Anke Krueger

**Carbon Materials
and Nanotechnology**

Related Titles

D.M. Guldi, N. Martin (Eds.)

Carbon Nanotubes and Related Structures

Synthesis, Characterization, Functionalization, and Applications

2010

ISBN: 978-3-527-32406-4

K.E. Geckeler, H. Nishide (Eds.)

Advanced Nanomaterials

2010

ISBN: 978-3-527-31794-3

C.S.S.R. Kumar (Ed.)

Nanomaterials for the Life Sciences

10 Volume Set

2011

ISBN: 978-3-527-32261-9

S.C. Tjong

Carbon Nanotube Reinforced Composites

Metal and Ceramic Matrices

2009

ISBN: 978-3-527-40892-4

C. Hierold (Ed.)

Carbon Nanotube Devices

Properties, Modeling, Integration and Applications

2008

ISBN: 978-3-527-31720-2

O. Tabata, T. Tsuchiya (Eds.)

Reliability of MEMS

Testing of Materials and Devices

2008

ISBN: 978-3-527-31494-2

Anke Krueger

Carbon Materials
and Nanotechnology

WILEY-VCH Verlag GmbH & Co. KGaA

The Author

Prof. Dr. Anke Krueger
Universität Würzburg
Institut für Organische Chemie
Am Hubland
97074 Würzburg
Germany

The Translator
Dr. Stefan Brammer

All books published by **Wiley-VCH** are carefully produced. Nevertheless, authors, editors, and publisher do not warrant the information contained in these books, including this book, to be free of errors. Readers are advised to keep in mind that statements, data, illustrations, procedural details or other items may inadvertently be inaccurate.

Library of Congress Card No.: applied for

British Library Cataloguing-in-Publication Data
A catalogue record for this book is available from the British Library.

Bibliographic information published by the Deutsche Nationalbibliothek
The Deutsche Nationalbibliothek lists this publication in the Deutsche Nationalbibliografie; detailed bibliographic data are available on the Internet at http://dnb.d-nb.de

© 2010 WILEY-VCH Verlag GmbH & Co. KGaA, Weinheim

All rights reserved (including those of translation into other languages). No part of this book may be reproduced in any form – by photoprinting, microfilm, or any other means – nor transmitted or translated into a machine language without written permission from the publishers. Registered names, trademarks, etc. used in this book, even when not specifically marked as such, are not to be considered unprotected by law.

Printed in Great Britain
Printed on acid-free paper

Cover Design   Schulz Grafik-Design, Fußgönheim
Typesetting   Toppan Best-set Premedia Limited
Printing and Bookbinding   T.J. International Ltd., Padstow, Cornwall

ISBN: 978-3-527-31803-2

Contents

Preface *XIII*

1 Carbon – Element of Many Faces *1*
1.1 History *1*
1.2 Structure and Bonding *6*
1.2.1 Graphite and Its Structure *7*
1.2.2 Diamond and Its Structure *9*
1.2.3 Structure of Other Carbon Allotropes *10*
1.2.4 Liquid and Gaseous Carbon *13*
1.3 Occurrence and Production *13*
1.3.1 Graphite and Related Materials *13*
1.3.2 Diamond *17*
1.4 Physical Properties *20*
1.4.1 Graphite and Related Materials *21*
1.4.2 Diamond *23*
1.5 Chemical Properties *24*
1.5.1 Graphite and Related Materials *26*
1.5.2 Diamond *29*
1.6 Application and Perspectives *29*
1.6.1 Graphite and Related Materials *30*
1.6.2 Diamond *30*
1.6.3 Other Carbon Materials *30*
1.7 Summary *31*

2 Fullerenes – Cages Made from Carbon *33*
2.1 History – The Discovery of New Carbon Allotropes *33*
2.1.1 Theoretical Predictions *33*
2.1.2 Experimental Proof *34*
2.2 Structure and Bonding *36*
2.2.1 Nomenclature *36*
2.2.2 The Structure of C_{60} *37*
2.2.3 Structure of Higher Fullerenes and Growth Mechanisms *40*

Carbon Materials and Nanotechnology. Anke Krueger
Copyright © 2010 WILEY-VCH Verlag GmbH & Co. KGaA, Weinheim
ISBN: 978-3-527-31803-2

2.2.4	Structure of Smaller Carbon Clusters	45
2.2.5	Structure of Heterofullerenes	46
2.3	Occurrence, Production, and Purification	47
2.3.1	Fullerene Preparation by Pyrolysis of Hydrocarbons	48
2.3.2	Partial Combustion of Hydrocarbons	49
2.3.3	Arc Discharge Methods	50
2.3.4	Production by Resistive Heating	51
2.3.5	Rational Syntheses	52
2.3.6	Enrichment and Purification	54
2.3.7	Preparation of Heterofullerenes	56
2.4	Physical Properties	57
2.4.1	Properties of C_{60} and C_{70}	57
2.4.1.1	Solubility	57
2.4.1.2	Spectroscopic Properties	59
2.4.1.3	Thermodynamic Properties	63
2.4.1.4	Solid C_{60}	64
2.4.2	Properties of Higher Fullerenes	65
2.5	Chemical Properties	66
2.5.1	General Considerations on Fullerene Chemistry	67
2.5.1.1	Typical Reactions of Fullerenes	67
2.5.1.2	Regiochemistry of Additions to Fullerenes	68
2.5.1.3	Secondary and Multiple Additions	68
2.5.2	Electro- and Redox Chemistry of Fullerenes	72
2.5.2.1	Electrochemistry of Fullerenes	72
2.5.2.2	Reductions of Fullerenes	74
2.5.2.3	Oxidations of Fullerenes	76
2.5.3	Inorganic Chemistry of Fullerenes	77
2.5.4	Endohedral Complexes of Fullerenes	82
2.5.4.1	Metallofullerenes	82
2.5.4.2	Endohedral Compounds with Nonmetallic Elements	86
2.5.5	Organic Chemistry of Fullerenes	87
2.5.5.1	Hydrogenation and Halogenation	87
2.5.5.2	Nucleophilic Addition to Fullerenes	93
2.5.5.3	Cycloadditions	98
2.5.5.4	Photochemistry	103
2.5.5.5	Radical Chemistry of Fullerenes	105
2.5.5.6	Fullerenes in Polymeric Materials and on Surfaces	107
2.5.6	Supramolecular Chemistry of Fullerenes	112
2.5.7	Polymeric Fullerenes and Behavior under High Pressure	116
2.5.8	Reactivity of Further Fullerenes	117
2.6	Applications and Perspectives	118
2.7	Summary	121
3	**Carbon Nanotubes**	**123**
3.1	Introduction	123

3.2	The Structure of Carbon Nanotubes	126
3.2.1	Nomenclature	126
3.2.2	Structure of Single-Walled Carbon Nanotubes	127
3.2.3	Structure of Multiwalled Carbon Nanotubes	135
3.3	Production and Purification of Carbon Nanotubes	140
3.3.1	Production of Single-Walled Carbon Nanotubes	140
3.3.1.1	Arc Discharge Methods	140
3.3.1.2	Laser Ablation	142
3.3.1.3	The HiPCo Process	144
3.3.1.4	Pyrolysis	146
3.3.1.5	Chemical Vapor Deposition (CVD)	147
3.3.2	Production of Multiwalled Carbon Nanotubes	150
3.3.2.1	Arc Discharge Methods	150
3.3.2.2	Laser Ablation	153
3.3.2.3	Chemical Vapor Deposition (CVD)	154
3.3.2.4	Decomposition of Hydrocarbons – Pyrolytic Methods	156
3.3.2.5	Production of Double-Walled Carbon Nanotubes	158
3.3.3	Strategies for the Rational Synthesis of Carbon Nanotubes	159
3.3.4	Structure and Production of Further Tubular Carbon Materials	163
3.3.4.1	Bamboo-Like Carbon Nanotubes	163
3.3.4.2	Cup-Stacked Carbon Nanotubes	164
3.3.4.3	Carbon Nanohorns (SWNH)	165
3.3.4.4	Helical Carbon Nanotubes (hMWNT)	166
3.3.5	Arrays of Carbon Nanotubes	168
3.3.6	Purification and Separation of Carbon Nanotubes	171
3.3.6.1	Removal of Impurities from Carbon Nanotube Materials	171
3.3.6.2	Evaluating the Purity of Carbon Nanotube Materials	173
3.3.6.3	Cutting of Carbon Nanotubes	176
3.3.6.4	Separation of Carbon Nanotubes by their Properties	177
3.3.7	The Growth Mechanism of Carbon Nanotubes	180
3.3.7.1	Arc Discharge Methods	180
3.3.7.2	CVD-Methods	185
3.4	Physical Properties	186
3.4.1	General Considerations	186
3.4.2	Solubility and Debundling of Carbon Nanotubes	187
3.4.3	Mechanical Properties of Carbon Nanotubes	190
3.4.4	Electronic Properties of Carbon Nanotubes	194
3.4.4.1	Band Structure and Density of States of Carbon Nanotubes	194
3.4.4.2	The Mechanism of Electric Conduction in Carbon Nanotubes	202
3.4.4.3	Field Emission from Carbon Nanotubes	204
3.4.5	Spectroscopic Properties of Carbon Nanotubes	206
3.4.5.1	Raman and Infrared Spectroscopy of Carbon Nanotubes	206
3.4.5.2	Absorption and Emission Spectroscopy of Carbon Nanotubes	209
3.4.5.3	ESR-Spectroscopic Properties of Carbon Nanotubes	212
3.4.5.4	Further Spectroscopic Properties of Carbon Nanotubes	213

3.4.6	Thermal Properties of Carbon Nanotubes	216
3.4.6.1	Specific Heat Capacity of Carbon Nanotubes	216
3.4.6.2	Heat Conductivity of Carbon Nanotubes	216
3.5	Chemical Properties	217
3.5.1	General Considerations on the Reactivity of Carbon Nanotubes	217
3.5.2	Redox Chemistry of Carbon Nanotubes	220
3.5.3	Functionalization of the Caps or Open Ends of Carbon Nanotubes	224
3.5.4	Side Wall Functionalization of Carbon Nanotubes	226
3.5.4.1	Covalent Attachment of the Functional Groups	226
3.5.4.2	Noncovalent Attachment of Functional Units	240
3.5.5	Composite Materials with Carbon Nanotubes	246
3.5.5.1	Composites with Covalent Bonding of the Polymer	248
3.5.5.2	Composites with Noncovalent Attachment of the Polymer	249
3.5.5.3	Nanotube Composites with Different Polymers	250
3.5.5.4	Nanotube Composites with Other Materials	254
3.5.6	Intercalation Compounds and Endohedral Functionalization of Carbon Nanotubes	255
3.5.7	Supramolecular Chemistry of Carbon Nanotubes	263
3.6	Applications and Perspectives	266
3.6.1	Electronic Applications of Carbon Nanotubes	267
3.6.1.1	Nanotubes as Tips in Atomic Force Microscopy	267
3.6.1.2	Field Emission	268
3.6.1.3	Field Effect Transistors	269
3.6.2	Sensor Applications of Carbon Nanotubes	271
3.6.2.1	Physical Sensors	271
3.6.2.2	Chemical Sensors	271
3.6.3	Biological Applications of Carbon Nanotubes	273
3.6.3.1	Recognition of DNA Sequences	273
3.6.3.2	Delivery of Drugs and Vaccines; Gene Therapy	274
3.6.4	Materials with Carbon Nanotubes	275
3.6.5	Further Applications of Carbon Nanotubes	277
3.6.5.1	Heterogeneous Catalysis	277
3.6.5.2	Hydrogen Storage in Carbon Nanotubes	278
3.6.5.3	Carbon Nanotubes as Material in Electrical Engineering	279
3.7	Summary	280
4	**Carbon Onions and Related Materials**	**283**
4.1	Introduction	283
4.2	Structure and Occurrence	284
4.2.1	Structure of Carbon Onions	284
4.2.2	Structure of Faceted Carbon Nanoparticles	289
4.2.3	Occurrence of Carbon Onions and Nanoparticles	290

4.3	Preparation and Mechanisms of Formation	*291*
4.3.1	Arc Discharge Methods	*291*
4.3.2	CVD-Methods	*293*
4.3.3	Preparation of Carbon Onions by Ion Bombardment	*294*
4.3.4	Chemical Methods	*296*
4.3.5	Transformations of Other Carbon Species	*298*
4.3.5.1	Thermal Transformations of Soot-Like Structures	*298*
4.3.5.2	Irradiation of Soot-Like and Other sp^2-Hybridized Carbons	*300*
4.3.5.3	Thermal Transformation of Diamond	*303*
4.3.5.4	Irradiation of Diamond Materials	*304*
4.3.6	Further Methods to Produce Carbon Onions	*305*
4.3.7	Growth Mechanisms of Carbon Onions	*307*
4.3.7.1	Growth Mechanisms of Carbon Onions Obtained by Electron Irradiation	*307*
4.3.7.2	Growth Mechanisms of Carbon Onions Obtained by Thermal Treatment	*309*
4.4	Physical Properties	*313*
4.4.1	Spectroscopic Properties	*313*
4.4.1.1	IR- and Raman Spectroscopy	*314*
4.4.1.2	X-Ray Diffraction	*315*
4.4.1.3	Absorption Spectra of Carbon Onions and Related Materials	*316*
4.4.1.4	EEL-Spectra of Carbon Onions and Related Materials	*317*
4.4.1.5	Further Spectroscopic Properties of Carbon Onions and Related Materials	*318*
4.4.2	Thermodynamic Properties	*319*
4.4.3	Electronic Properties	*320*
4.5	Chemical Properties	*321*
4.5.1	Reactivity and Functionalization of Carbon Onions and Carbon Nanoparticles	*321*
4.5.2	Conversion into Other Forms of Carbon	*323*
4.6	Applications and Perspectives	*324*
4.6.1	Tribological Applications	*325*
4.6.2	Applications in Catalysis	*326*
4.7	Summary	*327*
5	**Nanodiamond** *329*	
5.1	Introduction	*329*
5.1.1	Historical Background to the Discovery of Nanodiamonds	*329*
5.1.2	Natural Occurrence of Nanodiamond	*331*
5.2	Structure of Nanodiamonds	*332*
5.2.1	The Lattice Structure of Nanodiamond	*332*
5.2.2	The Surface Structure of Nanodiamond	*333*
5.2.3	Diamond or Graphite? Stability in the Nanometer Range	*336*
5.2.4	Agglomeration of Nanodiamond	*338*

5.3	Preparation of Nanodiamond	340
5.3.1	Detonation Synthesis	340
5.3.2	Shock Syntheses of Nanodiamond	344
5.3.3	Further Methods of Nanodiamond Preparation	346
5.3.4	Deagglomeration and Purification	349
5.4	Physical Properties	351
5.4.1	Spectroscopic Properties of Nanodiamond	351
5.4.1.1	Raman Spectroscopy	351
5.4.1.2	Infrared Spectroscopy	354
5.4.1.3	X-Ray Diffraction and EELS	356
5.4.1.4	Absorption and Photoluminescence Spectroscopy	358
5.4.1.5	Further Spectroscopic Properties	360
5.4.2	Electronic Properties of Nanodiamond	362
5.4.3	Mechanical Properties of Nanodiamond	365
5.5	Chemical Properties	367
5.5.1	Reactivity of Nanodiamond	367
5.5.2	Surface Functionalization of Nanodiamond	370
5.5.2.1	Hydrogenation	370
5.5.2.2	Halogenation	371
5.5.2.3	Oxidation of Nanodiamond	373
5.5.2.4	Reduction of Nanodiamond	374
5.5.2.5	Silanization of Nanodiamond	374
5.5.2.6	Alkylation and Arylation of Nanodiamond	375
5.5.2.7	Reactions on sp^2-Hybridized Domains on the Nanodiamond Surface	376
5.5.2.8	Further Functionalization of Nanodiamond	377
5.5.2.9	Composites and Noncovalent Interactions with Nanodiamond	380
5.5.3	Transformations of Nanodiamond into Other Forms of Carbon	382
5.6	Applications and Perspectives	382
5.6.1	Mechanical Applications	382
5.6.2	Thermal Applications	384
5.6.3	Applications as Sorbent	384
5.6.4	Biological Applications	385
5.6.5	Further Applications and Perspectives	385
5.7	Summary	386
6	**Diamond Films**	**389**
6.1	Discovery and History of Diamond Films	389
6.2	Structure of Diamond Films	391
6.2.1	General Considerations on the Structure of Diamond Films	391
6.2.2	The Surface Structure of Diamond Films	394
6.2.2.1	Structure of the (111)-Plane	394
6.2.2.2	Structure of the (100)-Plane	396
6.2.2.3	Structure of the (110)-Plane	398

6.2.3	Defects and Doping of Diamond Films	399
6.2.4	Structure of Further Diamond-Like Film Materials	402
6.3	Preparation of Diamond Films	403
6.3.1	CVD Methods for the Preparation of Diamond Films	403
6.3.1.1	Hot Filament CVD	404
6.3.1.2	CVD at Simultaneous Electric Discharge	405
6.3.1.3	Microwave CVD	405
6.3.2	Growth Mechanism of Diamond Films	407
6.3.3	Preparation of UNCD	410
6.3.4	Further Methods of Diamond Film Production	412
6.4	Physical Properties of Diamond Films	413
6.4.1	Spectroscopic Properties of Diamond Films	413
6.4.1.1	Infrared and Raman Spectroscopy	413
6.4.1.2	Optical Properties of Diamond Films	416
6.4.1.3	XRD, XPS, and EELS of Diamond Films	418
6.4.2	Electronic Properties of Diamond Films	420
6.4.2.1	Electric Conductivity of Diamond Films	421
6.4.2.2	Field Emission from Diamond Films	423
6.4.3	Mechanical Properties of Diamond Films	424
6.4.4	Thermal Properties of Diamond Films	428
6.5	Chemical Properties of Diamond Films	428
6.5.1	Considerations on the Reactivity of Diamond Films	428
6.5.2	Covalent Functionalization of Diamond Films	430
6.5.2.1	Hydrogenation of Diamond Films	430
6.5.2.2	Halogenation of Diamond Films	430
6.5.2.3	Oxidation of Diamond Surfaces	432
6.5.2.4	Radical and Photochemical Reactions on Diamond Surfaces	433
6.5.2.5	Cycloadditions on Diamond Surfaces	436
6.5.2.6	Further Reactions on Functionalized Diamond Films	438
6.5.3	Noncovalent Functionalization of Diamond Films	440
6.5.4	Electrochemistry of Diamond Films	440
6.6	Applications and Perspectives	443
6.6.1	Mechanical Applications	443
6.6.2	Electronic Applications	444
6.6.3	Chemical, Electrochemical, and Biological Applications	446
6.6.4	Further Applications of Diamond Films	447
6.7	Summary	448
7	**Epilog** 451	
8	**Further Readings and Figure References** 455	
8.1	Further Readings 455	
	Carbon in General 455	
	Fullerenes 455	
	Carbon Nanotubes 456	

Carbon Onions 457
Diamond Films 457
Nanodiamond 458
8.2 Figure References 458

Index 467

Paper Cutout DIY Kit 475

Preface

On first sight, it might seem overdone to write an entire book on a single element, all the more since it is not the first one on the subject "carbon," and most likely will not be the last. Nevertheless, carbon deserves our particular attention for being among the most significant elements on earth and the essential one for life on it. The chemistry of carbon gave rise to an own branch of science: Organic Chemistry. Yet in the present book this will only be considered when organic chemistry takes place on new carbon materials.

This textbook is intended as an overview on developments concerning carbon that caused, within the last 20 years, a virtually explosive expansion of research on new carbon materials. Starting from the discovery of fullerenes, a whole new field of research opened up between chemistry, physics, and materials science, which exclusively deals with this single, but fascinating element, its properties, and applications.

The present text is based on a one-term lecture that I have given several times by now at the University of Kiel, Germany. Contents are selected in a way to enable graduate students of chemistry and physics as well as future materials scientists to pick the relevant and useful from the topics at hand. The textbook is conceived as concomitant reading for a lecture and for self-studying alike. There are attached copy templates of models facilitating the spatial understanding of different carbon structures. Chapter 8 gives reference to further books and review articles to simplify a more thorough ingress into the field.

The book would not have become what it is without the help of many. I express my thanks to Prof. Dr. Henning Hopf and Dr. Torsten Winkler for critically reviewing the German version of the manuscript and to Prof. Dr. Florian Banhart and Dr. Carsten Tietze for looking through selected chapters of the original German text. Thanks as well to the librarians of the chemical institutes in Kiel for their aid in collecting the relevant literature. Numerous figures and pictures were supplied by authors and publishers, which is gratefully acknowledged. I am also thankful to Ulrich Sandten and Kerstin Hoffmann of Teubner Publishers for their support to the German version. For the support in producing this English version, I thank Lesley Belfit from Wiley-VCH for thorough reading of the translated text. Finally, I thank the Fonds der Chemischen Industrie (German Fund of the Chemical Industry) for support by a *Liebig* fellowship for junior faculty.

Carbon Materials and Nanotechnology. Anke Krueger
Copyright © 2010 WILEY-VCH Verlag GmbH & Co. KGaA, Weinheim
ISBN: 978-3-527-31803-2

Without the help and understanding of my husband Dr. Stefan Brammer this book would have never come into existence. I thank him for the never-ending support, motivation, proofreading, translating, and everything else which helped to finish this project.

Comments and suggestions to improve future editions of the text are highly welcome; still I wish that all readers of this book may have fun and gain knowledge while discovering the many facets of Element 6.

Kiel, January 2010 *Anke Krueger*

1
Carbon – Element of Many Faces

The sixth in the periodic table of elements is, at the same time, among the most important ones. With about 180 ppm, carbon is only 17th on the list of terrestrial elements' frequency, situated even after barium or sulfur – for comparison, the second-most frequent element, silicon, is about 1300 times as abundant as carbon. Still the latter is essential for the assembly of all organic matter. It is predestined for this central role especially due to its mid position in the periodic system and its associated ability to form stable substances with more electropositive and more electronegative reaction partners. Yet in the present text the organic chemistry resulting from these various bonding possibilities will only be mentioned if it is employed to modify carbon materials or, to put it in other words, the element itself as a material will be in the focus.

Another feature that gave reason to write this book was the occurrence of various allotrope modifications with in parts completely opposite properties. For the time being, this renders carbon one of the most interesting topics in materials science and research. Be it fullerenes, nanotubes or nanocrystalline diamond phases – they all are subject to intensive investigation and promise a multitude of applications, for example, in electronics, medicine, and nanotechnology.

Nevertheless, it is hardly possible to assess the development within the last two decades and to perceive its impact on the multitalented carbon's perspectives in chemistry, material science, and physics, without an understanding of its long-known modifications, mainly graphite and diamond. Hence, the first chapter summarizes the essential facts on the "classical" modifications and their properties, for only a solid comprehension of basic concepts and principles enables us to understand the properties of "new" carbon materials and to develop new ideas.

1.1
History

Elemental carbon in different shapes played a role in human life long before the term "element" was even coined. Charcoal and soot have been known and utilized for various purposes since ~5000 BC. They were mainly obtained from wood and employed, for example, for metallurgic processes such as the production of iron.

Carbon Materials and Nanotechnology. Anke Krueger
Copyright © 2010 WILEY-VCH Verlag GmbH & Co. KGaA, Weinheim
ISBN: 978-3-527-31803-2

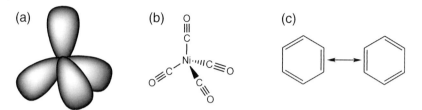

Figure 1.1 (a) sp³-Hybrid orbitals of carbon, (b) nickel tetracarbonyl, (c) mesomerism of the aromatic compound benzene.

The first application of graphite is documented from the late central European Iron Age (La Tène period). It was excavated near Passau (Bavaria) and employed to blacken pottery. Further examples of carbon being used for artistic purposes include pigment black from charcoal for cave painting, for example, as seen in the caves of Rouffignac (France).

The term "graphite" reflects its use as a pigment: it is derived from the Greek word *graphein*, meaning *to write*. Pencils, that became fashionable in the Middle Ages, originally were made from graphite, and ancient Egyptian papyri bear hieroglyphs written with ink were made from soot. The latter also became important for written records and works of art in East Asia from the beginning of modern times. Moreover, carbon in the shape of medicinal charcoal is employed in therapeutics to treat gastrointestinal diseases by its degassing and adsorbing effects. Charcoal is also a constituent of black powder – it is mainly black alder wood (*frangula alnus*) being charred at relatively low temperatures for this purpose.

In the 19th century, the exploration of the element took a rapid progress, and many important insights were obtained. Shortly after Berzelius made a distinction between inorganic and organic matter in 1807, it became obvious that carbon played a central role in organic substances. His definition of organic compounds only to occur in living organisms had to be revised soon (Wöhler conducted his important experiments to synthesize organic urea from definitely inanimate ammonium acetate in 1828 already). Nevertheless, his work laid the path for extensive research that largely influenced the development of bonding theories for complex molecules.

F. v. Kékulé interpreted benzene to be a cyclic entity in 1865. The concept of carbon as a tetrahedrally, four-fold coordinated atom was presented independently by J. H. van t'Hoff and J. A. Le Bel in 1874 and revolutionized the interpretation of the element's chemical activity (Figure 1.1). Since then, fundamental discoveries on this ubiquitous element multiplied. L. Mond and co-workers published the first metal carbonyls in 1890, and in 1891, E. G. Acheson for the first time achieved artificial graphite via intermediate silicon carbide (*carborundum*), which itself had been unknown then, too.

The development proceeded in the early 20th century. Next to the first preparation of a graphite-intercalation compound (C_8K, 1928), A. S. King and R. T. Birge found the element's composition from the isotopes ^{12}C and ^{13}C in 1929. Radioac-

Figure 1.2 The blue *Hope* diamond kept in the National Museum of Natural History in Washington counts among the world's most famous diamonds (© Kowloonese 2004).

tive ^{14}C was detected in 1936 by W. E. Burcham and M. Goldhaber. This enabled the development of radiocarbon dating for the age determination of the organic matter by Libby, who was awarded a Nobel prize for that discovery in 1960. One year later, the mass of isotope ^{12}C was established as basis for the standard atomic masses (^{12}C = 12.0000 amu). When NMR spectrometers with ^{13}C-Fourier transformation became widely available in the 1970s, structural analysis of organic molecules was immensely facilitated.

From 1985 on, the research on carbon gained new impetus from the first observation of fullerenes, and in 1991, carbon nanotubes were presented as another new allotrope modification. The whole field has been in a very dynamic state ever since, and the number of publications has grown to more than 50 000 in the meantime, demonstrating the large interest of specialists from various areas of research.

Diamond, whose name is derived from the Greek words *diaphanes* (*translucent*) and *adamas* (*invincible*), has also been known for long (Figure 1.2). It was first discovered in India around 4000 BC. The oldest, and at the same time one of the biggest diamonds surviving to our days (186 carat) is the *Kooh-i-Noor*. It was presumably found in India about 3000 BC and is kept in the Tower of London today. From about 600 BC on, diamonds from India came to Europe. As early as that, magical powers were ascribed to the stone due to its distinct appearance and resistant nature. Diamond as a decorative gem did appear later, though. In ancient Rome, the stone was known and appreciated as well, and Pliny the Elder (23–79 A.C.) already did mention its use as a tool. Still, the stones had to be imported from southeast Asia, which made them an extremely rare material, and even after reports on the first finding of diamonds on the island of Borneo (Indonesia today), India for its favorable situation remained the most important source until modern times.

It was only when the Indian mines were exhausted in the 18th century that new sources of the coveted stone had to be found. In the year 1726, diamond for the first time was found out of Asia while searching for gold in the then Portuguese colony of Brazil. This caused a veritable "diamond rush" that lasted for several years. But after some time, these sources began to run dry as well (today, there

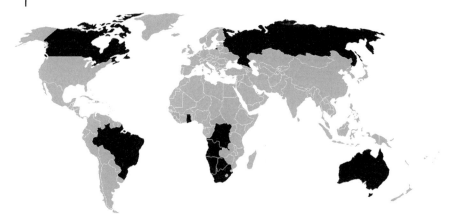

Figure 1.3 World-wide deposits of diamond (dark shading: countries with diamond resources). The classical African export countries experience increasing competition from discoveries in Canada and Australia.

are still black diamonds, so-called *carbonados*, mainly processed to be tools for oil drilling). Then, in 1870, the first so-called *Kimberlite* pipe was found near the town of Kimberley in South Africa, which within shortest time turned that country into the world's main supplier of diamond.

Discoveries of diamond increased in other African countries too. From the beginning of the 20th century, partly copious deposits were found, for example, in Congo, Namibia, Ghana, Guinea, Sierra Leone, and the Ivory Coast, which made Africa the virtually sole source of diamonds until the 1950s. It was not before 1950 that the large-scale exploitation of Siberian mines began and the former Soviet Union started supplying the world market with big amounts of diamond. Mainly the deposits in Jakutia near the city of Mirny, 4000 km east of the Ural Mountains, furnish good yields of jewelry and industrial grade diamonds to the present days (Figure 1.3).

Today, more or less abounding deposits have been found on all continents including Europe, for example, near Arkhangelsk (Russia). Especially the Kimberlite pipes discovered in northwestern Australia in the 1970s were very yielding, and by now, Australia is among the world's biggest suppliers. In recent years, also Canada became a big player on the world market after having prospected diamond sources and starting their exploitation (Figure 1.4).

However, a part of the diamonds offered today does not originate from the production controlled by a few big companies. These so-called blood diamonds are hauled in central Africa under inhumane and degrading conditions, and the money from their sale serves local warlords to subsidize a multitude of civil wars. In 2001, a treaty under the auspices of the UN was signed to make the raw material's origin traceable and thus prevent the selling of conflict diamonds, but despite this certification, the source of many stones remains unclear because documents are frequently faked and the diamond itself does not give information on its place of discovery.

Figure 1.4 The Ekati-Billiton mine in Canada (© Billiton Corp.).

Another milestone in the history of diamond was its first synthetic production. A team of Swedish researchers of the ASEA Corp. first succeeded in this task by using a pressure chamber, but their work did not obtain wide attention (refer to Section 1.3.2). A lot of attempts had been made before to produce the coveted stone in the lab, with approaches ranging from alchemistic folly to serious experiment. Among the noteworthy attempts there is the work of the French chemist H. Moissan (awarded the Nobel prize in 1906 for the discovery of fluorine). He saturated liquid iron with carbon and precipitated parts of the carbon as small, colorless crystals on shock-cooling. Nevertheless, diamond formation could not doubtlessly be proven. In 1880, J. B. Hannay published his work on heating a mixture of hydrocarbons and lithium in a closed iron container, which resulted in crystals with a density of $3.5\,g\,cm^{-3}$ and a carbon content of 98%. However, doubts on whether these samples really originated from his decomposition experiment could never be cleared out. C. Parsons experimented for more than 30 years on the manufacture of synthetic diamond. He failed to reproduce the experiments of Moissan and Hannay, but he also tested the addition of iron to a mixture according to Hannay. Although convinced at first to have made diamond, he later retracted this thesis after further, thorough investigation. He even doubted that it had then (1943) ever been possible to make synthetic diamond. Yet there had been experiments by C. V. Burton in 1905, consisting in the precipitation of carbon from oversaturated lead–calcium alloys by removing the calcium. Eighty years later, these experiments could be reproduced, and at least in the X-ray powder diffractogram, the product did show signs of a diamond-like crystal structure.

In 1955, the age of industrial diamond began when General Electrics started the first large-scale production in the United States. Six years later, P. S. DeCarli and J. C. Jamieson introduced a method employing the shock-wave energy of an explosion to generate pressure and thus to change graphite into diamond. From the mid-1960s on, an entirely new type of material became available besides the classical, more or less bulky diamond. Obtained by chemical vapor deposition, it does exist in the shape of thin films and serves, for example, as surface coating. Today, several other diamond materials are known which will be discussed in Chapters 5 and 6.

It was not before the end of the 18th century that the fact of diamond, graphite and soot being the very same element was acknowledged. The German chemist C. W. Scheele demonstrated graphite to be a kind of carbon, known in his mother tongue as Kohlenstoff (*coal-matter*). The french name *carbone* for the element was coined in 1789 by A. L. Lavoisier, who deducted it from the latin *carbo*, meaning *charcoal*. And only in 1796, S. Trennant could prove that diamond as well is a form of carbon.

By now, the phase diagram of carbon has grown out to be rather complex, with a multitude of allotrope modifications which in their turn are surrounded by additional high-pressure and high-temperature phases. All of them have distinct, and in parts even opposite properties, which put carbon among the most versatile and many-sided of elements in materials science.

1.2
Structure and Bonding

The carbon atom bears six electrons – two tightly bound, close to the nucleus, and the remaining four as valence electrons. The electronic configuration is $1s^2$, $2s^2$, $2p^2$, accordingly (Figure 1.5). This implies a bivalence, which in fact does only exist in a few structures (carbenes), though. In the vast majority of its compounds, carbon is tetravalent. Yet in more recent times, even higher coordination numbers were found in a series of molecules, for example, in Al_2Me_6, in different carbaboranes, or in the octa-coordinated carbon atom in $[Co_8C(CO)_{18}]^{2-}$ (Figure 1.6).

The preferred tetravalence may be explained with the hybridization model: the energetic difference between 2s- and 2p-orbitals is rather low compared to the energy released in chemical bonding. Therefore, it is possible for the wavefunc-

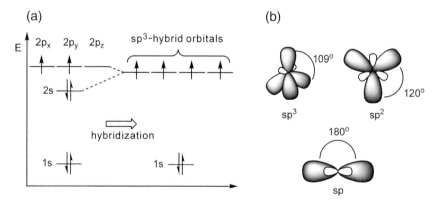

Figure 1.5 (a) Diagram of atomic orbitals and sp^3-hybridization, (b) hybrid orbitals of carbon.

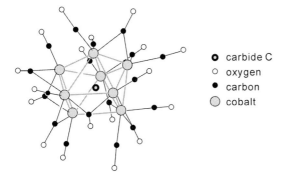

Figure 1.6 Hypervalent carbon. In the complex [Co$_8$C(CO)$_{18}$]$^{2-}$ the isolated carbon atom has eight coordinating neighbors.

tions of these orbitals to mix and form four equivalent hybridized orbitals. These sp^3-hybrid orbitals are directed toward the four corners of a tetrahedron circumscribed to the carbon atom. Likewise, the 2s-orbital may mix with a lower number of 2p-orbitals to form sp^2- or sp-hybrid orbitals, respectively. Figure 1.5b displays the hybrid orbitals and their spatial arrangement. A carbon atom may accordingly form bonds with one to four partners. Depending on the degree of hybridization, the resulting compounds show different structural features: sp-hybridized C-atoms form linear chains, whereas sp^2- and sp^3-hybridization give rise to planar structures and to three-dimensional tetrahedral networks, respectively. In sp- or sp^2-hybridized C-atoms there are two or one p-orbitals not taking part in hybridization. These can form additional π-bonds that, in contrast to the aforementioned σ-bonds, do not exhibit rotational symmetry. Their existence shows in the bond lengths and enthalpies. For example, the C–C distance in a double or triple bond is 133.4 or 120.6 pm, respectively, as compared to 154.4 pm in a single C–C-σ-bond. An analogous trend is observed for bonding enthalpies.

1.2.1
Graphite and Its Structure

The structure of graphite had been elucidated from 1917 on by Debye, Scherrer, Grimm, Otto, and Bernal. It is characterized by a succession of distinct so-called graphene layers that spread over an *xy*-plane. They are stacked in a *z*-direction, and there are only weak van der Waals interactions among them (Figure 1.7).

Within one graphene layer, the carbon atoms are situated at the corners of regular hexagons that constitute a two-dimensional lattice. From each C-atom there are three σ-bonds leading along the hexagons' edges, which correspond to sp^2-hybridization. As a consequence, only three out of four valence electrons participate in hybridization. The remaining electrons, contained in the p$_z$-orbitals, do also interact: they form a π-cloud that is delocalized over the entire graphene layer. The π-electrons thus behave like a two-dimensional electron gas, and by their

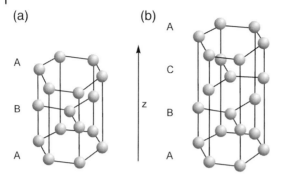

Figure 1.7 The sequence of layers in hexagonal (a) and rhombohedral graphite (b).

mobility, they cause material properties like an anisotropic electrical conductivity (Section 1.4.1). Within one graphene plane, the distance between adjacent atoms is 141.5 pm. This corresponds to a bond order of 1.5 and twice the covalence radius of an aromatic carbon atom (C–C distance in benzene: 139 pm), and it is a clear indication of the π-bond contribution. The distance between adjoining graphene sheets, on the other hand, is 335.4 pm (about twice the van der Waals radius) because there is only weak van der Waals interaction. That is why the planes of graphite are easily shifted in parallel against each other.

There are two different modifications, the hexagonal ("normal") or α-graphite, and the rhombohedral β-graphite. The latter is quite frequent in natural graphites and may be transformed into the thermodynamically more stable hexagonal form by the action of heat. The individual graphene layers in the hexagonal structure are stacked according to a sequence ABAB with the atoms of a layer B being situated above the centers of the hexagons in a layer A, and vice versa (Figure 1.7a). The third layer is congruent with the z-projection of the first one. The hexagonal unit cell, containing four carbon atoms, has a $P6_3/mmc$ (D_{6h}^4) symmetry and the dimensions 2.456 × 2.456 × 6.708 Å. In rhombohedral graphite, the layers are stacked in an ABCABC order (Figure 1.7b), giving rise to a bigger unit cell (2.456 × 2.456 × 10.062 Å) that also contains four carbon atoms nevertheless. Rhombohedral material is converted into the α-form above 1025 °C, whereas the reverse transformation may be achieved by milling. The enthalpy of formation is just about 0.06 kJ mol^{-1} higher for rhombohedral than for hexagonal graphite.

In reality, however, there is a wide range of graphitic materials apart from these ideal manifestations. They feature a perfect arrangement of carbon atoms within single graphene sheets, but the interplanar distance is about 344 pm due to stacking disorders. As a consequence, there is virtually no more interaction between the layers, and the orientation of individual planes no longer has an influence on the effective forces. Usually, the layers are irregularly turned around the z-axis and shifted against each other in the xy-direction. These structures are called *turbostratic*.

1.2.2
Diamond and Its Structure

In diamond, each carbon atom has four immediate neighbors situated at the corners of a tetrahedron that is circumscribed to the atom in question. All carbon atoms are sp^3-hybridized, and the bond length uniformly is 154.45 pm.

There are two modifications of diamond, the cubic and the hexagonal type (*Lonsdaleite*), with the first being the more abundant one. Another structural distinction of diamonds is made by their content of nitrogen: type Ia contains nitrogen as platelets with an approximate composition of C_3N. Type Ib, on the other hand, has nitrogen evenly distributed throughout the crystal. Diamond virtually void of nitrogen is classified as type IIa, but it rarely ever occurs naturally. Type IIb, which in contrast to the other variants is semiconducting, does not contain nitrogen either, but it bears a certain concentration of aluminum.

The crystal lattice of "normal" diamond is face-centered cubic with a lattice constant of 356.68 pm and eight atoms contained in the unit cell (Figure 1.8). The structure is not densely packed, but of the *sphalerite* kind, which is a penetration of two face-centered cubic lattices shifted against each other along the unit cell's space diagonal. An analogous structure is found in zincblende (ZnS). Upon heating to more than 3750 °C under a pressure of 1840 psi, cubic diamond turns into graphite.

Hexagonal diamond (*Lonsdaleite*) is extremely rare in nature. It was first discovered in a meteorite from Arizona in 1967. Still it is possible to obtain *Lonsdaleite* from graphite by exposing the latter to extreme pressure along the z-axis at ambient temperature. The hexagonal lattice (*Wurtzit* type) like its cubic relative is built from tetrahedrons of carbon, but these are arranged in a different way (Figure 1.9). The unit cell contains four atoms, and the lattice parameters are $a_0 = 252$ pm and $c_0 = 412$ pm.

As a visualization, starting from hexagonal graphite, one may fancy the freely mobile π-electrons alternately to form upward and downward σ-bonds with the respective atoms of adjacent layers. This causes a formal compression and undulation of the lattice (change of hybridization), but conserves the ABAB stacking of the imaginary planes. For the formation of cubic diamond, on the other hand, a slight

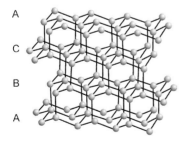
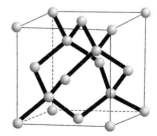

Figure 1.8 The lattice structure of cubic diamond and its elementary cell.

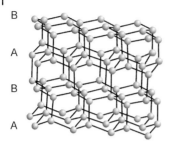

Figure 1.9 The lattice structure of hexagonal diamond. The arrangement of atoms in the horizontal crystal plane somewhat resembles a "wavy" graphite structure.

displacement of the third layer is required, which causes an ABCABC sequence and the interplanar distance of only 205 pm. To be sure, all of this is only a model made up to illustrate the lattice relation of the different carbon modifications. Actually, both cubic and hexagonal diamonds are isotropic, three-dimensional structures with every atom tetrahedrally surrounded by four neighbors.

1.2.3
Structure of Other Carbon Allotropes

There are a number of further carbon phases besides the "classics" graphite and diamond. In the 1960s, a white allotrope was found that was named *chaoite*. It was discovered in the Bavarian Ries (northeast of Ulm) in graphite gneiss that had molten under pressure. It may further be generated by heating pyrolytic graphite to about 2000 °C *in vacuo* (about 1.9×10^{-6} psi). The resulting hexagonal crystals grow as dendrites, the lattice parameters are $a_0 = 894.5$ pm and $c_0 = 1407.1$ pm, and the material's density of 3.43 g cm^{-3} is quite similar to that of diamond. However, the actual crystal structure has not yet been fully elucidated, but carbyne units (–C≡C–C≡C–C≡C–) are assumed to be at least partial constituents.

The so-called carbon(IV) does also seem to contain such carbyne structures. It may be obtained by heating graphite, preferably with a laser under argon at a pressure between 1.5×10^{-6} and 1.5 psi. The unit cell is of hexagonal shape with the parameters $a_0 = 533$ pm and $c_0 = 1224$ pm. The density of 2.9 g cm^{-3} is markedly higher than that of graphite. Further carbynoid allotropes are assumed to exist between ca. 2300 °C and the melting point of carbon.

While for the time being it is unclear for the aforementioned allotropes whether they are more of a graphitic or a diamond-like crystal structure, soot (or carbon black) definitely is a graphitic phase with an expanded interplanar distance of about 344 pm. Electron micrographs (Figure 1.10) show soot particles to be more or less spherical structures that form loose, grape-like aggregates. Within the individual particles there are very small domains (up to 3 nm along the hexagon planes and 2 nm perpendicular to them) that are graphitically crystallized. These are stacked in the manner of roof tiles, with a dominant turbostratic order in the piles. The particles' centers contain structures arising from polycyclic aromatic hydrocarbons (PAHs) attached to nucleation centers during formation of the soot

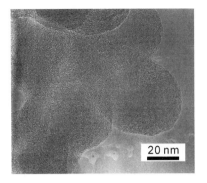

Figure 1.10 HRTEM image of soot particles (© M. Ozawa).

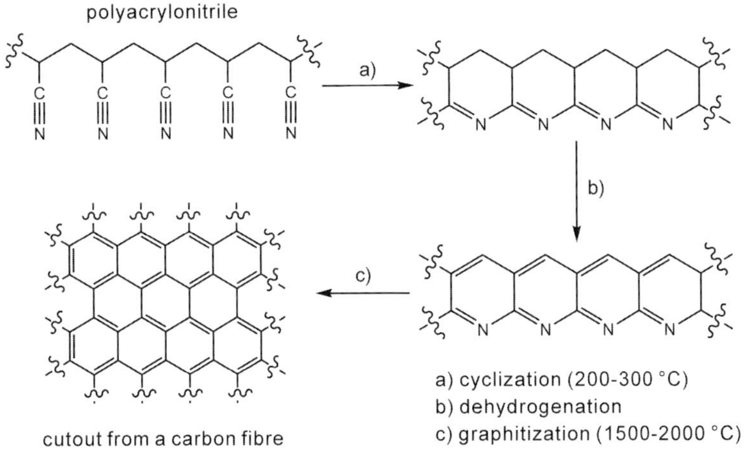

Figure 1.11 Conversion of polyacrylonitrile fibers into carbon fibers.

a) cyclization (200-300 °C)
b) dehydrogenation
c) graphitization (1500-2000 °C)

particles. By arranging with their molecular planes, they form the first pile, which is the starting point for particle growth. The lower the pressure during soot formation, the bigger the particles grow due to a then diminished number of condensation nuclei. Common sizes of soot particles range from 20 to 300 nm, which gives carbon black a considerable outer surface of about $100\,m^2\,g^{-1}$, and additional activation by heat may increase this value to even $1000\,m^2\,g^{-1}$. Owing to the loose stacking of subdomains, soot only has a low density of about $1.85\,g\,cm^{-3}$.

Carbon fibers are another graphitic manifestation of carbon; they only differ from the bulk material in the arrangement of graphitic elements that follow the fibrous habit here. The latter is caused by the manufacturing process: starting from polyacrylonitrile (PAN), an initially nitrogen-rich, fibroid structure is obtained by heating. This loses the nitrogen and graphitizes on further heating. The mechanism of this transformation is illustrated in Figure 1.11. Not only

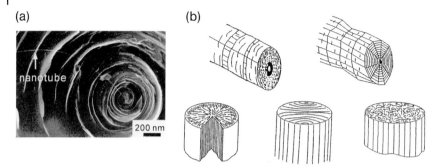

Figure 1.12 Carbon fibres; (a) microscopic image from out of the core protrudes a multi-walled carbon nanotube (bright line) (© Elsevier 1995), (b) schematic view of possible fiber structures (© Springer 1988).

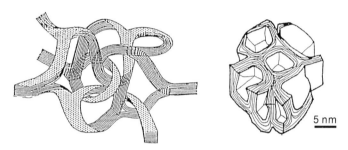

Figure 1.13 The structure of glassy carbon. The entangled graphitic ribbons give rise to extraordinary rigidity and isotropic materials properties (© Cambridge University Press 1976).

PAN, but also cellulose acetate or the tarry residues of oil refining (mesophase pitch) may serve as starting material. In the center of carbon fibers, structural elements that resemble multiwalled carbon nanotubes are frequently observed (see Chapter 3). Around these, graphitic domains are arranged in parallel with the fiber's axis with their specific structure depending on the temperature during preparation. Further thermal treatment after manufacture may change the respective morphology, but the orientation of the graphene layers in parallel with the fiber's axis is largely conserved. Figure 1.12 gives some examples of possible structures.

The so-called *glassy carbon* also bears some interesting structural features. It does contain ribbon-like graphitic domains of sp^2-hybridized carbon atoms stacked in layers, but these are irregularly twisted and entangled, resulting in a mechanically and chemically very resistant material. Figure 1.13 illustrates the respective structures. In contrast to graphite, glassy carbon is a material of low density ($1.5\,\mathrm{g\,cm^{-1}}$) with isotropic thermal and electrical properties.

1.2.4
Liquid and Gaseous Carbon

The melting point of graphite at ambient pressure of about 4450 K is extremely high – in fact it is the highest value measured for an element so far. Accordingly, its determination was anything but trivial as crucibles of any conceivable material would have molten before their content under examination. Eventually crucibles made from graphite itself were employed. Their center then was heated, for example, by means of a laser. The melting point of diamond slightly deviates from that of graphite, but the resulting liquid phase presumably is the same.

The boiling point of carbon is close to the melting point, so there is a considerable vapor pressure over the liquid phase. It is not single atoms evaporating, though, but small carbon clusters of up to several tens of atoms (a magnitude comparable to fullerenes). The most abundant species have structures of C_n with $n = 3, 2, 4$. The release of single atoms is restrained by the coincidence of high vapor pressure and the high C–C-bonding energy.

1.3
Occurrence and Production

1.3.1
Graphite and Related Materials

Above all, graphite is found in places where igneous rock (e.g., gneiss) and stratified rock border on each other. The graphitization of coal present in the stratified material was presumably initiated by ingressing magma and the corresponding high temperatures (300–1200 °C). There are different kinds of deposits bearing graphite in big pieces (lump and chip), in flakes, or as microcrystalline powder. The latter is frequently (and falsely) called "amorphous" due to its small particle size and the related aspect. For big, coherent lumps there is also a hydrothermal formation being discussed. Some of the deposits got a lenticular shape, and usually there are layers or pouches in the surrounding rock, which is another sign of the generation from coal seams.

China, India, Brazil, Korea, and Canada are the main suppliers, besides further sources of high-quality graphite like Sri Lanka or Madagascar. The overall production was about 742.000 t of natural graphite in 2003 (Table 1.1). It has also been

Table 1.1 Annual production of natural graphite in 2003 (according to US Geological Survey).

	China	India	Brazil	North Corea	Canada	Mexico	Czech Republic
Annual production of graphite in tons	450.000	110.000	61.000	25.000	25.000	15.000	15.000

mined in Germany of old, for example, in the Passau area (Bavaria), with an annual output of about 300 t.

The concentration of graphite in the deposits ranges from 20 to 50%. Depending on this content, there are different ways of processing. The crushed and milled rock is separated from the graphite by several steps of oil floatation. In some cases, there is also work-up with hydrofluoric or hydrochloric acid. Raw material from Sri Lanka, however, does not require floatation. It is simply sorted into graphite-containing and dead fractions before the muddy parts are washed away. The graphite portion is then treated with molten soda. The products obtained make different prices on the world market depending on their degree of crystallization and their purity (the latter usually is higher for well crystallized flaky graphite than for powdered graphite).

Yet the amounts from natural sources are far too low to cover the overall demand of graphitic carbon, and so a lot of the material is obtained synthetically. The world production is in the range of two-digit gigatons per year, with coal, oil or natural gas serving as carbon sources. Various graphitic materials such as soot, activated carbon, artificial graphite or carbon fibers, etc. are obtained by thermal decomposition of the starting materials at 600 to 3000 °C.

The properties of these graphitic products, such as degree of crystallization, particle size, or layer structure, depend on the manufacturing process and the kind of source material. Some products of low crystallinity exhibit a turbostratic structure of the graphene sheets. The higher the temperature during the decomposition of the primary material, the bigger domains of graphitic order are obtained, and the products' properties approximate those of original graphite. The individual products and the ways of their production are presented below.

Coke The annual production is in the range of hundreds of megatons as big amounts of this energy source are required to feed blast furnaces and for heating. It contains about 98% of carbon. Coke is obtained by strong heating of pit-coal (coking) with the specific kind of starting material leading to either *gas coke* or *metallurgical coke*. The latter is obtained by coking coal with a low content of gas. It is harder and more suitable for the furnace process than gas coke. In addition, there is *petrol coke* from the residues of distillation in oil refining. In comparison to other types of coke, this variant is well graphitized and hence employed to make *artificial graphite*.

Artificial Graphite This type of carbon is quite close to the natural material. It is formed by pyrolysis of carbon compounds at extremely high temperatures. In large-scale production, it is mainly petrol coke being graphitized at 2600–3000 °C. The process starts by preheating to about 1400 °C to rid the petrol coke of volatile components. Subsequently, molded parts bound with pitch are made and preburned at 800–1300 °C. This *artificial carbon* is then embedded in coke in an electric furnace and graphitized by resistant heating. The molded parts are covered with sand in this step, which is not only for thermal reasons: the silicon contained therein obviously has a catalytic effect on the formation of graphite by forming intermediate silicon carbide. At the prevailing temperatures, the latter has a considerable vapor pressure of silicon, and the carbon formed during the consequent

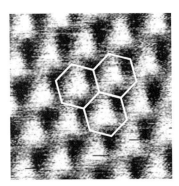

Figure 1.14 Atomic force microscopic image of highly ordered pyrolytic graphite (HOPG). The graphene lattice is indicated by the white lines. The different shading of carbon atoms results from the different situation in the atomic layer underneath (© A. Schwarz).

decomposition of carbide is obtained as graphite due to its lower chemical potential ($C_{microcryst.}$ + Si → SiC → Si + $C_{graphitic}$).

Pyrolytic Carbon Depending on the manufacturing conditions, a highly parallel order of graphene layers may be obtained here. Common pyrolytic carbon is made by thermolysis of hydrocarbons at about 700 °C and low pressure (about 0.15 psi). The product precipitates on a plain surface so that the graphene layers align with it. Thermolysis at 2000 °C and postgraphitization at 3000 °C yield a material called *pyrolytic graphite* that virtually matches natural graphite in its anisotropy of thermal and electrical conductivity. When, in addition to extreme temperatures, also shear forces are applied to the precipitated carbon, the almost completely parallel *highly ordered pyrolytic graphite* (*HOPG*) is obtained (Figure 1.14). The deviation from perfect parallel order is less than 1% here.

Carbon Fibers Pyrolysis of organic fibers under tension yields the aforementioned carbon fibers (Section 1.2.3), the hexagonal planes of which are oriented in parallel to the initial strain. The most common starting material is polyacrylonitrile (PAN). The covalent connection of atoms along the fiber's axis provides the material with high strain resistance and a large modulus of elasticity.

Glassy Carbon If unfoamed organic polymers are employed instead of PAN fibers, another variant of graphitic carbon is obtained. It is characterized by an entanglement of graphitic ribbons (Figure 1.13) that confer extreme hardness to the material. Furthermore, its physical properties are isotropic on a macroscopic scale as there is no preferred spatial order of the graphitic domains – they are rather arranged at random. The term *glassy carbon* originates from its aspect that resembles black glass. It can only be processed with diamond tools, and also chemically it is very stable. It may be applied, for example, as material for electrodes or crucibles.

Activated Carbon With a surface of 300–2000 $m^2 g^{-1}$ and a pore diameter of 1–5 nm, activated carbon is an outstanding adsorbent for a multitude of substances. It is produced by gentle heating of organic precursors like wood, turf or coconut peel, but also of pit-coal etc. The activation (to obtain the required porous structure) results either from hydrothermal surface reaction with water vapor or from impregnating the starting material with agents that cause oxidation and dehydration (e.g., ZnCl$_2$, H$_3$PO$_4$, or NaOH). Their residues are washed from the pores after charring. The annual production is some 100.000 t worldwide.

Soot (Carbon Black) Gigatons of carbon black are made each year mainly because it is required as filling agent by the tire industry. Several kinds of soot are distinguished depending on the respective starting material. In general, there are combustion soot from incomplete combustion and split soot obtained by thermal decomposition. The most important among the combustion soots are *furnace black* (95% of C), that is produced by burning mineral oil and chilling the resultant gases with water, and *channel black* obtained by precipitating on cooled iron surfaces the carbon from a shining flame of anthracene oil. Split soot can be made starting from natural gas, methane or acetylene, the latter yielding *acetylene black* with its particularly high content of carbon (98–100%). Soot like activated carbon possesses a high specific surface due to its particle structure (Figure 1.10). The small and irregularly arranged graphitic domains cause the material to be practically isotropic.

Natural Deposits of Carbon-Containing Materials Anthracite, pit-coal, brown coal, turf, wood, mineral oil, and natural gas are possible natural sources for the production of carbon materials. They all have different carbon contents and partly result from a long-lasting series of geological events. Within millions of years, a process called carbonization first changed organic matter to humic acids by anaerobic decomposition, later to turf, and then, under elevated pressure and exclusion of air, to brown or pit-coal. The carbon content increased in the course of this process, and the content of water concomitantly was reduced. Brown coal was mainly formed during oligocene and miocene (Figure 1.15) and contains up to 45% of water. The carbon content of the dehydrated material is about 70%. Pit-coal usually originates from the older age of *Carboniferous*; it bears less than 20% of water and about 90% of carbon after drying. The coal of highest quality, called anthracite, is a very hard, shiny gray material with less than 15% of residual water and more than 90% of carbon when dried. The different percentages of carbon correspond to the respective temperature and pressure during formation; the progressing coalification does not reflect the geological age – some of the oldest coals known are pieces of brown coal poor in carbon.

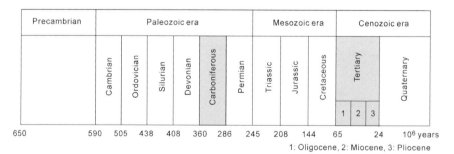

Figure 1.15 Chronological table of geological ages. The periods of coal formation are shaded gray.

1.3.2
Diamond

Natural deposits were the sole source of diamond until the mid-1950s as there was no synthesis of artificial material available – despite a variety of approaches.

The geological process of diamond formation has not been fully elucidated to the present day, but it usually is thought to be a transformation product of other carbons, for example, graphite. At a depth of more than 150 km from the earth's surface, at a pressure of more than 4.5 GPa and temperatures between 900 and 1300 °C, these change into the then-stable diamond. In volcanic eruptions several hundreds of millions of years ago the material was carried from the upper earth's mantle to the surface through chimneys of molten rock, reaching a speed of up to Mach 2 (700 m s^{-1}). The eruptive process was too fast for the retransformation into graphite to take place in spite of the reduced pressure and the lasting high temperature. The magma cooled down in the volcanic chimneys and formed igneous rock (Figure 1.16). The transporting rock of magmatic origin most frequently bearing diamond is *Kimberlite*, thus called after the South African town of Kimberley where the first of such chimneys was detected. But diamond is also found in sedimentary rock that was formed by withering of the original stone – it simply escaped erosion due to its hardness. Subsequent washout carried it away and deposited it at the future site of detection. Off the coast of Namibia, there are even marine deposits resulting from such transport processes.

Diamond is won either by washing it out of the sedimentary parent rock or by exploiting the Kimberlite pipes. The latter produces gigantic craters in the former volcanic chimneys that are dug continually deeper into the ground (Figure 1.4). In some mines also underground working takes place. The crushed and not-too-finely ground rock is washed out and the fraction holding diamond is led over

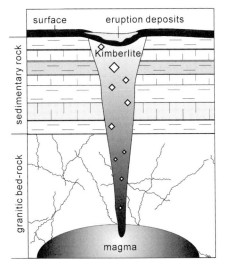

Figure 1.16 Scheme of a Kimberlite pipe. Due to the fast cooling of molten rock, carbon is conserved in the shape of diamond and conveyed to the earth's surface as such.

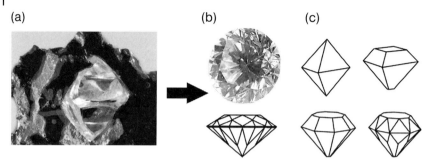

Figure 1.17 (a) Raw diamond in parent rock (© US Geological Survey), (b) jewelry diamond in brilliant cut and (c) evolution of cut types from simple octahedral stones to brilliants.

conveyor belts covered with a layer of fat. The diamond adheres to it, while other rock is carried on and discarded. Especially in large-scale industrial exploitation, there are also X-ray sorters in use. The current annual production is about 150.2 megacarat (30 t) with Australia (6.6 t), Africa (16 t in total), and Russia (4.8 t) being the main suppliers. Stones of gem quality are obtained from the respective diamond-rich fractions by assessing the raw diamonds' color, crystallinity and distribution of inclusions. Still a big part (an average of 50%, depending on the site of exploitation) does not fulfill the requirements for jewelry and thus is employed to make cutting and abrading tools. However, this amount is far from sufficient to satisfy the demand for industrial diamonds.

The selected raw diamonds are cleaved along suitable crystal planes to obtain stones for jewel manufacture. These are processed in a way to bring out the brilliance and its particular refraction of color (Figure 1.17). For grinding and polishing, diamond powder is used as no other material is hard enough to cause abrasion on diamond. The markedly varied hardness of its crystal planes is employed for this process: the polishing powder contains irregularly shaped fragments that represent material of every required hardness to abrade each crystal plane. The main task in diamond cutting is to accentuate the stone's luster. As much of the raw material as possible shall be preserved in doing so while achieving attractive gems nevertheless. Hence a multitude of cuts has been developed over the centuries, culminating in the so-called brilliant *full cut* that possesses at least 32 facettes and a plate on the upper half, and at least 24 facets and an optional calette on the lower. The stone's perimeter is circular. By this treatment the greatest possible dispersion of light is achieved (Figure 1.17b).

The quality of gem diamonds is judged by the *4 C-method*, comprising *color*, *clarity* (purity; size and kind of inclusions), *cut* (the quality of it), and *carat* (weight). 1 carat equals 0.2 g. The measure originates from antiquity – it is defined to be the mass of a seed from the carob tree. These tend to weigh exactly 0.2 g with surprising reproducibility.

As already suggested by the criteria of quality assessment, there are also colored or *fancy diamonds*. They are generated by substitution of foreign atoms to single

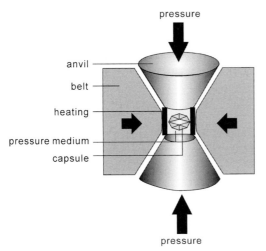

Figure 1.18 Press for the production of artificial diamond. The capsule in the heart of the press contains the carbon, a seed crystal and the catalyst.

lattice positions (Section 1.2.2). The black *carbonados* on the other hand are polycrystalline objects with inclusions of graphite, yet their natural color mostly is an unattractive brown that renders these stones only suitable to industrial purposes. Pure colors are rare, and the respective stones are very expensive.

Ways to produce artificial diamond had been sought for long. The first successful attempt in 1953 was largely ignored by the public, and to the present day it is not the Swedish scientists H. Liander and E. Lundblad of the ASEA company (Almänna Svenska Elektriska Aktiebolaget) being reputed for the diamond synthesis, but researchers from the General Electric company – even though they presented their effective procedure but in 1955. They employed the high pressure/high temperature transformation of graphite (*HPHT-synthesis*) in a hydraulic press generating several gigapascals (Figure 1.18) while the temperature was maintained at about 1500 °C. The addition of a catalyst accelerates the process; molten metals from the subgroups of iron or chromium (groups VIB and VIIIA, mainly iron and nickel) are suitable. In parts, the catalytic effect presumably consists in forming unstable metal carbide intermediates. Furthermore, graphite is more soluble in the molten metal than diamond. The graphite is covered with a thin film of liquid metal in which it solves to saturation. As graphite is thermodynamically unstable under the given conditions, it changes into diamond. This now precipitates from the melt as the latter is already supersaturated regarding the less soluble product.

Today about 540 megacarat (108 t) of industry-grade diamond are synthesized per year. In terms of cost per amount, the production of artificial diamond can easily compete with the hauling of natural material and so the first clearly surpasses the latter in the annual output. Main suppliers in 2003 have been Russia (16 t), Ireland (12 t), South Africa (12 t), Japan (6.8 t), and Belarus (5 t) (according to US Geological Survey)(Table 1.2). Recently, China entered the market on a large scale as well.

Table 1.2 Annual production of natural diamond in 2003 (accordng to US Geological Survey).

	Australia	Botswana	Congo	Russia	South Africa	Canada
Annual production of diamond in megacarat	33.1	30.4	27.0	24	12.7	11.2
of these: stones of gem quality in megacarat	14.9	22.8	5.4	12	5.1	n.a.[a]

a) n.a.: not available.

Industrial diamonds typically are colored gray to brown; they have numerous defects and a maximum size of 1 carat. It has been possible, however, to synthesize jewelry-grade diamonds as well. These may additionally be colored by deliberate doping. Still, for the time being, the considerable price for crystal-clear synthetic diamonds prevents a wide commercial use. The largest synthetic diamond to date, weighing 14.2 carat, was made in 1990. Synthesis starts with carbon strongly depleted of the isotope ^{13}C to distinguish artificial from natural stones.

Today there are several other ways of diamond synthesis besides the HPHT method. For example, it is possible to utilize the pressure of a shock-wave generated in an explosion. This process mostly yields powdery products with particle sizes in the range of micrometers (1 mm at max.) that may be employed for industrial purposes as well. Moreover, very small diamonds (5–20 nm) can be made by reacting explosives in confined containers. Diamond films are produced on various substrates by chemical vapor deposition (*CVD method*) using methane as a carbon source. Detonation synthesis and vapor deposition will be described in detail in Chapters 5 and 6.

1.4
Physical Properties

The variety of modifications is an outstanding characteristic of carbon. This becomes more than obvious from its phase diagram that has grown into the most complex figure considering ever new scientific results. Besides the long-known variants existent at normal pressure, there are also high-pressure phases. They are widely unexplored because their extreme conditions of existence render them almost impossible to study. Figure 1.19 exemplarily shows a simplified phase diagram.

Carbon is a typical main group element due to its properties. It is a nonmetal with the ground state electron configuration [He]$2s^2 2p^2$. The electronegativity of 2.55 on the Pauling scale is quite close to that of adjacent elements in the periodic table, for example, P (2.1), B (2.0), or S (2.5). The first energy of ionization is 1086.5 kJ mol^{-1} (refer to Table 1.3 for further values).

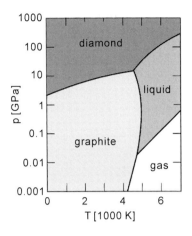

Figure 1.19 The phase diagram of carbon, not including fullerenes and carbon nanotubes (© Wiley-VCH 2000).

Table 1.3 Energies of ionization of carbon.

Ionization energy	1st	2nd	3rd	4th
kJ mol^{-1}	1086.5	2351.9	4618.8	6221.0

Besides the most abundant isotope ^{12}C (98.89%) there are also the isotope ^{13}C (1.11%) and the radioactive ^{14}C (traces). The molar weight of carbon (12.011 g mol^{-1}) results from this isotopic composition. ^{14}C is employed to determine the age of archeological objects (radiocarbon dating). The stable isotope ^{13}C is a valuable tool for molecular structure elucidation by NMR spectroscopy because its nuclear-spin quantum number is $I = \frac{1}{2}$.

In any further discussion of physical properties the considerable differences between modifications must be accounted for, so it is sensible not to describe the element's characteristics, but those of the respective allotropes.

1.4.1
Graphite and Related Materials

α-Graphite is the thermodynamically most stable form of carbon under normal conditions. Its standard enthalpy of formation is 0 kJ mol^{-1}, accordingly. The application of pressure and simultaneous heating convert it into diamond. β-Graphite can be obtained directly from milling α-graphite, whereas a straightforward transformation into further modifications like glassy carbon, fullerenes (Chapter 2), or nanotubes (Chapter 3) does not succeed. Table 1.4 compares some of the most important properties of graphite and diamond.

The density of graphite varies strongly depending on the material's origin and dispersity. It is 2.26 g cm^{-3} for ideal graphite, but it may be as low as 1.5 g cm^{-3} in

Table 1.4 Comparison of physical properties of graphite and diamond.

Characteristic	Graphite	Diamond
Color	Black with metallic luster	Colorless
Refractive index $n_{D(546\,nm)}$	2.15 ∥; 1.81 ⊥	2.43
Density	2.266 (exp. 1.5–2.2) g cm^{-3}	3.514 g cm^{-3}
Combustion enthalpy	393.5 kJ mol^{-1}	295.4 kJ mol^{-1}
Hardness (Mohs)	1 ∥; 4.5 ⊥	10
Band gap	0 eV	5.5 eV
Specific resistance	0.4–0.5 × 10^{-4} Ω cm ∥; 0.2–1.0 Ω cm ⊥	10^{14}–10^{16} Ω cm

∥: parallel to graphene layer.
⊥: perpendicular to the planes (along z-axis).

poorly graphitized, powdery samples. Carbon black has a density of 1.8 g cm^{-3} that might be increased by thermal treatment – volatile hydrocarbons are expelled, and values of up to 2.1 g cm^{-3} may be achieved. The density of glassy carbon is about 1.5 g cm^{-3}.

The graphite's anisotropy regarding many characteristics is a noteworthy feature. It is caused by the layered structure and affects electrical and thermal conductivity as well as mechanical characteristics like modulus of elasticity or tensile strength. The first is 5.24×10^5 N cm^{-2} in parallel with the z-axis, but 18.77×10^5 N cm^{-2} in parallel with the xy-planes. The tensile strength perpendicular to the planes is two to five times higher than in parallel to them, and also the hardness according to Mohs is larger in the z-direction ($H_M = 4.5$) than parallel to the layers ($H_M = 1$).

In graphite possessing a more or less undisturbed lattice, the heat capacity at 25 °C is 0.126 J mol^{-1} K^{-1}, whereas in samples with a more defective lattice (soot, activated carbon, etc.) it is higher due to the imperfect stacking. Conduction of heat in graphite is mainly affected by lattice wave transmission along the graphene layers, which also reflects in the values of thermal conductivity. These pass a maximum of more than 4.19 W cm^{-1} K^{-1} at about room temperature. Once again the method of preparation substantially influences the exact value.

The electrical conductivity is caused by π-electrons that always belong to a specific graphene sheet and rarely ever change to another. Graphite thus is a good electric conductor in parallel to the lattice planes with a conductivity of 10^4 to 2×10^5 Ω^{-1} cm^{-1} (specific resistance 5×10^{-5} to 10^{-4} Ω cm). In the direction of the z-axis, on the other hand, the material behaves more like an insulator with a conductivity of 0.33 to 200 Ω^{-1} cm^{-1} (specific resistance 0.005 to 3 Ω cm).

The most particular among the optical properties of graphite and related materials is the approximation to an ideal black body: microcrystalline forms of carbon have the highest absorptivity. Soot, for example, absorbs up to 99.5% of the incoming radiation. This value is much lower for other, better crystallized carbons, and very thin crystals of graphite are even transparent. The big absorptive capacity and black color of polycrystalline carbons is caused by their low particle size. Within

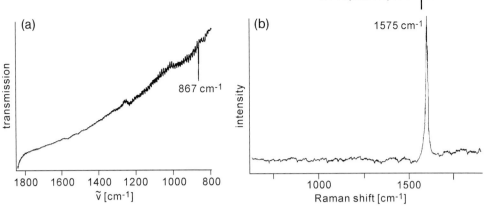

Figure 1.20 (a) IR spectrum of graphite (© APS 1985), (b) Raman spectrum of graphite (© AIP 1970).

the pores and in between particles the irradiated light is repeatedly reflected, leading to a stepwise absorption so it does not leave the surface again. The reflecting power of graphite is between 1 and 50%, depending on the wavelength. The UV spectrum exhibits a maximum at 260 nm related to the difference of energies between states of maximal density in valence and conduction bands (π-resonance). The IR and Raman spectra of graphite each show a solitary, characteristic signal (Figure 1.20) at 867 cm^{-1} and 1575 cm^{-1} respectively.

1.4.2
Diamond

Although the enthalpy of formation under standard conditions for diamond is 1.9 kJ mol^{-1} higher than that for graphite, the first does not spontaneously transform into the latter, but rather represents a metastable modification under normal conditions. In contrast to graphite it is not at all anisotropic regarding any property.

The density of diamond is comparatively high, 3.514 g cm^{-3}. Its hardness according to Mohs is 10, which is the highest value ever measured for a natural material. (Note that today there are several synthetic materials with a hardness higher than that of diamond; these include some of the polymeric fullerenes (Chapter 2) obtained under high pressure.) Still hardness is not equal for each crystal plane.

Furthermore, diamond exhibits the largest thermal conductivity among all naturally occurring materials. With 20 W cm^{-1} K^{-1}, it is about five times higher than that of copper. At the same time it expands only to small extents, which reflects in a coefficient of thermal expansion of 1.06×10^{-6} K^{-1}(mK^{-1}). The specific heat capacity at 25 °C is 6.12 J mol^{-1} K^{-1}.

Diamond without any defects is an electric insulator with a band gap of 5.5 eV. The specific conductivity of such isolating material is 8×10^{-14} Ω^{-1} cm^{-1}, whereas for semiconducting diamond of type IIb it is about 10^{-8} Ω^{-1} cm^{-1}.

The optical properties are the reason for diamond being used as a gem. Besides a high refractive index it features considerable dispersion: while the refractive index is 2.41 at a wavelength of 644 nm, it is already 2.54 at 300 nm and 2.70 in the UV at 230 nm. This gives rise to the so-called fire of the diamond – the varied refraction of different wavelengths and the associated ardent play of colors in cut stones. The good reflecting power ($R = 30–60$, depending on λ) is an additional contribution. Neat diamond is transparent both in the ultraviolet and the infrared range of the spectrum. Owing to their purity, type IIa diamonds are suited best to examine absorptive properties. The absorption edge at room temperature is situated at 230 nm and shifts toward higher wavelengths with increasing temperatures. Type I diamond shows two characteristic bands at $\lambda = 415$ nm and $\lambda = 503$ nm and the absorption edge lies at a markedly higher wavelength ($\lambda = 330$ nm) than in perfect diamond.

Samples of diamond sometimes also exhibit considerable fluorescence. Especially type I diamonds bearing impurities of nitrogen have a pronounced spectrum with two maxima already known from UV/Vis absorption. ($\lambda = 415$ nm: most likely from transitions without participation of foreign atoms or vacancies, but at defects generated from the breaking of C–C-σ-bonds; and $\lambda = 503$ nm: from transitions including foreign atoms on lattice positions.)

The Raman spectrum of diamond is very simple, with a single line at $1331\,cm^{-1}$ resulting from the vibration of the two face-centered partial lattices moving against each other. In the infrared spectrum, on the other hand, there are several bands with in parts distinctive features of types I and II, respectively. In the range from 1500 to $5000\,cm^{-1}$, there are two signals for diamonds of type IIa (~2400 and ~$3600\,cm^{-1}$) resulting from lattice vibrations. For type IIb, an additional band at $2800\,cm^{-1}$ is perceptible. Nitrogen-containing diamonds of type I show yet another signal at about $1200\,cm^{-1}$ caused by the nitrogen; this may be split into anything up to four maxima. In Chapters 5 and 6, the spectroscopic properties of diamond materials will be discussed in more detail.

1.5
Chemical Properties

Carbon is a fairly inert element and most of its modifications may only be reacted under rather harsh conditions. Nevertheless the entire Organic Chemistry, known exclusively to deal with its compounds alone, is founded on the chemistry of carbon. This apparent contradiction is resolved by the simple compounds being hard to obtain from the elements, but any further reaction being quite easy to achieve. They succeed in impressive variety, mainly due to the manifold ways of carbon bonding with itself (chains, rings, single and multiple bonds, etc.).

Carbon's chemistry is governed by its position in the fourth main group. In contrast to the higher elements in this group it does not tend to exert only two out of its four valencies. The maximum connectivity, as present, for example, in diamond, is 4. The octet rule is strictly obeyed in covalently bound carbon, yet there are several coordination compounds with up to eight bonding partners. On these occasions, carbon is engaged in three-center two-electron bonds (for example, $Al_2(CH_3)_2$). They are, however, only possible with more electropositive coordination partners. Due to its medium position in the periodic table, carbon reacts with oxygen as well as with hydrogen and may adopt any oxidation number from +4 to –4. Methane (CH_4) and carbon dioxide (CO_2) represent the extremes of this range.

Carbon monoxide and carbon dioxide, respectively, are obtained from reacting carbon with water or oxygen at a sufficiently high temperature. The latter determines the course of the reaction as well as the present amounts of oxygen or water vapor do. The different enthalpies of the oxidative steps may be explained by the destruction of the crystal lattice required in the reaction of solid carbon to give CO. No energy has to be applied for this process in the second step from CO to CO_2, so more heat is released here.

$$C_{(s)} + 0.5O_2 \rightarrow CO \quad (\Delta H = -110.6 \text{ kJ mol}^{-1}) \quad (1.1)$$

$$CO + 0.5O_2 \rightarrow CO_2 \quad (\Delta H = -283.2 \text{ kJ mol}^{-1}) \quad (1.2)$$

Strong heating of carbon with gaseous sulfur yields carbon disulphide (CS_2) in an endothermal reaction. This may be further reacted with elemental chlorine to give carbon tetrachloride.

With metals and elements such as boron or silicon (in general, with less electronegative elements) carbon forms carbides. Consequently carbon is the electron acceptor in these compounds. There are three different types: salt-like, metallic and covalently bound carbides.

Salt-like carbides are obtained with the electropositive metals from main groups 1 to 3, and with certain lanthanides and actinides. The characteristic feature of this type is the presence of carbon anions. There are methanides (containing C^{4-}), acetylenides (C_2^{2-}) and allenides (C_3^{4-}). Methanides have been described for aluminum and beryllium; they yield methane upon hydrolysis. The acetylenides, that produce ethine when reacted with water, hold isolated $[C\equiv C]^{2-}$-ions in their lattice. Elements from the first main group and subgroup form structures with a composition of $M_2(C_2)$, whereas it is MC_2 with those from the second main group and subgroup. With trivalent metals a stoichiometry of $M_2(C_2)_3$ is obtained (M = Al, La, Ce, Pr, Tb). The most important of acetylenides is calcium carbide. It is produced on a million-tons-scale to generate acetylene for welding or, by reacting it with atmospheric nitrogen, to give calcium cyanamide ($CaCN_2$), which is a valuable raw material for the fertilizer industry. As for the allenides, there are only Li_4C_3 and Mg_2C_3 known to date. These contain isolated $[C=C=C]^{4-}$-ions and release propyne upon hydrolysis.

Elements from subgroups 4 to 6 form metallic carbides that feature a series of distinct metallic properties (conductivity, metallic luster). Where the atomic diameter of the metal is more than 2.7 Å, the carbon atoms may sit in the octahedral gaps of the host lattice. If all of these sites are occupied, a compound with the composition MC (M = Ti, Zr, Hf, V, Nb, Ta, Mo, W) is obtained regardless of the metal's preferred valency. Usually they exhibit a cubic dense packing. Filling only 50% of the octahedral gaps yields carbides with a stoichiometry of M_2C (M = V, Nb, Ta, Mo, W), normally with a hexagonal dense packing. The different types of intercalary carbides show in parts remarkable properties: their melting points lie between 3000 and 4000 °C, chemically they are largely inert, and their hardness comes close to that of diamond. Especially tungsten carbide finds widespread application as material for heavy-duty tools. Furthermore there are metallic carbides of third period metals from the subgroups 6 to 8 whose atomic diameter is less than 2.7 Å. Their respective octahedral gaps are too small to harbor carbon atoms, and so the metal lattice is distorted on carbide formation. At the same time the carbon's coordination number increases, giving rise to a stoichiometry of M_3C (occasionally also M_3C_2, M_5C_2, M_7C_3, etc.; M = Cr, Mn, Fe, Co, Ni). Herein the carbon atoms usually are trigonal-prismatically surrounded by metal atoms. *Cementite* (Fe_3C, an important structural constituent of steel) is an exponent of this class of carbides.

Elements with an electronegativity similar to that of carbon yield covalent carbides. They also feature great hardness. Most of all, silicon carbide has several applications for mechanically stressed objects (grinding and cutting tools) or as material for high-temperature transistors, light-emitting diodes, infrared radiators, etc. It is produced, for example, according to Acheson by the direct reaction of coke and quartz sand in an electric furnace. Boron as well forms several covalent carbides with stoichiometries of $B_{12}C_3$, $B_{13}C_2$, and $B_{24}C$, which are employed for steel borination, as neutron absorber in nuclear reactors, or for vehicle armoring.

Carbon forms a multitude of compounds with the halogens. Typical stoichiometries are CX_4, C_2X_6, C_2X_4, and C_2X_2. Commercially important examples for this class of compounds include carbon tetrachloride, FCCs (fluoro-chloro-carbons) and PTFE (polytetrafluoroethylene, teflon). The FCC's significance has decreased, though, because of their detrimental effect on the ozone layer and the resulting ban of their use.

With nitrogen, carbon forms a series of nitrides with a composition of $(CN)_n$ (n = 1, 2, x, where x can be any large number except ∞) among which the cyanogen (n = 1) is stable at high temperatures only. Paracyan $(CN)_x$ is obtained from dicyan by polymerization.

1.5.1
Graphite and Related Materials

Graphite may be the thermodynamically most stable modification of carbon. Still it is chemically attacked more easily than diamond due to its layered structure and the comparatively weak interaction between the graphene sheets. Altogether the graphite's reactivity toward many chemicals is rather low nevertheless. With chlorine, for example, it does not react at all under usual conditions, and even with fluorine reaction occurs only at more than ~400 °C. Suitable performance yields the transparent, colorless carbon monofluoride CF (up to $CF_{1.12}$ due to additional fluorine atoms at marginal positions and defects), a chemically very resistant

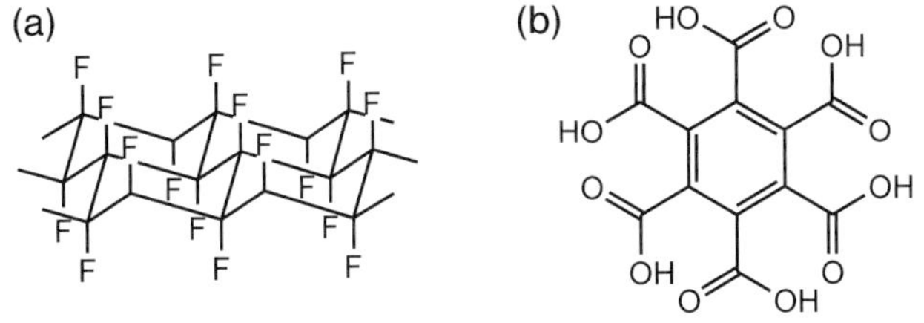

Figure 1.21 (a) Structure of the compound $(CF)_x$ resembles an undulated graphene sheet or the structure observed for *Lonsdaleite*, (b) mellitic acid.

insulator that is employed as dry lubricant. The individual layers of the graphitic lattice are conserved in $(CF)_x$ (Figure 1.21a), with the fluorine atoms alternatingly pointing upward and downward. This causes an interlayer distance of 6 Å, compared to 3.35 Å in graphite. On nonstoichiometrical fluorination ($CF_{0.8-0.9}$), the obtained graphite fluoride retains its electrical conductivity. Fluorination at temperatures >700 °C converts graphite into CF_4, whereas concentrated nitric acid turns it into mellitic acid $C_6(CO_2H)_6$ that contains planar C_{12}-units (Figure 1.21b). In the mixtures of HNO_3/H_2SO_4, graphite reacts with potassium perchlorate to give the so-called graphitic acid, named such for the existence of weakly acidic OH-groups. This compound is obtained as greenish-yellow, lamellar crystals that explode on heating. Aqueous bases do not perceptibly attack graphite, and the reaction with hydrogen succeeds only at high temperatures, for example, in an arc, mainly to yield acetylene. Graphite is not soluble in any of the common solvents; only molten iron does solve significant amounts.

A particularity within the chemical behavior of graphite is the formation of the so-called intercalation compounds. These result from embedding atoms or molecules in between single graphene sheets ("sandwich" structure). Their weak interaction enables the formation of this kind of compounds. There is a rearrangement of charges upon intercalation so the carbon layers are polarized (or even ionized) either negatively or positively, depending on the kind of intercalate (embedded atom or molecule). The exchange of electrons between the intercalate and the graphite's π-bands increases the number of charge carriers in the conduction or the valence band, respectively. Electron-withdrawing species cause holes in the valence band of the graphite, whereas electron donors partly fill the conduction band with electrons. Therefore graphite intercalation compounds exhibit an increased electrical conductivity.

The longest known among these compounds are the C_nK phases ($n = 8, 24, 36, 48$), with the first of them, C_8K, having been made in 1926 already. Depending on the stoichiometry, the potassium atoms are separated by one to four graphene layers (Figure 1.22) in these compounds, which are referred to as *steps*. The first of these steps deviates from the $C_{12n}K$-rule ($n = 2, 3, 4$) that holds for higher homologs – it represents the compound with the highest possible number of potassium atoms incorporated. It is noteworthy that the distance between graphene layers increases only by 2.05 Å, which after all is much less than could be

(a)

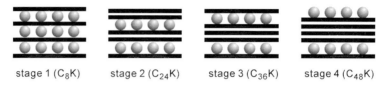

stage 1 (C$_8$K)   stage 2 (C$_{24}$K)   stage 3 (C$_{36}$K)   stage 4 (C$_{48}$K)

(b)

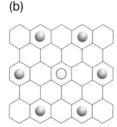

Figure 1.22 Intercalation compounds of potassium with different contents of metal are highly ordered structures. (a) Stacking of graphene layers and metal atoms (side view), (b) top view of the metal atoms arranged in the graphene lattice. The dark circles indicate positions occupied in all cases, while the position marked by the light circle is only occupied in C$_8$K.

expected upon the intercalation of metal atoms measuring 3.04 Å. This indicates an interaction of the potassium atoms with the clouds of π-electrons of the surrounding graphene sheets.

Actually, K$^+$ ions are formed – the released electron is transferred into the graphite's conduction band, causing the high electrical conductivity. C$_8$K is a versatile reducing agent in organic synthesis. Besides potassium, other alkaline or alkaline earth metals may form ionic intercalation compounds as well. Their composition is C$_n$M (e.g., M = Rb, Cs, Ca with n = 8, 24, 36, 48 or M = Li, Ca, Sr, Ba with n = 6) with the graphite being reduced. Sodium–graphite compounds, on the other hand, are hard to obtain. The golden-colored C$_6$Li is employed in efficient batteries.

As the valence and the conduction band exhibit similar energies, the graphite in intercalation compounds may also be oxidized when electron acceptors are embedded. The intercalation compounds with bromine or certain interhalides are important examples here: iodine and chlorine do not react with graphite, whereas with fluorine, the aforementioned carbon fluorides are obtained. In C$_8$Br, then, electrons from the valence band of graphite are transferred to the bromine. However, it is not present as isolated anions, but in the shape of polybromide ions. The Br–Br distance of 254 pm correlates quite well with that between the centers of the hexagons in the graphite lattice (256 pm). For other polyhalogenide anions, on the other hand, the interatomic distance (Cl–Cl: 224 pm and I–I: 292 pm) is too low or too high, respectively, to perfectly fit in these voids, and consequently no interstitial compounds are formed. Iodine monochloride (expected I–Cl distance: 255 pm) then again is incorporated. Halogen intercalation compounds are also good electric conductors. However, the charge transport occurs via holes (defect electrons) instead of electrons, similar to the mechanism in p-doped semiconductors.

Apart from graphite halogenides, there are also intercalation compounds with sulfuric, nitric, phosphoric, perchloric, and trifluoroacetic acid. They all exhibit a considerably larger distance between their graphene layers, and the graphite donates electrons to the respective intercalate. With sulfuric acid, for example, graphite hydrogensulfate is obtained, and the reaction with nitric acid yields graphite nitrate. Also metal salts and oxides like $FeCl_3$, $AlCl_3$, SbF_5, CrO_3, CuS, etc. may be embedded between the graphene sheets. The chromium oxide compound CrO_3-graphite is a selective oxidant to convert secondary and tertiary alcohols into the related keto-compounds (*Lalancette* reagent), whereas $AlCl_3$-graphite is employed as a selective Friedel–Crafts catalyst.

1.5.2
Diamond

Diamond is extremely hard to subject to a chemical reaction. Due to its perfect crystal structure, it is only attacked at very high temperatures. It is true that diamond powder is inflammable, yet bigger lumps are only ignited in an oxygen blower at more than 800 °C. The oxidation rate depends on the size and surface characteristics of the single particles; the temperature of combustion varies between 750 and 880 °C. In a stream of oxygen at 900 to 1200 °C, diamond reacts completely to carbon dioxide, whereas graphite yields a CO/CO_2 mixture.

No reaction takes place with acids or bases, appreciable attack only occurs with chromosulfuric acid, producing carbon dioxide. With fluorine at more than 700 °C, diamond reacts to give carbon fluorides or to be surface-fluorinated. The hydrogenation of the diamond's surface succeeds in a stream of hydrogen at high temperatures, yet the entire sample is not converted. The action of iron at more than 1150 °C yields an alloy. With nitrogen, no appreciable reaction is observed even at high temperatures, and with water vapor at about 1000 °C, diamond weakly reacts forming carbon monoxide. Highly oxidating melts of $KClO_3$ or KNO_3 do not attack diamond (with the exception of black, polycrystalline material). Altogether, little is known about "diamond chemistry" in the true sense of the meaning, because this modification of carbon largely resists any chemical transformation.

1.6
Application and Perspectives

The main part of the carbon employed is consumed for fueling and in the steel industry. The combustion of fossil raw materials alone absorbs the largest fraction of the hauled carbon, but besides that there is a multitude of applications of its different modifications.

1.6.1
Graphite and Related Materials

The applications of graphite are as manifold as the existing variants of graphitic materials. Large amounts (natural or artificial) are used for electrodes in arc furnaces to run different electrolytic processes like production of certain metals or alloys (sodium, aluminum, steel, etc.), and chlorine or carborundum are likewise obtained by electrolysis on graphitic electrodes. Graphite further is needed for carbon brushes in electric engines. In addition, graphitic substances are employed in arc lamps, as materials for molds or to line ovens and apparatus. In chemical facilities, graphite rendered gas-tight by impregnation with plastics is utilized, and in nuclear reactors, rods of graphite are used as moderators due to their large neutron-capture crosssection. Provided the distinctly anisotropic thermal conductivity at highly parallel orientation of the crystal lattices, the respective material may be used in making heat shields for spacecrafts, brake blocks or the like. Graphite is also employed as a dry lubricant due to its low hardness. In particular for chemically aggressive environments, it usually is a better choice than, for example, hydrocarbon-based agents. The leads of modern pencils contain graphite as well, and finally, it is used as carrier for certain stationary phases in reversed-phase high performance liquid chromatography (RP-HPLC).

1.6.2
Diamond

While its brilliance and fervent iridescence make it one of the most coveted gems, diamond is also most profitably applied to a number of far less decorative purposes. This is due to its great hardness, chemical and mechanical resistance, large thermal conductivity, and good electrically insulating properties. A considerable part of the natural and artificial diamond produced is employed for grinding, cutting or drilling instruments. Their use pays off even at their high price because there are less down-times due to failure or change of tools. Furthermore so-called wire drawing stones are made by drilling a hole through diamond with the desired diameter. They are used to draw wires from hard metals. In apparatus construction, bearings for fast-rotating axles can be made from diamond. Finally, diamond windows are employed for special spectroscopic applications as they are transparent over a wide spectral range, in particular for the UV where many of the materials commonly used for spectroscopic windows let pass only little or no light at all.

1.6.3
Other Carbon Materials

Apart from being a popular filler for elastomeric materials, soot is an excellent black pigment. Several properties make it downright ideal for printing purposes: it is light-fast and insoluble in almost any common solvent as well as it exhibits low particle size and great color depth and strength. Yet more than 90% of the carbon black produced is used as a filler – about two-third of it for tires, and the rest for other things made from rubber. An average tire for cars contains about one-third of soot that serves not only to increase the wear- and tear-resistance of the rubber (caoutchouk), but also to blacken it. Other plastics, lacquers and inks for writing or printing are likewise colored by the addition of carbon black.

Owing to their chemical, thermal, and mechanical resistance, carbon fibers are an ideal material for highly durable composites. Compared to steel, they are four times lighter at similar strength, which in addition is preserved up to temperatures of more than 2000 °C. Certain threads, ribbons, and fabrics are made from pure carbon fibers, as well as there are carbon-reinforced plastics which are employed to manufacture, for example, components of aircrafts or cars, sporting devices, implants, or filters for dusts and aerosols.

Activated carbon is a versatile adsorbent, for example, to decolorize sugar or to free spirit from fusel oil. Gases may be purified and toxic substances may be removed from breathing air (application in gas masks) same as flue gases are desulfurized using activated carbon.

Glassy carbon is chemically very inert even at elevated temperatures because there is virtually no porosity at all and its structure does not allow for intercalation. It is only attacked by oxygen and oxidizing melts at more than 600 °C. Accordingly, it is employed in ultratrace analysis (no memory effects in crucibles of glassy carbon) and in the semiconductor industry.

1.7 Summary

Under standard conditions, there are several modifications of carbon which may be all traced back to the basic graphite and diamond.

Box 1.1 Graphite.

- Most stable form of carbon under standard conditions ($\Delta H_0 = 0\,\text{kJ}\,\text{mol}^{-1}$).

- Two modifications: hexagonal α-graphite and rhombohedral β-graphite.

- The carbon atoms are sp^2-hybridized, σ-bonds with three adjacent atoms within one layer (bond angle 120°). Additional delocalized π-bonds within these layers. Only weak van der Waals interaction between the graphene sheets.

- Considerable anisotropy of properties like electrical conductivity, modulus of elasticity, etc due to the layered structure.

- Despite chemical inertness several compounds are known, above all intercalation compounds with alkali metals or halogens.

Besides crystalline graphite, there are a number of related materials that are less crystallized and usually exhibit a finer particular subdivision. Their properties differ (in parts considerably) from those of the parent system, which is mainly due to the smaller particle size and the disordered lattice.

Box 1.2 Diamond.

- Metastable modification of carbon. Appears as cubic or as hexagonal diamond (*Lonsdaleite*).

- Each C-atom is tetrahedrally connected to four adjacent atoms by σ-bonds. The C-atoms are sp^3-hybridized.

- Greatest hardness and highest thermal conductivity among all natural materials. Electrical insulator, yet semiconductance achievable by doping.

- Chemically extremely inert, is only attacked by aggressive reagents like chromosulfuric acid.

Proceeding from the facts presented in this chapter, we will now discuss the "new" carbon materials, starting with the fullerenes. These have first been discovered in 1985, many years after their theoretical prediction.

2
Fullerenes – Cages Made from Carbon

This chapter deals with cages made from carbon atoms which, for being molecular allotropes, lastingly changed the scientific view of the element. Besides their esthetical structure, they feature interesting physical and chemical properties that will be discussed herein.

2.1
History – The Discovery of New Carbon Allotropes

2.1.1
Theoretical Predictions

The idea of cage-like carbon structures is by far not as new as one might believe. First theoretical considerations date back to the year 1966. The scientist D. E. H. Jones, who chose himself the pseudonym Daedalus, published theoretical treatises on fullerene-shaped objects. However, no attention of any kind was bestowed upon his speculations about hollow structures entirely made from carbon atoms.

A first theoretically substantiate paper on C_{60} was published by E. Osawa in 1970. He let himself be inspired by his son's soccer ball while reflecting on superaromatic π-systems and postulated an analogous structure with icosahedral symmetry for the C_{60}-molecule, predicting its stability from Hückel calculations. He saw that corannulene, synthesized shortly before by Barth and Lawton, had to be a partial structure of that cage (Figure 2.1). His publications as well were granted due attention only after the experimental discovery of fullerenes.

The same was true for some other theoretical approaches (especially Hückel calculations by the Russian scientists Bokhvar and Galpern correctly predicted the π-system of the C_{60}-molecule.

Even though they are not theoretical treatise themselves, the buildings of the American architect Richard Buckminster Fuller (1895–1983) were some structural inspiration nevertheless. His geodetic domes, self-supporting cupolas consisting of various polygons, show similar structural features like the carbon cages, and it was to his honor that the latter were named "fullerenes" or "buckminsterfullerenes". Still it had already been in the early twenties that the first geodetic dome

Carbon Materials and Nanotechnology. Anke Krueger
Copyright © 2010 WILEY-VCH Verlag GmbH & Co. KGaA, Weinheim
ISBN: 978-3-527-31803-2

Figure 2.1 Corannulene and C_{60}. The curvature is induced by the five-membered rings.

Figure 2.2 With his geodetic domes, R. Buckminster Fuller provided the architectonic inspiration for the naming of the fullerenes. The US pavilion for the Montréal EXPO76 is just one example (© Montréalais).

was realized by the German architect W. Bauersfeld, but he failed to label his structure with a catching name, and so today such edifices are inevitably connected to the name of *Fuller*. It was mainly in the sixties and seventies when a number of buildings were erected taking up Fuller's concept. One of the most impressive examples is the US-American pavilion at the 1967 World Expo in Montréal (Figure 2.2).

Microorganisms of the class of diatoms represent another example of a structural relation. These algae grow in structures that correspond to 100.000-fold enlarged fullerene cages (Figure 2.3) with the polygons on their surface exhibiting the same distribution pattern as observed for fullerenes.

2.1.2
Experimental Proof

In 1985, the world of the element carbon was deemed an exhaustively explored and mature field of research. For the most important allotropes, the essential properties were known as well as the interconnections between them, and apart from Organic Chemistry, the elemental chemistry of carbon was thoroughly examined, too. Then came the day when a single signal at m/z 720 in a mass spectrum induced a lasting change in the view on a then slightly neglected element. The era of "new" carbon allotropes had begun.

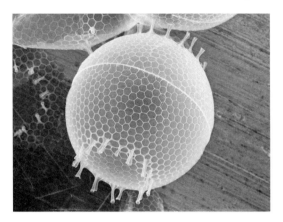

Figure 2.3 The diatom *Stephanopyxis turris* is an example of a biological structure similar to the fullerenes (© M. Sumper).

A year before that, first experimental results had already suggested the existence of certain carbon clusters. In experiments on laser-evaporation of graphite, signals attributable to the clusters of 30 to 190 carbon atoms had been detected in the time-of-flight mass spectrum. Most peculiar was the fact that only even numbers of atoms had been observed. Still for the time being, the clusters' structure had not been elucidated.

H. W. Kroto, J. R. Heath, S. C. O'Brien, R. F. Curl, and R. Smalley then made the decisive breakthrough in the discovery and identification of fullerenes when attempting to simulate stellar conditions, namely those in the class of Red Giants. Their experiment consisted in focusing a pulsed laser beam on a target of graphite under a stream of helium and subsequently analyzing the resulting particles in a mass spectrometer. As expected, they mainly detected molecules with compositions of HC_7N and HC_9N like they are also prevalent in space, but on modifying the experimental conditions, they suddenly found the spectra dominated by a signal at m/z 720. Yet another peak at m/z 820 was observed, which relates to C_{70}. Their conclusion that they had found a C_{60}-molecule with a highly symmetric cage structure was awarded with a Nobel prize for Kroto, Smalley, and Curl. Still this first evidence did not constitute a suitable method to supply sufficient quantities for an investigation on physical and chemical properties of the new substances.

It was only in 1990 when the groups of W. Krätschmer and D. R. Huffman succeeded in isolating macroscopic amounts of the most abundant fullerene C_{60} by the arc evaporation of graphite or by resistance heating of graphite electrodes, respectively. In fact, their initial intention, too had been to produce in the laboratory substances that exist in interstellar dust, but at suitable conditions, their IR-spectra exhibited four sharp, characteristic signals that related to those predicted for the C_{60}-molecule. Later on, C_{70} and the higher fullerenes were also obtained.

The first description of fullerenes containing one or more atoms within the cage followed soon after detection of the fundamental structure. These so-called endohedral fullerenes played a crucial part in the isolation of higher fullerenes.

2.2
Structure and Bonding

2.2.1
Nomenclature

Before turning toward the structural features of fullerenes, it is necessary to find a convention allowing for their unambiguous description in a simple and conceivable way. Unfortunately, the IUPAC nomenclature not really has a catching suggestion here; the systematic name for the most common fullerene consisting of 60 carbon atoms is:

Hentriacontacyclo-$[29.29.0.0^{2,14}.0^{3,12}.0^{4,59}.0^{5,10}.0^{6,58}.0^{7,55}.0^{8,53}.0^{9,21}.0^{11,20}.0^{13,18}.$
$0^{15,30}.0^{16,28}.0^{17,25}.0^{19,24}.0^{22,52}.0^{23,50}.0^{26,49}.0^{27,47}.0^{29,45}.0^{32,44}.0^{33,60}.0^{34,57}.0^{35,43}.0^{36,56}.0^{37,41}.0^{38,54}.$
$0^{39,51}.0^{40,48}.0^{42,46}]$*hexaconta-1,3,5(10),6,8,11,13(18),14,16,19,21,23,25,27,29(45), 30,32(44),33,35(43),36,38(54),39(51),40(48),41,46,49,52,55,57,59-triacontaene.*

Obviously this proposal is neither practicable nor instructive, and so a much simpler notation has been conceived which nevertheless describes all essential structural features. This system is based on the assumption that all fullerenes are cage-like structures. This allows for the omission of any information on the interconnectivity of rings, provided the symmetry of the object is included in its name. A fullerene's essential characteristic is the number of carbon atoms contained, so this figure must be found at a prominent position in the name. Finally, information on the cage's constituent rings is required, which are five- and six-membered in the majority of cases. With these data (number of C-atoms, symmetry and kind of rings), any fullerene can be described sufficiently. For example, the following name results for C_{60}: (C_{60}–I_h) [5,6]-fullerene. The first term indicates the number of carbon atoms and the symmetry by way of the point group (icosahedral symmetry I_h in this case), the second specifies the object to consist of five- and six-membered rings, and the name "fullerene" stands for a cage-like structure. The name of the next higher homolog is formed accordingly: (C_{70}–D_{5h}) [5,6] fullerene.

As will be seen later, there are also fullerenes with one or more carbon atoms being substituted by other elements. In fullerene nomenclature these molecules are described by prefixing to the name the designating syllable for the respective hetero atom, for example, aza[70]fullerene $C_{69}N$, bis(aza[60]fullerene) $(C_{59}N)_2$, bora[60]fullerene $C_{59}B$, or phospha[60]fullerene $C_{59}P$.

In certain compounds, the so-called endohedral fullerenes, atoms or small molecules are incorporated into the cage without forming a covalent bond to the fullerene scaffold. The notation $M@C_n$ (pronounced M *at* C_n) is used to indicate the position within the fullerene, with M representing the atom or molecule inside the cage. In this way, hetero- and endofullerenes can clearly be distinguished, for

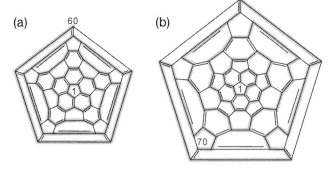

Figure 2.4 The Schlegel diagrams of C_{60} (a) and C_{70} (b). The lines shaded in grey indicate the numeration of carbon atoms along a spiral trace.

example, it is evident from the terms $N@C_{60}$ and $C_{59}N$ where the respective nitrogen is situated. Indeed, the IUPAC suggests the notation iMC_n (pronounced [n] fullerene-incar-M), but this is unusual in praxi, and the @-notation will be used in this book as well.

Now with functional groups attached by chemical reaction to one ore more of the carbon atoms, the hitherto existing nomenclature does not suffice any more. An additional, unambiguous way of numbering the cage's atoms is required. The basic idea in this is drawing a helical trace along all cage atoms of the fullerene, which are depicted in a so-called Schlegel diagram (Figure 2.4). This shows a projection of the cage in which the bonds are expanded radially around an axis of rotation of the fullerene until all atoms lie within a common plane. As a reference for this projection the axis of highest possible order is chosen. It must allow for drawing a helical line that begins at the atom or ring where the axis of rotation intersects the fullerene's shell on the side facing the spectator. Accordingly, the outer pentagon of the Schlegel diagram in Figure 2.4(a) corresponds to the five-membered ring on the rear side of the C_{60} through the center of which the respective axis passes. Schlegel diagrams may also serve to express the chirality of fullerenes, the notation for such systematically numbered fullerenes then is $^{f,s}C$ and $^{f,s}A$ for *c*lockwise or *a*nticlockwise helical traces, respectively.

2.2.2
The Structure of C_{60}

The structure of the most important fullerene, C_{60}, shall now be discussed in detail. The major difference, compared to classical modifications of carbon, is the molecule being a more or less punctiform, discrete unit instead of a structure repeating three dimensionally through space.

If in a graphene sheet a few six-membered rings are replaced by five-membered ones, the layer will be forced out of its plane and into a bent shape (Figure 2.5a). Placing the pentagons at suitable positions, a spherical structure comprising of 60

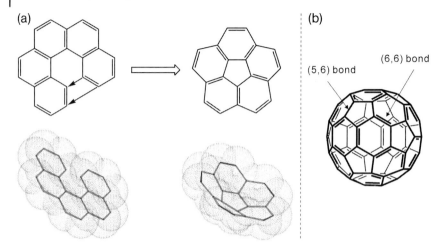

Figure 2.5 (a) A previously planar aromatic structure is getting curved by the closure of a five-membered ring; (b) C_{60} contains 12 five-membered and 20 six-membered rings showing the characters of radialenes or cyclohexatrienes, respectively. The bonds situated between two adjacent six-membered rings are called (6,6)-bonds, those between a five- and a six-membered ring (5,6)-bonds.

carbon atoms is obtained – the buckminsterfullerene that resembles a classical soccer ball. Please find in the annex the grid of a C_{60} molecule that may be assembled (following the instruction) for a better spatial understanding.

⇨ **Construction of a (C60–Ih) [5,6]-fullerene (refer to the last page of the book)**

Other but five-membered rings do also induce a curvature of graphene layers, but they cause by far larger tension in a closed cage, and so structures with four-, seven-, or eight-membered rings are energetically unfavorable (*NB* spherical giant fullerenes in carbon onions, Chapter 4).

The 60 equivalent carbon atoms of C_{60} occupy the corners of a polyhedron that exhibits icosahedral symmetry (32 faces, 60 corners, 90 edges) and may be considered a truncated pentagon–dodecahedron. The object belongs in point group I_h and features 6 S_{10}-rotary alternating axes (including 6 C_5-axes of rotation), 10 S_6-rotary alternating axes (including ten 10 C_3-axes of rotation), 15 C_2-axes of rotation, and 15 mirror planes each of which contains two C_5- and two C_2-axes of rotation. The molecule is easily identified spectroscopically due to this high symmetry (refer to Section 2.4). It is composed of 12 pentagons and 20 hexagons, with each five-membered ring being surrounded by six-membered ones alone. We will later find this to be the typical feature of stable fullerenes.

Theoretical as well as experimental studies proved that not all C–C bonds in C_{60} are of equal length, they rather alternate with the (5,6)-bond between five- and six-membered rings being longer than the (6,6)-bond between adjacent hexagons. The double bonds are thus localized in the six-membered rings, with the bond lengths of 139 pm and 145 pm roughly correlating to those in conjugate polyolefins. The diameter of a C_{60} molecule is 0.702 nm. The bond lengths meas-

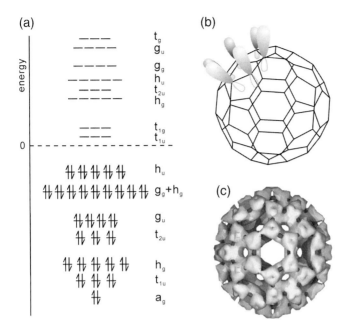

Figure 2.6 Scheme of molecular orbitals for C$_{60}$ (a) and radial orientation of π-orbitals on the fullerene surface (b). The AICD-plot indicates the calculated anisotropy of current density that may be correlated to the delocalization of π-electrons (c) (© F. Köhler). The character of C$_{60}$ as a moderately conjugated polyolefin is confirmed here.

ured gave rise to the Kekulé structure depicted in Figure 2.5b. The pentagonal rings show characteristics of a [5]-radialene, while the hexagons each correspond to 1,3,5-cyclohexatrienes.

Still the stability of C$_{60}$-fullerene is partly based on its three-dimensional aromaticity. Actually the valence bond formula in Figure 2.5 does show the alternating bonds, but not the delocalization of π-electrons. The molecular orbital model serves better to this purpose, here it becomes clear why C$_{60}$ experiences a particular stabilization: C$_n$-molecules are the most aromatic when (a) the bonding π-orbitals are fully occupied and (b) these orbitals have a low energy as well as a big energetic distance to the first unoccupied orbital (Figure 2.6a). Another approach follows the so-called *leap frog rule* that predicts those fullerenes to be especially stable that possess three times as much carbon atoms as a conceivable fullerene. In this way the marked stability of C$_{60}$ may be deduced from C$_{20}$, its smallest closed-cage relative.

In principle, the bonding situation in C$_{60}$ does not differ from that in other carbon skeletons bearing delocalized π-electrons; only the scaffold of σ-bonds is forced out of its usual angle of 120° by the surface curvature. The actual degree of hybridization is sp$^{2.278}$, which means a considerable shift toward tetrahedrally coordinated carbon atoms. The π-electrons of the individual atoms are situated in the p$_z$-orbitals perpendicular to the cage's surface (Figure 2.6b). These give rise to a cloud of π-electrons enclosing the carbon cage with a diameter of about 1.05 nm. The s-fraction of the π-orbitals is nothing less than 0.085. The delocalization of

Figure 2.7 Two examples of further theoretically possible structures of C_{60}. The neighboring five-membered rings are accentuated by bold lines.

π-electrons in C_{60} is less than in real aromatic structures for two reasons: on the one hand, the atomic lattice does not extend over a plane, but in three dimensions, and on the other hand, the molecule's shape causes a less efficient overlap of orbitals (Figure 2.6).

The electronic structure of C_{60} is characterized by a fivefold degenerate HOMO (h_u) and a threefold degenerate LUMO (t_{1u}), divided by a gap of 1.8 eV (or 2 eV in other publications; Figure 2.6a). The LUMO can accept up to six electrons, so C_{60} may form anything from mono- to hexa-anions in redox reactions. The interaction with metals of the first and second main groups includes *n*-doping, too.

Actually C_{60} is an excellent electron acceptor due to a particularly low energy of reorganization in electron-transfer reactions and a good delocalization of electrons and charges. For the same reasons, a retransfer from the fullerene (charge compensation) is restrained in comparison to normal electron acceptors (e.g., quinones), so charge separation is strongly favored in donor–acceptor systems containing fullerenes.

The question remains why there is only a sole icosahedral structure of C_{60}, while there is a theoretical number of 1812 isomers. Even with a restriction to accept five- and six-membered rings alone, there is still a variety of different conceivable closed-cage isomers. All of them, however, would contain five-membered rings not exclusively surrounded by six-membered ones as in the structure actually detected (Figure 2.7). Now, adjoining pentagons cause elevated tensile stress within the molecule, thus destabilizing the respective isomers in comparison to I_h-C_{60} (refer to Section 2.2.3). This is the smallest possible fullerene with all pentagons being isolated.

2.2.3
Structure of Higher Fullerenes and Growth Mechanisms

According to Euler's theorem on closed polyhedrons (refer to Section 4.2.1), every fullerene composed of five- and six-membered rings contains 2 (10 + M) C-atoms, and the resulting cage is made of exactly 12 penta- and M hexagons. The smallest conceivable fullerene accordingly is C_{20} (M = 0). While this molecule could not be isolated experimentally so far, the related hydrocarbon $C_{20}H_{20}$ is known. It features the expected pentagon–dodecahedral structure. With M increasing, the number of possible cage structures rises, for example, for the C_{78}-fullerene there are already more than 20 000 theoretical isomers.

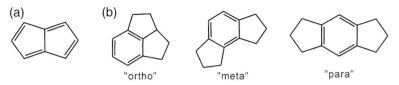

Figure 2.8 (a) The unfavorable pentalene structure shown here can only be avoided with isolated five-membered rings. (b) The "meta"-arrangement of five-membered rings is preferred for its avoidance of both neighboring pentagons and double bonds within five-membered rings.

The actual number of isomers observed is much smaller – there is, for instance, exactly one each for C_{60} and C_{70}. Other fullerenes, like C_{62-68}, do not form at all, even though they are theoretically possible. It is obviously impossible for such carbon cages to adopt a structure exclusively made from hexagons and pentagons with the latter being isolated.

The above observations may be summarized to state some rules that quite reliably predict the formation of certain, stable fullerenes:.

- Open-shell structures are energetically unfavorable and thus avoided.

- Structures with isolated pentagons are preferred to those containing adjacent five-membered rings. In this way, pentalene-like units with eight π-electrons are avoided – these would otherwise lead to a resonance destabilization (Figure 2.8a). Also the strain of the cage would increase if the elements causing the curvature were not evenly distributed over the entire surface. The unnatural bond angles of the involved carbon atoms would considerably destabilize the molecule. This is the so-called isolated pentagon rule or *IPR*.

- Another consequence of the IPR is that the structures best approximating a spherical shape are most stable.

- The number of double bonds situated in five-membered rings is minimized. As a result, five-membered rings connected via a six-membered one in *"meta"* position are preferred to the respective *"para"* variant (Figure 2.8b). *"Ortho"*-oriented pentagons are already disfavored according to the IPR.

The number of possible isomers is drastically reduced by these rules, and it becomes clear why many C_n molecules do not adopt a cage-like structure. The "magic" numbers for stable fullerenes obeying the rules above are n = 60, 70, 72, 76, 78, 84, ... , and really all of them (except for C_{72}) can be affirmed experimentally. Possibly any C_{72} initially created is transformed *in situ* into the even more stable C_{70}.

Higher fullerenes are by far less abound than the archetypical compound C_{60}. The formation of these cage-like compounds is kinetically controlled: among the fullerenes, C_{60} is by no means the molecule with the least enthalpy per carbon atom. Normally the larger, thermodynamically more stable homologs (refer to Section 2.4) should be more abundant than C_{60} with its higher curvature-induced

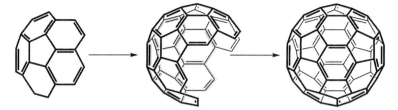

Figure 2.9 The formation of a fullerene cage. The tendency toward minimizing the number of unsaturated bonding sites causes the generation of five-membered rings and thus an increasing curvature and final closure of the structure to be a cage-like molecule.

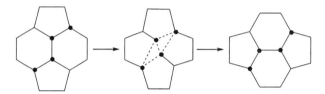

Figure 2.10 The five-membered rings are arranged in an energetically favorable way by rearrangements via four-centered Hückel transition states.

strain. Yet it seems to resist further transformation due to its perfect and inert structure. Once formed, it represents a local minimum on the energy hypersurface and will not react with further carbon clusters. Structures with about 60 carbon atoms will also form C_{60} by assimilating or expelling single atoms or small clusters, respectively. Only when markedly exceeding the number of 60 atoms, another energetic minimum like, for example, C_{70} is attained. As only graphene fragments with a smaller defect-induced curvature may grow beyond a critical size and close to higher fullerenes, the unexpected abundancy of C_{60} becomes explicable.

The *mechanism of fullerenes growing* from small carbon clusters is itself the subject of lively discussion. The only undisputed fact therein is that it is not a thermodynamically determined process, but a kinetically controlled, radical one. It also seems established by now that the initial clusters tend to minimize the number of their unsaturated bonding sites ("dangling bonds") by rapidly connecting them. Thus five-membered rings are formed in the developing graphene network. From trapping experiments various intermediate polycyclic aromatic compounds are known to appear that represent the partial structures of the final fullerene. Some of them may already be curved by the presence of five-membered rings. Further growth will then "coil up" the structure and finally lead to the closure of a cage (Figure 2.9). This ending benefits from the number of dangling bonds being reduced to zero in a single step. If the fast saturation of free bonding sites brings about a structure with unfavorably positioned pentagons, a better distribution of strain energy is established by processes like, for example, Stone–Wales rearrangements (refer to Sections 3.4.3 and 3.5.1). These are concerted reactions passing a four-centered Hückel transition state (Figure 2.10).

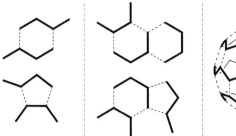

Figure 2.11 The mechanism of fullerene formation by spirocyclization. Indicated to the left and in the middle: possible reaction schemes for the formation of six-membered (top) and five-membered rings (bottom). To the right the bold line illustrates a hypothetical polyyne chain that is closed in several (simultaneous) cyclization steps to become the fullerene cage (© Royal Soc. 1993).

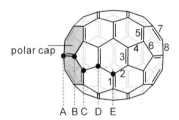

Figure 2.12 The structure of C_{70}. The closest similarity to C_{60} exists in the vicinity of the poles, which is also the region showing the highest curvature. The distinguishable C-atoms are indicated by capital letters, the different bonds by numbers.

An alternative growth mechanism proposes an initial formation of long polyyne chains that react in a spirocyclization, completely or partially to give fullerene-like structures (Figure 2.11). Sequential cycloadditions of smaller units (C_2 to C_4) are yet another conceivable mechanism, for example, C_{60} might entirely be assembled by cycloadditions of C_4-units (as diyne). Again the incorporation of pentagons would give rise to bowl-shaped intermediates that finally would close to be cage-like structures.

The next higher stable fullerene to C_{60} is C_{70}. Its shape resembles a rugby ball, and it contains twelve isolated pentagons as well. The structures differ in a row of additional six-membered rings inserted along what could be called the waist line of C_{70}, thus increasing the number of hexagons from 20 to 25 (Figure 2.12) and, at the same time, reducing the molecule's symmetry to D_{5h}. Furthermore, the bond lengths vary stronger than in the highly symmetrical C_{60} (Table 2.1). In C_{70} there are five distinct kinds of carbon atoms and eight different C–C bonds (Figure 2.12). The structure exhibits domains of varying curvature, strongly resembling C_{60} around its poles where the inflection is highest. The belt of hexagons in the equator region is the least strained.

The more carbon atoms a fullerene contains, the greater becomes the number of actual isomers. Yet, as mentioned before, not all theoretical structures are found, but only a selection of them, namely those that obey the rules stated above. Hence, there is exactly one isomer of C_{76}, 5 of C_{78}, 24 of C_{84} and 46 of C_{90} (compared

2 Fullerenes – Cages Made from Carbon

Table 2.1 Theoretical and experimental bond lengths in C_{60}.

Method	Hartree–Fock (HF STO-3G)	Møller–Plesset (MP2)	NMR	Neutron diffraction	X-ray diffraction
Length of (5,6)-bond	146.5 pm	144.6 pm	144.8 pm	144.4 pm	146.7 pm
Length of (6,6)-bond	137.6 pm	140.6 pm	137.0 pm	139.1 pm	135.5 pm

See A. Hirsch, M. Brettreich, *Fullerenes*, Wiley-VCH, Weinheim **2004**.

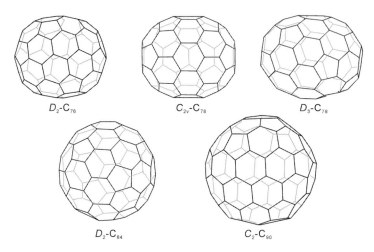

Figure 2.13 Some examples of higher fullerenes.

to a theoretical >20 000 for C_{78}). Figure 2.13 shows some essential examples of higher fullerenes and their respective point groups.

⇨ **Construction of a (C_{76} D_{5h}) [5,6]-fullerene (chiral element highlighted; refer to last page of the book)**

A noteworthy feature of C_{76} is the helical arrangement of carbon atoms on its surface. It is thus a chiral molecule, which also reflects in its symmetry of D_2. As there is no chiral information applied in the making, naturally a racemic mixture is obtained which may, nevertheless, be separated by chiral chromatography. There are other higher fullerenes that exhibit chiral structural isomerism as well, for example, D_3–C_{78} or D_2–C_{84} (Figure 2.13).

Besides these big fullerenes, there are also the so-called giant fullerenes, which are postulated to be constituents of the carbon onions (Chapter 4). Their structure is still subject to lively discussion because it could not be elucidated beyond all doubt why they usually show a perfectly spherical shape – assuming the IPR also to hold for giant fullerenes, one would expect markedly faceted or more or less icosahedral objects (Figure 4.4). Obviously the spherical structure is caused by further defects in the graphitic network. These might be pentagon–heptagon defects (Stone–Wales defects) incorporated in the lattice instead of two neighboring six-membered rings and leading to a better distribution of the curvature. Yet giant fullerenes have not been isolated as independent objects by now. Cages of that size presumably are unstable and collapse to form more compact structures.

2.2.4
Structure of Smaller Carbon Clusters

According to Euler's theorem, the smallest possible polyhedron consisting of pentagons and/or hexagons alone is the pentagon–dodecahedron (N = 2 (10 + M) with M = 0). For 20 carbon atoms, structural isomers as for instance depicted in Figure 2.14a might be conceived. These may be classified as ring-, bowl-, or cage-shaped objects. Quantum mechanical calculations show the bowl and the fullerene to be more stable than the monocyclic structure. In 1993, Prinzbach and co-workers published indication of the existence of cage-like C_{20}, but they could not doubtlessly prove the postulated structure. Detection took place by mass spectrometry, so the molecular mass could be determined, but the structure was not fully elucidated.

The hydrocarbon dodecahedrane ($C_{20}H_{20}$), on the other hand, is well known. Its three-dimensional structure is outlined in Figure 2.14b. It may be considered a per-hydrogenated fullerene. Since the initial description by Paquette, the chemistry of $C_{20}H_{20}$ has been studied in detail, and even 12-fold substituted derivatives are known. The synthesis according to Prinzbach starts from isodrin to obtain a derivative of pagodane. This is reacted to give an ester of dodecahedrane-1,6-dicarbonic acid, which in a sequence of saponification, Barton reaction toward the 1,6-dibromide and reduction is transformed into dodecahedrane. In experiments to characterize multiply substituted derivatives by mass spectroscopy, the spectrum of $C_{20}Cl_{16}$ revealed a successive loss of all chlorine atoms down to the decaene (m/z 240). This may not be conclusive evidence of the existence of a C_{20}-fullerene, but there is indication that its formation is possible at least at the conditions given in a mass spectrometer.

While in theory fullerene-like structures can be made up for any even number of carbon atoms (except for 22), the classical methods of fullerene production do

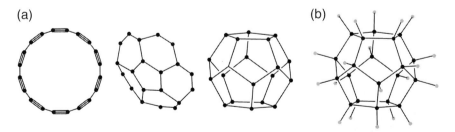

Figure 2.14 (a) Possible structures of a molecule C_{20}, (b) the structure of dodecahedrane $C_{20}H_{20}$.

not yield detectable amounts of cages with less than 60 atoms. The strain energy in intermediate precursors of these structures obviously is big enough to make them either decompose or form larger, less curved species. As mentioned before, C_{60} is the smallest conceivable fullerene with all-isolated pentagons, so lesser homologs would inevitably contain bonds with two bordering five-membered rings. This would be an unfavorable situation due to the distorted bond angles and the associated increase of strain energy.

Depending on the size of a carbon cluster, different structural characteristics are energetically preferred. Very small, positively charged species ($n < 7$) mainly exist as linear chains. Next, there are monocyclic compounds, bi- and polycycles and, finally, cage-like structures. Depending on the number of atoms contained, these may stabilize either by expelling C_2-units or by Stone–Wales rearrangements. Calculations on C_{20}, for instance, revealed that even if energy rich, the cage-like structure is still the most favorable of isomers.

From about 50 carbon atoms on, fullerene-like structures become dominant in the analysis of carbon plasma. These cannot be isolated, but at least there are publications on highly substituted C_{50+x}-compounds that will be presented in Section 2.5.8.

Some of the medium-sized carbon clusters, especially C_{36}, and their structure have likely been discussed for quite some time. The most extensive investigations dealt with the question of whether stable fullerenes with less than 60 atoms exist. Calculations suggested C_{36} with a cage-like structure of D_{6h}-symmetry to be stable (Figure 2.15a), and indeed experiments did at least indicate the existence of a C_{36}-fullerene. It is generated in an electric arc (refer to Section 2.3.3, $P = 7.7$ psi (400 Torr) helium, $I = 100$ A) with the fullerene soot directly being analyzed by time-of-flight mass spectrometry. Besides C_{60}, the expected peak for C_{36} is detected at m/z 432. After removal of those fullerenes soluble in toluene (C_{60} and C_{70}), the reaction with potassium in liquid ammonia provides a depleted fullerene soot that exhibits a prominent signal for 36 carbon atoms in its mass spectrum. In the solid state, C_{36} does not exist as isolated molecules, but as a three-dimensionally crosslinked structure (Figure 2.15b). The D_{6h}-symmetry is determined from the ^{13}C-NMR spectrum, and in addition there is a slight Jahn–Teller distortion observed for the C_{36}-cage.

2.2.5
Structure of Heterofullerenes

The so-called heterofullerenes are obtained by substituting one or several carbon atoms by other elements (Figure 2.16). The most prominent exponents of this class contain nitrogen or boron instead of carbon atoms, but other heterofullerenes are known as well, for example, with niobium, silicon, germanium, phosphorus, or arsenic incorporated in the cage.

By substituting single atoms, for example, with N, a heterofullerenyl radical is obtained from C_{60}. Usually this will form a dimer, which indeed is the species that has first been detected. The bond connecting two azafullerenes is formed between

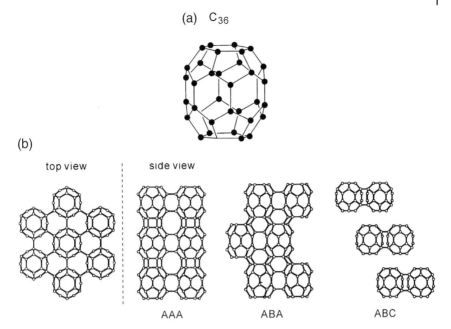

Figure 2.15 (a) The structure of C_{36}, (b) structures of polymeric C_{36} in solid state. Different stacking orders (AAA, ABAB, ABC) may be observed in the z-direction (© Wiley Interscience 2000).

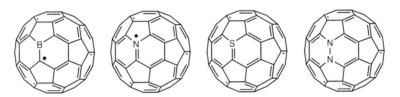

Figure 2.16 Different heterofullerenes.

two carbon atoms directly adjacent to the respective nitrogen. An alternative stabilization as $C_{59}N^+$ is possible.

For higher heterofullerenes the possible incorporation of heteroatoms at different positions of the cage must be considered. Consequently different isomers are obtained, and the subsequent radical combination yields mixed dimers as well.

2.3
Occurrence, Production, and Purification

Although discovered in the laboratory at experimental observation, fullerenes supposedly also exist in nature. Several publications claim the presence of C_{60} in interstellar space, and there is indication that structures at least fullerene-like are present in the mineral *shungite* that may be found in Karelia (Russia). This substance, entirely consisting of carbon, exhibits a deformed, wavy graphitic structure (Figure 2.17) that is said partly to contain fullerene-like inclusions. There are, however, no known deposits of fullerenes or fullerene-like substances that might serve as a source for this material.

The production of fullerenes may be achieved in different ways which all of them still have one thing in common: they are more of preparations than syntheses in the true sense of the meaning. The starting materials may be various other modifications of carbon, but hydrocarbons are suitable as well. The methods differ in the choice of primary materials and in the conditions applied for their conversion.

In general, processes generating low-molecular carbon clusters in an arc or plasma have to be distinguished from such with thermal processes like combustion or pyrolysis supplying the building blocks for fullerene growth. The most important variants of both basic forms are discussed below.

2.3.1
Fullerene Preparation by Pyrolysis of Hydrocarbons

Polycyclic aromatic hydrocarbons (PAH) are suitable for the preparation of fullerenes because they already contain structural elements of the fullerene cage. In particular, such hydrocarbons consisting of five- and six-membered rings should rather easily transform into fullerenes upon thermal treatment as a part of the curvature is preformed already.

Practically this method is performed with naphthalene, corannulene or higher polycyclic aromatic compounds. These are heated to about 1000 °C in an atmosphere of inert gas (usually argon). At these conditions, mainly C_{60} and C_{70} are generated with concomitant cleavage of hydrogen. (These cyclodehydration

Figure 2.17 *Shungite*, a carbon mineral (a), shows an undulated, graphite-like structure (b).

reactions also play a role in certain proposals for rational syntheses of fullerenes (refer to Section 2.3.5).) Hydrofullerenes like $C_{60}H_{36}$ are observed as minor products. With about 0.5%, the overall yields of fullerenes are not particularly high.

2.3.2
Partial Combustion of Hydrocarbons

When hydrocarbons are not completely incinerated, carbon black is obtained. Under suitable conditions this may contain not only particles of classical soot, but also fullerenes. The trace existence of fullerenes in lamp black was first proven by mass spectrometry, but after working out appropriate protocols for the reaction, smoking flames can by now be used for the production of weighable amounts. Benzene is the most common source of carbon for this process. It is mixed with oxygen and argon and burned in a laminar flame (Figure 2.18). The resulting mixture contains soot, polycyclic aromatic compounds and a certain fraction of fullerenes that make up 0.003–9.0% of the soot's total mass. Other hydrocarbons like, for example, toluene or methane may be employed as well.

Furthermore, the combustion of hydrocarbons is the very method which upon choosing suitable reaction parameters can yield an increased portion of C_{70} in the fullerene soot. At best, it may constitute up to 80% of the fullerene obtained. The gas pressure seems to be most influential for the C_{70}/C_{60} ratio: the fraction of C_{70} becomes larger with increasing pressure. The concentration of oxygen and the temperature of combustion are further important parameters.

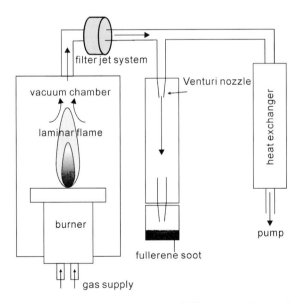

Figure 2.18 Combustion synthesis of fullerenes in a laminar flame.

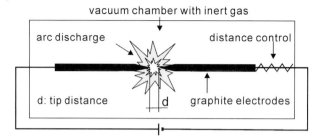

Figure 2.19 Fullerene synthesis by arc discharge.

By now the method is also used to prepare large amounts of C_{60}. A plant producing about 500 t of fullerene per year by benzene combustion has recently gone operative in Japan.

2.3.3
Arc Discharge Methods

When high voltage is applied to two graphite electrodes, an arc flashes over between them (provided they are not positioned too far from each other) and an arc develops. The temperature in the resulting plasma is sufficient to vaporize the graphite. In fullerene synthesis the required plasma is generated between two pointed graphite electrodes that barely touch each other (contact arc). The resultant particles rise from the plasma zone and deposit on the reactor walls (Figure 2.19). The yields of fullerene are about 15% with C_{60} constituting about 80% of the fullerene material. Graphite "soiled" with other elements (B, Si, or Al) turns out to be a suitable material for electrodes if the portion of higher fullerenes are increased. Finally, electrodes made of coal can be used in these syntheses too; only do the yields drop to 4–6% of the employed carbon.

Usually electrodes with a diameter of less than 6 mm are used for the arc method because the output of fullerene decreases on larger dimensions. This effect is caused by the fullerene's sensitivity toward radiation: while traveling toward cooler parts of the apparatus, the fullerene molecules are exposed to very intensive UV-rays emitted by the arc plasma. They get excited, and the resulting triplet state with a lifetime in the range of μs gives rise to an increased reactivity toward other carbon clusters C_n. The latter can be explained with the *open-shell* structure of the excited fullerene.

$$C_{60}(S_0) + h\nu \rightarrow C_{60}^{*}(T_1) \xrightarrow{+C_n} C_x \text{ (insoluble carbon)} \quad (2.1)$$

As larger electrodes generate larger plasma zones, they also emit more radiation. Hence more of the initially formed fullerenes react with other carbon clusters present in the apparatus, and the yields of cage-like structures decrease rapidly.

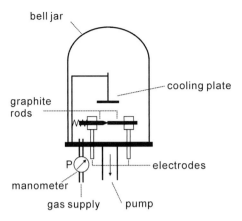

Figure 2.20 The first synthesis providing larger amounts of fullerenes succeeded in an apparatus for resistance heating of graphite rods presented by Krätschmer *et al.* Its essential feature is the electrodes just touching each other in one point.

2.3.4
Production by Resistive Heating

Another method of fullerene preparation has been presented by Krätschmer and co-workers. It is based on the strong heating that occurs at point contacts in an electric circuit. This is due to the small cross-section and the related high current density. The apparatus shown in Figure 2.20 bears two graphite electrodes that touch at just one point. Under tension this contact heats up to about 2500–3000 °C and the electrode tips start glowing. Smoke is generated that rises up and precipitates on the cool reactor walls or on an optional cooling finger. Usually electrodes with a diameter of about 3 mm are employed in this process, one of them with a pointed end to attain the smallest possible contact area. Thicker graphite electrodes tend to be unsuitable as the glowing zone expands over the whole electrode, thus diminishing the evaporation of graphite from its tip. The surrounding atmosphere has to be of inert gas (preferably helium) as even nitrogen may lead to undesired side reactions. The pressure inside the reactor must be thoroughly controlled to ensure optimal yields of fullerenes: with the pressure too low, the generated carbon radicals will diffuse from the reaction zone too quickly to be available for fullerene growth. Under excessive pressure, on the other hand, too many radicals are produced per time unit. Very fast growth of carbon clusters is observed, but the structures do not close on themselves to be cages. A pressure ranging from 140 to 160 mbar of helium has proven best. The method is easily performed on a laboratory scale as only little and not very expensive equipment is needed. At suitable conditions, total yields of fullerene may range up to 15% (with reference to the employed graphite). After purification, mainly C_{60} is obtained ($C_{70}:C_{60} = 0.02:0.18$).

There are further methods as well to heat graphite up to the required temperature. Laser oven methods (refer to Section 2.5.4), inductive heating, or the exploitation of solar energy in a parabolic-mirror oven have to be mentioned here.

2.3.5
Rational Syntheses

A rational synthesis of the fullerene skeleton, that is, a stepwise assembly from organic precursors, means a considerable challenge; the target is a molecule bare of any functional groups (except for double bonds). Any strategy thus has to include the preceding synthesis of partial structures that may be condensed to become a cage-like molecule in a way that removes any remaining substituents. Pyrolytic reactions suggest themselves here. Accordingly, the five-membered rings have to be introduced in advance at appropriate positions to achieve the required curvature.

Recently, there have been numerous attempts on the directed synthesis of C_{60} or other carbon cages. Most of these, however, dealt with the assembly of partial structures of C_{60}. L. T. Scott's preparation of circumtrindene **II** starting from trichlorodecacyclene is just one example. The sequence begins with reacting 2-chloronaphthaline to give 2-acetyl-8-chloronaphthaline which in four steps is further transformed into 8-chloroacenaphthenone (Figure 2.21). The titanium-catalyzed cyclotrimerization yields the C_3-isomer of 3,9,15-trichlorodecacyclene **I**. In the last step, the cleavage of HCl is effected by flash vacuum pyrolysis (FVP). The resultant bowl-shaped product represents 60% of the C_{60}'s structure. Compared to the pyrolysis of a mere hydrocarbon, the generation of alkyl radicals by halogen abstraction is advantageous because ring closure by bond formation preferably takes place at these sites, and the number of isomers emerging is markedly reduced.

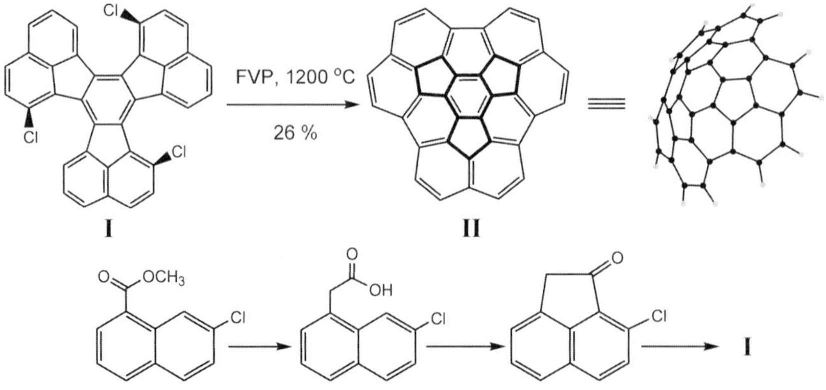

Figure 2.21 Synthesis of circumtrindene (**II**) from trichlorodecacyclene (**I**). The bowl-shaped structure represents 60% of the fullerene cage.

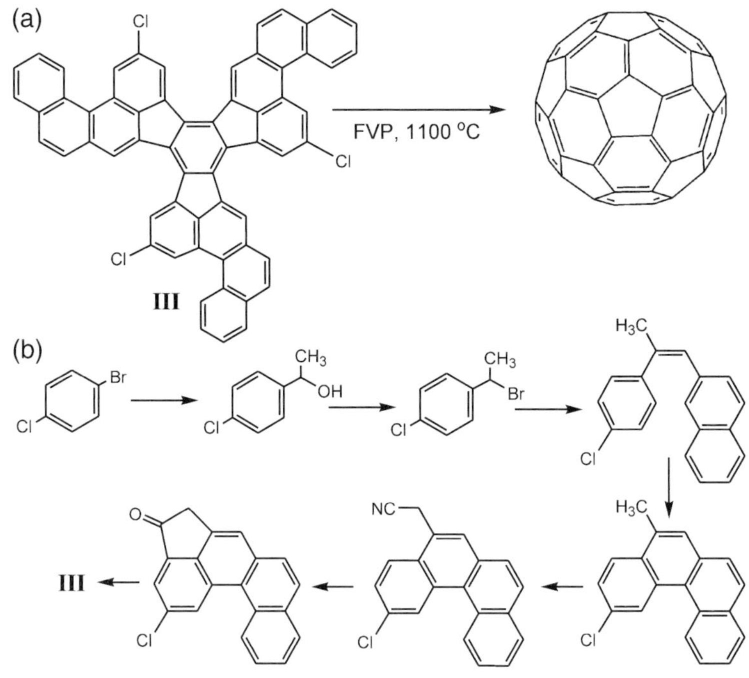

Figure 2.22 C_{60} is obtained by the pyrolysis of compound **III** (a). The strategy for its synthesis is outlined in (b).

A similar approach, with all 60 carbon atoms of the target fullerene present in the starting molecule, has also been presented by L. T. Scott. Again the last step is a flash vacuum pyrolysis which in this case transforms the open starting substance $C_{60}H_{27}Cl_3$ **III** into the C_{60}-cage (Figure 2.22). Evidence was given that the reaction proceeds directly from the hydrocarbon, without intermediate formation of smaller carbon clusters in the gas phase. The overall yield of the twelve-step sequence is about 1%, so this route of synthesis will never replace the cheap arc production of fullerene. Yet the feasibility of constructing geodetical domes and cages by rational synthesis could be demonstrated.

Other concepts include the reaction of cyclophane structures to give C_{60}. These molecules feature a big number of triple bonds, so the carbon content is very high in the starting material already. Figure 2.23 shows a synthesis suggested by Y. Rubin: starting from tris-(bromoethynyl)-benzene that is obtained by NBS-bromination of 1,3,5-triethinylbenzene, Gabriel coupling generates an intermediate which represents one half of the desired cyclophane. The latter then is obtained by Hay coupling. Even though at least partial dehydrogenation was observed in the mass spectrum, a transformation into C_{60} could not be achieved, which is possibly due to the structure's flexibility. A halogenation strategy for directed positioning of bonding sites as in Scott's examples or an increased number of connecting chains might be helpful here.

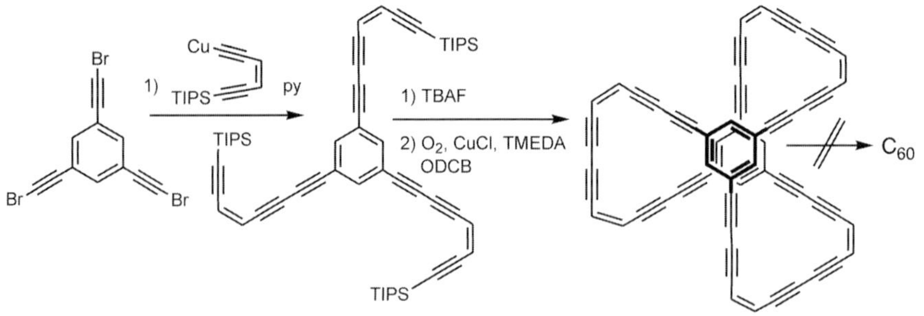

Figure 2.23 For the time being, it has not been possible to synthesize C_{60} by a dehydrogenation of carbon-rich cyclophanes.

This synthetic route is also suitable for the production of heterofullerenes as the use of cyclophanes containing nitrogen, for example, pyridines, enables the controlled incorporation of heteroatoms. In this way the nitrogen analog of C_{60}, the molecule $C_{58}N_2$, could first be detected in a mass spectrum.

All of these synthetic routes are most unlikely ever to compete seriously with the other methods of preparation because the reactions' complexity and the low overall yields do not allow for a large-scale performance. Still these methods are promising approaches for the directed synthesis of specific fullerenes, to accurately introduce heteroatoms or to obtain unusual endohedral fullerene complexes.

2.3.6
Enrichment and Purification

One thing is common to all practicable methods of preparation: they are not directed syntheses of single fullerenes with defined sizes, but rather yield a mixture of different carbon cages. Consequently, the separation and purification of fullerenes become necessary. As the concentration of C_{60} and C_{70} is much higher than that of the larger homologs, the latter are much harder to isolate.

In addition to the desired species, the fullerene soot contains a multitude of insoluble components like soot particles or graphitic fragments. Moreover, there may be polycyclic aromatic hydrocarbons also. The separation from insoluble contaminants may be affected by extraction because the fullerenes, and most of all the small ones, exhibit significant solubility in some organic solvents (refer to Section 2.4.1.1). Toluene is most commonly used for this process. Other possible solvents include chlorinated aromatic species, carbon disulphide, benzene and hexane. As an alternative, the fullerene part of the raw material may be separated by sublimation. In any case a mixture of the generated fullerenes is obtained which has to be resolved into its constituents. In most cases chromatography is used to this purpose.

As indicated on the flowchart in Figure 2.24, a preseparation into lower and higher species is already achieved by the extraction of the pristine soot. The process

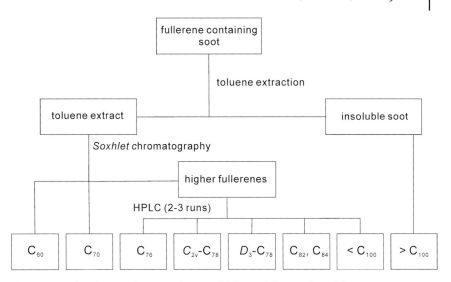

Figure 2.24 Flow diagram for an enrichment of different fullerenes from fullerene soot.

takes the advantage of the lower solubility of the big fullerenes ($n > 100$) in toluene. These will then be extracted from the depleted soot with chlorinated aromatic solvents. In a second step, the extract of small fullerenes ($n = 60–100$) in toluene is chromatographed on a stationary phase of alumina. Pure C_{60} and, in a second fraction, pure C_{70} are obtained. It is true now that the fullerenes favorably dissolve in toluene, but this solvent does not provide sufficient resolution in chromatography, and so the latter is run with mobile phases of hexane or hexane/toluene mixtures. Unfortunately the process consumes large amounts of solvent due to the fullerenes' low solubility in hexane. Therefore a combined process of extraction and chromatography was developed. It is called Soxhlet chromatography and requires significantly less solvent. The fullerenes C_n ($n = 76$ up to about 100) that remain on the chromatographic column after elution of C_{60} and C_{70} are washed out with solvent mixtures and further separated by repeated HPLC. Thus C_{76}, C_{78} (as resolved isomers) and other fullerenes can be obtained. Reversed-phase materials (silica gels rendered unpolar by functionalization with long alkyl chains) proved efficient for this step, but mixtures of silica and graphite have been employed too, and gels of polystyrene turned out to be the most effective.

The separation of C_{60} may further be attained by absorptive filtration on a mixture of charcoal powder and silica gel. Pure toluene can be used here which distinctly increases the solubility, thus accelerating the process and markedly reducing the solvent consumption. Apart from C_{60}, it has by now become possible to obtain C_{70} on a multigram scale as well as to make higher fullerenes available at least in amounts sufficient for physical examination. Highly purified materials can be produced by sublimation of the above substances. Chiral fullerenes (refer to Section 2.2.3) may be split into enantiomers by suitable methods, for example,

HPLC on a stationary phase based on amylose. The latter has been successfully applied to D_2–C_{76}.

2.3.7
Preparation of Heterofullerenes

Usually a synthesis of heterofullerenes by simply reacting the carbonaceous starting material with a source of the respective heteroelement does not succeed. The reaction of graphite with boron nitride or cyanogen $(CN)_2$, for example, does not yield the desired heterofullerene. Other methods starting from organic derivatives of fullerenes have to be considered instead.

Epi-iminofullerenes and ketolactames with opened fullerene cage have proven very suitable to the synthesis of azafullerenes. The ketolactames are generated by oxidizing with singlet oxygen the (5,6)-fulleroids which in turn are obtained from the [3+2]-cycloaddition of azides to C_{60} with subsequent nitrogen cleavage. On addition of a large excess of p-toluenesolfonic acid to a solution of the keto compound in boiling o-dichlorobenzene, the azafullerene $C_{59}N$ is generated by the cleavage of formaldehyde. The product dimerizes due to its open-shell structure (Figure 2.25). If hydroquinone is added in excess to the last step of azafullerene synthesis, it will act as a reductant and the reaction yields the hydroazafullerene (Figure 2.25).

Epi-iminofullerenes are obtained from the respective carbamates and transform into protonated derivates of azafullerenes at the conditions of desorptive chemical ionization in a mass spectrometer.

Bora- and further heterofullerenes (C_nSi, C_nGe) may be produced, among others, by arc discharge methods. To this end, suitably impregnated rods of graphite are employed in typical fullerene reactors. For the preparation of phosphafullerenes, the simultaneous vaporization of carbon and phosphorus in a radio frequency oven proved its worth. The vaporization of both elements at different positions in the oven, corresponding to different temperatures, was found to be crucial in this

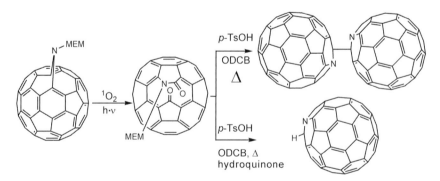

Figure 2.25 Synthesis of derivatives of azafullerene.

process. Some heterofullerenes contain radioactive nuclides. These compounds are generated by nuclear recoil upon deuteron bombardment. The heterofullerenes of arsenic $C_{59}^{71}As$, $C_{59}^{72}As$ and $C_{59}^{74}As$ are relevant examples here. The respective C_{70} analogs have been obtained with the same method.

2.4
Physical Properties

Contrasting the "classical" modifications of carbon discussed in Chapter 1, fullerenes are discrete molecules, which clearly reflects in their properties. Above all, the spectroscopic features show significant differences to those of other carbon allotropes.

2.4.1
Properties of C_{60} and C_{70}

Owing to their wide availability, C_{60} and C_{70} are the best explored fullerenes. By now, there are comprehensive collections of data covering all aspects of physical and chemical properties. Only the most important characteristics shall be mentioned here, more detailed information may be found in the copious specialized literature.

2.4.1.1 Solubility

Fullerenes are scarcely or not at all soluble in many common solvents (Table 2.2). Hence detailed knowledge about their solubility is important for any further examination of their properties. Only aromatic solvents like benzene, toluene, or chlorinated aromatic species solve significant amounts. Carbon disulfide can also be a potential candidate, but it is seldom used for its acute toxicity.

A striking observation is the high solubility of fullerenes in higher aromatic compounds (Table 2.3). Obviously the π–π interactions causing the solubilization of individual C_{60}-molecules are most effective here. A chlorinated aromatic solvent positively affects the solubility of C_{60}, too. In aliphatic solvents only minor amounts

Table 2.2 Solubility of C_{60} in different solvents at 25 °C.

Solvent	n-hexane	n-decane	CH_2Cl_2	$CHCl_3$	CCl_4	acetone	THF	MeOH	EtOH
Solubility of C_{60} in mmol l^{-1}	0.060	0.099	0.35	0.22	0.44	0.001	0.08	4.6×10^{-5}	0.0014

See M. V. Korobov, A. L. Smith, *Solubility of Fullerenes*, in: K. M. Kadish, R. S. Ruoff (editors), *Fullerenes*, Wiley Interscience, New York **2000**.

Table 2.3 Solubility of C_{60} in aromatic solvents and in CS_2 at 25 °C.

Solvent	CS_2	Benzene	Toluene	Chloro-benzene	1,2-dichloro-benzene	Benzo-nitrile	Piper-idine	1-methyl-naphthaline	1-chloro-naphthaline
Solubility of C_{60} in mmol l^{-1}	11.0	2.36	3.89	9.72	37.5	0.57	74.0	45.8	70.8

See M. V. Korobov, A. L. Smith, *Solubility of Fullerenes*, in: K. M. Kadish, R. S. Ruoff (editors), *Fullerenes*, Wiley Interscience, New York, **2000**.

of C_{60} are dissolved, although solubility improves with increasing chain length. Chlorinated alkanes as well do not solve marked amounts of C_{60}. Especially in polar aprotic and protic media like ethanol, acetone, or tetrahydrofurane, the solubility is virtually zero.

Due to its distinctly hydrophobic character, C_{60} is also practically insoluble in water. The concentration of a saturated solution is a homeopathic 3×10^{-22} mol l^{-1}. This value increases upon addition of amines as they may form donor–acceptor complexes with the fullerene. Colloidal solutions with particles measuring 0.22 µm and a concentration of 7×10^{-6} mol l^{-1} have been obtained by the aid of ultrasound. Also did the application of complexing host molecules like γ-cyclodextrines lead to a higher solubility. Surfactants serve to the same purpose. Strictly speaking, however, there are no real solutions in this case, but the fullerene molecules are located inside of micelles.

C_{60}, for being highly electronegative, may act as an electron acceptor in charge-transfer complexes. Accordingly, it is dissolved much better by solvents providing a free pair of electrons, and electron-rich aromatic species may, to the same effect, form complexes with their cloud of π-electrons. As a matter of fact, significantly higher solubilities are observed in aromatic media, especially such with electron-donating atoms (e.g., pyridine). The comparison of benzene to its saturated analog, cyclohexane, is an impressive example here: 1.4 mg l^{-1} vs. 0.036 mg l^{-1}. The solubility increases with the aromatic molecule's electron density (comp. toluene to benzene). This argumentation is also supported by the trend observed for halogenated solvents: the interaction with the halogen's free pair of electrons causes higher solubility.

Similar considerations hold of course for inorganic solvents: C_{60} is best solved in substances able to form donor–acceptor complexes. Examples may be found among the halogenides of elements from main group 4, for example, $SiCl_4$ (0.09 mg l^{-1}), $SnCl_4$ (1.23 mg l^{-1}), $SiBr_4$ (0.74 mg l^{-1}), or $GeBr_4$ (0.68 mg l^{-1}), etc., but also other Lewis-acids like $AsCl_3$ solve significant amounts of C_{60}.

The trends observed on C_{70} for the choice of a suitable solvent are in principle alike, Table 2.4 collects various solvents.

2.4.1.2 Spectroscopic Properties

First fundamental investigations on the photophysical properties of fullerenes were performed as early as 1991. A comprehensive picture of possible transitions and photophysical characteristics have been collected since then. Figure 2.26 summarizes the essential processes.

Solutions of C_{60} in organic media are colored dark violet, and those of C_{70} are deep red. These colors vary slightly with the individual solvents due to their different interactions with the fullerenes. Their specific structure imparts particular spectroscopic characteristics to C_{60} and C_{70} (Table 2.5). First, the absorption in the UV and visible range of the spectrum will be considered. The absorption spectra of the two smallest stable fullerenes are presented in Figure 2.27. Both feature several distinct bands at wavelengths between 200 and 400 nm and, although weaker, further absorption in the visible range. The latter causes the typical coloration of the solutions.

The spectra are dominated by allowed $^1T_{1u} \rightarrow {}^1A_g$ transitions in the UV region ($\varepsilon \sim 10^4$ to $10^5\,cm^2\,mmol^{-1}$). The additional bands in the visible range result from

Table 2.4 Solubility of C_{70} at 25 °C.

Solvent	Benzene	Toluene	n-Hexane	CH_2Cl_2	1,2-Dichloro-benzene	Tetralin	CS_2	Water
Solubility of C_{70} in mmol l^{-1}	1.55	1.67	0.015	0.095	43.1	14.6	11.8	1.6×10^{-10}

See M. V. Korobov, A. L. Smith, *Solubility of Fullerenes*, in: K. M. Kadish, R. S. Ruoff (editors), *Fullerenes*, Wiley Interscience, New York **2000**.

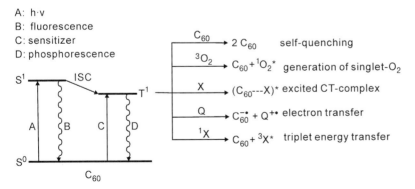

Figure 2.26 Transitions between different energy levels for C_{60}. Besides phosphorescence, C_{60} may undergo several other deexcitation processes from the first excited triplet state.

Table 2.5 Photophysical properties of C_{60} and C_{70}.

	E_S	E_T	τ_S	τ_T	k_{ISC}
C_{60}	193.0 kJ mol^{-1}	184.6 kJ mol^{-1}	1.2 ns	40 µs	1.5×10^9 s^{-1}
C_{70}	157.0 kJ mol^{-1}	146.5 kJ mol^{-1}	0.7 ns	130 µs	1.25×10^9 s^{-1}

E_S: singlet energy, E_T: triplet energy, τ_S: life time of singlet state, τ_T: life time of triplet state, k_{ISC}: velocity constant of intersystem crossing.
See A. Kleineweischede, J. Mattay in *CRC Handbook of Organic Photochemistry and Photobiology*, 2nd ed., CRC Press, Boca Raton, **2003**.

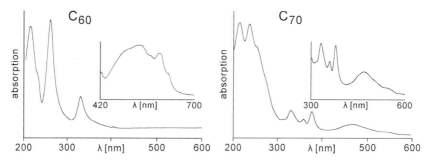

Figure 2.27 Absorption spectra of C_{60} and C_{70}. The detail inserts each show an enlargement of the respective absorption in the visible range of light that is responsible for the coloration (© ACS 1990).

transitions only weakly allowed due to the high symmetry of C_{60} in its closed-shell configuration ($\varepsilon \sim 100$ to 1000 cm^2 mmol^{-1}). C_{70} possesses markedly lower molecular symmetry and, accordingly, exhibits much stronger absorption at these wavelengths ($\varepsilon > 10^4$ cm^2 mmol^{-1}, Figure 2.27b).

The spectrum of C_{60} in hexane shows three major bands in the UV at 211, 256, and 328 nm. In addition, the latter has several shoulders to its long-wave side. These absorptions correspond to π–π* transitions across the HOMO–LUMO bandgap of 1.8 eV. The visible bands point to the vibrational structure of orbital-forbidden electronic transitions (singlet–singlet transitions), with the intensity of absorption being respectively low. The characteristic bands are situated at 492, 540, 568, 591, 598, 621, and 635 nm.

The structure of C_{70}'s absorption spectrum is explained in the same way. Here the allowed transitions lie at wavelengths of 215, 236, 331, 359, and 378 nm. Additional bands of medium intensity are observed at 469 and 544 nm, and further to the visible range there are several weak absorptions between 594 and 650 nm. The spectrum is altogether more structured than that of C_{60}, which reflects the higher number of different electronic transitions.

Upon a change to aromatic solvents, it is mainly the 328 nm band of C_{60} and the 331, 359, and 378 nm bands of C_{70} that experience a bathochromic shift. Most likely a charge-transfer interaction of the electrophilic fullerene with the solvent's π-system gives rise to this effect. Changing to more polar solvents, on the other hand, hardly any variation is observed because fullerenes do not possess a permanent dipole moment.

The fluorescence spectrum of C_{60} shows a minor emission at $\lambda = 720$ nm with a comparatively poor quantum yield Φ of about 10^{-5}–10^{-4}. This low fluorescence activity can be attributed to the short lifetime (~1.2 ns) of the excited singlet state and to the transition of lowest energy being forbidden for symmetry reasons. Again a stronger fluorescence ($\lambda = 682$ nm, $\Phi \sim 10^{-3}$) is found for C_{70}. The short lifetime of the first excited singlet state is caused by an efficient inter-system crossing toward the first excited triplet state (triplet quantum yield ~100%). Limited by triplet–triplet and ground-state quenching, this latter state lasts about 40 μs in C_{60}, but about three times as long in C_{70}. Despite a high rate of triplet formation, no (C_{60}) or only extremely weak phosphorescence (C_{70}) is observed.

Turning toward the absorption of longer wave radiation, that is, in the IR range, both fullerene molecules likewise feature characteristic properties. The FTIR spectrum of C_{60}, for instance, exhibits four bands at 1430, 1182, 577, and 527 cm^{-1}, exactly as they could be expected from symmetry considerations (Figure 2.28). The C_{60}-molecule possesses 180 degrees of freedom. Subtracting those

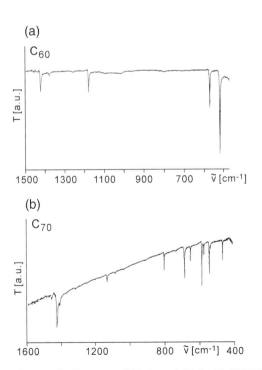

Figure 2.28 IR-spectra of (a) C_{60} and (b) C_{70} (© RSC 1991).

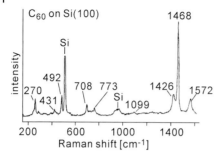

Figure 2.29 Raman spectrum of a film of C_{60} on a silicon substrate (© John Wiley & Sons 1996).

of translation and rotation, 174 modes of vibration result, but only 42 of them represent fundamental vibrations. Of these, in turn, only the four modes actually observed feature t_{1u}-symmetry rendering them IR-active. So the IR spectrum gives unambiguous evidence of the symmetry postulated for the substance under examination.

A much larger number of bands is found in the IR-spectrum of C_{70} because the molecule is less symmetric (Figure 2.28). Again the experimental observations are reflected in theoretical considerations on molecular symmetry. There is a total of 204 vibrational modes, including 122 fundamentals. 13 of these are IR-active, and 53 are Raman-active.

Actually, Raman spectra of fullerenes are very elucidating as well because their characteristic bands can be assigned to the respective symmetries. Ten out of the 42 fundamental vibrations of C_{60} are Raman-active, and accordingly there are ten clearly distinct bands in its first-order Raman spectrum. These signals belong to two A_g and eight H_g-modes (Figure 2.29). The A_g-modes are a radial breathing mode of the entire C_{60}-molecule at 492 cm^{-1} and a so-called pentagonal pinching mode at 1468 cm^{-1}, which is a simultaneous contraction of pentagons and expansion of hexagons. For the H_g-symmetric modes, ranging wider from 270 to 1575 cm^{-1}, the corresponding vibrations are much more complex to describe. Still a general statement would be that smaller wavenumbers relate to more radial vibrations.

Spectra in nuclear magnetic resonance (NMR) do also prove the symmetry of the particular fullerene under examination. As expected, the ^{13}C-NMR spectrum of C_{60} exhibits exactly one signal. This is found at 143.2 ppm (using d_6-benzene as a solvent; Figure 2.30, top). It is situated rather in the shift region of electron-deficient polyolefins than at typically aromatic values, which is an indication of the molecule's relatively low aromatic character. For C_{70}, the ^{13}C-NMR spectrum shows five signals with relative intensities of 1:2:1:2:1, thus sustaining the D_{5h}-symmetry (Figure 2.30, bottom). In d_6-benzene the signals are observed at 130.9, 145.4, 147.4, 148.1, and 150.7 ppm.

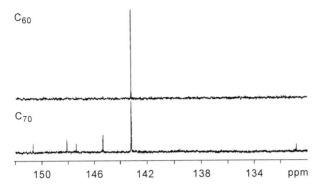

Figure 2.30 The ^{13}C–NMR spectra of C_{60} (top) and C_{70} (bottom, in a mixture with C_{60}) show exactly the number of signals expected for I_h- or D_{5h}-symmetry, respectively (© ACS 1990).

^{13}C-NMR spectroscopy is also an appropriate method to study solvent effects. A charge-transfer interaction, for example, induces a down-field shift of the ^{13}C-signal by changing the electronic environment between C_{60} and solvent. Yet an easy prediction of signal positions is not possible because other effects like the size, shape, and orientation of solvent molecules also influence the chemical shift.

2.4.1.3 Thermodynamic Properties

Comparing them to graphite or diamond from a thermodynamic point of view, fullerenes are a modification of carbon containing markedly higher energy. Graphite, for being the most stable form at ambient temperature, possesses by definition a standard formation enthalpy of 0 kJ mol^{-1}. For diamond, the value is just a little higher at 1.7 kJ mol^{-1} per C-atom. The latter is meta-stable at room temperature and transforms into graphite only upon heating. Fullerenes behave in a similar way. C_{60} as the smallest stable species has a standard formation enthalpy of 42.5 kJ mol^{-1} per C-atom. For C_{70}, the value is already diminished to 40.4 kJ mol^{-1}, and this trend continues toward larger fullerenes. Their formation enthalpy approximates that of graphite with increasing size and concomitantly decreasing curvature and strain. It would also be 0 kJ mol^{-1} then for a hypothetical fullerene of infinite dimensions.

The size of the carbon cage also governs other thermodynamic properties of the fullerenes in question, among them are, for example, the melting point or the bond energy per carbon atom. The latter is estimated to be 7.18 eV in C_{60} (experimental value: 6.98 eV), and for C_{70} 7.21 eV are calculated. Graphite, for comparison, has a bond energy of 7.37 eV. The melting points of the fullerenes altogether are higher than 4000 K, while for the boiling points no exact data could be obtained by now.

The phase diagram of carbon considerably gained complexity by inclusion of the fullerenes. These are situated close to the graphitic region, with some

2.4.1.4 Solid C_{60}

Solid C_{60} may be obtained both as a black powder or as well-crystallized solid. The latter possesses a face-centered cubic (fcc) lattice. The C_{60}-cages take positions at the lattice points, so they might be considered large pseudo-atoms in a classical fcc-lattice. For pure C_{60} there is a slight disorder in the orientation of single fullerene molecules, so the less symmetric space-group $Fm\bar{3}m$ results instead of the expected $Fm\bar{3}$ group. The lattice constant is 14.157 Å. The molecules freely rotate on their lattice positions, and actually they do so faster than they would in solution. The existence of two different bond lengths could be proven as well, which had strong impact on the discussion about the actual structure and aromaticity of C_{60}.

A successful X-ray analysis of pure C_{60} could only be performed as late as 1991, but the cage-like structure and the extraordinary symmetry of the molecule had already been demonstrated before by measuring several crystalline derivatives of the fullerene. The first fullerene compound characterized by X-ray analysis was the adduct with osmium tetroxide (Figure 2.31).

When irradiated with ultraviolet light, C_{60} forms a substance insoluble in typical "fullerene solvents." It has been identified to be a polymerization product consisting of fullerene molecules linked by [2+2]-cycloaddition. This reaction takes place even in the crystal because the free rotation of monomers enables sufficient approach and suitable orientation of the double bonds involved (refer to Section 2.5.4).

The spinning of the individual molecules ceases upon cooling and a lattice of space group $Pa\bar{3}$ with a lattice constant of $a = 10.041$ Å is formed. The influence of neighboring fullerene molecules has been determined for this structure: due to the molecules' spherical shape there are mainly van der Waals interactions, but in this case an additional anisotropic charge distribution is induced by the existence of different bond types. In the cooled crystal, a given, electron-deficient five-

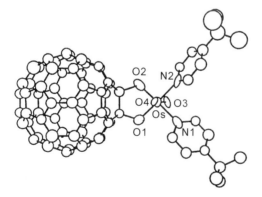

Figure 2.31 The first X-ray structure obtained for a fullerene compound. It is the addition product of C_{60} with $OsO_4(C_5H_4N-CH_3)_2$ (© AAAS 1991).

membered ring of one fullerene faces, in an almost parallel orientation, one of the rather electron-rich six-membered rings of an adjacent molecule, thus optimizing the electrostatic potential. Due to this fixation of single molecules on their lattice positions, the aforementioned polymerization no longer occurs below the temperature required for a free rotation.

2.4.2
Properties of Higher Fullerenes

The main problem in studying the properties of higher fullerenes is their poor availability. Even little amounts of the pure substances are hard to obtain. For the time being it still takes immense efforts to produce milligram amounts of the higher species. It is true that they result from the separation of C_{60} and C_{70} in the work-up of fullerene soot (refer to Section 2.3.6), yet further fractioning is hard to achieve as solubility decreases with growing cage sizes. Accordingly, the extractive and chromatographic methods for isolating single species become ever more challenging. The simultaneous occurrence of a multitude of structural isomers further complicates the isolation of substances with defined cage geometry.

Still it has been possible to determine spectroscopic data for some samples. Figure 2.32 shows the absorption spectra of the higher fullerenes C_{76}, C_{78}, and C_{84} as obtained from fullerene soot by HPLC methods. Solutions of the single substances appear yellow-green (C_{76}), brown (C_{2v}–C_{78}), golden-yellow (D_3–C_{78}), and

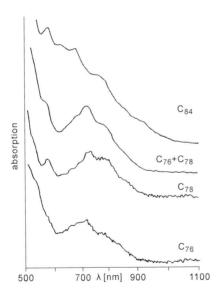

Figure 2.32 Absorption spectra of some higher fullerenes. The pure compounds also show very characteristic color in solution. The spectra of C_{76} and C_{78} were obtained after enriching by HPLC the respective substances from a mixed fraction containing both (© JCS 1991).

olive-green (C_{84}). The UV/Vis spectra also demonstrate that the onset of absorption lies already at wavelengths much larger than for C_{60} or C_{70}. This is indicative of a smaller HOMO–LUMO gap (only 1.2 eV in C_{84}) and suggests interesting electronic properties.

For several higher fullerenes the crystalline form has been isolated as well. C_{70}, for example, crystallizes in a face-centered cubic lattice that bears impurities, though, of a hexagonal phase (hcp). The cubic lattice constant is 14.98 Å. Like in C_{60}, the molecules exhibit a strong inherent motion that causes the crystal to be isotropic. It is only at low temperatures that the nonspherical shape of C_{70} shows effect: first a trigonal structure forms, that finally turns into a monoclinic one. X-ray analysis confirmed the D_{5h}-symmetry of the single molecules as well as the existence of eight different bonding types in C_{70}. Most striking are the distinct alternance of bonds about the belt region and the analogy of the polar region to the bonding situation in C_{60}.

The chiral C_{76} as a pure substance also crystallizes in a face-centered cubic form. Both enantiomers are evenly distributed throughout the crystal, and each C_{76}-molecule rotates freely at its lattice position. In contrast to C_{60} and C_{70}, lowering the temperature does not induce a phase transition into a less symmetric structure. The molecules rather seem to sit at their lattice positions in an unordered fashion, only with their rotation ceased.

For even higher fullerenes like C_{84}, usually several isomers crystallize simultaneously, rendering the structural determination of distinct variants hard to perform. X-ray analysis of these compounds becomes possible only after chromatographic separation. A disorder of individual molecules is observed in the higher fullerenes as well.

2.5
Chemical Properties

Up to now chemical properties have been studied especially on C_{60} and C_{70}, which is due to the starting materials being more easily available. Only endohedral functionalization, as presented in Section 2.5.4, is mainly performed with higher homologs. The other sections will chiefly deal with the chemistry of C_{60} (and C_{70}, where appropriate). The reactions of higher fullerenes are comparable to those of C_{60} as they all have similar structural features. Certain differences in chemical behavior are possible only with regard to a higher number of conceivable isomers and to a diminished reactivity which is caused by a lower degree of pyramidalization.

Chemically speaking, fullerenes are polyolefins or more or less aromatic systems. As already mentioned in Section 2.2.2, an alternation of (5,6)- and (6,6)-bonds, respectively, is observed. The latter feature double-bond characteristics and the π-electrons are just moderately delocalized. Hence the aromatic nature of C_{60} as well as of higher fullerenes is limited.

The electron affinity of fullerenes is noteworthy. They easily accept one or even several electrons. Obviously they are strong electrophiles, which markedly influences their reactivity. However, C_{60} can also act as an electron donor, yet this feature is far less pronounced than the acceptor properties.

2.5.1
General Considerations on Fullerene Chemistry

2.5.1.1 Typical Reactions of Fullerenes

For hollow objects like the fullerenes, a general distinction has to be made between outside and inside reactivity. Modifications to the outside are termed *exohedral functionalization*, and those to the inside are *endohedral*. Both variants are observed for the fullerenes. Classical fullerene chemistry deals with exohedral functionalization by one or more groups attached to the carbon atoms. Endohedral chemistry, on the other hand, studies compounds consisting of atoms or small molecules included in the cavity within the fullerene cage. The exohedral processes may further be divided into covalent and noncovalent interactions with the reaction partner.

Owing to its electrophilic character, and for being a polyolefin rather than a three-dimensional aromatic compound, C_{60} readily enters into addition reactions with nucleophiles. In doing so, an initial addition product $C_{60}Nu^-$ is formed. This is saturated by an electrophile E in the second step of the reaction, which usually results in a 1,2-addition, unless sterically demanding groups are added. In these cases a 1,4- or even 1,6-addition occurs. The latter is found in extreme situations, for example, when extremely bulky addends are employed (Figure 2.44).

Apart from the nucleophilic addition, also radical reactions can occur on the surface of the fullerenes. These act as a "sponge" for radicals which are easily added, yielding dia- or paramagnetic products. Dimers of fullerenes are frequently obtained as well in this kind of reaction.

Moreover, pericyclic reactions like [4+2]-, [3+2]-, and [2+2]-cycloadditions, to name the most prominent, are observed for C_{60} and also for higher fullerenes. In these cases, C_{60} may be reacted with electron-poor as well as with electron-rich species. The cycloaddition chemistry of fullerenes yields a wide spectrum of valuable derivatives that may be employed to the production of new, further functionalized materials (refer to Section 2.5.5.3).

The driving force of fullerene reactivity most of all originates from their curved surface. Any addition from the outside to one or more of the already prepyramidalized carbon atoms leads to sp^3-hybridization and thus reduces the strain in the fullerene molecule. An addition from the inside, on the contrary, would increase the strain and consequently is impossible. In addition the overlap of the participating orbitals would be insufficient (Figure 2.33). Endohedrally included atoms or molecules accordingly are not bound to the carbon skeleton, but they exist inside of it as more or less free entity (refer to Section 2.5.4).

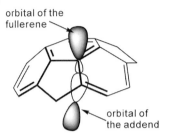

orbital of the fullerene

orbital of the addend

Figure 2.33 An attack from the inside is impossible due to poor orbital overlap and to an otherwise increasing strain.

As higher fullerenes exhibit a lower surface curvature, they are also less reactive toward potential addends because less energy is gained by stress relief. The same is true for multiple additions to C_{60} or its higher homologs. The supplemental reduction of strain energy decreases with each newly introduced addend. Other factors like the steric demand of the added groups become dominant from a certain degree of functionalization on, which leads to a limited number of addition steps even with an excess of reactands.

2.5.1.2 Regiochemistry of Additions to Fullerenes

The regiochemistry of additions to fullerenes is governed by a tendency to minimize the number of (5,6)-double bonds in the products. Hence in 1,2-additions the major products usually are those with the addends situated at both ends of a (6,6)-bond. Yet there are also adducts with opened (5,6)-bonds as these are still more stable than closed structures with addends at this bond (Figure 2.34a).

In higher fullerenes the surface curvature decisively influences the sites for reactions to occur. The strain at individual carbon atoms is by far higher around the molecular poles than in the so-called belt region, and so any addition reaction will preferably take place at the outer polar caps. Nevertheless, the regiochemistry of a specific addition is governed by the aspects discussed above. Accordingly, also here (5,6)-double bonds are unfavorable from an energetic point of view and will be avoided wherever possible.

The same rules in principle apply to additions in 1,4-position or at even more distant sites. Only the number of possible isomers increases with the distance between the added groups, and the formation of (5,6)-double bonds can no longer be evaded. For the latter reason these reactions, as compared to 1,2-additions, are energetically disfavored (Figure 2.34b), even though steric factors may still call for the 1,4- instead of the 1,2-addition. The formation of product mixtures is also a frequent observation.

2.5.1.3 Secondary and Multiple Additions

For all reactions on the fullerene, not only the stoichiometry, but also the regiochemical course of possible multiple reactions has to be considered. In principle, derivatives of C_{60} with 60 addends are conceivable, which would mean complete saturation of the carbon core. Still for the time being molecules like that have not

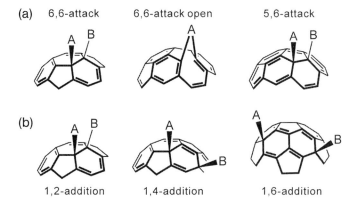

Figure 2.34 The regiochemistry of additions to fullerenes: (a) the attack on a (6,6)-bond is preferred for avoiding the formation of additional (5,6)-double bonds; (b) possible addition patterns on C_{60}, the 1,2-addition is the most favorable from a thermodynamic point of view.

been observed, which obviously is related to the space on the surface being insufficient to such a dense occupancy. However, there is a number of derivatives with up to 48 functional groups, and so any production of these substances is in peril of yielding an inseparable mixture that would contain a multitude of isomers with in parts varying numbers of addends. Consequently the stoichiometry must thoroughly be controlled in such reactions.

For the compounds generated, the degree of functionalization is governed by different factors. Firstly, the preparation of modestly functionalized species can only succeed with a sufficiently low amount of reactand, and secondly, the number of possible addition steps is limited by the steric and electronic conditions in the fullerene. With bulky addends, for example, a much lower number of addition steps is possible than with smaller residues. With the degree of functionalization increasing, the remaining six-membered rings more and more resemble benzene, which finally causes the addition of further reactand molecules to cease.

For the regiochemistry of multiple additions to C_{60} a complex picture is found. Starting from an initial 1,2-adduct, there are nine discernible sites for a second addition to another (6,6)-bond as depicted in Figure 2.35. The resulting regioisomers are termed *e′*, *e″*, *trans-1*, *trans-2*, *trans-3*, *trans-4*, *cis-1*, *cis-2*, and *cis-3*.

In reality, however, not all of these isomers are found in the respective second additions, and others are particularly frequent. Which ones these might be depends on the kind of addends. Provided they are very small, like, for example, hydrogen atoms, the second addition of another H_2-molecule takes place in *cis-1*-position. For larger addends, an attack at this site is impossible due to steric repulsion. In these cases, the equatorial isomers are found preferably, besides certain amounts of the *trans-3*-isomer.

An even larger number of isomers is conceivable if the formation of (5,6)-adducts is included in the considerations on possible regioisomers. These can no longer be described with the nomenclature of secondary addition outlined

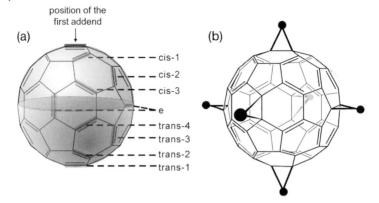

Figure 2.35 (a) Nomenclature for the second addition to fullerenes. There are three regions: cis, equatorial (e) and trans; (b) hexakis adduct of C_{60}.

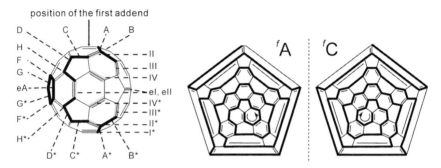

Figure 2.36 The nomenclature for the regiochemistry of additions must rely on a more complex system if (5,6)-adducts are to be considered, too. Moreover, the compound's chirality must be regarded for the derivatives concerned.

above. Here the fullerene's double bonds are identified by a code of Roman figures (6,6-bonds) and letters (5,6-bonds) instead. The Schlegel diagram of the respective basic skeleton is employed to make a distinction between enantiomers (Figure 2.36).

The regioselectivity of the second addition can at least partially be explained by electronic effects. The reactivity of certain double bonds in C_{60} is significantly influenced by the bonding of the first pair of addends. The *cis-1*-bond, for example, is markedly shortened, which, to a lesser extent, also holds true for the equatorial bonds. The bonds *cis-2* ad *cis-3*, on the other hand, are elongated. Only little effects are observed at the opposite side of the carbon cage. In comparison to all other bonds, the reactivity of the *cis-1*- and the *e*-bonds increases accordingly, whereas it is diminished for the (6,6)-bonds in the neighboring six-membered ring (*cis-2* and *cis-3*). With larger addends, the enhanced reactivity of the *cis-1*-bond nevertheless stands back to steric demands, thus turning the second addition at equatorial position the most favorable. Frontier orbital methods yield similar results for C_{60}. Calculations give particularly high coefficients for the frontier orbitals at the preferred *cis-1-* and *e*-positions in the monoadducts, while the respective bonds are found to shorten. A detailed discussion of orbital considerations in second additions may be found in the book of Hirsch and Brettreich.

The same fundamental aspects as in second addition are, in principle, also valid for further reactions of a bisadduct of C_{60} (or higher fullerenes). Again, only some of the potential reaction sites are favored, and from the multitude of possible isomers only a markedly reduced number will be observed experimentally. Still the structure determination of different isomers is quite difficult and, for the time being, has mainly been achieved for highly symmetrical species. Sufficiently large addends provided, the products of a *cis*-addition can be excluded beforehand, thus reducing the number of possible isomers. Upon equatorial orientation of functional groups, further addends are observed each to take another equatorial position that is still free. This results, for example, in highly symmetric hexakis-adducts that exhibit T_h-symmetry (Figure 2.35b). The increased reactivity at the respective positions may again be explained by the large coefficients of frontier orbitals in the starting molecule.

To gain better control over the formation of certain regioisomers or to obtain geometries hard to access, several strategies can be applied that enforce a definite spatial arrangement of addends. Reactive units may, for example, be connected by means of a linker whose length will then define the possible reaction sites for the second addition. Bridged derivatives result from this route (Figure 2.37). Topochemically controlled solid phase reactions have been employed, too; here the crystal lattice predetermines the structure of the addition product.

Yet another scheme exists besides the principles mentioned above. This is mainly encountered in the addition of radicals and certain nucleophiles like lithium or copper organyls. Upon reaction with an excess of the reagent, the favorable products are a hexakis- or a pentakis-adduct, respectively, that adopts the structure of a 1,4,11,14,15,30-hexahydrofullerene (Figure 2.38). The adduct's characteristic feature is a cyclopentadienyl unit whose double bonds are isolated from the conjugated π-system of the remaining fullerene skeleton. This structure is

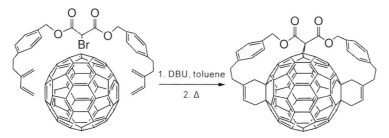

Figure 2.37 Example of a regioselective, template-directed second addition to a fullerene.

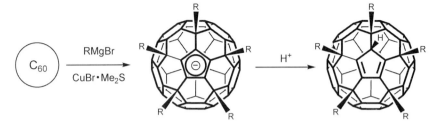

Figure 2.38 Synthesis of a hexahydrofullerene by way of the penta-substituted anion (the unaffected double bonds have been omitted for clarity reasons).

Table 2.6 Ionization potential (IP) and electron affinity (EA) of fullerenes (calculated for individual molecules).

Fullerene	C_{60}	C_{70}	C_{76}	C_{78}	C_{82}	C_{84}
IP in eV	7.8	7.3	6.7	6.8	6.6	7.0
EA in eV	2.7	2.8	3.2	3.4	3.5	3.5

Acc. to L. Echegoyen, F. Diederich, L. E. Echegoyen, *Electrochemistry of Fullerenes*, in: K. M. Kadish, R. S. Ruoff (editors), *Fullerenes*, Wiley Interscience, New York **2000**.

formed by a stepwise addition of radicals or nucleophiles, respectively (Figures 2.38 and 2.66). From the addition reaction with potassium fluorenide, for example, the tetrakis-adduct can be isolated and further reacted with different nucleophiles to yield the respective hexakis radicals. The cyclopentadienyl structure is even able to form η^5-complexes with transition metals (Figure 2.43).

2.5.2
Electro- and Redox Chemistry of Fullerenes

Judging from the molecular orbital diagram, C_{60} should easily accept electrons because it has a triply degenerate LUMO that lies energetically low. Analogous considerations hold for C_{70}, C_{76}, C_{78}, C_{82}, and C_{84}. Their calculated ionization potentials and electron affinities are collected in Table 2.6.

2.5.2.1 Electrochemistry of Fullerenes

The MO-model (Figure 2.6a) shows that C_{60} can theoretically accept up to six electrons in its triply degenerate LUMO. It has indeed been possible to show these one-electron processes experimentally and to determine their six reduction potentials by cyclovoltammetry (Figure 2.39). For C_{70}, the expectable six reductive

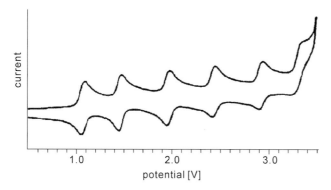

Figure 2.39 Cyclovoltammogram of C_{60}. The six oxidation steps are clearly to be seen (© ACS 1992).

Table 2.7 Reduction and oxidation potentials of C_{60} and C_{70} in volt.

	E_{red}^1	E_{red}^2	E_{red}^3	E_{red}^4	E_{red}^5	E_{red}^6	E_{ox}^1
C_{60}	−0.98	−1.37	−1.87	−2.35	−2.85	−3.26	+1.26
C_{70}	−0.97	−1.34	−1.78	−2.21	−2.70	−3.07	+1.20

See L. Echegoyen, L. E. Echegoyen, *The Electrochemistry of C_{60} and Related Compounds*, in H. Lund, O. Hammerich, *Organic Electrochemistry*, 4th ed., Marcel Dekker, New York **2001**.

steps have been found as well, with an additional observation being that from the third step on, the electron uptake occurs more easily than in C_{60} (Table 2.7). This effect can be explained firstly by the charges being farther distant from each other in the larger molecule and secondly by a better delocalization over more carbon atoms.

The choice of solvents has considerable impact on the reduction and oxidation potentials measured. The values vary up to 400 mV depending on the solution medium. Obviously factors like donor/acceptor properties, Lewis basicity, or the ability to form hydrogen bonds have large influence on the redox potential.

The higher fullerenes are more easily reduced and oxidized than C_{60}. The first reduction potential of C_{76}, for example, is about 100–200 mV (depending on the solvent) more positive than for C_{60}. Altogether the first one-electron reduction is facilitated with an increasing number of carbon atoms and the reduction potential shifts toward more positive values. Furthermore, with the LUMO degeneration removed, a wider variation is observed for the potentials of different reduction steps. Finally, the HOMO–LUMO-gap, which can be estimated from the difference between the first oxidative and the first reductive step, decreases with a growing number of carbon atoms.

The fulleride anions generated by electrochemical reduction may be reacted with electrophiles, yielding functionalized fullerenes like, for instance, alkylated derivatives.

The oxidation of C_{60} should be harder to achieve due to the high ionization potential and the completely occupied HOMO. The first oxidation step of C_{60} has experimentally been found at a relatively high potential of +1.26 V. Nevertheless, it has by now been possible to isolate a number of compounds with an oxidized C_{60} as well as a fairly stable salt of C_{60}^+, and the oxidation potentials to C_{60}^{2+} (+1.71 V) and C_{60}^{3+} (+2.14 V) could be determined, too. Likewise the electrochemical oxidation of C_{70} succeeded; the anodic potential was measured to be +1.2 V. The resulting radical cations are highly reactive and could be trapped with a variety of nucleophiles. The numerous derivatives of fullerenes naturally possess a varied electrochemistry of their own, yet a detailed discussion will be left out here to refer instead to the review literature on the topic.

2.5.2.2 Reductions of Fullerenes

Apart from electrochemical oxidation and reduction, the fullerenes' electron affinity can also be used in wet-chemical processes. The formation of alkali-metal fullerides is among the most important reactions with reducing agents. Here, the respective elemental metal and C_{60} are reacted in solid phase, for example, by heating in a closed ampoule. The stoichiometry of the resulting products is controlled by the ratio of fullerene and metal:

$$C_{60} + nM \rightarrow M_n C_{60} \ (n = 2, 3, 4, 6, 12; M = \text{Li, Na, K, Rb, Cs}) \quad (2.2)$$

Mixed alkali-metal fullerides can also be obtained in this fashion:

$$C_{60} + n_1 M^1 + n_2 M^2 \rightarrow M^1_{n_1} M^2_{n_2} C_{60} \cdot (M^1, M^2 = \text{Li, Na, K, Rb, Cs}) \quad (2.3)$$

Due to C_{60} being easily reduced, stoichiometries can range from $n = 1$ (CsC_{60}) to 12 ($Li_{12}C_{60}$). Among the respective compounds, those with a composition of M_3C_{60} are unique for displaying metallic character and, in addition, possessing supraconducting properties even at relatively "high" temperatures. Such fullerenes are known either with a single alkali metal or as a mixed compound, only Na_3C_{60} has not been observed experimentally because it disproportionates to give Na_2C_{60} and Na_6C_{60}.

The alkali-metal fullerides are intercalation compounds with the metal atoms embedded in the octahedral and tetrahedral gaps of the fullerene crystal. In the face-centered cubic crystal there are two tetrahedral and one octahedral gap per molecule of C_{60}, with radii of 1.12 Å and 2.06 Å, respectively. Resulting from a different occupation of these interstitial sites, four different structures are conceivable (Figure 2.40a).

- An MC_{60}-stoichiometry is obtained from the occupation of octahedral gaps alone, the compound has a *sodium chloride* structure.
- With half of the tetrahedral gaps occupied, another compound with a composition of MC_{60} results, yet its structure is of the *zinc blende* type.

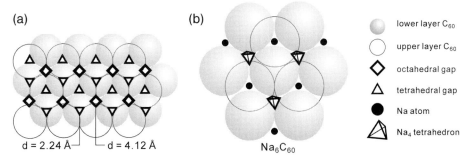

Figure 2.40 (a) The structure of the cubic face-centered lattice of C_{60} with intercalation sites; (b) example of the intercalation compounds Na_6C_{60} with Na_4-tetrahedra occupying the octahedral gaps of the fullerene lattice.

- The insertion of alkali metal into all tetrahedral gaps yields an M_2C_{60}-compound with *antifluorite* structure.
- A stoichiometry of M_3C_{60} results from the occupation of all tetrahedral and octahedral gaps, the compounds feature the *cryolite* structure.

The alkali atoms' size is responsible for the crystallographic properties of the fullerides. The atomic radius, for example, of potassium and rubidium is too big for the atoms to fit into the tetrahedral gaps. Nevertheless they do form compounds of the M_3C_{60}-type, crystallizing in the *cryolitic* structure. Yet their C_{60}^{3-}-anions are hindered in free rotation and face the metal atoms in the enlarged tetrahedral gaps with a six-membered ring. In addition, the entire crystal lattice widens slightly ($a = 14.24$ Å for K_3C_{60}, compared to $a = 14.157$ Å in C_{60}). Still the overall enlargement is not too pronounced because, among other reasons, the atoms situated in the octahedral gaps are a bit too small for this position and exert an opposite effect. Na_3C_{60} is unstable and disproportionates into Na_2C_{60} and Na_6C_{60}. Other fullerenes with a general composition of $M^1_{n1}M^2_{n2}C_{60}$ crystallize in different structures, for example, for $Na_2M^2C_{60}$ (M^2 = Rb, Cs) a primitive cubic lattice is found.

With the tetrahedral gaps occupied by lithium atoms like in Li_2CsC_{60}, no supraconductivity is observed even though the stoichiometry would formally give reason to assume it. Actually, with the lithium atom being small and, in comparison, weakly electropositive, there is stronger interaction between the C_{60}-molecules and the metal atoms. It causes partial hybridization of the $2p_z$-orbitals of C_{60} and the 2s-orbitals of lithium, which in turn prevents the system from assuming the fully charge-separated state of $(M_3^+)C_{60}^{3-}$. The actual charge of the C_{60}-molecule is slightly lower, and accordingly the t_{1u}-band is less than half filled.

Different kinds of structures are obtained especially for the intercalation of more than three metal atoms per C_{60}-molecule. Compounds of an M_6C_{60}-type (M = K, Rb, Cs) for instance crystallize as a space-centered cubic lattice with the metal atoms situated in the six equivalent tetrahedral gaps. The substances are isolators as the t_{1u}-band is completely occupied. However, Na_6C_{60} features some structural particularities that cause its high stability. The crystal structure contains Na_4-clusters situated in the octahedral gaps (Figure 2.40b). The remaining two sodium atoms occupy the tetrahedral gaps of the face-centered cubic crystal. With four metal atoms per C_{60}-unit, the actual structure depends on the kind of alkali atom, the lattice may be space-centered tetragonal (K_4C_{60}, Rb_4C_{60}) or space-centered orthorhombic (Cs_4C_{60}).

The question remains why M_3C_{60}-fullerides show metallic character, although isolating properties should be expected on strictly stoichiometric composition. Some authors explain this effect by crystal defects, especially by the absence of individual alkali metal atoms from their tetrahedral gaps (fraction about 0.07).

The production of metal fullerides succeeds as well with alkaline earth metals and rare earth elements. Among others, there are mixed alkali/alkaline earth fullerides with a composition of $M_{3-x}Ba_xC_{60}$ ($0.2 < x < 2$; M = K, Rb, Cs). Likewise may pure alkaline earth fullerides be obtained, for example, Ca_5C_{60} that becomes a supraconductor below 8.4 K. Barium and strontium give binary fullerides too, but with stoichiometric factors of 3, 4, and 6. Ytterbium and samarium may intercalate in the C_{60}-crystal as well, yielding supraconductive compounds with a composition of $M_{2.75}C_{60}$.

The generation of different anions of C_{60} succeeds also with other reducing agents like mercury or organic electron donors, for example, with the crystal violet radical, tetrakis(dimethylamino)ethylene, cobaltocene or decamethylnickelocene. These reagents generate salts of C_{60}^+. The sandwich complex $[Fe^I(Cp)(C_6Me_6)]$, possessing 19 electrons, further enables the reduction to the di- and trianion. The extent of reduction may be controlled by the stoichiometry of the complex and C_{60}.

2.5.2.3 Oxidations of Fullerenes

Oxidized fullerene species may be much harder to obtain, but a varied oxidation chemistry of C_{60} has by now been established nevertheless. The classical oxidation reaction with molecular oxygen takes place at elevated temperatures and yields $C_{60}O$ or $C_{70}O$, respectively, besides minor amounts of the unstable higher oxides (Figure 2.41). A mixture of these can also be obtained by electrochemical oxidation. The photo-oxidation is another possible reaction that is assumed to occur between singlet-oxygen and the first excited triplet state of C_{60}. The oxygen is added in an epoxidic manner to a 6,6-bond. Dimethyldioxirane was found another suitable source of oxygen, also yielding the monoxides plus a 1,3-dioxolan (Figure 2.41).

Oxidizing reagents that are also applicable to the epoxidation of double bonds, for example, MCPBA, enable further oxidation steps to di- and trioxides. Here again, different isomers are possible. Their number increases with the amount of oxygen incorporated. In addition to the epoxidic structure of oxides, another species is conceivable that features an opened ring. However, this structural isomer (depicted in Figure 2.41, bottom) has not been observed experimentally in direct oxidations. Yet the reaction with ozone generates ozonides that proceed to transform, for example, into such opened fullerenes or epoxidic structures (Figure 2.41).

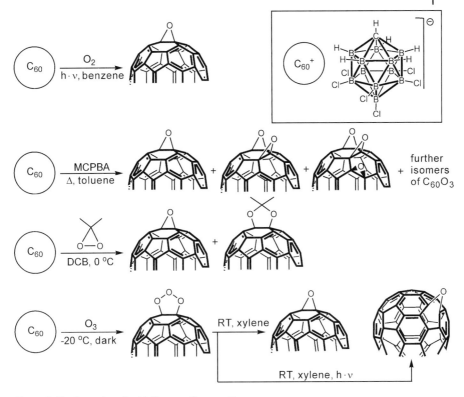

Figure 2.41 Examples of oxidation reactions on C_{60}.

Still more reactions with further oxidizing reagents are known. The osmylation described in Section 2.5.3 or reactions with strong oxidizing agents like SbF_5/SO_2ClF or $SbCl_5$ may serve as examples. The latter gives highly reactive radical cations that exist as dark green solutions. They are very unstable and decompose, partly destroying the fullerene cage, or react with nucleophiles present in the mixture. The stabilization of a C_{60}-radical cation finally succeeds by using the strongly oxidizing hexabromocarbazole radical and a stable CB_{11}^--conunterion. The latter possesses a cage-like structure with three-dimensional aromaticity. Both oxidizing agent and counterion feature very low nucleophilicity and accordingly do not react with C_{60}^+. A salt is generated that gives a dark red solution in 1,2-dichlorobenzene (Figure 2.41, inset top right).

2.5.3
Inorganic Chemistry of Fullerenes

Like in most other cases too, it is hard to distinguish inorganic from organic chemistry of fullerenes. This section primarily deals with aspects of structural chemistry, whereas reactions with metal-organic agents to organic derivatives of

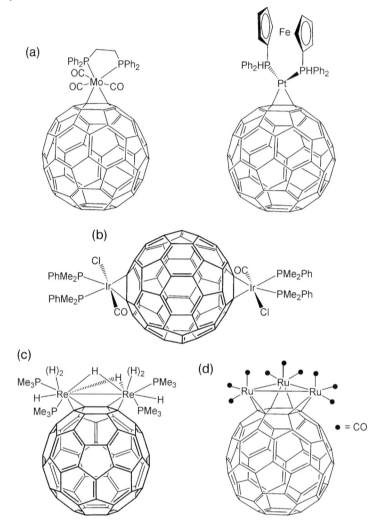

Figure 2.42 Examples of organometallic fullerene complexes with one metal center (a), two metal centers arranged at opposite sites (b) and neighbouring sites (c) and a complex comprising three metal centers in close proximity (d).

fullerenes will be discussed in the Section 2.5.5 on organic chemistry of fullerenes (Section 2.5.5).

Atoms of transition metals can coordinate the olefinic (6,6)-bonds present in C_{60} as these are already positioned suitably to complexation due to the curvature of the fullerene surface. Usually, the η^2-complexes typical also of other olefins are formed. The coordination to the metal center causes an extension of the respective (6,6)-bond and consequently reduces the fullerene's symmetry. At the same time, the affected carbon atoms' degree of hybridization is shifted further toward sp^3. Figure 2.42 collects some examples that exhibit in some cases totally different stabilities. The complex $(\eta^2-C_{60})Mo(CO)_3(dppe)$, for instance (Figure 2.42a), is fairly inert and disintegrates upon strong heating only, while other compounds decompose already at room temperature.

η^2-coordinated complexes are formed in particular with metals known for their affinity toward electron-deficient olefins, that is, with elements from subgroups 6–8 like Pt, Pd, Fe, Co, Ni, Ir, Rh, Ru, Os, Mn, and Re as well as Ta, Mo, and W. Titanium may also form olefin complexes with fullerenes. Especially the low-valent complexes of the aforementioned metals readily incorporate olefins in their coordination sphere.

With certain complexes it is also possible to obtain multiple coordination of the fullerene. One way to achieve this is the application of a large excess of the metal complex to be coordinated, for example, in the preparation of $(\eta^2-C_{60})[Ir(CO)Cl(PPh_3)_2]_2$. Due to the size of the ligands the two iridium centers on the C_{60} are situated in *trans-1*-position toward each other in this complex (Figure 2.42b). Furthermore succeeds the generation of bridged complexes with C_{60}. These may contain metal–metal bonds or bridging ligands. The structure of such complexes causes a close proximity of the transition metal atoms. An example is depicted in Figure 2.42c. The red complex $Ru_3(CO)_9(\mu_3-\eta^2, \eta^2, \eta^2-C_{60})$ represents an extreme, as all double bonds of one single six-membered ring are engaged in the coordination of the three metal centers (Figure 2.42d). This species should be distinguished from the η^6-complexes, which C_{60} does not form for its structure containing localized double bonds and for the associated low aromaticity of its six-membered rings. This may as well be deduced from considering the orbitals: their overlap does not suffice for η^6-coordination due to their radial orientation on the surface, which causes an ever growing distance of orbitals toward the outside. In a metallacyclopropane at η^2-complexation of C_{60}, on the other hand, the overlap is even stronger than in the reference molecule benzene.

The complex $Tl[C_{60}Ph_5]$ represents a particularity as the thallium cation is η^5-bound in this species. A five-membered ring entirely surrounded by phenyl groups serves as a coordinating unit (Figure 2.43).

These kinds of metal complexes with singly or multiply coordinating fullerenes are known as well for the higher homologs of C_{60}. The structural variation, however, is much wider and more intricate due to their lower symmetry. C_{70} for instance possesses four kinds of olefinic (6,6)-bonds. The shortest of them, that is, the ones

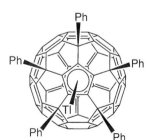

Figure 2.43 η^5-$Tl[C_{60}Ph_5]$ – a complex with fivefold coordination of the thallium atom which is η^5-complexed by the cyclopentadienyl structure.

with the most distinct double bond character, are situated at the polar caps of the carbon cage. In even higher fullerenes, a large number of possible isomers complicate the situation even more in addition to the existence of different olefinic bonds. Complexation on these fullerenes inevitably yields a multitude of structural isomers.

Inorganic compounds may also enter into σ-additions with fullerenes. Osmium tetroxide in the presence of pyridine, for example, adds to a (6,6)-bond of C_{60} forming two carbon-oxygen bonds (Figure 2.31). The stoichiometry of this reaction can be controlled by the amount of osmium tetroxide added. With an excess of OsO_4, the major product is the bisadduct that is obtained as a mixture of five out of the eight possible regio-isomers. These might be separated by HPLC.

In the presence of 4-*tert*-butylpyridine, an analogous adduct bearing additional ligands on the metallic centers is obtained. This substance has been the first derivative of C_{60} to be examined by X-ray analysis, thus confirming the cage-like structure of the fullerene molecule (Figure 2.31). The X-ray structure also reveals that the carbon atoms carrying the addend are slightly extricated from the fullerene cage, indicating a higher degree of pyramidalization.

C_{60} also participates in addition reactions with other inorganic compounds. The reaction with di-rhenium decacarbonyl, for instance, yields the respective 1,4-bisadduct of $Re(CO)_5$ (refer to Section 2.5.5.5), and with lithium alkyls the products of a 1,2-addition are obtained. With larger residues, however, the reaction no longer occurs in 1,2-position, but at more distant sites on the carbon cage. For example, $C_{60}(Si^tBu(Ph)_2)_2$ is a 1,6-adduct (Figure 2.44a).

Fullerenes possess a distinct ability to form intercalation compounds with other substances, especially with solvents. From benzene for instance, C_{60} crystallizes as a compound $C_{60} \cdot 4$ (C_6H_6). Similar crystallizations are known for further solvents like cyclohexane or CCl_4.

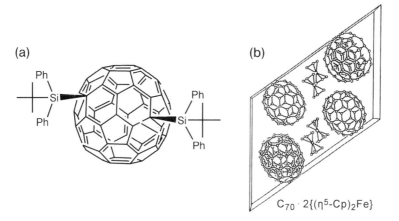

Figure 2.44 (a) Product of the 1,6-addition of a silicon compound to C_{60}; (b) co-crystallizate of C_{70} and ferrocene. An attack by the latter does not occur as its reduction power is too low (© Elsevier 1999).

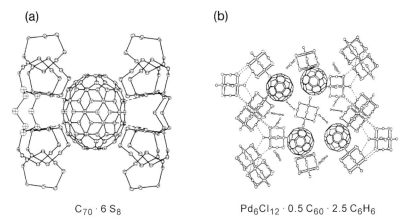

Figure 2.45 (a) Inclusion compounds of C_{70} with sulfur. The enclosure of the fullerene cage by the sulfur crowns is evident (© Neue Schweiz. Chem. Ges. 1993); (b) ternary cocrystallizate (© ACS 1996).

Contrasting many other substances however, fullerenes may also cocrystallize with molecules that cannot be numbered among the solvents, like hydroquinone, ferrocene or elemental phosphorus, to name some well-known examples. Although ferrocene is a reducing agent, its reducing power does not suffice to generate C_{60}^- or C_{70}^-, respectively. Upon crystallization, they form substances of the type $C_{60} \cdot 2\,[\eta^5\text{–}C_5H_5)_2Fe]$ (Figure 2.44b). On cocrystallization with elemental phosphorus, compounds with a composition of $C_{60} \cdot 2\,P_4$ are obtained. Here the phosphorous tetrahedrons are situated in the gaps of the C_{60}-lattice with a tetrahedral face oriented in parallel with a six-membered ring of an adjacent fullerene molecule. Slowly evaporating solutions of C_{60} in CS_2 yields a cocrystallizate of the fullerene with elemental sulfur S_8. The same is true for C_{70} and C_{76}. The respective complexes, for example, $C_{70} \cdot 6\,S_8$, feature a fullerene entirely surrounded by sulfur molecules (Figure 2.45a).

Even inorganic complexes like $(PhCN)_2PdCl_2$ react with C_{60} forming intercalation compounds. In this case a black, crystalline substance with the composition $C_{60} \cdot 2\,(Pd_6Cl_{12}) \cdot 2.5\,C_6H_6$ is obtained. The three components C_{60}, benzene, and Pd_6Cl_{12} (a cubic structure with the Pd-atoms on the faces and the Cl-atoms on the centers of the edges) join in a common crystal lattice (Figure 2.45b). Furthermore, a reaction of C_{60} with cobalt porphyrines has been described. In the products the fullerene is quite close to the center of the porphyrine system, but still too far away to call it a real chemical bond. In solid state the planar porphyrines and the fullerene cages stack over each other to form long, columnar aggregates.

Fagan studied the reaction of C_{60} with the complex $[Cp^*Ru(CH_3CN)_3]$ and found that it did not lose all three molecules of acetonitrile to form a η^6-complex as observed for aromatic systems, but rather gave a η^2-olefin complex by releasing just one ligand of acetonitrile. With some sufficiently nucleophilic hydrido complexes, C_{60} does not yield η^2-complexes, but a hydrometallation occurs instead. The zirconocene complex $Cp_2Zr(H)Cl$ reacts with C_{60} to give a red solution containing $Cp_2ZrCl(C_{60}H)$. From the acidic hydrolysis of this complex, isomerically pure 1,2-dihydro-C_{60} can be obtained (refer to Section 2.5.5.1).

2.5.4
Endohedral Complexes of Fullerenes

As seen before, the inside of the fullerene cages is empty, so it is self-suggesting to try to fill this cavity with atoms or molecules. Indeed the isolation of so-called endohedral fullerenes succeeded quite soon after the discovery of fullerenes themselves. In endohedral species, single or even several atoms are situated inside the carbon cage, so it is purely topological circumstances that prevent them from breaking out of these compounds.

2.5.4.1 Metallofullerenes

The preparation of metallofullerenes can be achieved in different ways. The methods most commonly used today are the arc-evaporation of impregnated graphitic rods and the laser-evaporation process (also called the laser-oven method). For the latter a rotating target of graphite and metal oxide (cemented with a pitch binder) is placed in an oven heated to about 1200 °C and irradiated with a doubled-frequency Nd:YAG laser under a stream of argon. The resulting fullerenes and metallofullerenes are swept along with the gas current and precipitated in the cooler zones at the end of the quartz tube. A temperature of not less than 800 °C is required for this process as neither empty nor filled fullerenes are formed below that value (refer to Section 2.3.4).

The arc-method uses the same procedure as successfully employed to produce empty fullerenes (refer to Section 2.3.3), with the sole difference that the anodes of graphite evaporated in a classical arc-apparatus are impregnated with metal oxides or carbides. (In the graphitic rods treated with metal oxides, the respective carbides will be generated, too, when heated to >1600 °C.) Besides empty species, the soot deposited on the cooler reactor walls contains various metallofullerenes.

Endofullerenes like Li@C_{60} can further be obtained by ion implantation techniques. High- energy ions of the desired element are targeted at a thin film of fullerenes. However, the produced amounts are small and accordingly the analysis of products is complicated.

Apart from the expected M@C_{60}, the examination of soot containing metallofullerenes also reveals carbon cages of unusual sizes. La@C_{74} and La@C_{82} are observed upon evaporation of a graphitic sample treated with La_2O_3. The C_{82}-metallofullerene exhibits the highest stability of all. It may, as an extract in toluene, even be stored under air without decomposing. La@C_{60}, on the other hand, cannot be isolated due to its instability.

A similar picture arises with other elements: The C_{60}-metallofullerenes are formed, yet they cannot be obtained in substance. Obviously, further factors influence stability in the generation of endofullerenes to the effect that smaller species

Table 2.8 Bonding energies ΔH_B of metallofullerenes.

Metallofullerene	Ca@C$_{82}$	Sc@C$_{82}$	Y@C$_{82}$	La@C$_{82}$
ΔH_B in kJ mol^{-1}	163	322	444	481

are less stable than bigger ones, with M@C$_{82}$ being the most abundant. Just recently it has been possible, however, to isolate a few selected C$_{60}$-metallofullerenes, for example, Er@C$_{60}$ or Eu@C$_{60}$.

In general, metallofullerenes are known for a multitude of elements by now, including alkali like Li or alkaline earth metals like Ca, Sr, and Ba, the elements of the scandium group Sc, Y, La and the lanthanides Ce, Pr, Nd, Sm, Eu, Gd, Tb, Dy, Ho, Er, Tm, Yb, Lu as well as titanium, iron, and uranium. The elements of main group and subgroup 3 are much easier to encapsulate than for instance those of main group 2, which reflects in the bonding energies of individual metallofullerenes (relative to M + C$_{82}$; Table 2.8). Endohedral compounds of heterofullerenes are know as well, it is possible, for instance, to enclose two lanthanum atoms in C$_{79}$N.

This is indicative of a more general finding: In addition to simple endofullerenes M@C$_n$ the encapsulation can yield species with several atoms confined to the same fullerene cage. The radius of the cavity inside C$_{82}$ is 0.4 nm, which is large enough to host up to three atoms of rare earth metals. The generation of M$_m$@C$_n$ becomes ever more probable with an increasing content of metal in the graphitic starting material. For scandium the complete series of Sc@C$_{82}$, Sc$_2$@C$_{82}$, Sc$_3$@C$_{82}$ and, just recently, even Sc$_4$@C$_{82}$ has been detected. Further on, Sc$_2$@C$_{84}$ was isolated, which is one of the best studied dimetallofullerenes besides La$_2$@C$_{80}$.

The formation of C$_{80}$-metallofullerenes is a particularly interesting because C$_{80}$ itself does not exist as an empty cage. The ground state of I_h-C$_{80}$ exhibits an antiaromatic open-shell structure explaining this instability. (There are only two electrons situated in the quadruply degenerate HOMO.) In a respective endofullerene however, the electronic structure is effectively influenced by electron transfer from the metal atoms to the carbon cage. The latter turns into a stable closed-shell C$_{80}^{6-}$-ion by accepting six electrons. This may be effected, for example, by two atoms of lanthanum, resulting in the formation of La$_2$@C$_{80}$. Today the metal atoms are known to circulate inside the icosahedral C$_{80}$-cage. The movement could be shown by the line broadening in the ^{139}La-NMR spectrum that is caused by the magnetic field resulting from this motion. The formation of C$_{80}$-compounds is speculated to happen by the way of the so-called shrink-wrap mechanism that explains the generation of smaller species by an expulsion of C$_2$-units from existing metallofullerenes and subsequent closure of the cage. This process may continue to cage sizes far smaller than C$_{60}$ and stops only at about 44 (but not less than 36) carbon atoms. Further shrinking leads to destruction of the cage-like structure. The size limit is determined, among others, by the dimensions of the encapsulated metal atom.

It soon became evident from ESR-measurements that the metal atoms exist as positively charged ions inside the fullerenes. Obviously, they transfer their valence electrons to the cage. For instance La@C_{82} really is La^{3+} + @C_{82}^{3-}. Yttrium expectedly behaves in an analogous manner, while scandium, due to its low-lying d-orbital, forms Sc^{2+} by transferring its 4s-electrons only. The tendency toward trivalent ions in Y and La is favored by their higher and more diffused d-orbitals. The charge status of lanthanide atoms bearing electrons in 4f-orbitals is subject to lively discussion. UV/Vis-spectra indicated that they are trivalent. Ce ($4f^1 5d^1 6s^2$) and Gd ($4f^7 5d^1 6s^2$) form trivalent ions accordingly, while elements such as Pr ($4f^3 6s^2$) and Nd ($4f^4 6s^2$) seem to transfer only two electrons. Eu ($4f^{14} 6s^2$) and Yb ($4f^7 6s^2$) each contribute their two 6s-electrons alone because a half or a fully occupied 4f-orbital is energetically favorable. Lutetium ($4f^{14} 5d^1 6s^2$) has a configuration similar to scandium, yet it transmits the 5d- and one 6s-electron because 6s-electrons are stabilized by relativistic effects. Summing up, the circumstances are rather complicated by backbonding and by orbitals lying in close proximity.

A major issue arising after the discovery of metallofullerenes was whether the metal atoms were really contained inside the carbon cage because in principle, they could just as well be located on the fullerene's outer surface. There were several approaches to answer this question. Firstly, endofullerenes were demonstrated to expel C_2-units in gas-phase fragmentation experiments (collision, laser impact, etc.), whereas exohedral compounds of the Fe(C_{60})-type yield an intact buckminsterfullerene by cleaving the metal atom. Secondly, the endohedral structure could be shown by electron microscopic examination (HRTEM, STM) of the solid phase. Conclusive evidence, however, could only be adduced from studies employing synchrotron X-ray diffraction. In these studies, significant electron density was detected inside the fullerene, making clear at the same time that usually the metal atoms are not situated at the center of the carbon cage (Figure 2.46). This method could even provide information about the charge of the encapsulated metal. The scandium atom in Sc@C_{82}, for instance, was found not to be trivalent, but the structure rather corresponds to Sc^{2+}@C_{82}^{2-}. The electronic properties of metallofullerenes are of special interest because they markedly deviate from those of empty fullerenes. The calculated ionization potentials and electron affinities for a series of metallofullerenes are collected in Table 2.9.

These values indicate that metallofullerenes can be both stronger electron donors and acceptors than their empty analogs C_{60} and C_{70}. This observation is

Table 2.9 Electron affinities (EA) and ionization potentials (IP) of metallofullerenes.

Metallofullerene	Sc@C_{82}	Y@C_{82}	La@C_{82}	Ce@C_{82}	Eu@C_{82}	Gd@C_{82}	C_{60}	C_{70}
EA in eV	3.08	3.20	3.22	3.19	3.22	3.20	2.57	2.69
IP in eV	6.45	6.22	6.19	6.46	6.49	6.25	7.78	7.64

See S. Nagase, K. Kobayashi, T. Akasaka, T. Wakahara, *Endohedral Metallofullerenes: Theory, Electrochemistry, and Chemical Reactions*, in: K. M. Kadish, R. S. Ruoff (editors), *Fullerenes*, Wiley Interscience, New York **2000**.

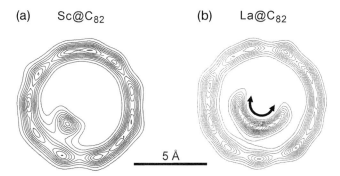

Figure 2.46 Distribution of electron density in endofullerenes. It turns out that, for example, lanthanum exerts a motion within the cage (b) whereas scandium remains fixed at its position out of the cage's center (a) (© Wiley Interscience 2000).

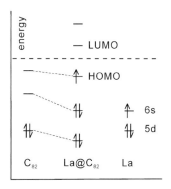

Figure 2.47 Molecular orbital diagram of La@C_{82} (© ACS 1993).

confirmed by measurements of the reduction and oxidation potentials. For La@C_{82}, the species best characterized, a first oxidation step was found at +0.07 V, corresponding to a moderate electron donor (comparable to ferrocene) and explaining its stability to air. Furthermore there are five reduction potentials at −0.42 V, −1.37 V, −1.53 V, −2.26 V and −2.46 V. Altogether, La@C_{82} is a better electron acceptor than an empty fullerene. As evident from Figure 2.47, the first reduction step above all is energetically favorable because the HOMO will be fully occupied by taking in an electron. The donor–acceptor characteristics of other metallofullerenes may be deduced from the MO-diagram as well. They turn out to be good electron acceptors just as well, yet they can also dispose of unpaired electrons and be oxidized. The 4f-electrons in Ce-, Pr-, and Gd-fullerenes do not seem to be involved in this process.

Metallofullerenes that could actually be isolated have been examined for their reactivity as well. Exohedral functionalization was achieved, for example, by photochemical reactions, Diels–Alder addition, etc. In the reaction with disiliranes, the thermal variant was found to be viable besides the photochemical addition observed with empty fullerenes. This effect is caused by the better donor–acceptor properties. Contrasting the noble-gas fullerenes discussed in Section 2.5.4.2, the electron transfer from an inward metal atom to the carbon cage largely influences the chemical properties of the respective endofullerene. Dimetallofullerenes may also react with disiliranes in a thermal way, which again is explained by the acceptance of electrons to the fullerene cage. However, the number of metal atoms inside the cage is not crucial as long as the reduction potential is sufficiently low (in $Sc_2@C_{82}$, for example, it is too high and the thermal reaction does not occur).

2.5.4.2 Endohedral Compounds with Nonmetallic Elements

Other elements but metals may also be enclosed in the cavities of fullerenes. Atoms of noble gases are particularly apt to this, but compounds with molecular hydrogen or nitrogen atoms included in the cage are known as well.

Customary methods of fullerene production are commonly performed in an atmosphere of helium, but still there are extremely few C_{60}-molecules containing helium atoms. However, charged fullerenes C_n^{m+} can incorporate atoms of noble gases when shot as molecular beam through a respective atmosphere. After reduction to a neutral particle, $He@C_{60}$ could for instance be detected in the mass spectrometer. Still caution is advisable as an empty molecule of C_{60} with four ^{13}C-atoms will give rise to the same peak as $He@^{12}C_{60}$, and so macroscopic amounts of the endohedral compound are required for unambiguous analysis. The production of different noble-gas endofullerenes $X@C_{60}$ (X = He, Ne, Ar, Kr, Xe) is feasible by heating C_{60} under high pressure (~3000 bar ~4.4 × 10^4 psi). Employing C_{70}, the same procedure also yields $X@C_{70}$. The conversion is not complete, however – really it is a rather small portion of the fullerenes incorporating noble-gas atoms (0.04–0.3%). While mass spectrometry was found unreliable to some extent, NMR spectroscopy may indeed serve to make a more accurate estimate on the amount of helium atoms in the carbon cages: Helium-NMR spectra can be recorded for the NMR-active ^{3}He-nucleus. Thus a content of 0.1% of $He@C_{60}$ has been determined.

Now the question arises of how the noble-gas atoms get into the fullerenes' cavity. The six-membered rings constitute the largest "passage" through the surface, but these would still have to be considerably widened to let a noble-gas atom pass. The activation energy for such a process would be accordingly high. By now it seems settled that at least one bond has to be broken to provide an entrance sufficiently large for the incorporation of atoms into the fullerenes. According to calculations it is energetically more favorable to break a (5,6)-bond and create an inlet through a nine-membered ring than to open a (6,6)-bond temporarily. The fullerene cage can re-seal then after the uptake or release of atoms, but it is also in danger of reacting otherwise. Deterioration products of C_{60} have been found in such experiments indeed.

In studies on helium endofullerenes, the respective signal in the ^{3}He-NMR spectrum experienced a high-field shift compared to the free helium atom (Figure 2.48). The shift is 6.3 ppm for $^3He@C_{60}$ and 28.8 ppm in C_{70}, indicating a diamagnetic ring current in the fullerene cage that is particularly pronounced in C_{70}.

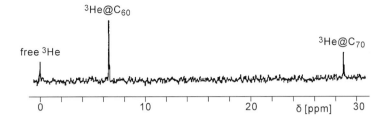

Figure 2.48 Compared to the signal of free helium, the signals of He@C_{60} and He@C_{70} in the ^{3}He-NMR spectrum show a significant high field shift (© Nature Publishing Group 1994).

Finally, other nonmetals too can be embedded in C_{60} and higher fullerenes. N@C_{60} may serve as an example here. Even though actually being reactive to the extreme, the nitrogen atom is so efficiently shielded by the carbon cage that it cannot react with its surroundings. Neither occurs a reaction with the inert inner surface of the fullerene. The production is achieved by shooting high-energy nitrogen ions at a film of fullerenes. The same process can also be employed to generate N@C_{70} or P@C_{60}.

2.5.5
Organic Chemistry of Fullerenes

2.5.5.1 Hydrogenation and Halogenation

For being polyolefins, the fullerenes present a multitude of double bonds that can in principle be transformed into C–C-single bonds by hydrogenation. Depending on the method applied, the reaction can lead to singly or multiply hydrogenated species. Still the directed production of defined hydrofullerenes is a demanding task, and structure determination sometimes is not trivial, especially for multiply hydrogenated compounds.

The reaction with molecular hydrogen is no suitable method to hydrogenate selectively one double bond of the fullerene cage, it rather yields a variety of polyhydrogenated compounds. The transformation does not succeed, however, without adding a catalyst or a radical promotor (like iodoethane) unless the process is run at extreme pressure (>3 GPa) and high temperatures (>700 °C). The radical hydrogenation mainly yields $C_{60}H_{36}$, whereas degrees of hydration up to $C_{60}H_{50}$ are observed for the catalytic variant with a ruthenium catalyst on activated carbon. Employing transition metals on alumina, the catalysts Pt, Co, Ni, and Pd mainly generate $C_{60}H_{36}$, while Ru, Rh, and Ir produce $C_{60}H_{18}$. Iron does not exhibit any catalytic activity.

In 1993 Rüchardt presented a very adroit method for the hydrogenation of C_{60} – a transferhydrogenation using 9,10-dihydroanthracene. Upon treating C_{60} with the molten reagent in a sealed ampoule at 350 °C, a colorless product is obtained within short time. Its empirical formula corresponds to $C_{60}H_{36}$. Continuing the reaction over 24 h, the product composition changes to $C_{60}H_{18}$. Only small amounts

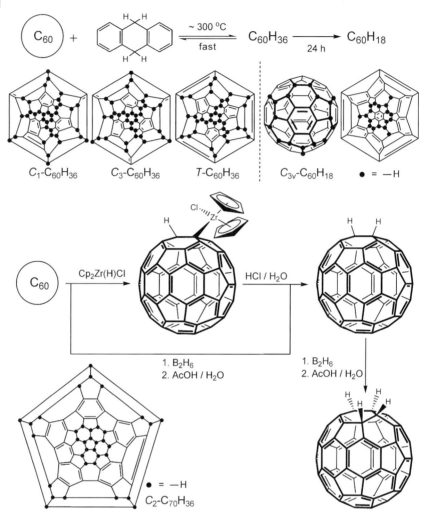

Figure 2.49 Depending on the method chosen, fullerenes with different hydrogenation patterns are obtained. Highly hydrogenated derivatives often feature remarkably symmetrical structures.

of impurities are found, which renders the method suitable to produce these hydrofullerenes on a preparative scale. Structure determination revealed exactly one isomer with C_{3v}-symmetry for the $C_{60}H_{18}$, while at least three isomers of $C_{60}H_{36}$ were isolated. These feature C_1-, C_3-, and T-symmetry (Figure 2.49). The application of 9,10-dideuterioanthracene led to the respective deuterated species.

Hydrogenation of C_{60} succeeds as well with certain base metals. The reaction with zinc in dilute hydrochloric acid gives the lowly hydrogenated compounds $C_{60}H_2$, $C_{60}H_4$, and $C_{60}H_6$. Other base metals generate hydrogenated fullerenes, too;

yet the product mixtures are far more complex and the yields are poor. The use of moist zinc–copper couple turned out the most attractive method. The stoichiometry of the hydrogenation product can easily be controlled between $C_{60}H_2$, $C_{60}H_4$, and $C_{60}H_6$ by varying the reaction time and the amount of reducing agent. When a solution of C_{60} in benzene is treated with zinc in concentrated hydrochloric acid, the hydrogenation progresses to the major product $C_{60}H_{36}$. Here the otherwise rarely observed $C_{60}H_{38}$ and $C_{60}H_{40}$ are found as byproducts.

The Birch reduction with alkali metals in liquid ammonia has been described for C_{60} already in 1990, just shortly after its isolation. A mixture of highly hydrogenated fullerene compounds $C_{60}H_x$ ($18 < x < 36$) is generated especially with lithium in ammonia in the presence of *tert*-butanol. Again $C_{60}H_{18}$ and $C_{60}H_{36}$ are the major products. Derivatives with even more hydrogen attached cannot be obtained by Birch reduction as they decompose at the conditions applied.

The classical approach to the directed hydrogenation of double bonds – the hydroboration – is feasible for C_{60}, too and allows for the selective preparation of 1,2-hydrogenated compounds. After reaction with a borane source (B_2H_6 or $BH_3 \cdot THF$), the second hydrogen atom is substituted for the borane residue upon acidic hydrolysis. The 1,2-addition is predetermined by the four-membered transition state in the formation of the hydroborinated intermediate. Hence the generation of further regioisomers of the dihydro compound is forestalled. A second hydroboration with subsequent protonolysis yields the 1,2,3,4-addition product $C_{60}H_4$ (Figure 2.49).

Another way to selectively obtain the 1,2-dihydrofullerene is to subject to acidic hydrolysis the addition product of $Cp_2Zr(H)Cl$ with C_{60}. Again the 1,2-addition pattern is determined by the addition of the zirconocene so no other regioisomers of the dihydrocompound emerge from the reaction (Figure 2.49).

A characteristic feature of the hydrogenated fullerenes is their acidity. In some cases its degree even matches that of organic acids like acetic acid. $C_{60}H_2$ features a pK_S-value of 4.7 and the compound $C_{60}(H)^tBu$ even one of 5.6.

C_{70} and higher fullerenes may as well be hydrogenated with the methods described above, yet in comparison to C_{60} their reactivity is lower. As for C_{60}, $C_{70}H_{36}$ is observed as major product for many methods of hydrogenation when applied to C_{70} (Figure 2.49).

How far now may a fullerene be hydrogenated while still conserving the cage-like structure? This question was posed soon after their discovery, and attempts were made to synthesize so-called fulleranes by perhydrogenation. For the time being, polyhydrofullerenes with up to 44 hydrogen atoms have doubtlessly been proven, and possibly there are even species with a higher degree of hydrogenation. Calculations for several highly hydrogenated derivatives of C_{60} revealed that considerable strain accumulated in the fullerene skeleton upon the introduction of additional hydrogen atoms and the concomitant generation of sp^3-hybridized carbon atoms. This turns the polyhydrofullerene more and more unstable with an increasing degree of hydrogenation until, finally, the cage-like structure collapses.

The extreme, $C_{60}H_{60}$, has not been proven experimentally by now. It would be a homolog to the dodecahedrane $C_{20}H_{20}$ initially described by Paquette, which also

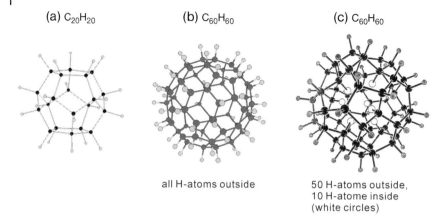

Figure 2.50 For the time being, dodecahedrane (a) is the only perhydrogenated carbon cage proven by experiment. $C_{60}H_{60}$ (b) could not be isolated so far, and calculations indicate a structure with ten internal hydrogen atoms (c) to be thermodynamically more stable (© AAAS 1991).

represents an exhaustively hydrogenated carbon cage. Figure 2.50 shows $C_{20}H_{20}$ as well as a hypothetical and a calculated structure of $C_{60}H_{60}$. It turns out that the icosahedral molecule depicted in Figure 2.50b is by far not the most stable isomer of $C_{60}H_{60}$. The total energy would rather be diminished if one or more hydrogen atoms were placed inside the fullerene cage. The energetically most favorable isomer still conserving the cage-like structure is one with ten inward hydrogen atoms. Thus the formerly spherical shape of the fullerene suffers considerable indentations (Figure 2.50c). However, this theoretical structure is not likely ever to be accessible to preparation because the inversion of an external hydrogen atom to the cage's inside inevitably faces a very high energetic barrier. It remains to be seen whether a "fullerane" can ever be isolated and, if so, what structure it might exhibit.

There is a close structural resemblance between the hydrogenated fullerenes and their halogen analogs. Like for hydrofullerenes, a large variety of halogenated derivatives can be prepared with some isomers like $C_{60}F_{18}$, being particularly stable. Compounds with different degrees of halogenation are obtained depending on the reaction conditions. Generally the halogen compounds are prepared by radical addition, while this approach fails for the introduction of iodine.

For being quite stable, the fluorofullerenes have been studied the most. They can be obtained by reaction with gaseous fluorine, xenon difluoride or further reactive fluorides like BrF_5 or IF_5. Normally, products with up to 48 fluorine atoms per C_{60} are found. Hyperfluorinated compounds ranging up to $C_{60}F_{78}$ (or even $C_{60}F_{102}$) have been observed upon application of drastic conditions, yet these are products with an opened cage. The reaction with heated metal fluorides mainly yields $C_{60}F_{18}$ and $C_{60}F_{36}$ that exhibit the same addition pattern like $C_{60}H_{18}$ or $C_{60}H_{36}$, respectively.

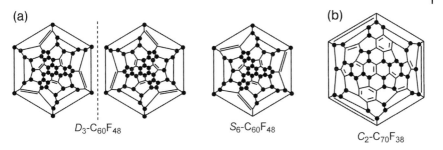

Figure 2.51 (a) $C_{60}F_{48}$ is obtained as a chiral D_3-isomer and as S_6-meso form; (b) $C_{70}F_{38}$ is found to possess C_2-symmetry.

Table 2.10 Composition of fluorination products of C_{60} with various metal fluorides.

Reagent	TbF_4, CeF_4, MnF_3	AgF	CuF_2, FeF_3	$KPtF_6$
Preferred composition	$C_{60}F_{36}$	$C_{60}F_{18}$	$C_{60}F_2$	$C_{60}F_{18}$

The direct fluorination with gaseous fluorine is hard to control, but conducted properly, it leads to $C_{60}F_{48}$ as major product. This is considered the upper margin of exhaustive fluorination as higher fluorinated derivatives most likely contain broken fullerene cages. Two symmetries are observed for $C_{60}F_{48}$, with one of them (D_3–$C_{60}F_{48}$) existing with two enantiomeric variants. The other (S_6–$C_{60}F_{48}$) represents the *meso*-form (Figure 2.51a). The structure also indicates why further fluorination of $C_{60}F_{48}$ is only possible when losing the structural integrity of the fullerene: The remaining double bonds are located in indentations of the cage which effectively shields them against further attack.

Other fluorofullerenes are obtained by reaction with metal fluorides. These may be less reactive than elemental fluorine, but consequently enable a much better selectivity in product generation. Depending on the metal chosen (either transition or rare earth metals might be considered), the composition of the fluorine compound ranges from $C_{60}F_2$ to about $C_{60}F_{36}$. Not only binary, but also ternary metal fluorides can be employed (Table 2.10).

Again the question arises on the limit to exhaustive halogenation. Due to its size, fluorine should be the halogen best suitable to a dense covering of the fullerene's surface. The compound undoubtedly characterized by now is $C_{60}F_{48}$, which is obtained by direct fluorination with F_2.

Halogenofullerenes can also be generated with chlorine, the next higher homolog. The synthetic methods are similar to those for fluorination. Both chlorine and halogenochlorides may be employed, but as with fluorine, complex mixtures are often obtained instead of separable compounds. Moreover chlorofullerenes are less stable than their fluoro-analogs, they are prone to hydrolysis

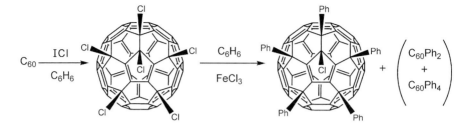

Figure 2.52 In a Friedel–Crafts-like reaction, the hexachlorofullerene $C_{60}Cl_6$ may be converted into a pentaarylated product.

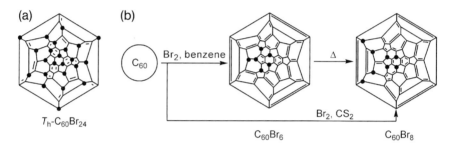

Figure 2.53 (a) The structure of $C_{60}Br_{24}$, (b) conversion of C_{60} into $C_{60}Br_6$ and $C_{60}Br_8$.

and reactive toward nucleophiles. With typical numbers ranging from 6–14 (partly even 26) chlorine atoms, the degree of halogenation achieved is lower than with fluorine.

The synthesis of a defined chlorofullerene succeeded using iodochloride and yielded $C_{60}Cl_6$ as the sole isolable product. The molecule features C_5-symmetry, and the substance is obtained as shining orange crystals. The compound may be employed in Friedel–Crafts-like reactions to yield a penta-arylated fullerene derivative (Figure 2.52). $C_{60}Cl_{24}$ is generated from C_{60} in a stream of gaseous chlorine, yet this highly symmetric compound can be obtained as well from reacting C_{60} with VCl_4 or $C_{60}Br_{24}$ with $SbCl_5$, respectively. The X-ray structure resembles that of $C_{60}Br_{24}$.

As indicated just before, C_{60} also reacts with bromine forming bromofullerenes. With liquid bromine, $C_{60}Br_{24}$ is obtained, which exhibits the highly symmetric structure (T_h) depicted in Figure 2.53a. The addition of bromine usually occurs as 1,4-addition. Owing to the spatial demand of the bromine atoms the 1,2-variant is rather unfavorable. The directed synthesis of bromofullerenes with a lower content of halogen can be achieved, too. $C_{60}Br_8$, for instance, is formed by reaction with elemental bromine in solution (in CS_2 or $CHCl_3$). Here the bromine atoms are arranged in two groups to both sides of the molecule's equator. The choice of solvents has an obvious impact on the stoichiometry of the products obtained as in CCl_4 or benzene; $C_{60}Br_6$ is generated instead of $C_{60}Br_8$. The hexabromide features the structure of the $C_{60}Cl_6$ mentioned above. Due to ecliptic interactions between cis-oriented bromine atoms it is less stable than the octabromide and consequently disproportionates into C_{60} and $C_{60}Br_8$ upon heating (Figure 2.53b). Altogether, the strength of the C–Br-bond is only moderate because it is stretched to more than 2.03 Å (normal: C–Br 1.96 Å) by the ecliptic interaction. This results in all bromofullerenes releasing bromine upon heating. The existence of (5,6)-double bonds after the addition of bromine further diminishes the stability.

Iodofullerenes have not yet been isolated. Presumably, they are unstable because the large iodine atoms cannot attack at sites close to each other, while an addition at farther distant positions would cause an extensive bond-rearrangement, including the generation of several (5,6)-double bonds. Furthermore, the carbon-iodine bond is markedly weaker than with other halogens.

Altogether, the stability of halogenated fullerenes increases with the halogen content, and it decreases in the series F > Cl > Br >> I. Successive reactions of the halofullerenes are limited, among others, by their in parts extremely low solubility in organic solvents. Still they tend to hydrolyze and react even with atmospheric humidity to give hydroxy derivatives or epoxides.

Halogenations have also been described for a variety of other fullerenes with both larger or smaller numbers of carbon atoms. Representatives of smaller fullerenes will be discussed in Section 2.5.8. For higher species, the same rules apply as for C_{60}, only the increased number of isomers and the accordingly complicated enrichment of individual species must be considered. The preparative methods largely resemble those for halogenated derivatives of C_{60}. Most results have been described for C_{70}. There are, for instance, characterized compounds with every halogen but iodine, like $C_{70}F_n$ (n = 34, 36, 38 [the C_2-isomer is shown in Figure 2.51b], 40, 42, 44), $C_{70}Cl_{10}$, and $C_{70}Br_{10}$, but derivatives of higher fullerenes like $C_{76}F_{36}$, $C_{78}F_{38}$, $C_{84}F_{40}$, or $C_{78}Br_{18}$ have been reported, too.

2.5.5.2 Nucleophilic Addition to Fullerenes

The reactivity of fullerenes toward nucleophiles is caused by their electron deficiency. Numerous reactions with nucleophilic reagents leading to useful and interesting derivatives have been described.

The nucleophilic addition to fullerenes corresponds to the common mechanism observed for electron-deficient olefins. A nucleophile A^- initially attacks the double bond and a reactive intermediate $C_{60}A^-$ is generated. This can be stabilized in several ways. The product $C_{60}AE$ results from the reaction with an electrophile E^+ (Figure 2.54a). These steps may proceed repeatedly to give compounds with a general composition of $C_{60}A_xE_x$. Furthermore, an intramolecular reaction is possible that yields bridging functional groups, and $C_{60}A_2$ is accessible by oxidative work-up of $C_{60}A^-$.

The alkylation by organolithium- or Grignard-reagents is a basic reaction between a fullerene and a nucleophile. According to the mechanism above, the $C_{60}R^-$-anion is generated in a first step. It is then stabilized with H^+ in an acidic

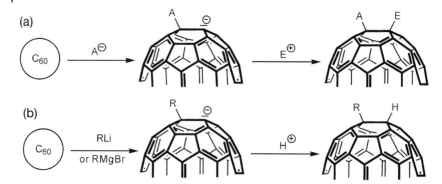

e.g. R = Me, ^{t}Bu, fluorenyl, (RLi); R = Et, ^{i}Pr, Ph, Me$_3$SiCH$_2$, octyl (RMgBr)

Figure 2.54 (a) General mechanism of the nucleophilic addition to a fullerene; (b) reaction of C$_{60}$ with organo-lithium or Grignard compounds.

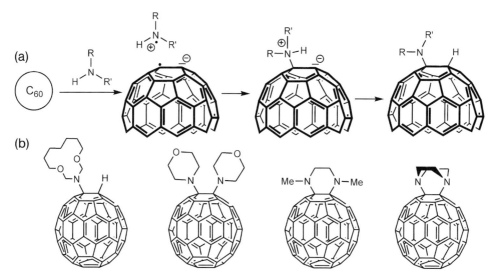

Figure 2.55 (a) The reaction of C$_{60}$ with amines yields aminofullerenes; (b) some examples of amino derivatives of C$_{60}$.

work-up yielding the hydroalkylated fullerene (Figure 2.54b). In analogous ways the reaction may be conducted with aromatic residues R or with acetylides. With a large excess of metal organyl applied, multiple additions of up to five alkyl groups can occur depending on the size of the residues.

Moreover, fullerenes easily enter into nucleophilic additions with primary or secondary amines. Here the first intermediate is an anionic complex with two electrons being transferred from the amine to the C$_{60}$. Subsequent recombination with Zwitterion generation and ensuing proton transfer yield the hydroaminated product (Figure 2.55a). Still an isolation of individual derivatives usually is impossible because the amines' high nucleophilicity leads to the formation of different multiple adducts. These tend to be inseparable, and just in some cases the selective isolation of the tetraamino-adduct succeeds. What is more, strict exclusion of oxygen is required to prevent the generation of dehydrogenated species and, to some extent, epoxides. Figure 2.55b collects some examples of isolated and characterized amino compounds. With diamines, C_{60} reacts to give bridged derivatives. The addition takes place on a (6,6)-bond. Here as well, dehydrogenated products are obtained as the amino protons present in the first stage are removed by oxidative elimination. Usually mono- or bisadducts are formed. Tertiary amines cannot react with C_{60} to yield analogous products. They give charge-transfer complexes instead, with some of them being considerably stable, for example, the C_{60}-complex with the *leuco*-form of crystal violet. Apart from that, they may also enter into photochemical or thermal radical processes (Section 2.5.5.5).

In reactions of alkyl cyanides with C_{60} an intermediate $C_{60}CN^-$-anion is formed which may then be saturated by various electrophiles (e.g., H^+, Me^+, etc.) A special feature of the cyanide addition is the generally exclusive production of the monoadduct. This turns the reaction valuable to the directed introduction of exactly one functional group.

When treating toluenic solutions of C_{60} with concentrated solutions of potassium hydroxide, a mixture of various hydroxylated products collectively termed *fullerenoles* is generated. These compounds are rather labile especially toward atmospheric oxygen, which renders their characterization complicated. What is more, these mixed compounds with their various numbers of hydroxyl groups can hardly be separated. Yet the introduction of single hydroxyl groups may be achieved by substituting halogen atoms in halofullerenes, for instance $C_{60}F_{35}OH$ or $C_{70}F_{37}OH$ can be obtained this way.

The Bingel–Hirsch Reaction One of the most valuable preparative methods applicable to fullerenes is the Bingel–Hirsch reaction. This is, in detail, the nucleophilic attack of a bromomalonate on the double bond of a fullerene in the course of which a cyclopropane ring is generated. The reaction exclusively yields the methanofullerene with a bridged (6,6)-bond. It has first been described by C. Bingel in 1993, and A. Hirsch subsequently optimized the conditions, so today it is one of the best controllable reactions of fullerenes and can be performed with a multitude of reagents and substrates.

The classical approach to this method employs the C–H-acidic diethyl bromomalonate that is reacted with sodium hydride, generating the actual reagent by deprotonation. The resulting nucleophile attacks a double bond of the fullerene to give the anionic intermediate. The negative charge in the latter is localized in α-position to the malonate, so this nucleophilic center in the C_{60}-skeleton can perform an intramolecular attack on the brominated carbon atom (S_Ni-reaction). Finally, the methanofullerene is formed by cleaving Br^- (Figure 2.56). The choice of a specific base is an important aspect in conducting the Bingel–Hirsch reaction. Sodium hydride is quite suitable as it does not enter into nucleophilic reactions. Primary and secondary amines on the other hand do not lead to the desired products because they are nucleophiles themselves and can attack C_{60} as such. Further examples of suitable bases are DBU (1,8-diazabicyclo[5.4.0]undec-7-ene), LDA,

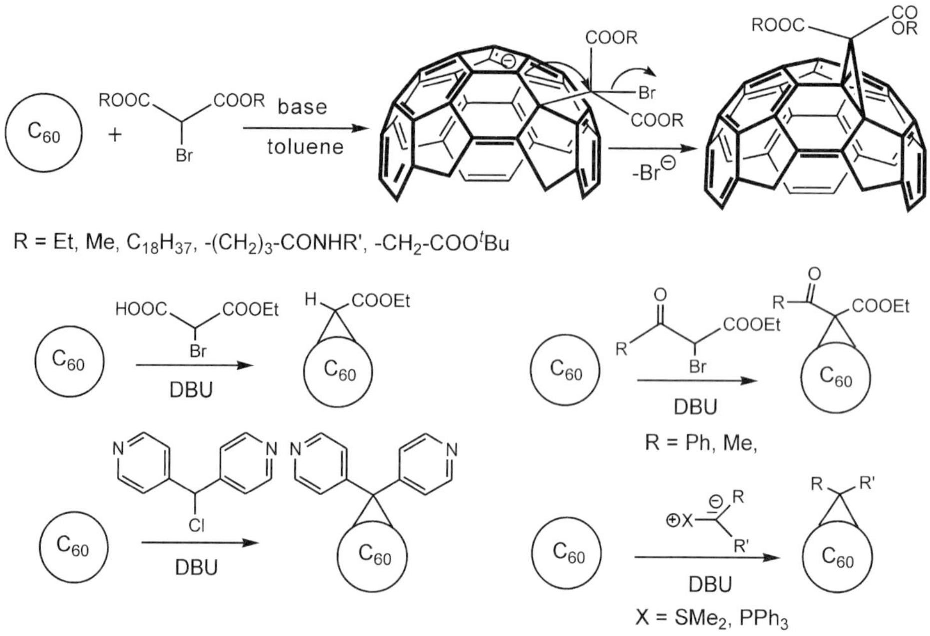

Figure 2.56 From a preparative point of view, the Bingel–Hirsch reaction counts among the most valuable transformations because it tolerates a wide variety of functional groups. Moreover, it provides access to interesting products of multiple additions taking place at defined positions.

pyridine, or triethylamine. A mechanochemical (solvent-free) variant employing a vibration mill and sodium carbonate as a base has also been reported. Saponification of the ethyl ester groups yields the methanofullerene dicarbonic acid.

Figure 2.56 collects some exemplary reactions of C_{60} with different bromomalonates and other applicable keto compounds. This collection shows that there is virtually no limit to the choice of side chains, which renders the Bingel–Hirsch reaction an attractive and flexible starting point for the synthesis of fullerene-containing materials. For example, the base-mediated conversion of a malonate and C_{60} into the respective methanofullerene can also be achieved. With the semi-ester of malonic acid instead of a malonate, the monosubstituted methanofullerene is obtained because the primary product is instantaneously decarboxylated (Figure 2.56). β-Ketoesters may also be reacted under Bingel–Hirsch conditions to give methanofullerenes.

In some cases the preparation of the bromomalonate may be difficult or even impossible. The reactive species may then be generated *in situ* by treating the malonate with CBr_4. The reaction of a malonate with iodine and DBU as a base also yields the respective methanofullerene. Today the *in situ* generation of α-haloesters and α-ketones find widespread application as it spares the efforts of a purification step before reacting them with the fullerene.

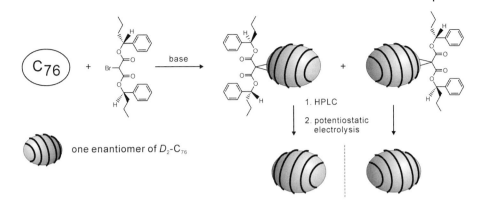

Figure 2.57 The enantiomeric resolution of D_2–C_{76} is achieved by chromatographic separation of the Bingel–Hirsch adducts and subsequent cleavage of the auxiliary group.

Due to the high reactivity of malonates toward C_{60}, an excess of the reagent will usually lead to multiply functionalized products. Now, as discussed in Section 2.5.1.3, there are several potential positions for a second or third reaction on the fullerene's surface, and so a strategy of regiochemical control is required to realize a certain addition pattern. One method uses interconnected reaction centers. The length of the respective linker then determines the possible sites for the second attack. For instance, with two malonates linked via a polyether chain, a crown-ether derivative, a substituted porphyrine or the like, the reaction with CBr_4 (or I_2) and a base will regio- and diastereoselectively yield the most favorable isomer depending on the length of the template linker.

It goes without saying that the Bingel–Hirsch reaction also takes place with larger fullerenes, but contrasting C_{60}, there are different double bonds, and so several regioisomers can arise. For the addition of diethyl malonate to C_{70} only the 1,2-adduct is observed experimentally. The reactivity of this 1,2-double bond by and large corresponds to that in C_{60} while the double bonds in the belt region of the C_{70}-molecule are less reactive, so here again the position most alike to C_{60} is attacked preferably. Similar results are found for even larger fullerenes (C_{76} and higher): a major product with the functionality attached to a strongly bent position of the carbon cage is obtained.

The inverse Bingel–Hirsch reaction is of interest, too. It serves to remove *Bingel* addends from a functionalized fullerene while re-generating the respective double bond. The reduction may be performed electrochemically as well as by reaction with magnesium amalgam or zinc/copper. Now, what is the benefit of removing a substituent once introduced with considerable effort? A first application of the retro-Bingel reaction has been reported for the separation of enantiomers of C_{76}: a pair of diastereomers is generated by functionalization with a suitable malonate. These can easily be separated (Figure 2.57), and the subsequent cleavage of the Bingel addend from both fractions yields the pure enantiomers of the smallest chiral fullerene. The Bingel addends may also act as protecting (and even directing) groups in further syntheses by occupying certain positions on the fullerene's surface and possibly by exerting directing effects on following additions. In an overall strategy employing protective groups, the Bingel addends are orthogonal to the pyrrolidines accessible by [3+2]-cycloaddition (Section 2.5.5.3). These will not be cleaved under the conditions presented above.

2.5.5.3 Cycloadditions

The cycloaddition to double bonds of the carbon cage is a major aspect in the organic chemistry of fullerenes. As mentioned before, C_{60} is an electron-deficient polyolefin with only moderate conjugation. These features clearly reflect in its behavior in cycloaddition reactions. In [4+2]-additions, for instance, it always acts as dienophile. A multifarious chemistry unfolded when exploring the whole range of possible cycloadditions. Today it allows for the attachment of virtually every functional group to the basic fullerene.

[2+1]-Cycloadditions It has been discussed in Section 2.5.5.2 already that cyclopropanation can also be achieved by adding nucleophiles. Yet the [2+1]-cycloaddition to fullerenes, and especially to C_{60}, further enables the generation of aziridines on the cage's surface.

The addition of a carbene can be achieved in the common way. The reaction will always take place on a (6,6)-bond. The respective diazirines represent a good source of carbene as they decompose to the latter by cleaving molecular nitrogen (Figure 2.58), but oxadiazoles are suitable carbene precursors as well. An important example of this type of reaction yields the dimethanomethoxyfullerene that can be further transformed into the fullerenecarbonic acid by treatment with trifluoroacetic acid (Figure 2.58). Halogen-substituted cyclopropanes are accessible, too, the cyclopropanation succeeds, for example, with trichloro acetate or with the mercury-containing carbene sources $PhHgCBr_3$ or $PhHgCCl_2Br$. The resulting dihalogenated methanofullerenes can themselves act as carbene precursor and enter into an interesting dimerization: It yields either C_{121} or C_{122}, depending on whether one methanofullerene attacks an unreacted C_{60} or two methanofullerenes react with each other via their methylene groups.

Nitrenes are the nitrogen analogs of carbenes and accordingly react with fullerene double bonds to give aziridine derivatives (Figure 2.58). Sources of nitrenes may be, for instance, amines treated with lead(IV)-acetate, various azides that are decomposed photochemically, or azidoformiates. The acylnitrenes generated by UV-irradiation of aroylazides form the expected cycloadducts at first, but these undergo a thermal rearrangement into an oxazole afterward.

In some cases though, the transformation of C_{60} into aziridine derivatives is not really a [2+1]-cycloaddition of the nitrene, but rather a [3+2]-addition of the azide (see there) with a subsequent nitrogen expulsion resulting in aziridine formation. However, this reaction also yields [5,5]- and [6,6]-opened compounds. An interesting finding is the possibility to prepare all eight of the regio-isomers (including *cis-1*) by twofold reaction with a nitrene.

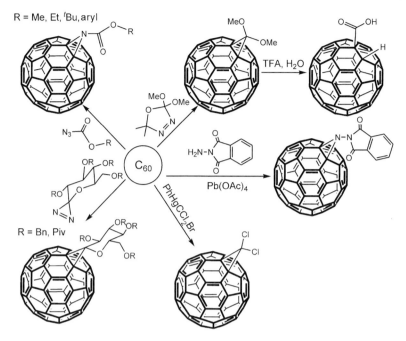

Figure 2.58 Examples of [2+1]-cycloadditions to C$_{60}$.

[2+2]-Cycloadditions As [2+2]-cycloadditions to fullerenes are photochemical reactions in the majority of cases, only a few selected examples will be presented herein. Further information can be found in Section 2.5.5.4.

Figure 2.59 collects several [2+2]-cycloadditions that can be carried out with C$_{60}$. The reaction with dehydrobenzene is instructive as it reveals electronic properties of the fullerene: The dehydrobenzene generated *in situ* from anthranilic acid reacts with C$_{60}$ exclusively in a [2+2]-cycloaddition, although in principle a [4+2]-reaction would also be possible, and dehydrobenzene usually enters into the latter when adding to electron-rich dienes. The nonoccurrence of this reaction clearly shows the electron-deficient character of C$_{60}$. For the same reason it never constitutes the diene part in a Diels–Alder reaction. Furthermore, the [2+2]-cycloaddition may be thermally effected, for example, the addition of long-chain cumulenes, allene amides or quadricyclan. The addition of ketenes as well occurs without irradiating the reaction mixture. Normally a reduced reactivity toward C$_{60}$ should be expected for the electrophilic ketenes, but in reality the products of a [2+2]-cycloaddition are even found to be the major product.

[3+2]-Cycloadditions 1,3-dipoles readily react with the (6,6)-double bond of C$_{60}$ that acts as a highly reactive dipolarophile. Five-membered rings are generated on the fullerene's surface this way, and these can bear a multitude of functional groups. Suitable 1,3-dipoles include, for instance, nitrile oxide, azomethine ylides, alkylazides, diazo compounds or trimethylenemethane (Figure 2.60).

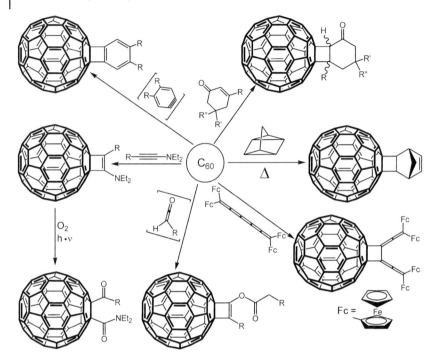

Figure 2.59 Examples of [2+2]-cycloadditions to C_{60}.

Depending on the reagent chosen, different products featuring three- or five-membered rings are obtained because the [3+2]-cycloaddition is often followed by a rearrangement. This may in parts include the opening of bonds in the fullerene cage. Consequently, the simple addition of a 1,3-dipole to C_{60} followed by formation of a three-membered ring could in principle yield four isomers (bridging of a (5,6)- or a (6,6)-bond, respectively, with both of them possibly being present as the closed or the opened variant). In fact the isomers (5,6-closed) and (6,6-opened) are not observed due to the unfavorable distribution of their double bonds. This problem has thoroughly been studied for the addition of diazomethane to C_{60} (Figure 2.60). Here the methanofullerene is formed by cleaving nitrogen from the intermediate. The choice of reaction conditions can favor the production of either the closed (6,6)- or the opened (5,6)-isomer. Cyclopropane-derivatives valuable for further syntheses may be obtained using singly or doubly substituted diazomethanes.

The [3+2]-cycloaddition of azides is also followed by the expulsion of molecular nitrogen and yields nitrogen-bridged fullerenes. The compound featuring a bridged and then opened former (5,6)-bond is the major product due to the stepwise mechanism of N_2-cleavage, while the minor product contains a bridged, but closed (6,6)-bond (Figure 2.60).

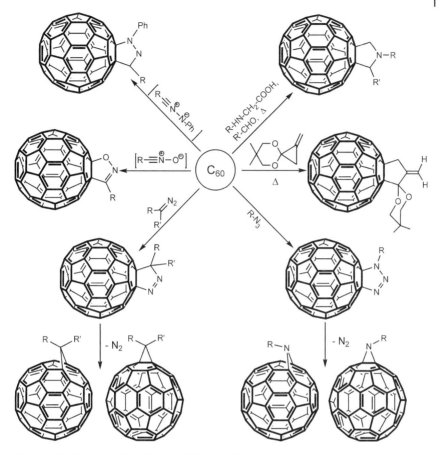

Figure 2.60 Examples of [3+2]-cycloadditions to C$_{60}$.

The reaction with azomethine ylides is an important class of [3+2]-cycloadditions to fullerenes because it leads to versatile pyrrolidine derivatives of C$_{60}$ (Figure 2.60). Immonium salts, aziridines, oxazolidines or silylated ammonium compounds are sources of the ylide structure. The sequence is often carried out according to Prato: in a first step an amino acid (like N-methylglycine) is reacted with an aldehyde or ketone. Subsequent reaction with C$_{60}$ then gives the desired pyrrolidine derivative. A multitude of these is accessible by choosing suitable functionalization of the nitrogen atom or the two carbon atoms in the ring (e.g., by choice of a functionalized aldehyde).

Reactions yielding hydrofuran derivatives have been reported as well. Formally these are [3+2]-reactions, still the mechanism of product generation has not yet been fully elucidated. Usually 1,3-diketones are attached to C$_{60}$ by oxidative addition in the presence of piperidine (Figure 2.61). Again the method is apt to supply a wide repertoire of functionalized fullerenes. A suitable choice of reagents

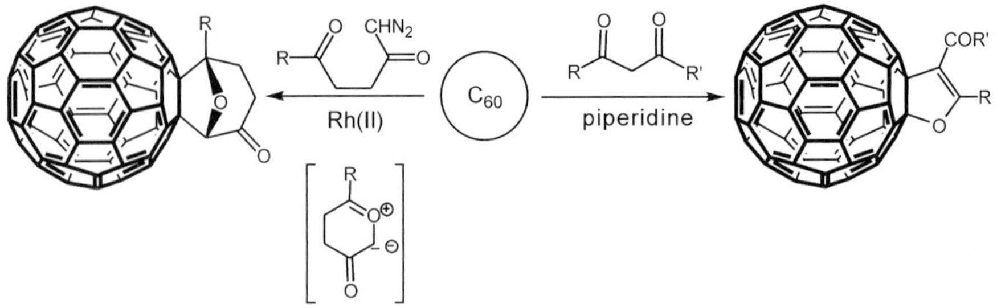

Figure 2.61 Carbonyl compounds as well may react with C_{60} in 1,3-dipolar cycloadditions and related reactions.

provided, other hetero-atoms like nitrogen or sulfur may as well be introduced into the annealed ring by oxidative addition. Even the synthesis of five-membered rings containing more than one hetero-atom succeeds this way.

Bicyclic compounds can be produced, for example, by adding to C_{60} the carbonyl-ylide generated from diazopentadione (Figure 2.61). In this case the carbon atom of the terminal carbonyl group may be functionalized in various ways, so again a large range of derivatives is accessible.

[4+2]-Cycloadditions The addition of dienes is one of the most valuable synthetic methods of fullerene functionalization as a wide variety of functional groups is tolerated. C_{60} always constitutes the dieneophile in Diels–Alder reactions, thus documenting its electron-deficient character. As a consequence [4+2]-reactions occur exclusively with dienes and always include a (6,6)-bond of the fullerene in the resulting six-membered ring. Figure 2.62 shows some examples. Typical dienes like cyclopentadiene or anthracene give the expected products with C_{60}. With the latter, however, a good part of the generated product decomposes again in a retro-*Diels-Alder* reaction due to the anthracene's limited reactivity. This observation is confirmed when treating fullerene with other poorly reactive dienes like 2,3-dimethyl-1,3-butadiene or 9-methylanthracene.

A strategy proposed by Müllen can help to overcome this problem of cycloreversion. It is based on the assumption that once generated by [4+2]-cycloaddition, aromatic rings will not be destroyed again save with considerable energetic effort. The *ortho*-quinodimethanes are a class of compounds most suitable to this reaction (Figure 2.27). They are obtained *in situ* from bis-bromomethylbenzenes. A bridged product comprising two fullerene cages linked via a benzene ring can be prepared using the respective bis-*ortho*-quinodimethane (from tetrakisbromomethylbenzene). The reaction with *ortho*-quinodimethane reagents yields a cyclohexane that is quite sturdily bound to the fullerene and further stabilized by the aromaticity of the neighboring ring, so multifarious functionalization may be performed at the distant end of the attached units without destroying the adduct.

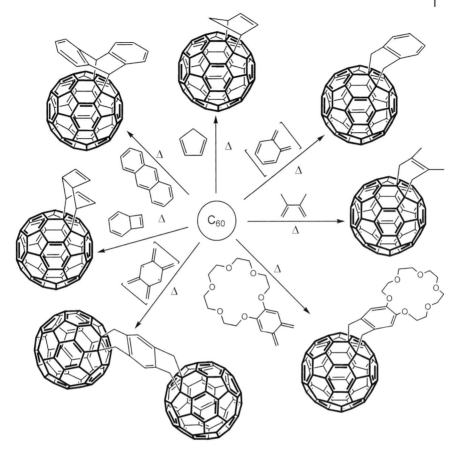

Figure 2.62 Examples of [4+2]-cycloadditions to C_{60}. The latter always reacts as dienophile.

2.5.5.4 Photochemistry

Fullerenes feature a rich photochemistry. Especially the [2+2]-cycloaddition to double bonds of the carbon skeleton has been studied down to detail. With the usual reagents, C_{60} enters into a [2+2]-cycloaddition forming the expected cyclobutane ring. The reaction most easily occurs with electron-rich alkenes or alkynes and sets out from the triplet state of C_{60}. The attachment of propenyl-anisol is an example of the cycloaddition of alkenes. The mechanism could be proven to be not a concerted, but a stepwise process including radical intermediates. The [2+2]-cycloaddition of alkyneamines is a classical reaction with electron-rich alkynes. With C_{60}, these yield an amino-functionalized cyclobutene ring that might subsequently be transformed into further interesting fullerene derivatives (Figure 2.59).

C_{60} may even react with itself in a [2+2]-cycloaddition forming a dimer as depicted in Figure 2.63. This is distinct from the product obtained in radical

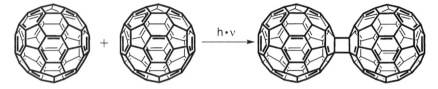

Figure 2.63 The photodimerization of C_{60} leads to adducts with a bridging cyclobutane ring.

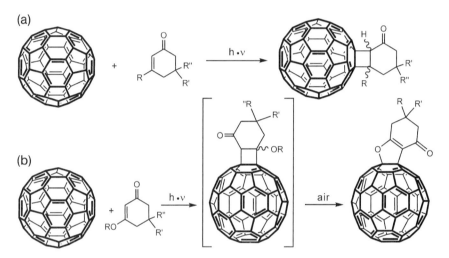

Figure 2.64 (a) Photochemical reaction of C_{60} with α,β-unsaturated carbonyl compounds; (b) in air a ring expansion occurs to yield the respective dihydrofurane.

dimerization by comprising a cyclobutane ring as connecting unit. For the time being, an oligomerization via [2+2]-cycloaddition has not been reported. However, the expected compounds would most likely be hard to dissolve, inferring problems in purification and characterization. The photodimerization may also be accomplished in the crystal state as the space between reacting bonds (3.5 Å) is markedly shorter than the maximum allowable distance of 4.2 Å for the reaction of two parallel double bonds.

Upon UV-irradiation ($\lambda \sim 300$ nm), α,β-unsaturated ketones and C_{60} form a cyclobutane ring too. The resulting product is a mixture of the *cis*- and *trans*-isomers (Figure 2.64a). Once again C_{60} reveals to be an electron-deficient olefin by entering into [2+2]-cycloadditions even with enones like 2-cycloheptenone that normally tend to be inert to this reaction. The requirement for UV-irradiation clearly indicates that an excited state of the enone reacts with C_{60}, and not vice versa like in the addition of alkynes. The latter occurs via an excited triplet state of C_{60} generated by absorption of visible light.

Besides [2+2]-cycloadditions, also the photochemical alkylation of fullerenes is of interest. It includes a photo-induced electron transfer, which due to the fullerenes' acceptor properties normally means the formation of a radical anion by reductive transfer. There are, however, examples of oxidative processes as well.

Furthermore, C_{60} facilitates the generation of singlet oxygen. The efficiency of this process is very high due to the large triplet quantum yield of the fullerene. Hence it can be used as a sensitizer for the oxidation of alkenes and dienes (Figure 2.26).

2.5.5.5 Radical Chemistry of Fullerenes

The fullerene skeleton can not only undergo nucleophilic additions, but it also enters into radical reactions very easily. The fullerene acts as a radical sponge in these reactions, and either dia- or paramagnetic derivatives are obtained depending on the number of transferred electrons.

A reaction of C_{60} with an alkyl radical yields a radical with the unpaired electron situated on the fullerene cage. ESR-measurements show the hyperfine coupling between the radical electron and the protons of the alkyl residue. Depending on the addend chosen, everything ranging from very reactive to rather stable radicals can be obtained. The latter is normally observed with sterically demanding groups like ^{t}Bu. The unpaired electron is delocalized within a certain area of the fullerene's surface. Figure 2.65 shows the resonance structures of a fullerene monoradical. The major part of spin density is centered at the C1-atom that stands in α-position to the alkyl residue. The second-highest value is found at the carbon atoms C3 and C3'. Alkylfullerenyl radicals tend to dimerize. The smaller the alkyl residue is, the more pronounced this effect becomes. The bond between the fullerene cages forms in close proximity to the site of the alkyl residues, which gives another hint on the whereabouts of the unpaired electron. The bonding strength is about 15 to 40 kJ mol^{-1}, so the dimers can easily be cleaved again, for example, by visible light.

A multiple addition of radicals is possible, too. Depending on whether an odd or an even number of radicals is attached, diamagnetic or paramagnetic compounds are obtained. Further radicals are always bound close to alkyl groups

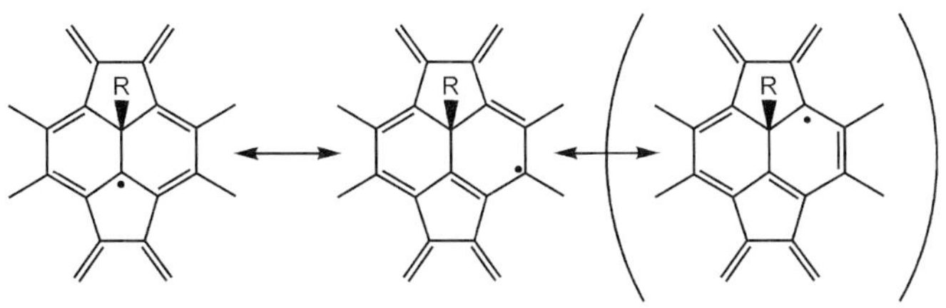

Figure 2.65 The fullerene monoradical experiences resonance stabilization with the major part of spin density being localized at the C-1-atom adjacent to the alkyl group (resonance structure to the left). For C-5, on the other hand, virtually no spin density is observed (to the right).

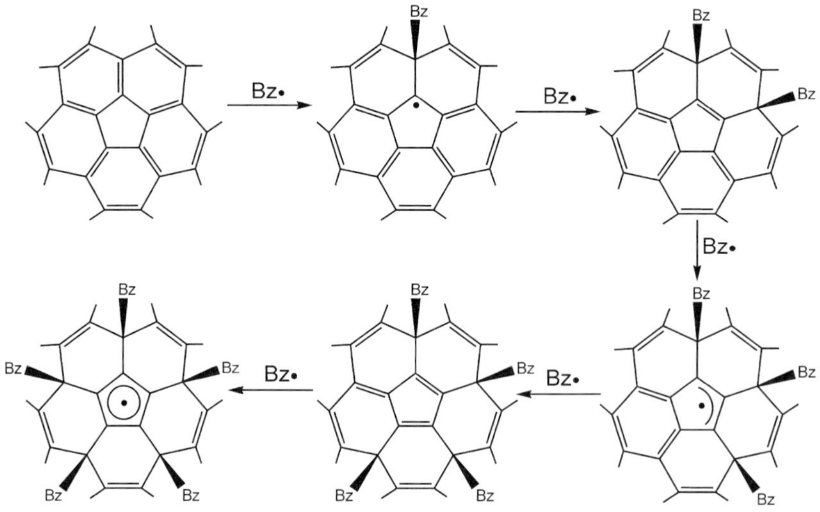

Figure 2.66 Successively reacting C_{60} with benzyl radicals leads to the radical pentaadduct.

already found on the surface. For instance, the pentaalkylfullerenyl radical depicted in Figure 2.66 is produced by the addition of five alkyl residues. The reason for this addition in close vicinity may be sought both in the electronic conditions and in the elimination of intermediate (5,6)-double bonds. These are always found in fullerene radicals with an even number of alkyl groups. Sufficiently large residues, for example, benzyl, sterically stabilize the unpaired electron in three- and fivefold alkylated radicals, while the small methyl unit does not show this effect. As a result, the formation of multiply substituted radicals is not observed for the latter.

Other radicals like trialkylsilyl or radical compounds with transition metals give the respective adducts with C_{60} as well. For instance, the attack of $(CO)_5Re\cdot$ photochemically generated from $Re_2(CO)_{10}$ yields $C_{60}[Re(CO)_5]_2$ favoring the 1,4-addition. The reaction of C_{60} with trialkylsilyl radicals very much resembles that with alkyl radicals. Compounds of the $(R_3Si)_2C_{60}$-type can be obtained by adding another radical to the initial $R_3SiC_{60}\cdot$. With bulky residues the reaction takes place at C_1 and C16. Tributyltin hydride and C_{60} form the radical $C_{60}H^{\bullet}$, and also other radicals such as $RO\cdot$, $RS\cdot$, or $(RO)_2(O)P\cdot$ react with C_{60}.

As mentioned before (Section 2.5.5.2), only primary and secondary amines are able to give a nucleophilic addition to fullerenes. Tertiary amines, on the other hand, exclusively enter into radical additions. Usually these are initiated photochemically, and they are very hard to control. However, the isolation of a triethylamine-adduct with C_{60} (see below) succeeded when oxygen was scrupulously excluded, while in its presence a completely different product featuring a pyrrolidine ring was generated. Moreover, the aerobic thermal reaction of C_{60} with NEt_3 has successfully been performed just recently. It yields the 3-N,N-diethylamino-5-

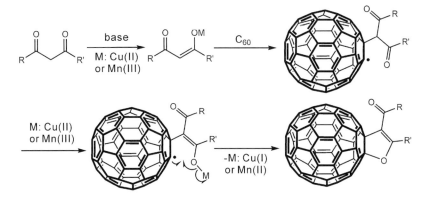

Figure 2.67 The reaction with 1,3-diketo compounds in the presence of copper or manganese ions also follows a radical mechanism.

methyl-cyclopenta-fullerene adduct that is otherwise produced at anaerobic photochemical conditions.

A fusion of C_{60} with a functionalized dihydrofuran is achieved by radical-mediated reaction of the fullerene with ketones. Again the process includes a radical intermediate with the highest spin density centered in direct vicinity to the site of initial attack. This turns ring closure into the most favorable reaction path (Figure 2.67). C_{60} may also join into radical polymerizations (refer to Section 2.5.5.6).

For radical reactions of C_{70}, the same principles hold as for those of C_{60}. Still it must be considered that due to the reduced symmetry of the fullerene cage there is a total of five distinct reaction sites. Therefore a mixture of isomers is inevitably obtained from additions of alkyl radicals to C_{70}.

2.5.5.6 Fullerenes in Polymeric Materials and on Surfaces

Fullerene molecules may be incorporated into polymers in a variety of ways. A first, basic differentiation must be made between noncovalently embedded fullerenes (either isolated or aggregated) on the one hand and molecules covalently attached to the polymer strands on the other. The interaction between dispersed molecules or particles of fullerenes and their polymer matrix is clearly electrostatic. Their production is very simple. The desired amount of fullerene is added to the polymerizable material as a solid or in solution, and the polymerization is initiated. Transparent films of C_{60}/PMMA are an example of this class of composites, they contain separated C_{60}-molecules. The characteristics of both the fullerene and the polymer are conserved.

Fullerene molecules covalently bound to the polymer have a larger impact on the properties of the resulting material. Possible modes of incorporation are summarized in Figure 2.68a. As the fullerenes might give multiple reactions, they can not only be attached end-on, but may also form further, more cross-linked compounds (Figure 2.68b).

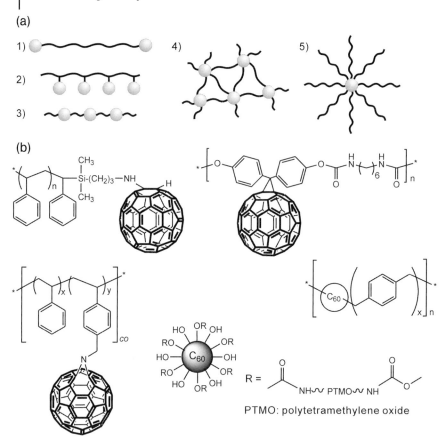

Figure 2.68 (a) Different kinds of polymer materials that may be obtained from different modes of fullerene incorporation; (b) examples of fullerene–polymer composites.

Some examples of polymer materials containing fullerenes shall be presented here. The number of coupling sites on the surface of the carbon cage is decisive for the structure of the emerging material. Types 1 and 2 may form when only one functional group is present; the fullerenes then are end-caps to the main or the side chain of the polymer. One of the few known cationic variants of binding a fullerene into a polymer, the reaction of C_{60} with poly-(9-vinyl-carbazole) in the presence of aluminum(III) chloride, results in such a type-2-composite.

Type 3, the so-called pearl necklace, corresponds to fullerenes lined up in the polymer as constituents of the main chain. It forms when each fullerene molecule is linked to the polymer at two sites. The reaction of C_{60} with p-xylylene is a classical example. In the emerging chains the fullerenes are statistically distributed in between oligomeric xylylene units. Composites with a structure slightly deviating from this "real pearl necklace" can be obtained from employing derivatives of C_{60} that carry functional groups with two reactive sites. These may be incorporated into a polymer chain, for example, by condensation, so this functional group becomes integral part of the main chain while the C_{60} itself does not.

The presence of functional groups on the fullerenes' surface is not necessarily needed for fusion with the polymer, as might be seen from the example of C_{60} reacting with *p*-xylylene radicals. Still in these cases the resulting polymer must be seen to inherently prefer a certain structure as otherwise a mixture with fullerenes in different binding modes might be generated. Especially types 1–3 are hardly ever obtained with uniform distribution and connectivity of the components. Prefunctionalized fullerenes, however, may serve to their production, and by now there has even been a series of attempts on attaching a defined number of polymer anions to nonfunctionalized C_{60}. The choice of a suitable solvent is crucial in these processes.

Cross-linked (type 4) and dendritic or starburst (type 5) polymers are usually synthesized starting from unfunctionalized C_{60}. Unlike the situation in classical cross-linked polymers, junctions are not established on single atoms, but the spherical C_{60} acts as a sort of interconnecting "super-atom". This extends the distance between the chains and concomitantly reduces the density of the resulting polymer. Besides cross-linked polymers, fullerenes with multiple reaction sites also allow for the production of composites with several chains originating from an individual C_{60}-molecule. In a way it serves as "nucleation center". Up to six strands can be linked to a C_{60} using reactive carbanions, for example, styryl or isoprenyl. Suitable carbanions generally include so-called living polymers, but species generated by later deprotonation might be considered as well. The latter approach has, for instance, been chosen to fix C_{60} on polyethylene films with terminal diphenylmethyl groups. Here deprotonation was achieved with butyl lithium. This reaction has also been realized for a series of other polymers. C_{60}-polystyrene is an example of type 5 structures that has been studied down to detail. It is synthesized by converting styrene into living polystyrene with *sec*-BuLi and subsequent reaction with C_{60}.

Apart from reactions with carbanions, C_{60} may also be included in polymers by radical polymerization in the presence of free fullerene. The method usually employs a radical starter. At first, however, this will not react with the monomer, but with the highly reactive C_{60}. Polymerization thus begins originating from the C_{60}-center and yields highly substituted derivatives. The chain length and the mass contribution of the fullerene can be controlled by tuning the concentration of the initiator and the amount of C_{60} present. For instance the copolymerization of styrene and C_{60} succeeds in the presence of an initiator.

Star-shaped structures may also be obtained using hydroxylated fullerenes, the so-called fullerenoles, and, for example, isocyanate-terminated polymers. These are linked by a condensation that yields urethane units (Figure 2.68b). The resulting material is highly viscous and soluble in common organic solvents.

The synthesis of dendrimers including fullerenes may be considered another aspect of polymer chemistry. The respective material falls under structural type 5. Yet in contrast to the star-shaped polymers mentioned above, they do not bear polymer chains originating from C_{60}, but regularly ramified structures. The fuller-

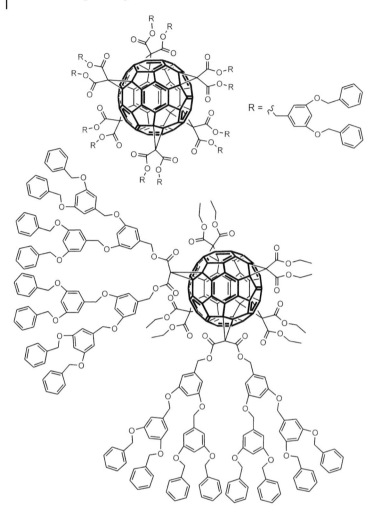

Figure 2.69 Examples of dendrimers with C_{60} constituting the core.

ene may be the core of such a dendrimer. The structure of the resulting compound then can easily be controlled by the initial functionalization of the C_{60}. As might be seen in Figure 2.69, the synthesis of unsymmetrical dendritic structures, too can be achieved, for example, by shielding with the dendrimer one hemisphere of the fullerene while the other stays available for further reactions.

Highly symmetric dendrimers with a core of C_{60} are obtained starting from six-fold cyclopropanated, T_h-symmetric derivatives of the fullerene (Figure 2.69). This central unit features twelve connective sites in shape of the acid functions, which leads to a dendrimer structured evenly in all directions. Other derivatives of C_{60} bearing four cyclopropanes are just as well suitable to dendrimer synthesis. These cores exhibit eight connective positions. Moreover six-fold cyclopropanated C_{60} enables the production of unsymmetrical dendritic derivatives like the one shown in Figure 2.69. These may subsequently be employed in making functional dendrimers that carry, for example, porphyrine or various alkyl chains attached to the remaining free bonding sites. The alkyl groups, for example, confer amphiphilic character to the resulting dendritic fullerene, rendering it suitable to the directed insertion into membranes or to the formation of monomolecular films on water surfaces. Micelles containing fullerenes can also be obtained this way.

Fullerene molecules bound to the surface of a dendrimer represent a fundamentally different mode of incorporation. Upon suitable decoration of the last dendritic generation with nucleophilic groups, well controllable amounts of the fullerene may be attached. The resulting substances are of interest, for example, for catalytic applications or to examine the fullerenes' properties as radical scavenger. The latter is largely facilitated by using these dendrimer composites for two reasons: firstly, the fullerenes do not exist in a free solution, but they are rather bound in a defined way, and secondly, the substances are soluble in common organic solvents despite their large fullerene content. Furthermore, the biggest part of irradiated light was found to be absorbed in the outer shell of the molecule due to the large number of fullerene units per dendrimer, and even what little of light reaches the core is transferred back to the fullerenes, which turns the center of such a molecule into a kind of "black box".

The covalent attachment of functionalized fullerene molecules to solid substrates is another subject of intensive study. These materials are interesting in particular for photovoltaic applications. Adducts of C_{60} and porphyrines bound, for instance, to indium-tin oxide (ITO) count into this class. Triethoxysilylisocyanides are used as an anchor group. Fullerenes may also be covalently bound to gold surfaces. The preparation of substrates entirely functionalized with C_{60} succeeds, for example, by adding to the fullerene the terminal amino groups of gold surfaces modified with 8-aminooctanethiole (Figure 2.70a). This material can also be made in a multilayer version: Fullerene molecules that have been fixed on the substrate before are modified by the addition of a bidentate primary amine in a way that the attachment of another layer can occur. Generally this covalent fusion of fullerene derivatives is an integral part in the development of fullerene-based optoelectronic devices. Selective extraction of fullerenes by means of a solid phase is feasible too, for example by silica bearing terminal amino groups. This so-called fullerene-fishing is once more based on the ease of the nucleophilic addition of amines to C_{60} (Figure 2.70b).

Apart from covalent bonding to the surface of substrates, fullerene molecules may also be arranged there employing other types of interaction. For instance, the adsorption of C_{60} on gold surfaces strongly depends on which compounds have been used to preimpregnate the substrate. Fullerene molecules adsorbed directly on a Au(III)-surface form a hexagonal-densely packed layer, while a markedly different arrangement is observed on surfaces modified with perylene or coronene. With the latter a honeycomb structure much resembling the hydrocarbon layer is found (Figure 2.70). The attraction between fullerene and aromatic compound is

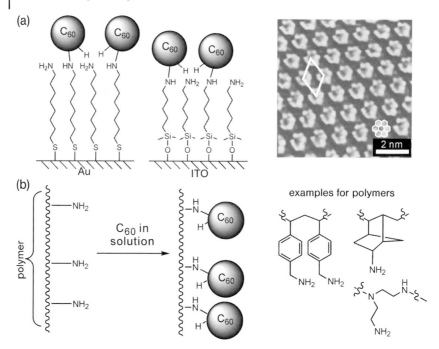

Figure 2.70 Immobilization of C_{60} on surfaces. (a) Covalent linking to gold or to indium-tin oxide and noncovalent attachment to gold modified with coronene (© RSC 2005); (b) attachment of C_{60} to polymers by nucleophilic addition of terminal amino groups.

based on π–π-donor–acceptor interactions. The adsorption of C_{70} to gold surfaces succeeds just as well, only the two possible orientations of the molecule (*tip-on*, *belt-on*) must be considered. These result in different packing. Both bonding types have been detected experimentally.

2.5.6
Supramolecular Chemistry of Fullerenes

Due to their size and shape, C_{60} and its higher homologs suit well to the formation of host–guest compounds. The fullerene molecule normally plays the role of a guest that is bound by a larger organic molecule. The inverse situation with atoms or molecules being included in the carbon cage is discussed in Section 2.5.4 about endohedral fullerenes.

The interaction between host molecule and fullerene is mediated by the π-electrons of the carbon cage. Very efficient complex formation is thus achieved by π–π-donor–acceptor interactions. The fullerene's distinct electrophilicity further causes preferred interaction with electron donors. Fullerenes participating in supramolecular structures have been reported soon after the discovery of the element's new modification, and some examples shall be presented in the following.

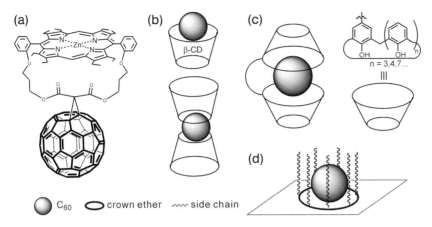

Figure 2.71 C_{60} may form supramolecular adducts with a variety of compounds, for example, with derivatives of porphyrine (a), cyclodextrines (b), calixarene (c) or crown ethers (d).

One of the most frequently reported interactions of fullerenes with organic compounds is that with porphyrines. These adducts are very interesting due to their donor–acceptor properties. Examinations focus on the compounds' application to the conversion of sunlight. The interactions between the bent π-surface of the fullerene and the planar porphyrine are mainly based on van der Waals forces. In metalloporphyrines cocrystallized with C_{60} the latter is inserted into the interstices of the chain-like or columnar arrangement of the first. The distance of a fullerene carbon from the plain of the porphyrine is 2.7–3.0 Å, which is exceptionally short for van der Waals interactions. The supramolecular exchange of these subunits is also observed with both of them being linked via a tether like in the so-called parachute adduct of a zinc porphyrine and C_{60} (Figure 2.71a). Furthermore, systems featuring a tweezer-like arrangement of two porphyrines have been prepared. These enable an even better complexation of the fullerene. In some of these compounds the C_{60}-molecule is enclosed in a kind of cage so its interaction with the solvent is rather limited and the solubility of the complex is exclusively controlled by the host.

The exchange between fullerenes and cyclodextrines has also been subject to intensive studies. The resulting adducts are soluble in water, which is of eminent importance especially for biological applications. The diameter of the void in α-, β-, and γ-cyclodextrine is 5.3, 6.5, or 8.3 Å, respectively, while the *van der Waals* diameter of a C_{60}-molecule is more than 10 Å. Hence on first sight the latter seems to be too large for the cavity available and one would not expect the interaction of fullerene with cyclodextrines to be very strong. However, stable complexes are formed from these species. The C_{60}-molecule does certainly not reach very deep into the hollow of the cyclodextrine, but obviously they enter into sufficient interaction nevertheless (Figure 2.71b). The stoichiometry may be 1:1 as well as 2:1, with two cyclodextrines embracing one fullerene in the latter case. This type is predominantly observed with very large carbon cages. As indicated before, the fullerenes are transferred into the aqueous phase by this complex formation. The varying strength of the interaction with different fullerenes may also be employed for chromatographic separations. This has, for example, been achieved for C_{60} and C_{70} using a stationary phase of silica modified with γ-cyclodextrine. C_{70} is bound more tightly there and consequently exhibits a longer retention time. Neat silica, on the other hand, does not effect any separation of the two fullerenes. The radical anion $C_{60}^{\cdot-}$ is also complexed by γ-cyclodextrine. There is only little influence of the interaction with cyclodextrines on the electronic properties of the fullerene. UV-spectra, for instance, show only minor changes in the position and intensity of absorption bands.

Complexation by plunging into the cavity of a host molecule does not only occur with cyclodextrines. Different calix[n]arenes (n = 4, 5, 8, etc.) may coordinate to C_{60} or higher fullerenes in a similar manner. Structures with one or two calixarene ligands are known here as well (Figure 2.71c). In complexes with two calixarenes these can be either independent or interconnected via a spacer. This must not necessarily be very long – the connection may be affected, for example, by a biphenyl unit that is constituent to both calixarenes. Again a tweezer-like coordination of different fullerenes is observed. Complexation constants of more than 76 000 M^{-1} have been found for calixarene complexes with C_{60}, which means a very strong interaction between host and guest molecules. The exact values depend on the size of the calixarene: the larger its void, the better it may enclose the fullerene and the higher a bonding constant is found.

Further complexes with C_{60} and C_{70} have been reported for crown-ethers. These plus possible side chains envelop the fullerene like a basket (Figure 2.71d). It has been possible to prepare films of aza-crown-ethers with their cavities filled with fullerenes; in addition, these were stabilized at their position by lipophilic side chains.

Chlathrates may, in a way, be regarded as supramolecular adducts in the crystalline state. C_{60} forms a series of these compounds which in most cases are obtained by simple cocrystallization of their constituents. Known examples are the inclusion compounds with hydrocarbons like pentane or nonane, etc., but chlathrates of C_{60} or C_{70} have also been found with hydroquinone in the presence of benzene. In these cases the fullerene molecules occupy interstitial spaces within the crystal structure of hydroquinone and form donor–acceptor complexes. Further inclusion compounds with fullerenes are known for ferrocene and other inorganic substances such as sulfur (S_8), white phosphorus (P_4) or complexes like $(PhCN)_2PdCl_2$ and Ph_3PAuCl (also refer to Section 2.5.3). For the complex of C_{60} with ferrocene the crystal structure of the first is found with the latter inserted into the given gaps. The whole structure can only exist because ferrocene is too weak a reducing agent to transform C_{60} into C_{60}^-.

A special host–guest geometry is realized in the so-called *peapods*. These are single-walled carbon nanotubes with fullerenes enclosed in their inner void so they are arranged like peas in their pod (also refer to Section 3.5.6). The embedded fullerenes can be C_{60} as well as higher homologs. These may themselves contain endohedral guests, which turns peapod-formation really into a super-supramo-

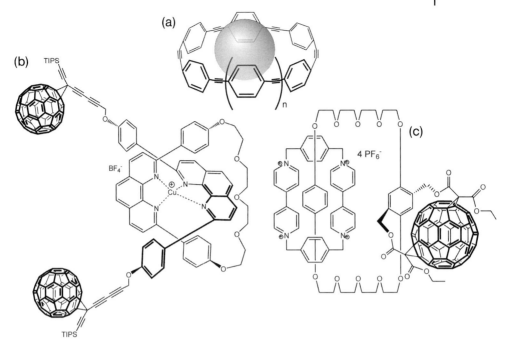

Figure 2.72 Examples of supramolecular derivatives of fullerene. (a) Complex with belt-shaped aromatic compounds, (b) rotaxanes, and (c) catenanes.

lecular process. C_{60}-peapods, however, consist of carbon atoms alone, so they are a supramolecular species made from a single element – a structural feature otherwise absent in supramolecular chemistry. Still they should not be considered an elemental modification of their own as simply two structural species distinct and demarcated from each other are combined within one adduct.

Similar arguments like for the peapods hold for C_{60} surrounded by a belt-like aromatic compounds (Figure 2.72a). Depending on their side chains the latter may confer some solubility in polar solvents. The interactions between belt and central fullerene are mainly of the π–π-type. They are exceptionally favorable due to the bent surface of both bonding partners. Inherently curved compounds like perchlorotriquinacene can as well coordinate to C_{60} by π–π interactions.

Besides supramolecular systems with the fullerene embedded as guest in some cavity, there is another class that features the carbon cage bound as constituent into larger species by functionalization, and these will then enter into supramolecular interactions. For example, for certain biological applications, it is of interest to find derivatives of fullerenes suitable to an exchange with DNA. The *N,N*-dimethylpyrrolidinium iodide of C_{60} may serve as an example here. It interacts with the phosphate groups of the DNA backbone. This causes remarkable changes of the tertiary structure of the double-stranded DNA that result in the latter wrapping around small aggregates of the fullerene compound. It is also possible of course to treat single-stranded DNA with fullerenes carrying oligonucleotides. This way, the fullerenes can be bound directly to the DNA system. Site-specific interactions of fullerenyl-oligonucleotides can even splice double-stranded DNA.

Metal-ligand exchanges may as well be employed to build up supramolecular systems with fullerenes. For instance, pyridines linked to a C_{60} via a cyclopropane can coordinate to platinum atoms generating a supramolecular cyclophane structure. The synthesis of rotaxanes containing fullerenes succeeds in a similar manner; here the C_{60}-molecules are positioned at the ends of the rod-shaped component to act as a stopper (Figure 2.72b). Rotaxanes and catenanes including fullerenes are attractive target molecules not only for aesthetic reasons – they also feature interesting electronic and photophysical properties predestining them to the development of molecular machines. The compound shown in Figure 2.72c, for instance, has distinctive redox characteristics. The copper coordinated by two phenanthroline moieties hardly influences the redox potential of the two terminal fullerenes, while the potential for the Cu^I/Cu^{II}-pair experiences an anodic shift of 0.3 V to be +0.865 V.

Finally, certain long-chain molecules are able to bind fullerenes by micelle formation and to finely disperse them in a medium. The examples reported for this kind of interaction mostly use surfactants like Triton X-100 or lecithin. The hydrophobic fullerene is incorporated into the micelle, partly in its center, but partly also in its periphery. Among other effects, this interaction leads to the solubility of such adducts in water. The embedding in long-chain, amphiphilic molecular units is also used to prepare artificial membranes holding fullerenes.

2.5.7
Polymeric Fullerenes and Behavior under High Pressure

The phase diagram of fullerenes is complex, and for the time being little has been published on the subject. Due to its availability and with its high symmetry simplifying the phase diagram, most studies focus on C_{60} so this has been examined in greater detail.

Under high pressures (6–18 GPa) the fullerene crystal undergoes an irreversible phase transition. The fullerene molecules are forced into tight proximity which leads to the formation of covalent bonds. These are 1.4–1.5 Å long, so the centers of two C_{60}-molecules cannot approach each other nearer than about 8.5 Å. The polymerization propagates throughout the crystal and yields an extremely hard, three-dimensionally interconnected material. This is able to scratch even diamond (Figure 2.73). At first, below 8 GPa, a two-dimensional polymer is generated, and it is only under higher pressures of up to 13.5 GPa and at sufficient temperatures (~800 K) that the three-dimensional polymer is obtained. The latter already bears a considerable fraction of sp^3-bonding. On even higher pressures the fullerene is transformed into diamond. Most fullerene polymers exhibit only a narrow band gap, and so electric conductivity is observed at slightly elevated temperatures.

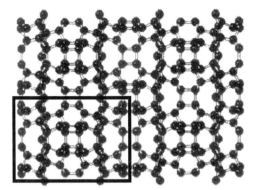

Figure 2.73 The structure of a high-pressure polymer of C_{60}. The unit cell is indicated by a rectangle (© Springer 2002).

2.5.8 Reactivity of Further Fullerenes

Fullerenes smaller than C_{60} inevitably contain neighboring five-membered rings. Therefore, they are particularly strained and highly energetic, which should reflect in an increased reactivity. Problem is, however, that these species cannot be isolated at all or in infinitely small quantities only, and any material actually obtained decomposes rapidly, especially in air.

Nevertheless, a derivative of the fullerene C_{50} has been isolated just recently. C_{50} is the first fullerene with a maximum of two five-membered rings each in adjacent position. In even smaller species groups of three five-membered rings must at least partly be immediate neighbors, which means an additional destabilization. C_{50} possesses a completely closed electron shell. This results in a more distinct aromatic character than in open-shell fullerenes.

The derivative of C_{50} is prepared directly from graphite and not via the unsubstituted parent fullerene. Mixing small amounts of carbontetrachloride (CCl_4) to the atmosphere in an arc discharge apparatus results, among others, in the generation of decachlorofullerene-[50] (Figure 2.74). The chlorine atoms are arranged around the equator of the cage-like structure (D_{5h}), thus stabilizing the molecule in comparison to the parent C_{50}, and so the chloro-compound could be isolated. Methoxy groups may be substituted for the chlorine atoms, yielding structures with the formula $C_{50}Cl_{(10-n)}(OMe)_n$. Moreover the compounds $C_{56}Cl_{10}$ and $C_{54}Cl_8$ have been detected as byproducts in the synthesis of $C_{50}Cl_{10}$, but as of now it is still unclear whether these really exhibit cage-like structures or a mere bowl-shaped constitution.

The preparation and reactivity of C_{36} have been studied, too. Regarding stability it is situated between fullerenes and rings and accordingly should show an interesting chemical behavior. Its most striking feature is the tendency to form intermolecular bonds, thus generating di- and trimers. According to different

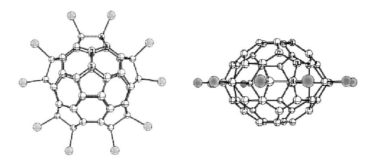

Figure 2.74 The calculated structure of decachlorofullerene[50] (© ACS 2004).

calculations, these may be connected by single C–C-bonds as well as by cyclic structures. Furthermore, the preparation of minor amounts of $C_{36}H_6$ and $C_{36}H_6O$ was reported, but it could not be clarified beyond all doubt whether these substances really are derivatives of fullerenes. The instability of C_{36} complicates more extensive studies on its reactivity, and only few results, obtained on samples of uncertain structure, have been published accordingly. Altogether, the successful synthesis of derivatives is more likely than that of the small parent fullerenes because the addition to bonds between neighboring five-membered rings markedly increases the stability of these structures.

Toward the other extreme, the reactivity of large fullerenes decreases with raising numbers of atoms. This effect is explicable by the lower degree of prepyramidalization and the reduced gain of energy from strain relief. The reactivity of very large fullerenes finally approximates the rather inert nature of graphene layers.

2.6
Applications and Perspectives

When fullerenes were first discovered, nobody thought of real commercial applications of the new material – the structure simply seemed too exotic to become available in macroscopic amounts. However, as the methods of producing C_{60} have continually been improved to the present day, the price per gram has been reduced to about €100 (highly purified lab quality), so in fact it is obtainable for various applications. Some of these shall be explained in short.

On one hand, C_{60} and its derivatives are considered promising materials for biological uses. The farthest progress has been made in experiments on its employment as sensitizer for the photochemical generation of singlet oxygen. The latter shall serve to the directed degradation of DNA or to the destruction of tumor tissue (photodynamic therapy). On the other hand, it is the electronic properties of the

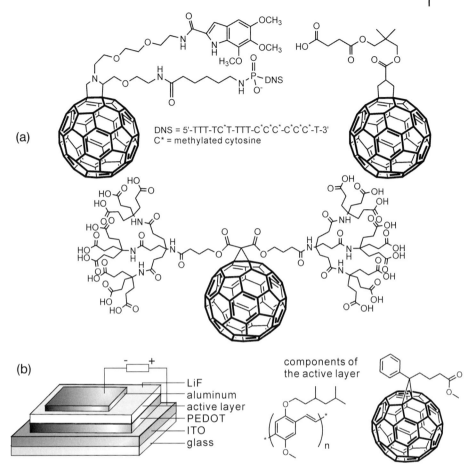

Figure 2.75 Functionalized fullerene materials may be employed in biological or electronic applications. (a) Examples of water-soluble derivatives of fullerene that might be suitable for medical uses; (b) solar cells with their active layer consisting of a C_{60}-composite.

cage-like molecule that are of interest for applications as redox-active substance or as surface film.

Especially the functionalized C_{60}-materials feature characteristics that turn them attractive for a number of purposes. In a first step the solubility of the fullerene in water or a physiological medium is achieved by functionalization with hydrophilic groups. This is crucial for exploring biological or medical applications. Selected examples are shown in Figure 2.75a. Some of these are highly efficient in the *in vitro* inhibition of HIV-protease or in the aforementioned photodynamic therapy. They are also very efficient radical scavengers, for example they easily react with hydroxyl or peroxide radicals that cause premature cell death. Hence fullerenes might serve as antioxidants to prevent apoptosis. The fullerenols

$C_{60}(OH)_x$ proved most efficient in this regard, yet the polymer composites discussed in Section 2.5.5.6 are very promising as well. They exhibit the same properties as derivatives not bound in a polymer, but might be employed in the production of practicable formulations, most of all when the fullerenes are linked to a biopolymer.

Its abundant redox and photochemistry, combined with low reorganization energy, render C_{60} a promising candidate for electro-optical and light harvesting applications. To such purposes the carbon cage will be functionalized, for example, with organic donor molecules. Furthermore, derivatized fullerenes are of particular interest for the preparation of photoconductive polymers, electroluminescent substances or materials that enable photoinduced electron transfer. Figure 2.75b depicts the model of a solar cell using fullerenes. For the time being most of the materials under examination for this application contain fullerenes noncovalently embedded in the polymer matrix. It turned out, however, that solar cells made from such composites suffer phase separation and aggregation of fullerenes. The material properties, especially the phase stability, are significantly enhanced by covalent attachment of the fullerene molecules to the polymer.

Chemical sensors with good application profiles can be obtained from linking C_{60} to a solid phase. Surface films including fullerenes are of special value here, with an example being the trace detection of oxygen in solutions using a suitably modified film of polypyrrole. Fullerenes decorated with silicon alkoxides may be bound to silica phases to supply new materials for high performance liquid chromatography (HPLC). These exhibit high selectivity for aromatic analytes in organic as well as in aqueous media.

Hydrofullerenes with various degrees of hydrogenation are candidate materials for hydrogen storage. $C_{60}H_{36}$, for instance, has a hydrogen storing capacity of 4.8% of its own mass. This value may be below the 6.5% required for an economical application, but still it is a considerable improvement compared to other materials. Hydrogenated fullerenes may also be useful in lithium ion accumulators because they significantly prolong the batteries' lifetime.

Fullerenes are also a valuable material for imaging methods in medicine. Heavily iodinated derivatives may, for instance, be used as X-ray contrasting agent with the advantage that their toxicity is markedly lower than that of substances commercially employed. Additionally, the so-called endofullerenes are attractive for applications as radio tracer or contrasting agent in magnetic resonance imaging (MRI). An outstanding advantage of such endohedral derivatives like Gd@$C_{82}(OH)_x$ and ^{166}Ho@$C_{82}(OH)_x$ is the encapsulation of the respective metal. The toxicity of the latter is markedly reduced this way as it is no longer in contact with the surrounding tissue.

Summing up, a wide repertoire of possible applications has been developed for the fullerenes, and most of all for C_{60}. But despite a massive decrease of prices for fullerene soot, the cost for the starting material still is the biggest impediment to a profitable use of the outstanding properties. Real commercial applications will only come into reach when C_{60} can be produced at prices significantly below the current level.

2.7 Summary

Altogether the fullerenes represent an entirely new modification of carbon that stands out for a unique structure and remarkable physical and chemical properties.

Box 2.1 Structure of fullerenes.

- Fullerenes are a modification of carbon with a cage-like structure.
- The surface curvature arises from incorporating five-membered rings into the hexagonal network of graphene layers.
- In the most stable fullerenes, the five-membered rings are evenly distributed across the surface and isolated from each other (isolated pentagon rule – IPR).
- The double bonds are preferably located in the six-membered rings; double bonds in five-membered rings are energetically unfavorable.
- C_{60} and C_{70} are the most important species among the fullerenes.

Fullerenes are produced, among others, from graphite reacted in an arc discharge. The resulting fullerene soot is extracted and purified by chromatography. The physical properties of fullerenes differ significantly from those of graphite and diamond, the classical modifications of carbon.

Box 2.2 Physical properties of fullerenes.

- Fullerenes exhibit strong electron affinity and act as "radical sponges."
- The solubility in organic media depends on the type of solvent; aromatic and halogenated aromatic substances are the most suitable.
- The spectroscopic properties of fullerenes are closely related to their respective symmetry. Structural information can be obtained from the number of bands, for example, in the IR-spectrum.
- In comparison to graphite, all fullerenes are more energetic, with increasing size, however, their standard enthalpy of formation approximates the value of graphite.

The reactivity of fullerenes reveals significant differences from the chemical behavior of graphite and diamond.

> **Box 2.3 Chemistry of fullerenes.**
>
> - Fullerenes behave like electron-deficient polyolefins and not like aromatic compounds.
> - They easily enter into addition reactions with nucleophiles.
> - Functionalization can easily be achieved by cycloaddition reactions.
> - Fullerenes add hydrogen or halogens and may be transformed into highly functionalized compounds.
> - 1,2- or 1,4-addition occurs depending on the size of the addends.
> - The regiochemistry of multiple additions is governed not only by steric factors but also by a tendency to avoid double bonds in five-membered rings. Thus a *cis*-1-arrangement is observed with small addends, while larger addends prefer *e*- or *trans*-3-positioning.
> - The Bingel–Hirsch reaction is an important tool in fullerene functionalization. It yields cyclopropanated fullerenes by means of deprotonated bromomalonates.
> - Fullerenes may incorporate guests in their inner cavity, these compounds are called endofullerenes.
> - Heterofullerenes are compounds with one or more heteroatom substituting for carbon atoms of the cage.

C_{60} and its homologs are promising candidates for applications in different fields.

> **Box 2.4 Possible applications of fullerenes and their derivatives.**
>
> - Solar cells
> - Composite materials with interesting electronic properties
> - Derivatives of fullerenes for photodynamic tumor therapy
> - Chemical sensors
> - Endofullerenes as contrasting agent in MRI with reduced side effects

3
Carbon Nanotubes

A sheet of paper may be rolled up with its edges connected in a butt joint to generate a tube. Performing this experiment hypothetically with a graphene layer results in a carbon tube. Such structures actually exist; they are entirely made up from carbon atoms and accommodate a cylindrical cavity. Different diameters provided, several of these tubes may fit one into another to make a multiwalled carbon tube. The diameter of both single- and multiwalled species measures on the nanometer scale, so they have been named *carbon nanotubes* (CNT). These fascinating objects represent another modification of the element carbon. They will be the subject of the ensuing chapter.

3.1
Introduction

Carbon fibers have been known for long. They are used in a variety of materials for mechanical reinforcement and to make the respective composites more resistant to different external influences. Carbon-reinforced materials for sports gear, like the frames of a tennis racket or a mountain bike, are but a few examples. Having a closer look on the employed carbon fibers in the electron microscope reveals that at least some of these long, thin objects possess a tube-shaped core measuring several nanometers across (Figure 1.12a). Then the outer layers of the fiber are arranged around this central cavity. As early as 1976, M. Endo and co-workers reported on concentric, tubular structures in the core of carbon fibers and postulated a catalytic growth mechanism.

These publications, however, were paid attention again only after the group of S. Iijima had found carbon nanotubes when studying different soot types from arc-discharge experiments for fullerene production. Upon examination in the transmission electron microscope, areas of equidistant lines symmetrically arranged around a central void were observed. The explorers soon realized that these stripes were the projection of tubular objects. Consequently the observed structures had to be tubes fitted one into another (Figure 3.1a). Just a little later the same group described single-walled carbon nanotubes (SWCNs) that had been obtained after modifying the preparative conditions (Figure 3.1b). It also became

Carbon Materials and Nanotechnology. Anke Krueger
Copyright © 2010 WILEY-VCH Verlag GmbH & Co. KGaA, Weinheim
ISBN: 978-3-527-31803-2

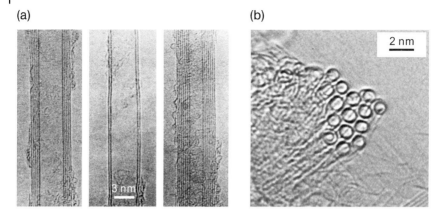

Figure 3.1 HRTEM-image of (a) multiwalled (© Nature Publishing Group 1991) and (b) single-walled carbon nanotubes (© F. Banhart).

evident soon that these nanotubes were closely related to the fullerenes that had been known since 1985.

Theoretical predictions of carbon nanotubes are largely unknown in the literature. Still there is a rich collection of publications on the structure of carbon fibers that include in their discussion the tubular core of the fibers. However, these facts were largely ignored until after the discovery of carbon nanotubes because it seemed unlikely ever to be able to obtain the isolated core of such a fiber.

Carbon fibers are close relatives to the nanotubes discussed in this chapter as they exhibit several similarities regarding structure and characteristics. For example, the graphene layers are oriented in parallel to the fiber axis. Still this does not mean that cross-sections with a concentric structure are found in all cases (Figure 1.12b). Nevertheless, definitely there are carbon fibers much resembling nanotubes after thermal treatment above 2000 °C, and it is only upon further heating to more than 3000 °C that graphitized, facetted fibers are obtained. The center of such carbon fibers often contains a multiwalled nanotube. Due to its higher mechanical strength, the latter is not destroyed when the fiber breaks, it may then be observed above the plane of fracture.

The carbon nanotubes also have a lot of structural features in common with the carbon cages described in Chapter 2. Contrasting some fullerene species, however, they do not occur naturally in any form, neither on earth nor in space, so they are a completely artificial form of carbon indeed. The question of whether nanotubes are a real modification will be discussed further below. In any case they exhibit bent graphene layers, but while the whole three spatial directions are affected in the fullerenes, the curvature is limited to two dimensions here. Hence, the incorporation of five-membered rings – indispensable to achieve closure of the bowl in fullerenes – is unnecessary for the construction of nanotubes. It suffices to bend the graphene sheet from its plane and make it a cylinder. Compared to fullerenes of equal diameter, nanotubes consequently bear less strain and their carbon atoms exhibit a smaller degree of sp^3-hybridization. However, considering the caps of closed carbon nanotubes reveals that these are partially made from fullerene fragments. Examples may be found in Section 3.2.2, the (5,5)- and (9,0)-nanotubes presented there are terminated by suitably oriented fullerene hemispheres. The cylindrical tube is situated between them. Hypothetically reducing its length to 0 leads to contact of the caps and establishes an intact molecule of C_{60}. Fullerenes may thus be considered as an extreme of closed nanotubes without cylindrical centerpiece.

The question arose before whether carbon nanotubes are an elemental modification in their own right. Modifications of a given material or element are, by definition, substances identical in chemical composition (pure carbon in this case), but different in crystal structure. This phenomenon is also called allotropy. The different lattices result in distinct properties and varied stabilities regarding temperature and pressure. At given values of these parameters there is usually but one modification that is thermodynamically stable. Suitable treatment may convert it into other allotropes. Graphite is the most stable form of carbon at standard conditions. However, this does not mean that any other modification of carbon spontaneously transforms into graphite at room temperature and normal pressure. The activation barrier for these conversions is high enough to let metastable materials exist under normal conditions. Diamond is the most prominent example of this effect.

Now, do nanotubes represent a modification of their own or not? There are two fundamentally opposite opinions on the matter, both of which shall be considered here. Nanotubes exhibit the same elemental composition as any other material from pure carbon and, at the same time, feature distinct electronic and mechanical properties. In this respect they might surely be considered as an autonomous modification of carbon. The question of different crystal structures, however, cannot unambiguously be answered. The bonding in a carbon nanotube is based on the same pattern as in graphite; in particular the structure of a graphene layer is obviously related to that of the carbon framework in a nanotube. Altogether they might simply be regarded as bent graphene sheets, and in multiwalled nanotubes the neighboring sheets are even connected via π–π-interactions, so basic structural features of graphite are present in carbon nanotubes as well. From this point of view, one might argue that nanotubes are a heavily distorted variant of graphite (or that graphite represents an extreme of nanotubes with infinitesimally low curvature). Each nanotube may further be described as a result of rolling up a graphene layer under a certain angle and with a certain diameter. As every one of these would crystallize in an individual structure, one might just as well argue that there is an infinite number of nanotube modifications (provided that their length and diameter could also be infinite). The same ambiguity must be stated for the fullerenes that exist in different sizes, too. Using a strict interpretation of the definition regarding distinct crystal structures, solids of C_{60}, C_{70}, and any other fullerene would each be a modification of their own.

3.2
The Structure of Carbon Nanotubes

A principal distinction can be made between single-walled nanotubes (SWNT) and multiwalled nanotubes (MWNT). Both classes comprise species of most different diameters and lengths. Besides dimensions, it is also the way the graphene layer is rolled up to be a tube that dominantly influences the properties of the resulting materials. Furthermore, there may or may not be caps at the tubes' ends, the respective structures then are called closed or open carbon nanotubes. The structural features of single-walled nanotubes will be discussed first in the following before the concept shall be extended to the multiwalled variants then.

3.2.1
Nomenclature

It is virtually impossible to name carbon nanotubes correctly according to IUPAC nomenclature for two reasons. Firstly, the number of carbon atoms constituting the nanotube is enormous – it might be tens or hundreds of thousands. Secondly, it would be extremely difficult to apply the classical nomenclature of organic compounds. A whole new system has consequently been established to classify carbon nanotubes. It is perfectly apt to distinguish different structures, and it even permits to infer certain properties of the species at hand from the components of its name. There are three classes of carbon nanotubes differing in the way the basic graphene sheet is rolled up:

- **Zig-zag carbon nanotubes**
  The graphene layer is rolled up in a way to make the ideal ends of an open tube be a zig-zagged edge (Figure 3.2a). It means that the rolling up is done in parallel to the unit vector $\vec{a}_1$ of the graphene lattice.

- **Armchair carbon nanotubes**
  In comparison to the zig-zag tubes, the graphene sheet is turned by 30° before rolling up. The perfect terminus is an edge consisting of the sides of the last row of six-membered rings.

- **Chiral carbon nanotubes**
  If the angle of turning the graphene layer before rolling up is between 0° and 30°, chiral nanotubes are obtained. They are characterized by a line in parallel with the unity vector $\vec{a}_1$ that spirals up around the tube. Consequently two enantiomeric forms exist for these species.

The naming of specific carbon nanotubes is based on using a pair of numbers that indicate the coiling direction and the perimeter of the tube. This pair of descriptors (n,m) results from geometrical considerations that are presented in the following section.

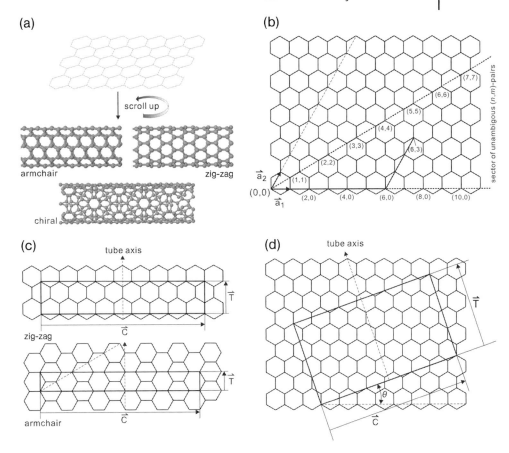

Figure 3.2 (a) Different types of carbon nanotubes result from formal rolling up of a graphene layer, (b) definition of the descriptors n and m and the parameter space, (c), (d) translational unit cells of a zig-zag, an armchair and of a chiral nanotube. Here, as well as in most figures to come, the double bonds are not shown for clarity reasons.

3.2.2
Structure of Single-Walled Carbon Nanotubes

Single-walled carbon nanotubes consist of a solitary graphene sheet which is rolled up to become a hollow cylinder that envelops the longitudinal axis of the tube (Figure 3.2). The size and orientation of the basic graphene layer are essential thereby for the structure of the nanotube under consideration. Figure 3.2 shows an uncoiled fragment of a carbon nanotube. It is evident that certain structural elements are lined up in a strictly periodic order. These ever recurring elements are called *translational unit cells*. They comprise the smallest repeating unit of a CNT.

There are two preferential orientations on a graphene sheet running in parallel with the unit vectors of the two-dimensional elemental cell of the hexagonal lattice (Figure 3.2). When the graphene layer is rolled up along $\vec{a}_1$ or along the bisector of the angle between $\vec{a}_1$ and $\vec{a}_2$, the resulting nanotubes are either of the zig-zag or the armchair type. The unit cells of these special CNT are exceptionally short (Figure 3.2c). Complexity increases quickly, however, considering CNT with larger unit cells that result from rolling up the graphene under a certain angle θ. Obviously there is need for a universally applicable procedure to describe the structure of any conceivable carbon nanotube in an unambiguous way.

The vector model has proven versatile to this purpose. It employs the unit vectors of the graphene's two-dimensional unit cell as a reference dimension. The vector $\vec{C}$ running in parallel to the coiling direction is a linear combination of integer multiples of the units vectors. It is an interconnection of two identical points on the graphene lattice (Figure 3.2). $\vec{C}$ describes a straight line that represents the uncoiled perimeter of the respective nanotube. It also defines the orientation of the nanotube as the tubular axis $\vec{T}$ is perpendicular to the tube's cross-section, which on its own part lies in a plane defined by the perimeter (or the coiled $\vec{C}$).

It becomes evident from close examination of the graphene sheet in Figure 3.2 that a lot of conceivable nanotubes are structural equivalents due to the high symmetry of the two-dimensional lattice. Consequently only $1/12$ of the potential carbon nanotubes actually possess distinct structures. These still mean an infinite number of tubes though (provided length and diameter may be infinite too), and thus an efficient nomenclature had to be established to describe the structural essentials. The characterization focuses on perimeter and orientation of the graphene as the length of individual tubes is of minor importance. The notation according to Dresselhaus was found very suitable here. It is based on the two standard vectors of the graphene lattice. A linear combination of them can describe every point of the graphene layer that bears a carbon atom. This means that the circumference of any chosen carbon nanotube is unequivocally defined by its perimetral vector $\vec{C}$. It connects its origin to the point that superimposes on it when rolling up the graphene sheet. $\vec{C}$ is defined by Eq. (3.1)

$$\vec{C} = n \cdot \vec{a}_1 + m \cdot \vec{a}_2 \, (n \geq m) \tag{3.1}$$

where $\vec{a}_1$ and $\vec{a}_2$ represent the unit vectors shown in Figure 3.2b. Each and every nanotube structure can thus be described unambiguously by the pair of descriptors (n,m). n and m indicate how many steps one has to progress on the hexagonal pattern in the directions of $\vec{a}_1$ and $\vec{a}_2$, respectively, to obtain the circumferential vector of the considered nanotube. The condition $n \geq m$ eliminates redundancies and ensures that every structure is represented exactly once only. Each pair (n,m) defines the diameter and orientation and, consequently, the chirality of the respective tube, which is sufficient for an unequivocal description. The appendix contains templates of graphene sheets with vectors $\vec{C}$ drawn in for the different classes of carbon nanotubes, these examples may serve to illustrate this descriptive system.

⇨ Construction of a (5,5)-, (9,0)- and (6,3)-Nanotube (refer to last page of the book)

The diameter of the carbon nanotube can easily be calculated from the absolute length of vector $\bar{C}$ that represents the circumference (Eqs. 3.2 and 3.3). The two unit vectors $\bar{a}_1$ and $\bar{a}_2$ measure 0.246 nm each ($\bar{a}$). Circumference and diameter thus are calculated to be

$$\text{Circumference of the nanotube (nm)} \quad |\bar{C}| = \bar{a} \cdot \sqrt{n^2 + nm + m^2} \quad (3.2)$$

$$\text{Diameter of the nanotube (nm)} \quad d = \frac{1}{\pi} \cdot |\bar{C}| = \frac{\bar{a}}{\pi} \cdot \sqrt{n^2 + nm + m^2} \quad (3.3)$$

For zig-zag nanotubes $m = 0$, while armchair tubes are defined by $n = m$. These two extremes are the borderline cases of the descriptor zone, their vectors stand at an angle of 30° to each other. Any other nanotube structure lies within these limits. n and m vary depending on the angle θ. The latter also describes the helical chirality found for each nanotube with different values of n and m (except for $m = 0$). θ indicates the degree of turning the graphene sheet from the zig-zag orientation before rolling it up. This angle can be calculated from the descriptors as well (3.4):

$$\text{Chirality angle } \theta(°) \quad \theta = \sin^{-1}\left(\frac{\sqrt{3}m}{2\sqrt{n^2 + nm + m^2}}\right) \quad (3.4)$$

However, θ does not state which enantiomer is at hand as the rotation of the graphene has the same absolute value for both of them, only the directions differ. Therefore some authors add a sign to the chirality angle to distinguish between enantiomers. The chirality of nanotubes may be illustrated by using the (6,3)-nanotube paper cutout DIY Kit found at the end of the book. A line along the direction of vector $\bar{a}_1$ is drawn in for better visibility. It winds around the resulting carbon tube in a left-handed helix when the model is rolled up with the imprinted graphene facing the outside. The other enantiomer is obtained by rolling up the tube with the imprint to the inside.

The length of carbon nanotubes is usually many times larger than their diameter. Hence the arrangement of six-membered rings recurs with a certain periodical frequency along the tubular axis. This circumstance is termed one-dimensional or translational unit cell. Unlike the unit cell in a three-dimensional crystal lattice, it corresponds to a cylindrical fragment of the structure, which renders it special for comprising the entire cross-section of the nanotube. The length of the unit cell in axial direction is defined by the absolute value of the translational vector $\bar{T}$ that runs in parallel with the tubular axis. Its value and direction can be determined by simple geometrical considerations from the planar representation of the graphene sheet related to a given nanotube. $\bar{T}$ is plotted at right angles to the circumferential vector $\bar{C}$, starting from the origin (0,0) and proceeding until an equivalent point of the graphene lattice is encountered. Both vectors spread a rectangle that corresponds to the uncoiled unit cell of the nanotube (Figure 3.2). The actual shape of the translational unit cell, however, is obtained only after rolling up. The respective unit cells are made discernable in the templates for constructing the different classes of nanotubes. It is evident that the length of the translational unit cell varies widely with the values of n and m.

The size of the elemental cell may as well be calculated from the descriptors. There are two possible formulas to be applied depending on the relation of the values for n and m. With the difference of n and m being an integer multiple x of $3g$ (g = biggest common divisor of n and m), the absolute value of $\bar{T}$ is obtained from Eq. (3.5):

$$\text{Length of the unit cell (nm)}: t = |\bar{T}| = \frac{\bar{a} \cdot \sqrt{3 \cdot (n^2 + nm + m^2)}}{3g} \cdot (n - m = x \cdot 3g)$$

(3.5)

If the difference between n and m is not an integer multiple of $3g$, Eq. (3.6) applies to calculate the length of the unit cell:

$$\text{Length of the unit cell (nm)}: t = |\bar{T}| = \frac{\bar{a} \cdot \sqrt{3 \cdot (n^2 + nm + m^2)}}{g} \cdot (n - m \neq x \cdot 3g)$$

(3.6)

Figures 3.2c and d show the unit cells of selected nanotubes. Obviously the unit cell of all zig-zag nanotubes is 0.426 nm long, and the armchair tubes also share a common length of 0.246 nm for their unit cells. It is a rule that for a given θ the perimeter has no influence on the length of the unit cell – chiral tubes with their circumferential vectors oriented equally on the graphene sheet just as well exhibit identical length of translational unit cells, regardless of their diameter.

The number of carbon atoms in the unit cell of a nanotube depends on its structure and may be anything up to several thousands. Again the respective mathematical expressions depend on the relation of n and m (Eq. 3.7):

$$\text{C-atoms in the unit cell}: \quad N_C = \frac{4 \cdot (n^2 + nm + m^2)}{3g} \cdot (n - m = x \cdot 3g) \quad (3.7a)$$

$$\text{C-atoms in the unit cell}: \quad N_C = \frac{4 \cdot (n^2 + nm + m^2)}{g} \cdot (n - m \neq x \cdot 3g) \quad (3.7b)$$

These values may grow large even for carbon nanotubes with typical diameters of 2–30 nm. A (95,51)-nanotube, for instance, has a diameter of 10.05 nm, its unit cell nevertheless measures 54.7 nm already and contains 65 884 carbon atoms.

Many of these translational unit cells are lined up in axial direction in real nanotubes. Their length thus can differ from their diameter by orders of magnitude. For this structural particularity (the expansion being wider than just a few nanometers in one direction only), carbon nanotubes may also be considered one-dimensional crystals, which largely influences their spectroscopic and electronic properties as will be discussed in more detail in the ensuing chapters.

After these systematic reflections on the construction of carbon nanotubes we shall now turn to their atomic structure. In nanotubes measuring much more in length than in diameter, all carbon atoms are situated on the corners of equally

3.2 The Structure of Carbon Nanotubes | 131

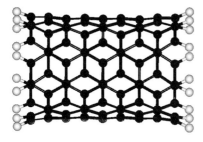

Figure 3.3 Finite nanotube with widening and nonequivalent bonds about the tube's ends.

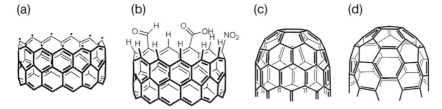

Figure 3.4 Different conceivable tube terminations: (a) dangling bonds, (b) terminal functional groups, (c) cap on a (9,0)-nanotube, (d) cap on a (5,5)-nanotube.

sized hexagons. The bond length is a uniform 1.425 Å. Regarding this detail, the carbon nanotubes do not differ from the arrangement in graphite that also features equivalent atoms. Consequently one might assume the π-electrons to be delocalized across the whole cylindrical structure.

In nanotubes with comparable diameter and length, on the other hand, the influence of the ends becomes apparent. In these cases the equivalence of bonds is broken. The cylinder is deformed toward the ends (Figure 3.3) and the degree of π-electron delocalization decreases in the entire system. It rather becomes possible to setup several resonance structures for these so-called finite nanotubes. They may include localized double bonds or six-membered rings without aromatic character (also refer to Figure 3.8). Alternating bonds are most frequent near the ends of the tube while their number decreases toward the center. However, calculations on different levels yield in parts contradictory values for the bond length in these short tubes. There is consensus that the alternation is between the bonds rectangular and those in a 30° orientation to the tube's axis. Still which one of these bonding types is the longer is discussed controversially. The longer the calculated nanotube gets, the less influence of this alternation is found.

Considering the ends of an open, single-walled nanotube reveals that the outermost carbon atoms must bear unsaturated bonding sites (Figure 3.4a). These so-called dangling bonds may be saturated by attaching either functional groups or a cap of carbon atoms. The orifice of the tube is retained in the first case (refer to Section 3.3.6), while in the latter a closed SWNT is obtained. Especially in short tubes these terminal groups exert marked influence on the structure, for example, the ends of the carbon cylinder are widened. However, this phenomenon is not

Figure 3.5 Shapes of caps on asymmetrically closed carbon nanotubes: (a) asymmetrical conical tip, (b) tip with inner cap, (c) beak-shaped tip.

confined to short nanotube segments. It is rather observed for tubes of any length, but its influence on the structure of the overall system remains negligible in longer species.

The shape of caps in closed SWNT is highly variable. Both approximately hemispheric caps and pointed structures are found, and even concave domains ("beak-shape") have been observed (Figure 3.5). However, the repertoire of potential caps is much less diverse for single-walled than for multiwalled nanotubes. According to Eq. (3.3), the (9,0)- and (5,5)-nanotubes given in the appendix have a diameter similar to that of C_{60}. Hence hemispherical fragments of that carbon cage should be suitable caps for these tubes. The C_{60}-molecule must be divided into different hemispheres depending on the type of nanotube (Figures 3.4c and d. The C_{60}-fragment required for the (9,0)-zig-zag nanotube must enable the formation of six-membered rings with the terminal carbon atoms of the tube, which means that it must exhibit a zig-zagged rim itself. To this end, the carbon cage is bisected at right angles to one of its trigonal axes of rotation. For the (5,5)-nanotube, on the other hand, the formal dissection of C_{60} is effected perpendicular to one of its pentagonal axes of rotation. Suitable caps for larger nanotubes can be identified in an analogous way. Fujita and Dresselhaus *et al.* have presented a very useful formalism for that. It is based on uncoiling the carbon nanotube to be a graphene sheet. In this model, pentagonal defects are generated by "cutting" 60°-segments from the lattice at the required positions so a five-membered ring and a convex curvature would arise there upon assembly (Figure 3.6, bottom right). To represent a fullerene in this projection, the position of five-membered rings is identified by color coding the respective six-membered rings and appointing numbers to them. Six-membered rings that would superimpose after assembly of the model (i.e., hexagons representing identical five-membered rings) are assigned the same ordinal (Figure 3.6). A multitude of caps for a given nanotube can be conceived this way.

This is indeed close to reality as actually hemispheric caps are observed rather seldom. In most cases conical structures are found. These bear defects at the tip itself as well as at the transition to the nanotube. They all have in common that the required curvature is brought about by pentagonal defects. The bevel angle of such conical caps depends on the number of five-membered rings present at their tips, as demonstrated by the values in Table 3.1. A cap with six pentagonal defects, for instance a hemispheric fullerene fragment, enables a direct transition to the tubular part, while a smaller number of five-membered rings causes a more or less conical shape.

Besides five-membered rings, the caps may also contain defects that cause an inverse curvature (Figure 3.5). Seven-membered rings are the most common

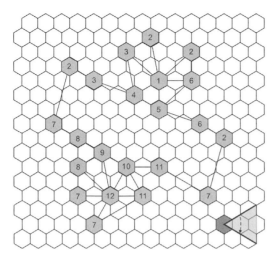

Figure 3.6 Determination of the structure of caps for nanotubes according to Dresselhaus and Fujita. The caps to both sides of the (7,5)-tube exhibit different geometries (© APS 1992).

Table 3.1 Relation between the bevel angle of nanotube tips and the number of five-membered rings.

Number of five-membered rings	1	2	3	4	5	6
Bevel angle of the conical tip	112.9°	83.6°	60°	38.9°	19.2°	0°

Acc. to P. J. F. Harris, *Carbon Nanotubes and Related Structures*, Cambridge University Press, Cambridge **2001**, p 75.

example. These defects lead to the aforementioned beak-shaped caps that exhibit considerably smaller diameters than the tubes themselves.

It has been mentioned before that the equivalence of all bond lengths (as confirmed for instance by STM measurements) led to the assumption of a completely delocalized system of π-electrons. Yet the reactivity of nanotubes (Section 3.5) suggests the character of a moderately conjugated polyolefin. This is, however, only a contradiction at first glance: The aromatic character is less pronounced than in the reference system benzene, but still the double bonds are not localized.

This raises the important question of nanotube aromaticity in general. A distinction must be made here between local aromaticity and superaromaticity. The latter arises from ring currents that circuit the perimeter of the respective tube. Figure 3.7 shows how the geometrical arrangement of carbon atoms in the nanotube should in principle enable this kind of aromaticity. Yet different calculations indicate that the actual currents are rather weak, which means that superaromatic behavior is largely negligible.

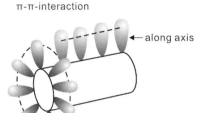

Figure 3.7 Depiction of the outer π-orbital lobes along the axis and around the perimeter of a carbon nanotube.

Aromaticity as such is a property that is hard to assess as it cannot be measured directly. It is true that there are several models to describe aromatic behavior, but they are partly incompatible with each other and yield rather different results. They always generate a theoretical value that must be compared to a reference system like benzene. Typical values employed to compare the aromaticity of conjugated systems include the topological resonance energy (TRE) and the nucleus-independent chemical shift (NICS) value. The latter indicates the absolute magnetic shielding of a test atom situated in the center of the system under consideration (negative NICS means aromaticity). Usually the TRE values are transformed into percentage resonance energies (%RE) to establish better comparability between molecules of different sizes. To this purpose the TRE value is multiplied by 100 and subsequently divided by the overall π-bonding energy of the graph-theoretical polyene reference.

Numerous calculations regarding the aromaticity of carbon nanotubes have been performed applying different procedures and theoretical levels. A general observation was that the aromaticity of nanotubes increases with growing diameters and approximates that of graphene with its perfectly even arrangement of carbon atoms. This is in unison with the negligible bond alternation in nanotubes studies by experiment—weaker conjugation and a more distinct localization of double bonds should reveal themselves by a better discernability of single and double bonds. The dependence of aromaticity on the diameter is explicable by the radial arrangement of π-orbitals. Their interaction is more efficient on large diameters than on thin nanotubes (Figure 3.7). Indeed one has to distinguish between the outer orbital lobes and those pointing inside the tube. The bigger outer lobes achieve a better overlap with increasing diameter, while those on the inside suffer the opposite effect. However, the latter ones play just a minor role in the considerations on aromaticity.

Recent calculations did further reveal that also the length of a finite nanotube has essential influence on its structure, which may adopt one out of three basic variants: The Kekulé-structure, the incomplete, and the complete Clar-structure (Figure 3.8). The first does not contain isolated double bonds, but is rather completely conjugated. The Clar-structures both feature p-phenylene moieties which, in the case of the incomplete Clar-structure, are flanked by isolated double bonds. This structural variety causes differences in electronic properties and thus in the

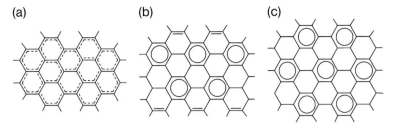

Figure 3.8 (a) Kekulé structure, (b) incomplete, and (c) complete Clar structure.

aromatic character. The calculations show that the three structures in Figure 3.8 alternatingly represent the energetic minimum when varying the tube's length. It is salient that the electronic properties and the reactivity of finite carbon nanotubes depend on the exact length of the tube.

If now the tube bears defects – either vacancies or sp^3-hybridized carbon atoms generated by the attachment of functional groups – the complete conjugation of π-electrons is broken and the aromatic character decreases. In detail, the defects give rise to finite sections within the tube that exhibit different conjugation patterns similar to those in short nanotubes. Then transition to another structural type is observed at the latitude of the defect.

3.2.3
Structure of Multiwalled Carbon Nanotubes

Multiwalled carbon nanotubes (MWNT) consist of a concentric arrangement of single-walled nanotubes with a usually constant distance of layers. There are examples with just two nanotubes fit one into another (so-called double-walled nanotubes, DWNTs) as well as species with many shells (more than 50). The latter measure many nanometers in diameter and may be so hard to distinguish from classical carbon fibers that only electron microscopy can reveal the difference. Common MWNTs, however, possess a smaller number of concentric tubes. The nomenclature of MWNT indicates the intercalation of the inner tubes by the notation $(n_1,m_1)@\ (n_2,m_2)@\ (n_3,m_3)@$...,starting from the central tube.

Usually a cylindrical void is found in the core of these tubes. Its diameter may be anything from 1 nm to several nanometers. The smallest possible hollow is determined by the size of the innermost SWNT, so diameters less than about 1.2 nm are rather unlikely to be found. The hitherto existing experimental results confirm this (although there are reports on SWNT with a diameter of 0.4 nm only). An upper limit to the size of the cavity theoretically does not exist. However, structures with very large hollows are increasingly unstable and, consequently, not very probable. Typical inner diameters of nanotubes with about 10 shells amount to several nanometers.

In principle, the same rules apply to the atomic construction of multiwalled nanotubes like they do to single-walled species. This is consequent as the first consist of the latter. Additional aspects are, however, which tubes are suitable to a concentric arrangement and what kinds of interaction exist between the components of a multiwalled carbon nanotube (MWNT).

The first issue to be solved is the question of intertubular distance and whether it is variable from one layer to another or constant in a system of telescoping nanotubes. Electron micrographs show that the space between individual tubes usually measures 0.34 nm, which is about the value found in turbostratically disordered graphite (Figure 3.1a). Larger gaps are observed for a few exceptional cases only. Firstly, there are often zones close to the tubes' ends where the inner layers already bend toward an internal cap while the outer ones still proceed straight ahead. And secondly, some nanotubes possess a different number of shells to both sides of the central cavity and partly exhibit an additional dilation of intertubular spacing.

The structure of "normal" MWNT will be considered here to begin with – "normal" meaning those with a constant distance between individual layers. This bears several consequences for the structure. An ABAB-sequence of layers like in graphite, for instance, can be maintained in very small areas only. The individual tubes are rather disorderly shoved one into another, rendering a direct interaction close to impossible. The reason of this effect is the introduction of additional carbon atoms into the lattice of the outer tubes. It is required because at a distance of 0.34 nm between individual shells their circumferences differ by about 2.14 nm (3.8).

Perimeter gain u' between $\quad u' = 2\pi(r_2 - r_1)$

Two tubes of an MWNT: $\quad u' = 2\pi \, 0.34 \text{ nm} = 2.14 \text{ nm}$ (3.8)

Considering an armchair nanotube reveals that a distance of 0.34 nm can almost exactly be realized: The most suitable extension of circumference causes a perimetral difference of 2.13 nm, which is quite close to the theoretical value. The smallest recurring unit along this direction is 0.426 nm long as results from the size of the unit vectors $\vec{a}_1$ and $\vec{a}_2$. Adding five of these building blocks generates an increase in circumference u' of (5×0.426) nm = 2.13 nm. Hence armchair nanotubes can form MWNT even under approximate retention of the graphitic ABAB-structure (Figure 3.9a).

Another situation is found for zig-zag nanotubes. Their smallest repetitive unit along the circumference corresponds to the unit vector $\vec{a}_1$ and thus measures 0.246 nm. The required perimetral difference of 2.14 nm cannot even approximately be achieved by adding integer multiples of that length. A number of nine additional six-membered rings fits in best under the given circumstances. Still the difference in perimeters is 2.214 nm, correlated to an intertubular distance of 0.352 nm. In each new tube added to the outside, 18 additional carbon atoms are inserted. These represent a kind of defect that causes a displacement of packing between layers. Thus the ABAB-structure of graphite is only retained in very small domains between these insertion sites (Figure 3.9b).

The constellation in chiral nanotubes is even more complicated. Only for a minority of cases an MWNT with the correct distance between shells can be gener-

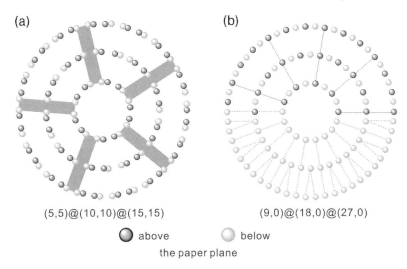

Figure 3.9 Projection of an MWNT made up (a) from armchair and (b) from zig-zag nanotubes; the domains of good π-overlap are indicated by bands and lines, respectively. The dashed lines show the addition of single carbon atoms in each new layer of the zig-zag tube impeding the π-overlap.

ated while keeping the same chirality angle for all constituent tubes. Two structural characteristics should consequently be expected: Firstly, an ABAB-packing of individual tubes will only be found in very small domains, and secondly, an MWNT will usually consist of nanotubes with different chirality angles. These results of theoretical consideration are supported by experimental findings. For instance, electron diffraction patterns of multiwalled carbon nanotubes often show signals of tubes with different geometry. This observation will later turn out to be an important indication of discrete tubes actually existing in multiwalled tubes. However, the presence of individual tubes with various orientations makes the prediction of MWNT properties extremely complicated, all the more as for the time being it is impossible to unequivocally assign the orientation of individual tubes.

The question remains which kinds of interaction exist between the single tubes of an MWNT. It must be answered separately for the terminal and the central regions. In the middle, far away from the ends of a multiwalled nanotube, the individuals take a constant distance of 0.34 nm while there is no direct covalent or ionic bonding between them. The interaction is thus limited to van der Waals exchange and to the zones where the packing enables π–π-interaction. The latter, as mentioned before, are usually rather small, so the overall interaction between two concentric nanotubes should be weak. This assumption is supported by calculations that predict a barrier of only 0.23 meV per atom for the rotation of a tube inside a bigger one and a respective translational barrier of 0.52 meV per atom. In effect, the inner tube should easily be rotated and shifted within the outer one due to this weak exchange.

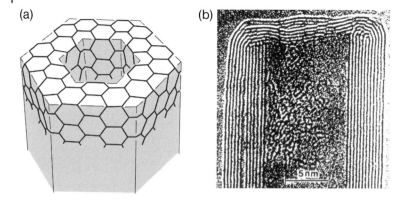

Figure 3.10 Toroidal nanotubes have one coherent surface (a), HRTEM-image of a toroidal nanotube (b) (© Cambridge University Press 1999).

In reality, however, this phenomenon is not observed because totally different conditions prevail at the ends of a multiwalled carbon nanotube. The interactions are no longer limited to π-stacking and van der Waals interaction. In some cases covalent bonds are found instead that interconnect the individual tubes. Caps may as well suppress a rotation of tubes against each other due to their asymmetric shape.

Especially open MWNTs feature remarkable structures owing to covalent linkage at their ends. Toroidal junctions are just one example; here the surface of the inner cavity in a way becomes part of the outer surface. Figure 3.10 shows a model illustrating this binding mode and experimental proof. It is evident that such an arrangement requires five-membered as well as seven-membered rings to generate the convex curvature of the outer edge and the concave one of the inner, respectively. This sort of semitoroidal structure usually contains six of these pentagon–heptagon pairs that formally may be formed by rearrangement from two six-membered rings.

During growth of an MWNT, it may further come to the incorporation of carbon atoms joined with two adjacent walls. This leads to an arrangement as depicted in Figure 3.11 that shows the common rim of a pair of nanotubes. As both tubes touch with their edges at this position, this kind of connection is called "lip–lip interaction." It plays a major role in the discussion of growth mechanisms.

Having described in detail the structure of MWNT, one important question remains: So far we have assumed the multiwalled nanotubes to be concentric arrangements of several SWNTs individually closed to themselves. But can we really be sure of that?

There are a number of arguments that favor this concentric principle for the major part of nanotubes at least. They include, for instance, the observation of internal single- or multiwalled caps and of secluded regions within multiwalled tubes. These features are incompatible with a spiral construction of MWNT as confined structures cannot be generated from such "infinite" arrangements, at

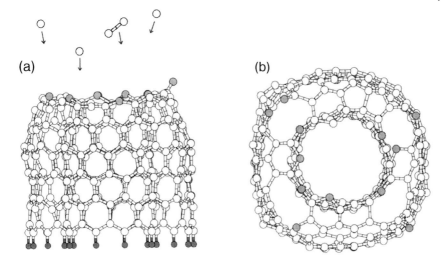

Figure 3.11 Lip–lip interaction at the open end of a multiwalled nanotube; side view with inserting carbon atoms and C_2-units (a) (© Springer 2000) and top view on the connected walls of the MWNT (b) (© Springer 1999).

least not without considerable effort. The reflexes of electron diffraction in the TEM also indicate a concentric composition of MWNT from independent tubes. In many cases, reflexes of chiral as well as of achiral tubes are observed for an individual MWNT. This would be impossible to a coiled-up species as it would consist of a single graphene sheet with determined orientation. It would necessarily have to exhibit the same chirality allover.

Therefore, multiwalled carbon nanotubes may henceforward be assumed normally to consist of concentric single-walled tubes. These are largely independent of one another and may belong to different structural types even within an individual MWNT. However, this does not mean that alternative structures do not exist. At least partially coiled species have been observed in some cases, and rolled up graphene layers play an important role in some hypotheses on the mechanism of nanotube formation.

The ways of tube termination and further defects like kinks are another essential aspect of MWNT structure. Like in the single-walled tubes, five-membered rings are crucial to the formation of these domains. As for the asymmetric conical structure, a single pentagon defect at the end of the innermost tube is presumed to cause the generation of such a cap. Sometimes the inner tubes end at a considerable distance from the actual tip. This gives rise to larger voids.

Besides cap formation, defects of individual tubes may influence the MWNT structure in other ways too. The simultaneous existence of five- and seven-membered rings, for example, makes a bent nanotube (Figure 3.12). In multiwalled systems, these so-called elbow structures are found in between encapsulated regions. The flexure thus is confined to a few graphene layers.

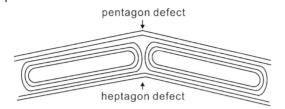

Figure 3.12 Scheme representing the structure of an encapsulated kink in a MWNT.

3.3
Production and Purification of Carbon Nanotubes

In recent years, the field of carbon nanotube production has seen a major progress. A multitude of techniques with different basic principles has been developed. They yield single- and/or multiwalled nanotubes depending on the method applied. All procedures have one thing in common: The hitherto available techniques always yield a product mixture as neither length nor exact geometry nor diameter can be completely controlled. Hence methods of purification and separation play an essential role in preparing highly homogeneous nanotube specimen. Technical applications furthermore require supply with sufficient amounts at acceptable prices. Here as well considerable progress has been achieved in recent times.

Most methods of nanotube production aim to vaporize a carbon source to allow for the growth of nanotubes from out of the gas phase. Only approaches of rational synthesis that intend a step-by-step construction of the lattice can do without this "detour" via the gas phase.

3.3.1
Production of Single-Walled Carbon Nanotubes

3.3.1.1 Arc Discharge Methods

The method developed by Krätschmer and co-workers for the preparation of macroscopic amounts of fullerenes can also be adapted to produce nanotubes. It has first been employed to make multiwalled nanotubes (Section 3.3.2.1), but by now SWNT production with this method succeeded as well. In doing so one must see to it that the growth phase of carbon structures is sufficiently long to prevent the cage closure of fullerenes and have cylindrical species growing instead.

The essential differences of an arc discharge apparatus for nanotube preparation are an increased pressure of inert gas and the distance between the electrodes. Contrary to fullerene production, these must under no circumstances touch each other. Figure 3.19a in Section 3.3.2.1 shows a scheme of the installation. The carbon source has to be added with a transition metal catalyst to obtain SWNT. Usually, this is achieved by drilling a hole through the anode in parallel with its longitudinal axis that is filled with a mixture of the catalytic metal and pulverized carbon. Catalyst will then continually be released while vaporizing the anode in the arc, thus enabling the generation of single-walled tubes. Metals such as iron, cobalt, nickel, rare earths, noble metals like platinum or combinations of these elements have successfully been employed in SWNT synthesis. It turned out that, as a rule, blends of two or more metals are more active than single elements. Mixtures of nickel with either cobalt or yttrium were found to be particularly suitable, while helium proved well as inert gas. It is applied at a pressure of 450 Torr/8.7 psi. The electrodes are placed at a distance of 3 mm from each other, which is readjusted continually during reaction. The current density of the arc is about 250–300 A cm^{-2}.

Like in MWNT production, a deposit is piled up on the cathode in SWNT synthesis, too (refer to Section 3.3.2.1). However, this matter firstly contains only a small fraction of the material evaporated from the anode, and secondly the concentration of single-walled tubes is relatively low. The cathode deposit consists of a hard, gray shell and a soft, black core. The shell contains sintered graphitic material, while the core exhibits a structure with aggregates in slightly columnar orientation (by far less pronounced than in MWNT synthesis, Section 3.3.7). The deposit's inner region bears SWNT, too, but to a small extent only. It rather contains graphitic nanoparticles, multiwalled tubes, graphite, and amorphous carbon. Particulate catalyst is found in the tips of some MWNT, and in the cathode deposit frequently an enrichment of one element from the catalytic mixture is observed. It is noteworthy, however, that no catalyst is found there when using a nickel–cobalt blend.

Yet contrary to the arc discharge synthesis of multiwalled nanotubes, the yield of carbon materials is not limited to the cathode deposit. Actually much larger amounts and concentrations of SWNT are found elsewhere in the reactor. Especially the deposits on the walls of the apparatus are rich in single-walled tubes. From these locations the precipitate can be peeled off like some textile material (Figure 3.13). Furthermore, there is a kind of collar around the edges of the

Figure 3.13 Felt-like material chiefly consisting of single-walled carbon nanotubes (© CRC Press 2005).

cathode that bears high concentrations of SWNT. Finally, the whole apparatus is interwoven with a cobweb-like fabric that largely consists of single-walled carbon nanotubes, too. All these indicate that the formation of SWNT is not confined to the space between the electrodes, close to the arc, but that the reaction zone extends far into the reactor chamber (refer to mechanism, Section 3.3.7).

The yield of single-walled carbon nanotubes is about 15%, while ca. 50% of the isolated carbon is obtained as amorphous material with the latter partly deposited on the surface of the first. The raw product also contains up to 20% of catalyst, which has to be removed before further use.

Single-walled nanotubes from arc discharge have diameters between 1.2 and 1.5 nm. This dimension may be controlled to some extent by the reaction temperature. Higher values generally give rise to larger average diameters which is explicable from the growth mechanism (refer to Section 3.3.7). The length of the tubes, on the other hand, depends on the catalyst chosen. Up to 20 µm are achieved when using a mixture of cobalt and nickel, while a nickel–yttrium blend rather yields nanotubes about 5 µm long. The tubes feature closed tips and seem relatively free of defects. Single-walled nanotubes made by arc discharge form strongly bound bundles.

In principle, the arc discharge preparation of single-walled carbon nanotubes is also applicable to the production of larger amounts. Apparatus apt to generate up to 100 g/h are equipped with electrodes 25 mm in diameter. Electrodes measuring up to 70 mm are feasible from a technological point of view, yet there is no proportional increase of yields with the growing diameter. It is crucial to adapt all other parameters, like pressure, cooling, reactor geometry, etc., to the larger turnover. Otherwise the growth period, that is, the lifetime of nanotube precursors and intermediates in the reaction zone, is either too short or too long.

3.3.1.2 Laser Ablation

Another method of destroying a graphite target employs a focused laser beam to locally supply the energy required for atomization. This technique as well has been used for the generation of fullerene clusters. The scheme in Figure 3.14a shows the essential components of a laser ablation apparatus.

It consists of an oven held at about 1200 °C with an internal quartz tube. The latter has an efficient cooling device (usually a cooling finger made of copper) situated at one end to collect the SWNT. However, nanotubes are also deposited on the reactor walls. Moreover, the installation must feature a part transparent to the laser light to enable vaporization of the graphite target by the focused laser beam. Helium and argon have been reported as carrier gases, but today argon has largely replaced helium as the latter has too high a cooling rate.

To obtain SWNT alone, a certain amount of catalyst (~1–2 at. %) must be added to the graphite target. Without this additive, multiwalled tubes are formed (Section 3.3.2.2). Larger amounts of the catalyst, on the other hand, coalesce during reaction to become bigger metal clusters that are inactive regarding the generation of nanotubes. Catalysts commonly applied include cobalt, nickel or mixtures including one of these metals. Ni/Co-blends provide particularly good yields of single-walled

(a)
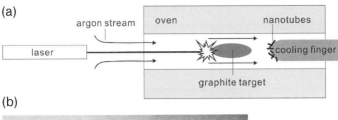

(b)
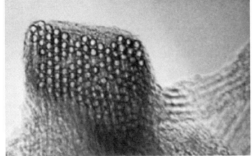

Figure 3.14 (a) Laser ablation apparatus for the production of SWNT. The catalyst is admixed to the graphite target. (b) HRTEM-image showing a bundle of SWNT. The hexagonal arrangement within the bundle is clearly to be seen (© AAAS 1996).

nanotubes. Other transition metals like platinum or copper, mixed with cobalt or nickel, generate SWNT too, yet the yields are considerably smaller.

The irradiation of high-power laser light causes local heating of the graphite target; it may reach several thousand degrees at the immediately exposed spot. This causes an increased number of carbon atoms and clusters as well as catalyst material to pass over into the gas phase. Thus conditions prevailing there are similar to those in an arc discharge apparatus. Consequently the nanotubes obtained from both methods closely resemble each other.

The concentration of catalyst metal at the surface of the graphite target increases during exposure to the laser, which reduces the rate of SWNT formation. Hence a technique simultaneously irradiating a neat and a catalyst-bearing target has been developed that leads to markedly better yields. The amount of SWNT obtained is further influenced by the temperature. The higher it is, the more nanotubes are produced – below 200 °C on the other hand, no SWNTs are observed any more. Usually the apparatus is run at a temperature of 1200 °C.

The resulting single-walled nanotubes contain very little defects and no catalyst particles. They also tend to be free of amorphous carbon surrounding them, which quite often is the case with other techniques (refer, e.g., to the arc discharge method). The diameters of individual tubes range from 1.2 to 1.7 nm with a very narrow size distribution, and their length may be several hundreds of nanometers. A peculiarity of these tubes is the formation of very stable bundles that results from individual SWNT arranging in parallel. The bundles usually exhibit overall diameters between 5 and 20 nm and a length of several hundreds of micrometers, so actually they might better be thought of as SWNT fibers. Although single-walled nanotubes show a general tendency to form bundles, this property is far less pronounced in samples from other methods of preparation. In many cases two-dimensional super-structures are observed within the bundles. Usually this is a kind of two-dimensional hexagonal lattice (Figure 3.14b), but more complex structures like circular arrangements have been found, too.

With the reactor being a confined apparatus, continuous operation of the process is impossible. It must be interrupted after complete consumption of the graphite target to replace the carbon source. Furthermore, the power consumption would be very high at large-scale applications, so this method is more likely to establish for the generation of pure, defect-free SWNT on the laboratory scale. Reactors yielding up to 5 g of SWNT per hour have indeed been presented, but larger amounts are improbable ever to be accessible using laser ablation because a number of unresolved problems (heat distribution, power consumption, continuous supply of target, etc.) impedes the development of larger apparatus.

3.3.1.3 The HiPCo Process

Another method to prepare single-walled nanotubes has been derived from a process that is also employed to produce classical carbon fibers. It consists in the reaction of a flowing, gaseous carbon source, for example, an alkane or CO, on a transition metal catalyst generated *in situ* by thermal decomposition of organometallic precursors like carbonyl complexes or metallocenes. The metal particles are very small and serve as nucleation centers for carbon fibers or nanotubes that start growing from these metal clusters (also refer to Section 3.3.1.5 on CVD methods). When using hydrocarbons as starting material, one must consider their potential pyrolysis. It usually sets on at about 700 °C and gives rise to amorphous graphitic structures. These must be separated then from the single-walled tubes, which means a considerable extra effort in SWNT production.

The preparation of single-walled nanotubes succeeds more easily by the so-called HiPCo-process that was published for the first time in 1998. The name is deduced from "high-pressure carbon monoxide" and signifies a crucial aspect of the method: Here the carbon source is not a hydrocarbon, but carbon monoxide that does not suffer pyrolysis at the relevant temperatures. The formation of carbon material is based on the Boudouard equilibrium (3.9):

$$\text{Boudouard equilibrium:} \quad 2\,CO \xrightarrow{[Fe]} CO_2 + C(SWNT) \quad (3.9)$$

The surface of iron clusters generated from iron pentacarbonyl catalyzes the reaction to carbon dioxide and elemental carbon. The latter is obtained in the shape of single-walled nanotubes that grow originating from the catalyst particles like in the production of carbon fibers. As obvious from the reaction equation, the reaction rate and the law of mass action depend geometrically on the pressure of carbon monoxide. As a consequence, the equilibrium is effectively shifted toward the products only at high pressures of CO. Still the reaction rate is rather low even at high pressure, and so a large fraction of the carbon monoxide passes the reactor without transformation. Furthermore, the entire iron carbonyl contained in the CO is thermally decomposed while the resulting iron is carried from the reactor,

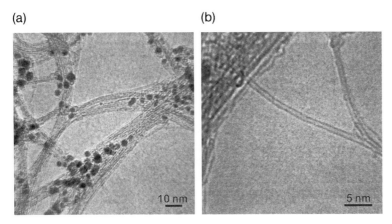

Figure 3.15 HRTEM-images of HiPCo-nanotubes with iron clusters (a) and after purification (b). The diameter of the individual tubes is about 1.0–1.4 nm (© Amer. Vac. Soc. 2001).

so new catalyst must be added continually. It comes in handy here that commercially available carbon monoxide contains small amounts (2–10 ppm) of iron pentacarbonyl, and indeed the dispensability of additional catalyst dosing has been reported for individual cases.

The active catalytic species are iron clusters generated from the pentacarbonyl. They consist of about 40–50 iron atoms at a diameter of ca. 0.7 nm. This corresponds roughly to the dimension of the smallest SWNT formed (also refer to Section 3.3.7). The iron particles generated from the catalyst attach to the outside of the carbon nanotubes and may contribute up to 7% to the overall mass of the product mixture (Figure 3.15). The metal clusters can, however, be removed by wet chemistry (Section 3.3.6) because they are not concealed deep inside the SWNTs, but usually surrounded by just two or three layers of carbon.

The reactor consists of a tube furnace with a mixture of CO and $Fe(CO)_5$ continually flowing through it. Once generated, the nanotube soot is carried out of the reactor by the gas stream. It is deposited either at cooler sites of the apparatus or on a dedicated cooling finger for subsequent work-up. A state-of-the-art laboratory scale HiPCo-generator (p_{CO} = 30 atm/446 psi, t = 24–72 h, flow rate ~ 300 slm, T = 1050 °C) can produce up to 450 mg of single-walled carbon nanotubes per hour. Continuous operation of the installation can easily be maintained for up to a fortnight.

The heating up of the initial gas mixture is an essential aspect of reaction control. $Fe(CO)_5$ starts decomposing at ~250 °C already, while the desired reaction proceeds with acceptable velocity only above ~500 °C. Hence, a compromise must be sought ensuring optimized yields of SWNT. To this end, both too fast or too slow a heating of the initial mixture have to be avoided. Otherwise only very small, re-evaporating iron clusters would be generated in the first case, while in the second, the iron clusters would be too big to allow for the growth of nanotubes. A suitable device was found here in a nozzle that enables the rapid injection of preheated carbon monoxide into a stream of CO and Fe(CO)$_5$. The heating up to acceptable reaction rates thus ensues very quickly. The reactor temperature is about 800–1200 °C, and the pressure of CO can be raised up to 30 atm/446 psi in a typical laboratory installation. The iron particles become surrounded by a coating of carbon shells during reaction and consequently loose their catalytic activity. Therefore, a continuous addition of iron pentacarbonyl to the gas mixture is required to maintain a constant amount of catalyst.

Typical SWNT samples from a HiPCo-apparatus are shown in Figure 3.15. It is striking that all of the individual nanotubes exhibit similar diameters of about 1.0–1.4 nm. Possible reasons for this surely include the nucleation on metal clusters generated *in situ* – these show a narrow size distribution themselves. The length of HiPCo-SWNT is about 1 μm. The individual tubes aggregate to form stable bundles, which can only be destroyed by drastic procedures (Section 3.4.2). The pressure of CO influences the average tube diameter. Increasing the first leads to a reduction of the latter to about 0.7 nm, which roughly corresponds to the diameter of a C_{60}-molecule.

The feasibility of continuous operation counts among the major advantages of the HiPCo-process over most alternative methods. The catalytic decomposition of carbon monoxide can thus be considered as a potential way of scaling up SWNT-production from a few grams to kilograms or even tons.

3.3.1.4 Pyrolysis

The thermal decomposition of organic compounds can also be employed to generate small carbon clusters or atoms. The borderline with chemical vapor deposition (CVD) as presented in the next section is not really fix. In both cases, the method is based on the thermal decomposition of organic precursors. Processes both with and without catalyst have been reported. Contrary to the chemical vapor deposition, however, the catalyst (if applied) is not coated onto a substrate, but the substance or a precursor is added directly to the starting material ("floating catalyst"). The resulting mixture is then introduced into the reactor either in solid or in liquid state by a gas stream. From this point of view the HiPCo-process could also be considered a pyrolytic preparation of SWNT, but due to its importance it is usually regarded as autonomous method.

Various alcohols turned out to be suitable raw materials for the pyrolytic production of single-walled nanotubes. For instance, the thermal decomposition of an ethanol/ferrocene mixture at 950 °C in a stream of argon yields long strands of single-walled tubes with an exceptional diameter of 2–3 nm (Figure 3.16). In this process, the liquid reactant mixture is finely dispersed into the argon flow through a nozzle. Alcohols generally give rise to single-walled nanotubes, while other carbon sources like acetylene preferably produce MWNT. Hence the OH-radicals from alcohol degradation have been surmised to reduce the generation of amorphous carbon and the subsequent growth of further walls on existing nanotubes. In more recent times, however, ferrocene-catalyzed syntheses of single-walled nanotubes from oxygen-free substances like hexane or benzene have been reported, so the aforementioned hypothesis can at least not be generally valid.

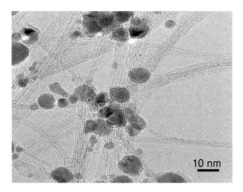

Figure 3.16 HRTEM-image of single-walled carbon nanotubes obtained by pyrolysis of ferrocene–ethanol solutions. The dark objects are iron particles (© Elsevier 2005).

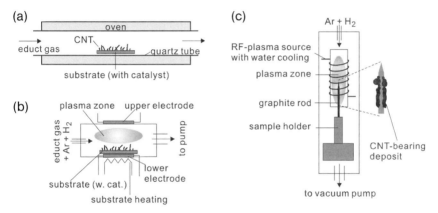

Figure 3.17 Different apparatus for the chemical vapor deposition of carbon nanotubes: (a) classical CVD-reactor with tube furnace, (b) bias-enhanced CVD reactor, (c) plasma-enhanced CVD-apparatus.

The decomposition of methanol or ethanol on an iron–cobalt catalyst deposited on a zeolite substrate represents a transition to the CVD methods (Section 3.3.1.5). Here as well, SWNTs of high purity are obtained. The yields can be up to 800% with reference to the catalyst employed (or 40% relative to the starting material).

3.3.1.5 Chemical Vapor Deposition (CVD)

It has been more than 20 years already since the first CVD has been developed to produce carbon fibers and filaments. This technique is also highly suitable for the preparation of various nanotube materials.

The apparatus consists of an oven surrounding a heat-resistant tube that can be evacuated. A substrate with particulate catalyst dispersed on it is placed inside the tube (Figure 3.17a). Then, a stream of the gaseous reactant mixture is led over the catalyst. Having finalized the reaction, the substrate with the SWNT grown on it is removed from the reaction chamber and the nanotubes are isolated. Typical temperatures for CVD range from 550 °C to 770 °C.

The most common carbon sources are hydrocarbons, and among these especially methane, with admixtures of hydrogen and inert gas, but comparable results can also be obtained with ethylene or acetylene. Due to the high reaction temperature, however, the choice of carbon sources is limited to hydrocarbons that do not undergo spontaneous decomposition between 500 and 1000 °C. A low partial pressure of the hydrocarbon may be applied to obviate this undesired process to some extent, but still the low-molecular species qualify best for the method.

The catalyst is usually coated onto its support in the shape of finely dispersed particles. The choice of the two materials is crucial to both product's quality and quantity. For example, the linkage of catalyst and support must be stable also at high temperatures. A large catalyst surface and a low tendency to aggregation further favor the growth of SWNT. Alumina and silica constitute typical substrates, but silicon itself or certain metals are suitable as well.

As for the catalyst, elements from the iron group have proven the most active. Hence, CVD-methods normally employ iron-, cobalt- or nickel-catalysts on alumina supports. The activity of these transition metals is due to their ability to solve small amounts of carbon at elevated temperatures. The carbon atoms released from methane can thus be solved in the catalyst or move on its surface by diffusion. Upon supersaturation then, they start to grow as nanotube from its surface. The catalyst also facilitates the thermal decomposition of hydrocarbons yielding carbon atoms and small clusters in the gas phase that can serve as building blocks for nanotube growth.

Control of the nanotubes' diameter by means of catalyst particle with defined dimensions is an enticing perspective. The size-selective synthesis of SWNT succeeded, for instance, with iron particles 3, 9, or 13 nm wide, respectively. The resulting nanotubes had average diameters of 3, 7, or 12 nm. Obviously, a uniform size distribution of the catalyst nanoparticles is crucial here. They are usually obtained by precipitation from organometallic precursors (e.g., iron pentacarbonyl).

Single-walled nanotubes prepared by vapor deposition are of high quality and, contrary to materials made by other methods, far less entangled. Hence the purification of CVD-SWNT is facilitated. Still the samples often bear defects, and they always contain considerable amounts of catalyst material, which has to be removed through a time-consuming purification procedure. The defects can be cured by annealing at high temperatures under an inert atmosphere (Section 3.3.6).

An upscaling of CVD-nanotube production is not easy to achieve since it is limited by the size of the substrates. What is more, these have to be taken from the reactor once the nanotubes attain the desired length. Installations working continually are therefore hard to conceive. Still the CVD-method has a right to exist also as a large-scale technical process; especially the deposition on structured surfaces, for example, integrated circuits, succeeds with CVD alone after applying the catalyst by a structuring process (inkjet printing, lithography, photoresists, etc.). This is why carbon nanotubes are manufactured by CVD if they are required in defined, high quality or as an ordered array on a substrate for certain applications. Nanotubes intended for bulk-utilizations like polymer reinforcement are prepared by other, less costly procedures (arc discharge, pyrolysis, HiPCo).

Hata recently reported on a method that allows for higher growth rates and a drastic improvement of yields by more than a hundred-fold to >50 000% with reference to the catalyst. It is, in detail, a CVD-method employing ethylene, which is decomposed on a catalyst (Fe, Fe/Al_2O_3, Co/Al_2O_3, etc., on silica substrate) while maintaining a strictly controlled water content of the reactant gas flow. The decisive advantage of this method consists in the water constantly removing amorphous carbon from the product mixture with the aid of its oxidizing power. The product's purity is thus enhanced and, what is more, the lifetime of the particulate catalyst is markedly prolonged as it is no longer covered with a growing layer of amorphous carbon. The product obtained is a dense film of parallel, vertically aligned nanotubes. These may reach a height of several millimeters after a growth period of as little as 10 min (Figure 3.18). The nanotubes are not soiled with metal particles or amorphous carbon and exhibit rather uniform diameters of 1–3 nm. The method is apt to produce structured nanotube films as required, for example, for field emission displays. To this purpose the catalyst is lithographed onto the substrate. The high growth rate then leads to sharply contoured structures of

(a)

(b)

Figure 3.18 Nanotube arrays generated by CVD with added water. (a) The high growth rates enable the production of macroscopic amounts within just a few minutes. (b) Various arrangements of nanotubes can be obtained by coating structured patterns of catalyst onto the substrate. The insert shows a detail of such a nanotube pillar (© AAAS 2004).

SWNT rising from the catalytic sites of the substrate (Figure 3.18). Larger amounts of SWNT films may easily be prepared by this method, but the limitation of substrate sizes to reactor dimensions inevitably remains.

3.3.2
Production of Multiwalled Carbon Nanotubes

The first MWNTs have been obtained as early as 1976 by iron-catalyzed pyrolysis of benzene. Apart from that, there is a number of methods to produce MWNT, which all of them differ in the way of generating small carbon clusters or atoms from the respective starting materials. They include arc discharge, laser ablation, chemical vapor deposition with and without plasma enhancement or the catalytic decomposition of various precursor compounds. It turned out that MWNTs from low-temperature syntheses bear more defects and, as a whole, are less ordered than those generated at high temperatures. However, these drawbacks can still be compensated by subsequent recuperation of defective samples at elevated temperatures.

3.3.2.1 Arc Discharge Methods
As early as 1991, carbon nanotubes have been prepared with an apparatus quite similar to the arc discharge reactor for fullerene production (Section 2.3.3). Here as well an electric arc is used to vaporize carbon introduced in the shape of graphite rods, and again deposition occurs on cooler parts of the installation. The decisive difference to fullerene preparation consists in the electrodes not touching each other during reaction and an arc evolving between them. The so-called contact arcing as employed in fullerene synthesis is not apt to produce carbon nanotubes. This is firstly because the fraction of soot bearing the highest concentration of multiwalled nanotubes is deposited on the cathode. Secondly, the plasma zone must have a certain extension to enable the growth of nanotubes as these, contrary to fullerenes and SWNT, are generated only within the plasma zone.

Figure 3.19a shows a typical installation to prepare carbon nanotubes in the laboratory. Usually a vessel of steel is employed as the reaction chamber. Glass vessels as used in fullerene synthesis are rather unsuitable because the mounting of graphite rods at a certain distance to each other is hard to achieve. The pressure of inert gas (helium, argon) is held at about 500 Torr/9.7 psi, which is markedly higher than the pressures employed (<100 Torr/1.9 psi) in the preparation of C_{60}. The current intensity is about 50–100 A. At even higher values, the method does not yield nanotubes, but a hard, sintered material useless for nanotube production. The voltage applied usually is about 20 V (DC). Such an apparatus may, for instance, be constructed with a consumer type arc welding unit.

Upon voltage application, an arc sparks over between the graphite electrodes. It transfers material from the anode into the gas phase, which then can deposit in shape of the desired nanotubes. A certain portion forms a deposit on the cathode, but condensation also occurs at other places within the reaction chamber. The graphite anode is used up in the course of the reaction, and so a continuous read-

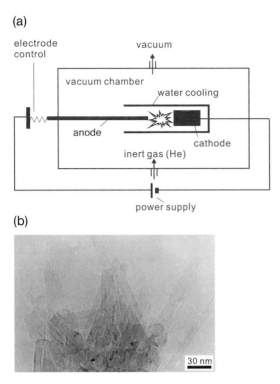

Figure 3.19 (a) Scheme of an arc discharge apparatus; (b) HRTEM-image of multiwalled carbon nanotubes prepared by arc discharge (© Elsevier 2002).

justment is required to maintain a stable arc. This consumption amounts to ≥1 mm/min. The dimensions of positive and negative electrode differ both in length and diameter. The latter is generally 6–12 mm for the anode, the adjustment of which is usually achieved by a stepping motor. Bearing and adjustment have to be designed and performed in a way to prevent contact with the cathode throughout the process. The cathode has a diameter of up to 12 mm and usually is laid out as a cylindrical unit, which enables the formation of a stable arc. Owing to the large evolution of heat during plasma generation, the cathode inevitably requires cooling; otherwise no satisfying precipitation of nanotubes will take place. Normally the anode is cooled as well. The cathode cooling has decisive influence on the quality of the soot deposited there. Being insufficient, the deposit will consist of hard material that contains only a few nanotubes. At adequate cooling, on the other hand, homogeneous deposits of nanotubes arranged in bundles are obtained.

The pressure of inert gas is another important parameter of the arc discharge method. The higher the pressure of helium in the combustion chamber, the better yields of nanotubes will be obtained while at the same time the fraction of carbon nanoparticles is reduced. The proportion of tubes and nanoparticles is about 2:1 in the best samples. At more than 500 Torr/9.7 psi, however, the overall yields decrease, so consequently typical reactors are operated at exactly this pressure. Procedures applying a more than atmospheric pressure of helium have been reported, yet a significant broadening of the diameter distribution is observed here. Alternative gases such as argon, nitrogen, or H_2/N_2-mixtures may be used instead of helium. Compared to the preparation of SWNT or fullerenes with the same method, the arc discharge synthesis of MWNT is much less sensitive to traces of oxygen in the inert gas. This is due to the reaction not taking place outside the plasma zone – SWNT and fullerenes, on the other hand, are more susceptible to oxygen attack as they cover considerable distances at high temperatures in the inert gas during formation.

The current intensity also influences yields and quality of the MWNT. If it is too high, the portion of very hard, sintered material will increase at the cost of nanotube yields. On the other hand, a current sufficient to sustain a stable arc and an extended plasma zone is required to achieve the continuous vaporization of carbon. Typical values range from 50 to 100 A.

The arc discharge method produces long, little bent MWNT featuring an asymmetric tip in most cases (Figure 3.19b). Their outer diameter usually ranges from 2 to 30 nm, and they consist of anything from a few to several tens of layers. The cylindrical inner cavity of these tubes measures 1–3 nm, their length is about 1 μm. The tubes are mainly found in the cathode deposit, while the soot deposited at the reactor walls predominantly contains nanoparticles and amorphous carbon. The latter is also often observed on the surface of MWNT from the cathode soot. The individual MWNT usually forms bundles due to strong van der Waals interaction. The single MWNT is of high quality, its electric conductivity is close to the theoretical limit of $(12.9\,k\Omega)^{-1}$ for ballistic electron transport (refer to Section 3.4.4.2), and current densities of up to $10^7\,A\,cm^{-2}$ can be achieved without appreciable superheating.

The deposit generated on the cathode consists of a very hard, gray shell of sintered graphite layers and an underlying soft, jet black core. This exhibits a columnar structure with a fluffy material filling the interstitial spaces. The columnar structures are oriented in parallel to the deposit's axis and may be up to 60 μm in diameter and several centimeters long. They contain graphitic and amorphous portions besides the nanotubes. The highest concentration of MWNT is found in the intercolumnar space, but in parts they also precipitate on the columns' outer walls (Section 3.3.6). Obviously, the cathode deposit has to be disassembled to isolate the MWNT. Furthermore, a stable arc turned out to be prerequisite to a good deposit quality with a large core portion. If, for example, the temperature exceeds a certain value, graphitization also proceeds into the core zone. In addition, the residual small amounts of nanotubes obtained then are heavily soiled. Still a distinct core zone with columnar structures is no guarantee for a high content of nanotubes – on the contrary, deposits seeming to be of high quality on first sight turned out to contain very low amounts of nanotubes upon detailed analysis.

Even the large-scale technical application of the arc discharge method could be established to produce multiwalled carbon nanotubes. The electrodes are designed

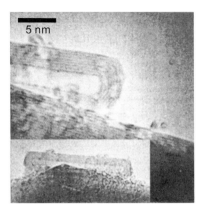

Figure 3.20 HRTEM-image of an MWNT obtained by laser ablation. The uniform shape of caps and the short length of these tubes are evident (© ACS 1995).

then as graphite rods several centimeters in diameter and the amperage is adapted to the conditions. Naturally the stabilization of the plasma in such an apparatus causes more problems than on the laboratory scale, and in comparison, the yields of MWNT decrease by about 25% indeed. This very energy-consuming process can be operated for a certain time. Continuous performance, however, is not feasible because the electrodes have to be replaced at intervals. Nevertheless, these reactors are apt to a production on the multikilogram scale.

3.3.2.2 Laser Ablation

MWNT can also be prepared by laser ablation. Contrary to the synthesis of single-walled nanotubes, no catalyst is added here, but a pure graphite target is vaporized by means of a focused laser beam. The resulting MWNT are precipitated at cooler positions within the reactor. Here as well, the operating temperature is about 1200 °C because the number of defects increases and the yields of MWNT decrease at lower temperatures. Below 200 °C then, no growth of carbon nanotubes is observed anymore. The process produces a considerable portion of amorphous carbon, fullerenes, and carbon nanoparticles besides the desired MWNT. These impurities have to be removed before further use. The yields of multiwalled nanotubes usually range about 40%.

Generally the MWNTs obtained possess 4 to 25 layers. They are a few hundreds of nanometers long, which makes them relatively short in comparison to tubes obtained from other methods. The structure of MWNT from laser ablation is virtually free of defects, and the individual tubes are closed with caps (Figure 3.20). The general question arises here how the emerging tubes can grow up to their actual length without prematurely being closed by caps – normally a fast introduction of five-membered rings and a subsequent closure of the tubes could be expected at the reaction conditions prevailing. The orifice is kept open, however, by so-called carbon–adatom bridges that interconnect neighboring layers ("lip–lip" interaction). This hypothesis is sustained, among others, by the observation that two adjacent tubes normally are closed at the same position (Figure 3.11).

Contrary to other methods, the laser ablation of solid carbon sources is not suitable to a large-scale synthesis of MWNT. The process cannot be conducted continuously, and the amounts of energy required render it uneconomic. Hence laser ablation apparatus are most likely to remain reserved to MWNT-synthesis in the laboratory.

3.3.2.3 Chemical Vapor Deposition (CVD)

Like in the preparation of single-walled carbon nanotubes, the chemical vapor deposition of MWNT consists in the generation of small carbon clusters or atoms from precursor compounds. The products precipitate in the shape of different carbon materials with the reaction conditions determining the specific structure of the substance.

The simplest CVD method is the thermal decomposition of carbonaceous compounds in a heated quartz tube bearing a substrate that is coated with catalyst particles. The reaction oven is heated at 500–1000 °C as no catalytic activation of the degradation is observed below 500 °C. The gaseous starting material is led over the substrate where a part of the molecules is decomposed and precipitates as nanotubes. Suitable carbon sources include methane, ethylene, acetylene, and ethane. The substrate is allowed to cool to about 300 °C after reaction to prevent damage to the nanotube structures, which probably would be done at higher temperatures. The growth rate of the nanotubes varies from a few nanometers to several micrometers per minute, depending on the reaction conditions. Besides the method of substrate-bound catalyst, one may also employ the "floating catalyst" process. In this case, a precursor of the catalytic species is added to the reactant gas to generate the active catalyst *in situ* within the reaction zone. It might be necessary to connect a second oven in series to ensure complete decomposition of the activated reactant–catalyst complex. Heat application may also be achieved by inductive heating or infrared irradiation. The products resulting from all of these CVD variants do not markedly differ from each other.

Furthermore, it is possible to promote reactant degradation by a glow discharge that may, for example, be generated by means of a plasma source (Figure 3.17b). Different techniques have been established over time. They include a direct current plasma that might be complemented by a hot filament or microwave irradiation. Inductive or capacitive radio frequency plasma as well is commonly employed for nanotube production. Contrary to the thermal methods presented above, much lower temperatures may be applied in the "plasma-enhanced CVD" (PECVD). This is of consequence especially for the deposition of nanotubes on structured, heat-sensitive substrates. However, a certain minimum temperature is indispensable also for the PECVD technique as otherwise catalytic decomposition of the precursor compounds does not occur to a sufficient extent. Measuring the actual temperature of the substrate though is a challenge of its own. Therefore, it is usually the temperature of the heating stage carrying the substrate that is reported in literature. This is easily determined with a thermocouple. The real temperature of the substrate is presumably much higher due to plasma heating and ion bombardment.

The potential starting compounds are the same as for thermal CVD. The plasma induces the decomposition of the initial hydrocarbons. However, a large portion of the resulting material consists of amorphous carbon. To suppress this off-catalyst reaction, the reactant gas is mixed with an inert carrier gas. This may be a noble gase (normally argon), hydrogen, or even ammonia. At an overall pressure usually between 1 and 20 Torr (0.02 and 0.39 psi), the content of starting material is about 20%. A performance at elevated temperatures turned out problematic regarding the generation of a stable plasma. Different types of reactors have been developed for PECVD (Figure 3.17), which shall be presented here in short:

In a classical plasma reactor, the plasma is sustained by a constant voltage applied between two electrodes. Above a certain minimum value, breakdown through the gaseous reactant–carrier mixture occurs and a glow discharge emerges. Reactive species like electrons, carbon atoms, ions, and radicals are observed within the discharge zone. Particle density in the plasma core zone is particularly high, while only a weak electric field exists there. The distance of the electrodes has to be adjusted with regard to the pressure inside the specific reactor. As a rule of thumb, lower pressure calls for a larger distance of the electrodes.

The substrate covered with catalyst may be attached to either the anode or the cathode. Generally an additional heating of the substrate is needed to achieve satisfying growth rates, but also the introduction of a tungsten filament into the plasma stream is employed as an alternative to the auxiliary heating. The method is called "hot filament CVD" (also refer to Section 6.3.1). Still a voltage of more than 300 V between the poles is required; but compared to other samples, the nanotubes prepared this way show less damage from ion bombardment.

Radio frequency enhancement was found to be the method of choice for plasma stabilization. A voltage oscillating at radio frequency (13.5 MHz normally) is applied here between a grounded and a live electrode. The glow discharge then takes place at lower voltages, thus reducing damage to already existing nanotubes. The ionization of reactive species as well is facilitated by the alternating field. The coupling of the plasma can be effected either by the same voltage difference applied to the substrate, or by external sources like induction coils or the coupling of microwave radiation via antennas or wave guides (e.g., 2 kW, 2.45 GHz). The method is then called "microwave assisted plasma-enhanced CVD." Using an external source to feed the glow discharge allows for a totally independent application of direct or alternating current to the substrate.

The one thing in common to all CVD methods is their need of a catalyst for reactant decomposition. Depending on the application, it may be coated onto a substrate or added directly to the reactant gas ("floating catalyst"). The simplest approach is the deposition of the respective transition metal on substrate plates (e.g., of SiO_2, Si, SiC, etc.) by physical methods like electron gun evaporation, thermal vaporization, pulsed laser deposition, ion beam-, or magnetron sputtering. The use of masks and lithographical techniques enables a structured deposition as known from semiconductor electronics for decades. For the time being, iron, cobalt, nickel, and molybdenum have been employed as catalysts. It turned our favorable in some cases not to coat the catalyst directly onto the substrate, but to deposit a catalytically inactive intermediate layer (e.g., Al, Ir, Ti, Ta, W) first and have the particulate catalyst placed on top of it. Alloy formation then causes an advantageous tighter attachment of the catalyst particles, and the heat transfer characteristics are also affected positively.

Besides physical deposition of the catalyst, the substrate can be prepared in a chemical way as well. To this purpose, a structure-bearing polymer (e.g., Pluronic 123) is mixed with an aqueous solution of inorganic salts. The resulting gel is applied to the substrate, and subsequent drying, reducing, and calcinating yield separated metal nanoparticles on the support that can serve as catalyst for nanotube production. This chemical deposition is, however, associated with large efforts as it requires several reaction steps to generate the active catalyst on the substrate. Furthermore, the procedure is too lengthy for surveying tests in the optimization of catalyst composition. Thus physical methods of substrate preparation gain increasing general acceptance.

For the direct injection of the catalyst or one of its precursors, the choice of compounds is naturally limited to vaporizable substances. The respective transition metal carbonyls and metallocenes proved their worth here. They decompose upon heating in the CVD apparatus and release the elemental metal in the shape of small, catalytically active clusters.

Small and uniform size of the individual particles is a prerequisite of a good catalyst, regardless of its preparation. The diameter of the metal particles codetermines that of the emerging nanotubes (Section 3.3.6), and consequently a narrow size distribution of the catalyst particles causes low variability in the tubes' diameters. The yields of nanotubes decrease if the metal particles employed are too large because these clusters preferably acquire a thin graphitic coating that quickly makes them loose their catalytic activity.

By the choice of starting material, one can even control whether single- or multiwalled nanotubes are generated. Reacting, for instance, benzene with a metallocene catalyst will give rise to multiwalled carbon nanotubes, while the use of acetylene leads to SWNT.

3.3.2.4 Decomposition of Hydrocarbons – Pyrolytic Methods

Multiwalled carbon nanotubes may be prepared from organic precursor molecules. These serve exclusively as carbon source here. The decomposition takes place on a catalyst at high temperatures, and even some methods capable of doing without a catalyst have been developed.

The conversion of acetylene on an iron catalyst on SiO_2-support is a typical example. In this process, acetylene is thermally decomposed by leading it over a bed of catalyst within a quartz tube heated at about 700 °C (500–1000 °C, generally). Apart from the desired MWNT, there are also larger, fibrous structures and layers of amorphous graphene observed. These tend to coat the catalyst particles. The bamboo-like nanotubes (Section 3.3.4) usually obtained from this method are often covered with amorphous carbon too and, in parts, they are considerably curved. In addition to these bent species, there is also a spiral or helical structure

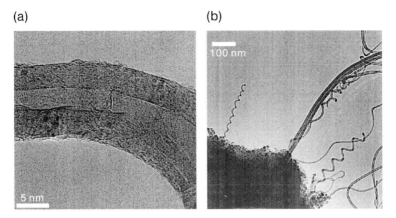

Figure 3.21 (a) MWNT generated by pyrolysis of acetylene on a cobalt catalyst at 720 °C. (b) The tubes are heavily bent and often show helical structures (© ACS 2001).

(Figure 3.21). Most of them bear a comparably wide, cylindrical hollow. It is ca. 10 nm in diameter when using cobalt on a support of graphite as catalyst. The outer diameter of the multiwalled nanotube is about 30 nm. Including the fairly thick layer of amorphous carbon usually observed as a coating, the structure roughly has an overall diameter of 130 nm.

The length of the tubes can range up to many micrometers and depends on the reaction time. The diameter of the MWNT can be controlled as well by coating suitably sized catalyst particles onto the substrate. Generally, the samples obtained are soiled with other structures, mainly with graphite-coated catalyst particles and so-called carbon nanoparticles (refer to Chapter 4).

Carbon nanotubes can also be produced by pyrolyzing camphor at about 900 °C under argon in the presence of an iron catalyst. Using ferrocene for the latter, the procedure yields vertically arranged, parallel MWNTs with diameters of 20–40 nm and a length of up to 200 μm. In comparison to other CVD methods, this approach requires only about 10% of the usual amount of catalyst, and consequently the nanotubes obtained feature much less contamination with metal particles. The oxygen contained in the camphor further causes *in situ* oxidation of the amorphous carbon generated. The yields amount to ca. 90% of the product mixture and the nanotubes produced exhibit a high degree of graphitization.

The chemical dehalogenation of perfluorinated hydrocarbons like perfluorocyclopentene, perfluoronaphthalene, or perfluorodecalin leads to carbon materials, too. At first, unstable carbyne phases are generated, which may be transformed then into nanotubes. The same is true for the decomposition of 1,3,5-hexatriyne that forms nanotubes as well. The yields, however, are rather low, with only 1–2% of the resulting material actually being MWNT. In addition, fullerenes and onion-like carbon particles are obtained.

A method to prepare nanotubes in a fashion similar to carbon fiber production consists in the pyrolysis of carbon-rich polymers. For example, the thermal treatment of polyvinylpyrrolidone (PVP) on a templating membrane of aluminum oxide yields multiwalled carbon nanotubes. These contain, however, a certain portion of nitrogen, and also C–N bonds have been shown.

An interesting procedure combining both the carbon source and the catalyst in one precursor molecule is the thermal decomposition of the iron phthalocyanine complex. The process is conducted at about 950 °C in a stream of argon and hydrogen. The complex disintegrates into its components. The organic ligand is degraded to give atomic carbon or little clusters, while the iron forms elemental particles. From their surface, it is where the growth of the multiwalled, normally bamboo-like nanotubes (Section 3.3.4) originate.

3.3.2.5 Production of Double-Walled Carbon Nanotubes

Double-walled carbon nanotubes are an attractive target for synthetic approaches due to their defined number of tubes. A possible way to prepare them is a variation of the arc discharge method. It employs a graphite electrode about 8 mm in diameter. Similar to SWNT production, it bears a bore hole filled with a catalyst, but compared to SWNT synthesis, this is a far less active material. Normally, Fe_2CoNi_4S is employed, which may be obtained from fusing the elements under argon at 700 °C. As for the inert gas, helium is replaced by argon that is added with hydrogen to facilitate the arc-over. The overall pressure is about 380 Torr/7.35 psi. In any other respect, the parameters are chosen with reference to typical SWNT syntheses by arc discharge.

In this production method, again a cathode deposit and a number of other carbon deposits are formed. While the first is of no great significance here, the latter contain partly large amounts of DWNT. Like in SWNT synthesis, a certain portion is precipitated on the reactor walls (~20% DWNT), and a "collar" evolves on the edges of the cathode (~50% DWNT). Especially rich in double-walled nanotubes, however, is a thick film covering the sides of the cathode. This may bear up to 70% of double-walled nanotubes. Carbon nanoparticles and amorphous carbon are found as by-products, while fullerenes are not observed at all. The DWNT obtained measure 3–5 nm in diameter. The distance of their graphene layers is 0.39 ± 0.02 nm, which is a considerable expansion compared to the usual value. The tubes' diameter can be controlled to some extent by the reaction temperature with higher values favoring larger dimensions. The double-walled tubes are closed and only loosely bundled, and their caps are normally free from catalyst particles.

Besides small carbon clusters generated in the reaction zone anyway, the presence of hydrogen further gives rise to light hydrocarbons that contribute to the deposition of DWNT as well. Hence, in principle, this is a floating catalyst CVD performed *in situ*. It has indeed been applied in a multitude of experiments for the deliberate production of double-walled nanotubes. Normally, acetylene is employed as carbon source because apparently it suits best to surround an existing nanotube with a second layer of amorphous carbon (refer to Section 3.3.6).

The CVD-synthesis of DWNT is conducted at conditions similar to those in the preparation of other nanotubes by deposition from the gas phase. The catalyst is

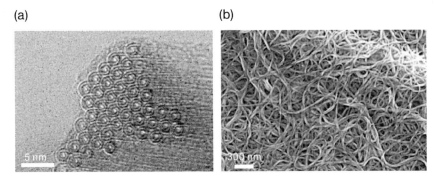

Figure 3.22 (a) Double-walled nanotubes prepared by chemical vapor deposition. (b) The DWNTs form a dense film (© Nature Publishing Group 2005).

either coated onto a substrate or fed directly into the stream of acetylene. Still the techniques commonly applied use an iron–cobalt catalyst. This inevitably gives rise to a mixture of different nanotubes from which the DWNT have to be isolated in tedious purification.

Today, however, another method has been developed that allows for the almost exclusive formation of double-walled nanotubes. It is based on using a co-catalyst that supplies markedly increased amounts of activated carbon (atoms, clusters) per time unit. The limited number of catalyst particles provided, this leads to a part of the carbon depositing on the growing tubes as an amorphous film, which promotes the generation of double-walled tubes. As co-catalyst, molybdenum has proven its worth. It is positioned at the entrance of the oven where it is the first thing a stream of methane and argon (1:1) is led over at 875 °C. Only then will the reactants contact the iron catalyst placed in the center of the oven. In the subsequent purification, metal particles are removed by treating with hydrochloric acid and amorphous carbon by partial oxidation with air. The product is obtained as a solid, black film that contains more than 95% of DWNT. These feature highly uniform diameters and form trigonally packed bundles (Figure 3.22).

3.3.3
Strategies for the Rational Synthesis of Carbon Nanotubes

If the directed preparation of specific single-walled carbon nanotubes succeeded, a multitude of studies would become possible regarding properties which, for the time being, are hardly or insufficiently accessible to examination. Moreover developmental materials for applications inevitably requiring a certain geometry of the tube would become available.

To the present day, however, a rational synthesis of SWNT regrettably has not been realized, yet there are a number of approaches that shall be presented here. As for the first method, it is not a rational synthesis in the true sense of the meaning. Still, the pyrolytic decomposition of the cobaltocarbonyl complex

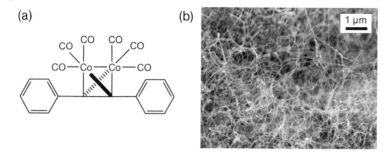

Figure 3.23 (a) Complex of tolane with cobalt carbonyl; (b) MWNT obtained by thermal decomposition of this complex (© K. P. C. Vollhardt).

of tolane (diphenylacetylene) (Figure 3.23) at 650 °C has been one of the first attempts to convert solely organic precursors into nanotubes. On closer inspection, the resulting sooty material turns out to contain considerable amounts of nanotubes.

The particularity of this method consists, on one hand, in applying a comparably low temperature for thermolysis of the starting material, and, on the other hand, in the organocobalt complex itself carrying the catalyst with it. For the latter reason, the metal is present not in the shape of catalytically active nanoscale clusters, but as solitary cobalt atoms promoting the growth of nanotubes. The structure of the tubes prepared by Vollhardt and co-workers using this method coincides with that obtained from other procedures based on the catalytic decomposition of hydrocarbons. They are multiwalled nanotubes which do, however, not represent the exclusive product of the pyrolysis. The resulting soot rather contains further carbon structures and, above all, amorphous carbon.

The choice of the organic starting component largely influences the kind of carbon species generated. This becomes evident, for example, in the pyrolysis of metal complexes of some dehydro[n]annulenes: Apart from the desired nanotubes, a series of other structures like carbon onions, graphite, or amorphous carbon is observed.

It seems advisable to choose for starting material a compound with as many precondensed aromatic rings as possible to promote the formation of larger structural units. This synthetic sequence as well is finalized by pyrolysis of the organic precursors (Figure 3.24). Large, disk-shaped aromatic compounds like hexa-*peri*-hexabenzocoronene can be modified with peripheral alkynyl substituents to enable complex formation with dicobaltocarbonyl. Depending on the temperature, two different nanotube structures are generated upon decomposition: the bamboo-like species described in Section 3.3.4 at about 800 °C, and straight, pervious MWNT above 1000 °C. In most cases the tubes are closed, and catalyst particles are found both in a part of the tubes' ends and in the compartments of the bamboo structure. It is remarkable that the conversion rate of carbon is almost 100% and neither other carbon structures nor amorphous carbon are observed. At an internal diam-

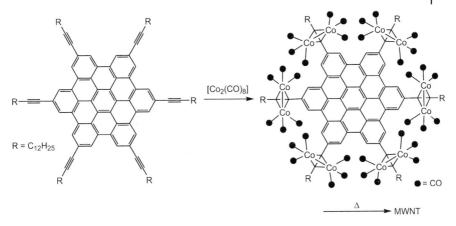

Figure 3.24 Highly condensed aromatic compounds may be converted into multiwalled nanotubes. Complexing eventual triple bonds with dicobalt octacarbonyl promotes this process.

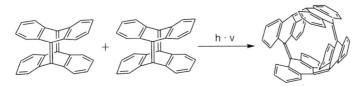

Figure 3.25 Photochemical ring-opening metathesis yielding the fragment of a picotube.

eter of about 15 nm and an outer one of ca. 33 nm, the size distribution of nanotubes produced this way is quite narrow because the catalyst is generated *in situ* and therefore evenly dispersed in the sample. The length of the nanotubes obtained ranges from several hundreds of nanometers to a few micrometers.

Studies on a rational synthesis of small, tubular aromatic compounds may also be considered under the aspect of directed nanotube synthesis. The resulting tubes would have diameters markedly below 1 nm, so they should better be termed "picotubes." Their structure distinctly deviates from cylindrical shape about the ends (Figure 3.25) as the cleavage of the terminal hydrogen atoms is problematic. For the example shown, a [4,4]-armchair nanotube would be formed if it succeeded some day. Besides this structural intricacy, another problem arises with increasing size of the molecules synthesized: The higher the degree of condensation of the aromatic systems, the lower their solubility gets, until finally a reaction in solution becomes practically impossible. Therefore, one has to rely on pyrolytic and/or solid phase methods in these cases.

The template-directed synthesis of nanotubes represents another approach. Suitable precursors (hydrocarbons, polyaromatic compounds, fullerenes, or other nanoscale carbon particles) may be subjected to pyrolysis while situated in pores

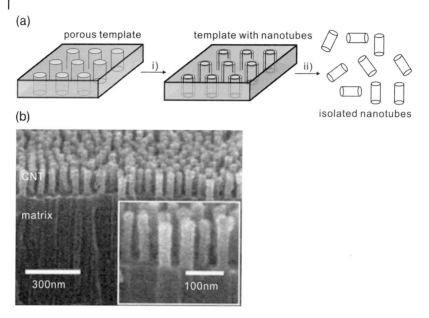

Figure 3.26 (a) Scheme indicating the generation of nanotubes inside a porous template; (b) SEM-image of a zeolitic template with protruding ends of nanotubes. The matrix has partly been removed by etching (© Wiley-VCH 2005).

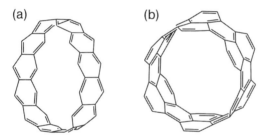

Figure 3.27 Belt-shaped aromatic compounds are equivalent to carbon nanotubes of minimal length. (a) The shortest possible zig-zag nanotube, and (b) hypothetical armchair nanotubes made of pyrene units.

of adequate size (Figure 3.26). The spatial confinement enables the growth of nanotubes adapting in diameter to the template. This concept has already been set in action in the preparation of double-walled nanotubes by coalescing C_{60}-molecules inside the so-called peapods (Section 3.5.6), but porous inorganic materials like zeolites might as well be employed. These offer the additional opportunity of chemically removing the template to isolate the nanotubes generated.

The so-called beltenes constitute an extremely curtailed model of a nanotube. Being belt-like aromatic compounds, they reproduce correctly an essential structural feature of the nanotubes, that is, the annular conjugation along the tube's perimeter (Figure 3.27). With only one benzene ring in width, however, they lack

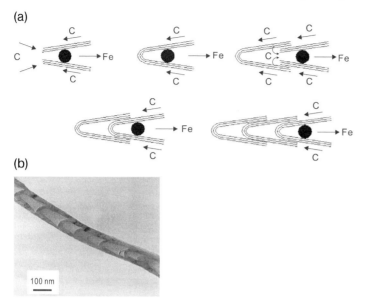

Figure 3.28 Mechanism of formation of bamboo-like nanotubes (a) (© Elsevier 2000) and HRTEM-image of such nanotubes (b) (© Elsevier 2002).

the expansion in parallel to the tubular axis. Therefore the π-system in the z-direction is rather short, which causes the electronic properties to differ dramatically from those of a proper nanotube. Yet if the construction of beltenes from condensed aromatic compounds like pyrene succeeded, a rational synthesis of predefined nanotube fragments would come much closer to reach.

3.3.4
Structure and Production of Further Tubular Carbon Materials

Apart from the hitherto described single-walled and multiwalled carbon nanotubes, there is a variety of further structures with tubular or related geometries. These include bamboo-like and so-called cup-stacked nanotubes as well as nanohorns and helical nanotubes. Like normal SWNT and MWNT they consist of curved graphene layers, but they feature additional structural peculiarities that will be considered in the following sections.

3.3.4.1 Bamboo-Like Carbon Nanotubes
Typical MWNTs consist of several pervious SWNT of different diameters. A rather distinct structure is observed for the so-called bamboo-like nanotubes (Figure 3.28). They contain a core of individual sections enveloped in some pervious nanotubes. Their diameter is roughly 30–120 nm, and the length reaches far into the micrometer range. The walls consist of up to several tens of individual tubes.

The production of bamboo-like structures can be achieved on different ways, for instance, by arc discharge between electrodes of graphite or carbon bearing metal particles (usually iron). The use of carbon instead of graphite causes a slightly varied growth mechanism due to the structure of the starting material: besides atomic carbon, larger fragments may be incorporated into the growing tube as well. This is especially due to the presence of loosely bound polycyclic aromatics and to the absence of an extended lattice in carbon. The growth mechanism of bamboo-like nanotubes from either graphite or carbon electrodes is based on interplay between the tube growing from carbon adsorbed at the catalyst and the movement of the metal particle. The iron is at least partially molten at the high temperature in the arc, and indeed small iron droplets are the catalytically active substance. They are carried along at the tip of the elongating nanotube and have been observed there experimentally (Figure 3.28). The size of the droplets determines the diameter of the emerging tube.

Chemical vapor deposition can also be employed to produce bamboo-like structures. To this end, the catalyst (normally from the iron group; Fe, Co, Ni, or mixtures thereof) is deposited on a substrate and pretreated there by etching. Subsequent catalytic reaction with acetylene at ca. 800 °C yields the desired nanotubes. The finding that no catalyst particles are detected inside the resulting nanotubes calls for a mechanism that is not based on these particles being pushed upward by the tube, but rather on them staying fixed on the substrate surface ("bottom growth"). Finally bamboo-like nanotubes have also been observed as a product in the pyrolysis of polyaromatic compounds (Figure 3.24) in the presence of cobaltocarbonyl.

3.3.4.2 Cup-Stacked Carbon Nanotubes

The foregoing case comprises nanotubes closed to the exterior and provided with a pronounced internal structure. The so-called cup-stacked nanotubes, on the other hand, exhibit a very particular surface. They consist of columnar stacks of hollow, truncated cones (Figure 3.29). This structure entails an open external shell

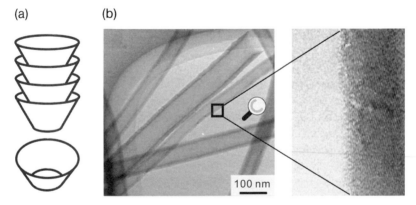

Figure 3.29 Cup-stacked carbon nanotubes consist of individual truncated cones. Model of a cup-stacked CNT (a) and HRTEM-image of these nanotubes (b) (© AIP 2002).

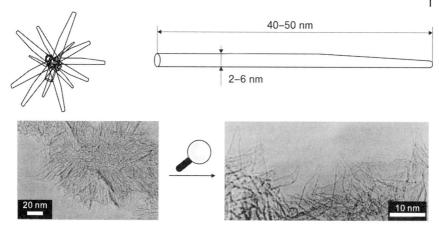

Figure 3.30 Carbon nanohorns often form agglomerates the shape of a dahlia or a sea urchin (© Elsevier 2004). The scheme is to illustrate the dimensions (top).

that should be highly reactive due to the large number of unsaturated bonding sites along the rims of the individual cups. Strictly speaking, this type of tube is not really a carbon nanotube, but rather a superstructure of so-called nanocones. These cone-shaped units of suitably rolled up graphene layers have been shown by electron microscopy, too (Figure 3.29).

The bevel angles of these graphene cones (cups) usually comply with the values given in Table 3.1 for different numbers of pentagon defects at the apex of a conical structure. The individual cups, however, are no closed cones, but, as a rule, they are open structures. The resulting tube consequently features a pervious cavity which, in addition, is in contact with the environment. Its diameter normally measures less than 100 nm, while the tubes themselves have diameters of 50–150 nm and a length of up to 200 μm. The bevel angle of the cone ranges from 40° to 85°, while the length of the cups is influenced by the catalyst particles. The unsaturated bonding sites on the inner and outer walls of cup-stacked nanotubes usually react with the rims of neighboring cups to form loops, or they cause the accumulation of, for example, amorphous carbon on the outer shell.

The preparation is achieved using the floating catalyst method. The precursor of the catalyst (in this case ferrocene or $Fe(CO)_5$ with H_2S as co-catalyst) is vaporized together with the carbon source (natural gas mainly consisting of methane here). The resulting carbon nanofibers then may be graphitized at about 3000 °C.

3.3.4.3 Carbon Nanohorns (SWNH)

Besides the comparatively long species discussed so far, short single-walled carbon nanotubes can be prepared as well (Figure 3.30). Generally they measure 2–6 nm in diameter and about 50 nm in length. A conical tip with an internal angle of ca. 20° is situated on one end. The structure resembles a small container not unlike to an Eppendorf tube. The term *nanohorn* has been established for these single-walled tubes closed on one side. Several of them cluster to form structures that have been compared to dahlia blossoms in the literature. SWNHs are of interest especially for their large inner surface and their very regular structure. What is more, they can be produced with high purity and good yields. The properties of SWNH suggest a series of possible applications. They may, for instance, be filled and can very effectively store, e.g., hydrogen or methane. SWNHs have also been demonstrated to possess catalytic properties, for example for the direct decomposition of methane to hydrogen, and they have further been employed as material for electrodes in fuel cells.

They are produced, for example, by laser ablation from pure graphite in an atmosphere of argon at normal pressure and without a catalyst added. The method is apt to prepare gram amounts of a material that is more than 90% pure. SWNHs are further available from different arc discharge methods. They may, for instance, be generated using an arc welding machine or a pulsed arc. The latter procedure achieves higher purity and a narrower size distribution of SWNH by preheating the graphite electrodes to 1000 °C.

3.3.4.4 Helical Carbon Nanotubes (hMWNT)

It was not long after the discovery of carbon nanotubes before specimens with coiled or helical shapes were observed. These helical nanotubes (*h*MWNT) are often found as by-product in the synthesis of multiwalled nanotubes, yet by now a directed preparation has succeeded as well. Their extraordinary shape turns the helical nanotubes into an interesting subject of research as the expected electronic properties like inductive effect under electrical current should enable an application, for example, as "nanocoils."

Sundry structural variants exist that differ in the way of winding. On the one hand there are objects resembling a microscopic spiral spring. On the other hand, there are those that are more similar to a distorted thread (Figure 3.31). Moreover, structures consisting of several spirally entangled tubes have been reported. The

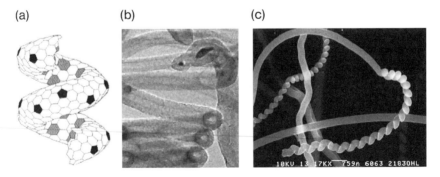

Figure 3.31 Helical nanotubes feature both five- and seven-membered rings (a) (© AIP 2002). There are spiral (b) (© ACS 2003) as well as twisted structures (c) (© APS 1993).

design of hMWNT is derived from that of straight nanotubes. However, it contains defects that give rise to the twisting. As mentioned in the chapters on fullerenes and on the caps of closed nanotubes, the curvature of a graphene layer is caused by the presence of rings with more or less than six carbon atoms. The frequently occurring five-membered rings effect a convex deformation of the graphene layer, while it is a concave one with seven-membered rings. The latter would be required to sit on the inside and the first on the outside to obtain a spiral structure. Hence one may assume to find pentagonal defects mainly on the outside and heptagonal defects chiefly on the inside of the carbon spiral (Figure 3.31). Close inspection of high-resolution electron micrographs of helical nanotubes also reveals that at highly curved positions the inner walls are partly no longer in parallel with the outer wall, but detach from it and buckle to the inside, which indicates the structure to be strained.

The production of hMWNT is possible on different ways. For instance, they are the main product of converting pyridine or toluene in a CVD apparatus in a stream of hydrogen at about 1100 °C in the presence of iron pentacarbonyl. But also the reduction of diethyl ether at ca. 700 °C on zinc yields up to 80% of helical nanotubes.

The mechanism of formation of helical structures is largely unexplored, yet first hypotheses claim that an uneven distribution of active sites on the catalyst surface leads to a different rate of extrusion along the tube's rim. Consequently, a different number of carbon atoms is incorporated, which results in a curvature of the emerging tube. Ellipsoidal catalyst particles, randomly oriented on the substrate, can also cause unlike growth rates on different sides of the tube and thus the formation of a spiral. The incorporation of a C_2-unit on one side of the tube generates pentagon and heptagon defects. These are separated upon insertion of further units and migrate until they reach the "knee" position of the tube (Figure 3.31). The hMWNTs represent a line-up of such "knees."

Apart from the hMWNT, larger helical objects made from carbon have been reported, too. These so-called carbon microcoils (Figure 3.32) as well exhibit a distinctly spiral shape. They are produced by thermal degradation of acetylene at

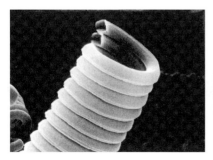

Figure 3.32 The microcoils are larger helical objects consisting of tubular carbon structures (© MRS 2000).

800 °C using traces of a sulfur or phosphorus catalyst. Their diameter ranges from 1 to 10 μm and the length from 100 to 500 μm. Microcoils thus already constitute a transition to the spiral carbon fibers. They can be graphitized by subsequent thermal processing at 3000 °C. It confers to them a herring-bone pattern with extended domains of graphene layers oriented in parallel.

3.3.5
Arrays of Carbon Nanotubes

For a variety of CNT applications, it is desirable to place the growing tubes in parallel on the substrate. There is a number of approaches to this aim that shall be presented in the following.

The requirement may be divided in two subsections to begin with. These are parallel arrangement of the tubes, and the site-selective deposition on the substrate. The formation of so-called arrays (patterns of carbon nanotubes on a two-dimensional matrix) can be achieved, for example, by a modified CVD process. To this end, the catalyst particles are positioned on the surface by lithographic methods or by direct "writing." Selective etching from a continuous catalyst film on a substrate has the same effect. Examples include the deposition of single iron particles on a silicon substrate partially protected by a photoresist or the vapor deposition of a nickel film, followed by plasma etching employing a copper mask. In the subsequent deposition of carbon from the gas phase, nanotubes can only grow where a catalyst particle is situated. This technique allows for the preparation of two-dimensional arrays that contain columns consisting of many individual nanotubes (Figures 3.18b and 3.33).

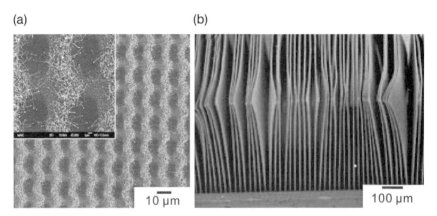

Figure 3.33 Carbon nanotubes may grow in arrays following to the structure of the underlying catalyst layer. Both (a) columnar or (b) lamellar structures can be obtained (within the lamina shown in (b), the individual tubes are aligned in parallel to each other and in perpendicular to the substrate's surface). More complex arrangements are possible, too (© Elsevier 2004, © AAAS 2004).

Structuring the substrate alone, however, does not suffice if parallel alignment of the carbon nanotubes is desired; the orientation must rather be induced by an external effect. The parallel arrangement can be achieved either during or after preparation.

Different external influences may serve to the alignment of existing nanotubes. For instance, the high pressure filtration of an SWNT-colloid under simultaneous application of a magnetic field of 7–25 T yields a membrane of parallel nanotubes. The embedding of single- or multiwalled nanotubes in nematic liquid crystals also leads to a spontaneous alignment of individual tubes. Reorientation of the liquid crystals by means of an electric field then may enforce a rearrangement of the tubes as well. Furthermore, the extrusion of SWNT-dispersion succeeded in oleum with subsequent drying. Positively charged nanotubes are generated here that are surrounded by a layer of sulfate and bisulfate ions. They align in parallel to the flow when extruded from a nozzle. A similar method is employed for spinning a mixture of SWNT and polyvinylpyrrolidone (PVP). The resulting composite materials exhibit a high degree of parallel arrangement of the nanotubes contained. Thermal treatment then removes the polymer while the nanotubes are conserved in their orientation. This technique is apt, for example, to arrange nanotube networks on a substrate, in this way generating, for example, electric conductors (Figure 3.34).

Besides these methods for the alignment of existing nanotubes, there is also a number of procedures to induce a parallel arrangement immediately during the production of the tubes. One way of predetermining a growth direction consists in the application of a template. With this method both single- and multiwalled nanotubes can be produced in an aligned and two-dimensionally arranged manner (Figure 3.26). To this purpose, an organic starting material is decomposed inside the pores of a zeolite or another mesoporous silica or alumina. Carbon sources for the synthesis of nanotubes inside or on zeolites may be hydrocarbons added from the outside as well as organic residues from zeolite production. The resulting nanotubes are forced to grow in a distinct direction due to the porous structure.

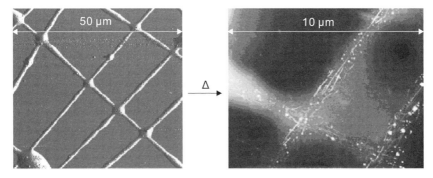

Figure 3.34 Conductor paths consisting of carbon nanotubes can be generated by heating a CNT–polymer composite on a substrate (© ACS 2004).

In some cases, the bundles of nanotubes grow beyond the zeolite's surface and form a pattern that is defined by the pore distribution. An example for this type of MWNT arrangement is found in the chemical vapor deposition of acetylene on anodic alumina with a pore size of ca. 25 nm. The resulting multiwalled nanotubes can be isolated by chemical removal of the template with chromic and phosphoric acid (Figure 3.26).

The directed preparation of DWNT on a support of silicon carbide is feasible by selective evaporation of the silicon atoms in the uppermost layer of the substrate. The carbon atoms released thereby organize themselves in the shape of double-walled nanotubes that grow up vertically on the surface of the substrate. Here the alignment is caused by the close proximity of neighboring tubes that does not allow for a deflection in other directions.

A high density of nucleation centers on a substrate can also be achieved by an accordingly dense arrangement of the catalyst particles in a CVD process. Examples include nanoparticles of iron or cobalt that are precipitated from solution on a silicon support or the pretreatment of the catalyst film with ammonia. Likewise does the thermal decomposition of iron phthalocyanine generate enough iron particles for the subsequent growing nanotubes to interfere with each other, thus forcing them into a vertical orientation.

A word of warning is indicated however: Many of the nanotube samples designated as parallel in the literature may seem like an array of parallel tubes on first sight in the scanning electron microscopy (SEM), but increasing the magnification until individual nanotubes become visible reveals that the alleged parallel arrangement is a fallacy (Figure 3.35). The tubes feature a lot of defects, kinks, and bends, and above all it is their closeness that keeps such arrays from being an unordered network of entangled nanotubes.

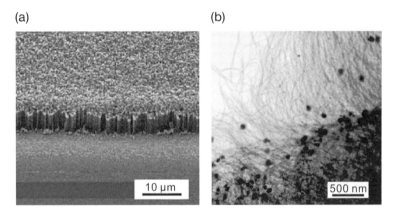

Figure 3.35 Carbon nanotubes largely aligned in parallel obtained by PECVD (a) and detail of a nanotube film (b) with the tubes just seeming parallel for the prevailing crowding. This picture, however, reveals a comparatively unordered internal structure (© Elsevier 2004).

Besides the arrangement enforced by spatial confinement, parallelism of carbon nanotubes can also be achieved by applying an electric field during the growth phase. This concept is realized in the PECVD. The alignment is effected by an electric field between substrate and electrode. The method yields thin MWNT with about four walls that are oriented in parallel to the surface of the substrate which is coated with iron particles.

3.3.6
Purification and Separation of Carbon Nanotubes

3.3.6.1 Removal of Impurities from Carbon Nanotube Materials

Regardless of the method used for their production, carbon nanotubes always contain a number of impurities. The most prominent among them are metallic catalyst particles, amorphous carbon, carbon nanoparticles containing metals, fullerenes, and polyaromatic fragments of graphene sheets. Hence, purification is indicated prior to further examination or use. A variety of methods at least partially fulfilling this requirement has been reported. For the time being, however, it has not been possible to obtain samples completely pure from an analytical point of view, neither of single- nor of multiwalled nanotubes, so any material available on the market contains varying portions of the aforementioned impurities.

Contrary to fullerenes, the carbon nanotubes are structures with much more variable parameters. Consequently, even neat samples will never be analytically pure, but exhibit a certain distribution of nanotubes differing, for example, in length. In comparison to fullerenes, the solubility of carbon nanotubes in common organic or inorganic solvents is markedly reduced due to this latter variety and to the fact that the tubes are very much longer than their diameter. Hence methods of purification based on the solubility of the crude mixture, like high-performance liquid chromatography (HPLC) or extraction, are not applicable to nanotubes in a straightforward way.

Besides the removal of foreign material, there is another crucial aspect in the purification of carbon nanotubes. Any of the procedures presented in Sections 3.3.1 and 3.3.2 yields tubes with different structure indices (n,m), that is, varying in diameter and electronic properties. Considerable efforts thus have been made recently to further separate these mixtures and to isolate nanotubes with uniform diameter, or at least with common electronic properties (Section 3.3.6.4).

A treatment with different mineral acids is usually apt to remove catalyst particles as these normally consist of base metals. Hydrochloric or nitric acid are most frequently used for this purpose (Figure 3.36). The first, however, can only attack on directly accessible metal particles, whereas concentrated nitric acid leads to quite an efficient removal of metal because the oxidizing power of the acid enables it to attack on and even to open the closed tips of nanotubes. This ensures direct contact with the metal particles. Yet at the same time, the acid reacts with other defects in the nanotubes' walls, which generates functional groups containing oxygen, like COOH, C=O, etc., on their surface. Upon prolonged acidic action and

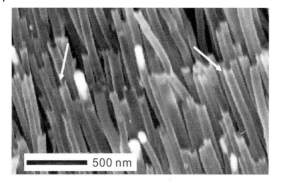

Figure 3.36 Carbon nanotubes opened by an acid treatment (examples indicated by the arrows). Some of the tubes are still closed by catalyst particles (bright ends, © Elsevier 2000).

at high density of defects, the tubes are markedly reduced in length. This process is termed "cutting" of the nanotubes and will be discussed in more detail at the end of this section. With MWNT the acid treatment is easier to perform than with single-walled species as the structure of the latter renders them more sensitive to the acid's oxidizing power. Even short action causes a considerable number of defects in the sidewalls of the SWNT, which is quite unfavorable for some applications. Even complete destruction of the tubular structure may be observed on longer reaction of excessively concentrated mineral acids with SWNT. In MWNT, on the other hand, the defects are mainly generated in the outer wall, leaving the structure as a whole more or less unaffected. The acid treatment can reduce the metal content down to below 1%, depending on the structure of the material purified. The lowest values, however, are only achieved if all nanotube tips have been opened, as a reliable dissolution is possible solely for those metal particles that can make direct contact with the acid through an orifice. Samples still containing closed nanotubes after acid treatment may bear up to 8% (m/m) of metal.

Other methods of oxidative purification include the treatment with super-critical water (hydrothermal procedure), thermal oxidation in air and plasma oxidation in the presence of water. All of them make use of the different reactivity of amorphous carbon and nanotubes, with the latter being more stable than the unordered material. The difference, however, is not very big, and so an extensive removal of amorphous carbon inevitably causes massive losses of nanotubes as well. Consequently the application of oxidizing methods always represents a compromise between complete elimination of impurities and retaining the largest possible amount of product.

The hydrothermal process, that is, the treatment of the sample with water at high pressure and temperature (often in the supercritical range), proved a suitable method for nanotube purification. Water evolves surprising reactivity in redox processes at these conditions. It is then capable of dissolving from the samples both amorphous carbon and metallic impurities (especially in the presence of acids during reaction). Choosing the conditions too harsh, however, will do severe damage to the structure of the tubes, and new variants of carbon will arise from graphitic fragments and residual amorphous material. These products include, for example, onion-like species that deposit on the remaining nanotubes. Very aggressive conditions during hydrothermal treatment (high pressure and temperature) trigger the conversion of single-walled nanotubes into other carbon forms. At 600 °C, mainly MWNT and graphitic particles are formed.

A general problem in oxidative purification is the poor accessibility of impurities situated within bundles of nanotubes, so these tend to remain in the sample. Hence it is important to achieve not only the purification, but also a debundling (Section 3.4.2) of the nanotubes. The oxidative methods further fail in the treatment of carbon nanotubes deposited on a substrate in a parallel alignment, as firstly they cannot be suspended, and secondly the spatial arrangement is sensitive to harsh conditions. Another problem is the cutting of the tubes that occurs at defect positions in strongly oxidizing acids, which causes an even wider length distribution of the samples after purification than before. If the catalyst particles are situated on the tips of the individual tubes, they may be dissolved in an immersion bath. As an alternative, the metals can first be transformed into the respective oxides by oxidation in air. Subsequent treatment with less concentrated acids can dissolve these oxide particles from the sample, which means preserving the nanotube arrays.

Besides the oxidative procedures, also some other methods have been established that can serve to the purification of nanotube samples. These include, for instance, the separation of dispersions by microfiltration. Auxiliary ultrasound may be applied to prevent the nanotubes from clogging the filter by depositing on it. Especially particles with a size smaller than the nanotubes, like amorphous fragments, are removed this way. Chromatographic separations may be suitable to functionalized carbon nanotubes. They are based on the interaction of attached functional groups with the stationary phase or on the different size of particles. It goes without saying that this method is limited to soluble samples. The general insolubility of nanotubes, on the other hand, may even be utilized in the extractive removal of polyaromatic hydrocarbons and fullerenes from the sample. Toluene, for instance, can remove the majority of such impurities.

Usually a combination of different and, possibly, complementary methods is employed for the purification of carbon nanotubes. A typical flow diagram of a routine procedure is shown in Figure 3.37. It comprises several filtration and centrifugation steps, which confers particular significance to the quality of the filter material and to the performance of centrifugation and subsequent decantation. Decanting the supernatant in particular requires much experience. Inappropriate performance may lead to irreproducible results and a marked decrease in sample quality.

3.3.6.2 Evaluating the Purity of Carbon Nanotube Materials

Considering the multitude of methods used to produce and purify nanotubes, the question arises of how to evaluate the quality of the purified materials. There are several analytical procedures, yet each of them features certain drawbacks. What

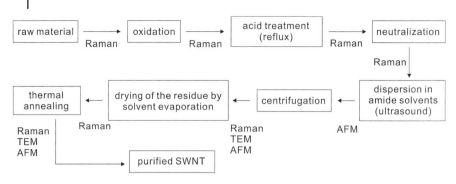

Figure 3.37 Flow chart of carbon nanotube purification. Indicated below the arrows are suitable analytical methods for quality control (© ACS 2004).

is more, the sample quality for the time being is not determined according to common standards, and so a 90% pure material from one source may represent a totally different quality than the 90% product of another supplier. Some analytical methods, however, allow at least for an estimate on the actual quality, and combining several of them gives a good clue of the real degree of purity. Hence an extensive analysis of samples should always be insisted on instead of accepting results from just a few of the techniques presented below.

Electron microscopic methods provide direct proof of nanotubes, of their structure and of the presence or absence of impurities like nanoparticles or metal residues. Evidential pictures of the samples produced can be obtained especially by high-resolution transmission electron microscopy (HRTEM; Figure 3.1). With the sample homogeneity being limited, however, the amount actually examined is too small (on the picogram scale!) to provide a representative overview of a macroscopic amount, even if testing many different samples. Quite the same is true for the SEM that serves to study surfaces. Again only infinitesimal amounts are examined, which renders the method unsuitable to routine quality control. Furthermore, catalyst particles and amorphous carbon are visible to SEM in exceptional cases only. Analysis with this method alone will therefore lead to a significant over-estimation of purity. A sample considered 100% pure on first sight from SEM, for instance, may really bear up to 30% of catalyst that would go undetected on exclusive SEM examination. For samples containing carbon particles and amorphous material next to nanotubes, the quality of the image further depends on the dispersion of the specimen. Samples only dispersed by ultrasound to be smaller aggregates, for example, seem to be much purer than an untreated control, which obviously cannot reflect reality (Figure 3.38). Consequently, the evaluation of sample purity cannot be based on electron microscopic methods alone.

Further techniques must be applied that include larger portions of the material into examination. These may be thermogravimetry as well as spectroscopic methods like infrared or Raman spectroscopy. They provide valuable data to evaluate sample quality and purity, but still they are insufficient to serve as stand-alone procedures.

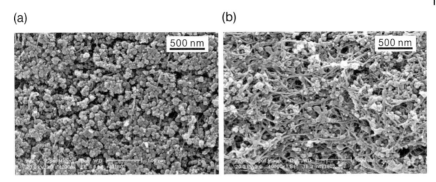

Figure 3.38 SEM-image of a nanotube before (a) and after ultrasonication (b). On first sight an improvement of sample quality seems to be achieved, which, however, does not reflect reality (© ACS 2005).

The thermogravimetric analysis (TGA) provides information especially on the content of metallic catalyst particles in the sample. When performed in air, the carbon species will be oxidized to give volatile gaseous products way below 1000 °C, whereas the metal portion is – in parts or completely – transformed into its oxides and remains in the crucible as such. Analyzing this residue thus allows for quite an accurate assay of the metal in a nanotube sample. It may be up to 30% (m/m) depending on the quality of the material under test. The portion of amorphous carbon, on the other hand, can only be insufficiently determined by TGA.

The spectroscopy in the near infrared (NIR) region has developed into another important method of examining sample purity. It is apt to determine the purity relative to a reference sample that is assigned 100% purity. As such a material actually does not exist by now, the best sample available is employed as reference. The S_{22}-interband transition turned out the signal of choice to be observed because it is less sensitive toward unintentional doping than the S_{11}-transition. The sample is prepared as suspension in N,N-dimethylformamide (DMF), thus ensuring better homogeneity than in a powdery specimen. The ratio V of the area A_{S22} associated to the S_{22}-signal and the integral A_T below the featureless baseline serves as the measure of sample quality (Figure 3.39). The slope of the baseline is attributed to the π-plasmons of both the nanotubes and the amorphous carbon and thus represents the portion of amorphous material in the sample. The observed spectral region is adapted to the position of the S_{22}-signal of the respective sample as the exact wavenumber depends on the nanotube's diameter. Typical values for the band range from 7750 to 11750 cm^{-1}. The value V obtained is then divided by the respective ratio V_{ref} of the reference. The result indicates the relative purity of the sample regarding its portion of amorphous carbon. However, the method is error-prone in the presence of incompletely deaggregated material. With these species an increased light scattering (Mie type) occurs as the particle size ranges about the irradiated wavelength. Therefore again, the sample preparation is crucial to obtain reproducible results.

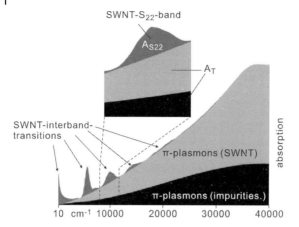

Figure 3.39 Determination of sample purity from the IR-spectrum of a nanotube specimen. The ratio V is calculated from the areas A_{S22} and A_T and divided by the ratio V_{ref} (© ACS 2005).

Raman spectra as well provide information on sample purity. The D- and G-band of the spectrum are analyzed to this purpose. The D-band, situated at about 1200–1400 cm^{-1}, arises from unordered, sp^3-hybridized carbon, whereas the G-band at ca. 1500–1600 cm^{-1} originates from the C–C-stretching modes in the nanotubes (Section 3.4.5.1). Due to the lack of a standard, the integration of the G-band alone does not supply information on the amount of nanotubes present in a sample, but the ratio of D- to G-band suits well to determine the portion of amorphous carbon. Yet the low sensitivity of the D-band causes small areas under its signal, which results in considerable uncertainty and scatter of the values.

3.3.6.3 Cutting of Carbon Nanotubes

The possibility to influence the length of the carbon nanotubes in a sample has been mentioned before in the context of some purification methods. There are several procedures including, among others, oxidative methods ("cutting"). They are of course no real cutting, but a chemical shortening of nanotubes. They make advantage of the increased reactivity at defect sites like holes, kinks, and end-caps that causes the oxidative attack preferably to take place at these positions. Suitable reagents include concentrated nitric acid, mixtures of HNO_3 and H_2SO_4, fluorine or elemental oxygen. Oxidative cutting usually leads to a considerable loss of material (up to 90%). At first, it is indeed the carbon atoms at defect sites being oxidized and attached with groups containing oxygen, but further reaction then leads to the generation of CO_2. Another time-tested reagent besides those mentioned above is the so-called *piranha* water, a 4:1-mixture of concentrated sulfuric acid and 30% hydrogen peroxide. This solution is, however, only able to cut at existing defects, whereas new defects cannot be generated this way. Hence the treatment with *piranha* water at room temperature is suitable to produce short nanotubes without the otherwise inevitable losses of carbon material. The nanotubes obtained have an average length of 500 nm, which usually means a significant reduction of the initial value.

A shortening of carbon nanotubes can also be achieved by mechanical means. Individual nanotubes, on one hand, may be cut very accurately using electron or ion beams, while on the other hand ultrasonication or ball milling can be applied to cut larger amounts. Here as well the energy is introduced via interaction with sidewall defects or with the tubes' reactive termini. The destruction preferably takes place at these positions as they feature reduced mechanical stability. Extending the ultrasound treatment or the milling for too long, however, is at the risk of destroying the tubular structure and producing an unordered carbon material with some nanotube debris. The length of the nanotube fragments obtained by mechanical cutting is a few hundreds of nanometers, too.

Another method to prepare segments of carbon nanotubes consists in embedding the tubes in a polymer matrix (e.g., epoxy resin) and subsequent cutting on an ultramicrotome. Nanotubes measuring less than 100 nm in length can be produced this way. Finally nanotubes with a confined length distribution may be obtained by choosing a certain growth period in controlled CVD (Figure 3.18).

3.3.6.4 Separation of Carbon Nanotubes by their Properties

With the directed synthesis of individual species of carbon nanotubes remaining unfeasible in the foreseeable future, the available mixtures have to be separated for their components before any application requiring specific kinds of tubes. This task turned out rather complex, and a conclusive solution is still pending, but some partial successes have been achieved nevertheless. These at least allow for the enrichment of certain nanotubes.

The nanotubes obtained from purification may largely be free of metallic particles and amorphous carbon, but they still constitute a most heterogeneous mixture of different carbon nanotubes with a wide variety of properties. The characteristics observed on such a mixture always represent an average of the entire sample, but by no means the actual properties of individual tubes. However, the uniformity of the employed carbon nanotubes is a prerequisite to many of their envisaged applications. Considerable efforts are made to separate, for instance, tubes with certain diameters or properties (metallic or semiconducting) from the rest of the sample. Some methods to approach this aim of completely resolving a nanotube mixture will be presented in the following. Carbon nanotubes may be sorted by different criteria, for example, the length of individual tubes, the presence of bundles or single tubes, their diameter, their chirality angle and last, but not least, by their electronic properties. There are various methods of separation depending on the distinctive feature chosen.

A sorting by length, for example, after oxidative cutting of a nanotube sample, can take advantage of the different masses of short and long tubes. Centrifugation and subsequent decanting of the supernatant separate the small tubes still in suspension from the larger ones in the precipitate. However, the removal of the liquid phase requires much experience to avoid stirring up large particles on the one hand, and at the same time, to leave behind as little liquid as possible. Using filters that retain longer nanotubes is another means of separation. Still this process demands a special filter membrane with pore diameters below 1 μm, and single long tubes may pass through the filter even then if oriented in an unfavorable way. Hence a sample will never be completely free from unshortened material. Size exclusion chromatography (SEC) as well suits the separation of carbon nanotubes by length. The stationary phase consists of glass with controlled porosity (controlled pore glass, CPG); a mean pore diameter may be, for example, 300 nm. A surfactant solution of sodium dodecylsulfate (SDS) in water is employed as mobile phase that serves to hold the eluted tubes in stable suspension. This method does not only make a selection by length, but it can also part the tubes from amorphous or nanoparticulate carbon and from residual catalyst as these exhibit dimensions rather distinct from nanotubes. Single tubes and bundles, on the other hand, are not separated, but during chromatography, a partial debundling of the sample on the column is observed.

Compared to the separation by length, a selection by thickness is much more difficult because in general there is no bimodal size distribution, but a continuous range of tube diameters. Macroscopic amounts of carbon nanotubes with defined diameters will hence be available only when a rational synthesis from low-molecular units succeeds. The choice of starting materials would then provide control over the size of the resulting nanotube (Section 3.3.3). Still there are some methods that allow at least for some enrichment of certain diameters. One of them is the diameter-selective oxidation of certain nanotubes with hydrogen peroxide under exposure to light of different wavelengths. Depending on the irradiated energy, chemical attack preferably takes place on a narrow fraction of diameters. Light with a wavelength of 488 nm, for instance, causes the oxidation of tubes measuring 1.2 nm across, while at 514 nm tubes with a diameter of 1.33 nm react. Closer investigation reveals that only semiconducting tubes are attacked by this oxidation, and that the energy of the irradiated light correlates to the S3-bandgap of the species preferably attacked.

The reaction with nitronium salts like NO_2BF_4 or NO_2SbF_6 contrarily prefers metallic nanotubes as these feature a high electron density close to the Fermi level. The nitronium ions intercalate into the nanotube bundles, where charge-transfer from the SWNT to the ions and subsequent selective destruction of the metallic tubes take their course. The nanotube fragments are removed by filtration. The respective residue then contains the semiconducting tubes that did not interact with the nitronium ions. However, the method is reliable for nanotubes with rather small diameters (≤ 1.1 nm) alone as only here the differences between metallic and semiconducting species are big enough to allow for a selection.

Other chemical transformations are suitable as well to discriminate between different types of tubes in a sample. The reaction with diazonium salts, for example, selectively functionalizes metallic nanotubes. Their solubility is enhanced by the surface modification so they may be separated from the unaltered, insoluble semiconducting tubes. Subsequent removal of the functional groups and annealing at elevated temperatures yield nanotubes with most of them being metallic conductors. As metallic nanotubes possess higher electron density close to the

Fermi level, they are in general rather attacked by electrophilic compounds. These reactions include, among others, the addition of carbenes or nitrenes. With the conductivity of the semiconducting species suppressed by a command voltage, metallic nanotubes can selectively be functionalized even on substrates in the electrochemical treatment with aryl diazonium salts. In this setup, only the metallic tubes are able to conduct the required current. The functionalized, former metallic tubes turn into isolators during the reaction, and the resulting sample exhibits purely semiconducting properties. This method is of broad interest, for example, regarding the production of field effect transistors.

Noncovalent interactions with belt- or tube-like host molecules might also be suitable to a separation of carbon nanotubes by diameter. Cyclodextrines or belt-shaped aromatic compounds could be named as examples here. They may not have proven their applicability as selective complexing agent yet, but considering their geometry reveals favorable dispositions for a discriminative interaction with certain nanotubes. Supramolecular arrangements with carbon nanotubes are also discussed in Section 3.5.7.

A very attractive method of separating metallic and semiconducting SWNT has been presented in 2003. Namely, it is the alternating current dielectrophoresis that is characterized by the metallic tubes moving in the medium and forming a parallel pattern on the electrode while the semiconducting species do not move at all (Figure 3.40a). It is crucial though to prepare stable suspensions of the nanotubes

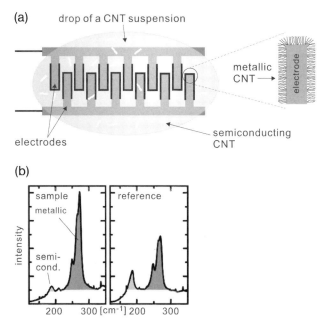

Figure 3.40 Separation of metallic and semiconducting carbon nanotubes by AC dielectrophoresis. (a) Scheme of the installation, and (b) Raman spectrum of a metallic sample in comparison to the starting material as reference (© AAAS 2003).

to be separated with this method, and for the time being, the amounts processed are no more than 100 pg, obtained from 100 ng of crude material. A scrutiny on the tubes reveals that the deposit on the electrodes contains about 80% of metallic nanotubes. The Raman signal of the RBM (radial breathing mode, Section 3.4.5.1) of semiconducting species decreases in the course of the separation, and a change of both the peak shape and height of the signals at high and low wavenumbers is observed for the G-band (Figure 3.40b). There is a significant difference between metallic and semiconducting nanotubes regarding this parameter: Semiconducting tubes give rise to a symmetrical peak with high intensities at high frequencies, whereas with metallic species the G-band is asymmetric and exhibits a balanced intensity in the high- and low-frequency ranges of the band. The reduction of the intensity at large wavenumbers and the increasing asymmetry of the band clearly indicate the enrichment of metallic nanotubes.

Another quite simple method of selectively destroying the metallic tubes consists in using the heat that evolves while passing a strong current through the sample. Semiconducting species do not heat up that much and consequently evade destruction because a given voltage induces a markedly lower current through them. However, this procedure is only applicable to single tubes, but not to bundles, as in the latter case the heating of adjacent metallic individuals would also affect the semiconductors.

Besides separating existing mixtures, it is an explicit objective to prepare nanotubes with defined dimensions in the first place. The template-assisted synthesis can be employed here for large diameters. It consists in nanotubes generated by vapor deposition growing from the pores of a uniformly porous template. Ceramic materials like alumina or silica proved their worth here (Figure 3.26). Yet the method can be applied only as far as sufficiently large cavities exist and the building units are not hampered in diffusion. That is why usually just nanotubes with a diameter of more than 100 nm are produced this way. Another limiting factor is the controlled preparation of the porous matrix, still the family of zeolitic materials provides a wide choice of suitable structures.

3.3.7
The Growth Mechanism of Carbon Nanotubes

The growth mechanisms of carbon nanotubes vary depending on the preparative method applied. Some of them have not been fully elucidated to the present day. Yet a detailed knowledge of this mechanism is crucial for the directed synthesis of certain types of nanotubes to establish optimal conditions for a controlled growth. Some of the popular growth mechanisms and hypotheses concerning them are presented below.

3.3.7.1 Arc Discharge Methods
One of the first syntheses of carbon nanotubes succeeded by using an arc discharge. The mechanism of this growth type, however, turned out a complex system of single processes that can only be co-ordinated by controlling various parameters

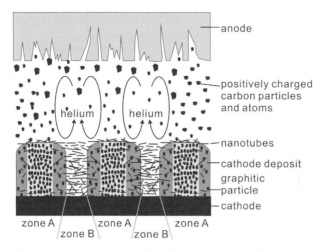

Figure 3.41 Zone structure of the deposit on the cathode in the arc discharge preparation of carbon nanotubes.

like pressure, temperature, arc intensity, anode composition or reactor geometry. Their individual impacts are partly unknown in detail. Successful syntheses therefore originate from empirically optimized setups. In general a deposit is formed on the surface of the cathode. It will exhibit a core–shell constitution as described in Section 3.3.2.1 if suitable reaction conditions are chosen (Figure 3.41). This structure is important for nanotube growth because a kind of reaction cycle can evolve in this way: protruding from the column (zone A) there are a few nanotubes that have been generated at first. Under the voltage applied they emit electrons into the plasma (field emission). The electrons ionize carbon clusters or atoms, giving rise to a flux of carbon ions above zone A. At the same time the helium is sucked in and pushed aside. This induces a circulation that keeps up a constant supply of carbon atoms and clusters that attach to and elongate the pieces of nanotubes already formed.

These tubes are found mainly in zone B in between the columnar structures. The growth mechanism itself differs depending on whether the process is performed with or without catalyst (e.g., scooter mechanism or lip–lip interaction; see below and Figures 3.11 and 3.43). MWNTs are formed without catalysis by transition metals, whereas for the production of single-walled tubes a catalyst is needed. Nanotubes already formed tend to revaporize from the columns because the system heats up strongly. Hence an efficient electrode cooling is required. From the anode, on the other hand, tiny fragments of graphitic material vaporize. Normally these crystallites are a few nanometers large and positively charged. For the latter reason they move toward the cathode and become degraded down to small C-clusters (usually C_3) and carbon atoms while passing the plasma. However, this process is only successful up to a certain size limit for the crystallites: above this value, the time of passage through the plasma does no longer suffice to allow for complete degradation. The "surviving" particles will then be incorporated into the cathode deposit, thus diminishing its quality. Atomization is the limiting step in graphite vaporization and deposition and restrains the maximum possible yield of MWNT to about 30%.

In the synthesis of single-walled nanotubes, on the other hand, it is not the vaporization of starting material, but the diffusion of carbon units through the catalyst that is found to be the rate-determining step. The formation of SWNT advances according to the so-called dissolution–precipitation model (DP-model) that postulates a sequence of three steps as discussed below:

At first, carbon is solved in the surface of heated catalyst metal. The solution of carbon in metals is a common phenomenon with steel being just one example – here a certain admixture of carbon reduces the brittleness of the product. Solving carbon in elements from the iron group is a strongly exothermic process, which causes the catalyst particles to heat up (up to 1300 °C) and to be molten even in some centimeters' distance from the arc itself. This is also why the zone of nanotube growth extends far into the reactor during the production of SWNT. With some ten nanometers, the catalyst particles are distinctly larger than the nanotubes obtained. Therefore, no residual catalyst is normally found inside SWNT from this procedure. The second phase in the DP-model is the diffusion of carbon species to other sites on the catalyst surface. Usually there are so-called hot spots, that is, exceptionally reactive centers on the metal particle, where nanotubes preferably grow.

In the third step, condensation of the fluid carbon material occurs – in this case as single-walled carbon nanotube. The process is endothermal (about $40\,kJ\,mol^{-1}$), which gives rise to a temperature gradient between the reactive zones and the remaining surface. It is, however, not steep enough to induce a directed transport of carbon material toward the sites of consumption. Indeed, it is rather a concentration gradient at work here. The continuous removal of carbon from the nucleation centers causes a depletion at these sites that is compensated by diffusion (phase 2). Normally, this step is rate-determining. Still the best growth rates are achieved with all three steps proceeding at about the same speed, so a pseudostationary state can evolve.

The caps of the individual tubes are first thing that is formed in the precipitation of the SWNT. A structure consisting of five- and six-membered rings is generated first. Ideally it contains six five-membered rings, which results in the formation of a hemisphere. Provided a very high temperature in the reaction zone, large amounts of carbon will then be supplied by diffusion. A relative excess of six-membered rings arises from this feed. With the number of five-membered rings remaining constant, the caps will inevitably acquire larger diameters and thus ensues the formation of thicker SWNT at higher temperatures. Further growth then takes place at the rim of the cap which, in the process, is pushed upward by the carbon newly incorporated into the structure.

Certain nanotubes have been observed to form more frequently than others in the arc discharge process. In general, armchair-SWNTs grow better than zig-zag or chiral species. On one hand, this is due to the extraordinary thermodynamic stability of, for example, (10,10)-nanotubes, but kinetic reasons contribute as well to the effect. In achiral nanotubes, and especially in those of the armchair type, the replacement of metal atoms by carbon is much easier due to the orientation of the lattice structure (see below). What is more, several SWNTs at a time will emerge from very reactive sites. With the prevailing temperature being constant for all tubes nucleating in these zones, their respective diameters will also be more or less the same. Consequently, they may form a symmetric packing, which is why the most stable bundles are observed in these cases.

The preparation of DWNTs in an arc as well features some particularities. With the use of argon as an inert gas, considerably higher voltages are required to establish a stable arc discharge. The breakdown can be facilitated by an admixture of hydrogen which, however, also serves to another purpose: In close vicinity to the arc's plasma zone, particulate graphite and carbon atoms from anode vaporization are converted into small hydrocarbons like acetylene or ethylene. Their concentration is further augmented by the hydrogen attacking the electrodes' surface. Convection and other processes transport them from the immediate arc zone to deposit them at different sites where they decompose again into the elements. This process requires a certain amount of energy. It further turned out that polyynes play a central role in the transfer of carbon atoms.

The actual growth of DWNT takes place on the existing catalyst particles. They determine the diameter of the resulting tubes. Normally, they are much smaller than those used in SWNT preparation. Hence, in comparison to the latter, the diffusion rate inside the catalyst particles must markedly be reduced to avoid an oversupply of carbon at the reaction site. Otherwise an uncontrolled nucleation of nanotubes and related structures would occur. The catalysts actually employed contain a certain amount of sulfur and exhibit much lower reactivity than the classical catalysts for SWNT synthesis. Fe_2CoNi_4S is a typical material of this kind. Initially, single-walled nanotubes are generated on the catalyst particles. Due to their limited reactivity, however, not all of the carbon available can form single-walled tubes, but a considerable portion remains as amorphous material. This can precipitate on the outer surface of already existing SWNT. Starting from there, the catalyst converts the unordered carbon into a second, concentric nanotube that encloses the first (Figure 3.42). The inner tube is no longer affected because the thermodynamic stability of amorphous carbon is distinctly lower than that of a firmly built tube.

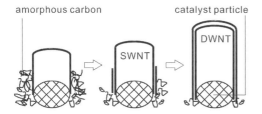

Figure 3.42 Growth mechanism for double-walled carbon nanotubes (DWNT).

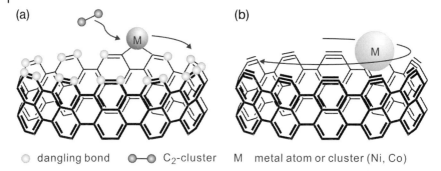

○ dangling bond   ⊙—⊙ C$_2$-cluster   M metal atom or cluster (Ni, Co)

Figure 3.43 Metal-catalyzed growth on the open end of a nanotube according to the scooter mechanism: (a) insertion of C$_2$-units into carbon–metal bonds, (b) complexing of the metal atom by a cyclic polyyne.

The question remains now for all types of carbon nanotubes whether they grow on their closed or their open end. There is wide consensus, especially regarding the arc discharge method with its high temperatures, that the growth of MWNT takes place on an open end. This would consequently exhibit a multitude of unsaturated bonds on each of its walls. However, these would be energetically unfavorable even under the prevailing conditions. Neighboring walls are therefore considered to interact via bridging carbon atoms. This mechanism, which is also termed as "lip–lip" interaction, markedly reduces the number dangling bonds, but at the same time allows for a facile insertion of further building blocks for tube growth. Figure 3.11 shows a model of this type of tube termination.

Single-walled nanotubes as well demand an explanation of the atomic level of how the tubes' growth and the closure with a tip proceed. The interplay between metal catalyst and the carbon atoms involved plays a major role in the considerations on this process. At sufficiently high temperatures, the catalyst exists in the shape of atoms or small clusters. These are situated at the open end of the growing tubes. Now there are different models of how the attachment is realized in detail. One theory is that metal atoms take the place of carbon atoms and that growth is achieved by the insertion of C$_2$-moieties or larger fragments from the gas phase. Due to the continuous incorporation of carbon, the position of the metal atoms is not static, but they rather move back and forth along the rim of the nanotube. Hence this way of growth has been named "scooter" mechanism (Figure 3.43a). The metal further causes the tubes' ends to remain open for instance by inducing rearrangements. These may correct defects like five-membered rings that would otherwise lead to an early closure of the orifice. Now with growth advancing, and provided the continuous supply of catalyst atoms from the electrodes, further metal atoms concentrate on the tubes' ends. They finally aggregate and coalesce to be small clusters. From a certain critical size on, the bonding strength with the carbon atoms becomes insufficient and the metal particle peels from the tube's tip. From this moment on, newly formed five-membered rings will no longer be corrected. The tubes' ends are soon to close and growth ceases. This kind of mechanism also explains why SWNTs from arc discharge synthesis do not contain catalyst particles. Small amounts of metal finely dispersed on an atomic level, however, are not detectable with the analytical methods commonly applied and might indeed be present.

A second hypothesis on the attachment of metal atoms postulates the existence of a cyclic polyyne on the rim of the growing nanotube where the metal atom moves along (Figure 3.43b). Yet this kind of structure is logical only for armchair nanotubes as they alone can form a respective polyyne without building up too much strain.

3.3.7.2 CVD-Methods

The mechanism of nanotube formation in chemical vapor deposition features characteristics rather distinct from those found for the synthesis by arc discharge or laser ablation. Contrary to the latter, a solution of small carbon clusters in and subsequent diffusion through catalyst particles play a minor role in the deposition from the gas phase. The employed hydrocarbons decompose directly on the surface of the catalytic particle. The carbon, therefore, becomes immediately available for nanotube growth.

The growth takes place in several steps that comprise the following processes:

- diffusion of starting molecules through the boundary layer of the catalyst particle;
- adsorption of the reactive species to the catalyst surface;
- formation of elemental carbon and gaseous by-products as well as growth of the tube;
- desorption of the gaseous by-products;
- transport of the gaseous by-products through the boundary layer.

Depending on the type of CVD-method employed, the individual steps contribute to the process to varying extents. In thermal CVD, the starting compound alone constitutes the reactive species, furthermore only the carrier gas and/or H_2 are found. In PECVD, on the other hand, various radicals, higher hydrocarbons, atomic hydrogen, and different ions have been detected as well.

In situ studies of the growth mechanism performed during the deposition of nanotubes from the gas phase revealed that carbon preferably attaches to steps on the crystal surface of the catalyst particles, thus initiating the formation of graphene layers. The continually changing shape of the metal particle also contributes to this process as it is generating new active centers all the time.

Two growth models are postulated depending on whether or not catalyst particles are found in the tips of the tubes obtained. In the so-called tip-growth, the metal is situated at the nanotubes' tip, whereas in the "base-" or "bottom-growth" it remains stuck to the substrate. Experimental examples have been observed for both mechanisms. The tip-growth variant ensues from the catalyst being pushed upward by the growing tube. This may only occur upon rather weak adhesion of the metal particle to the substrate, and its size must not exceed a certain limit

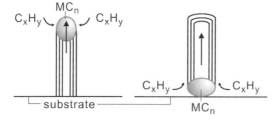

Figure 3.44 Different growth mechanisms for the deposition of carbon nanotubes from the gas phase: tip growth (the catalyst is moving upward) or bottom growth (the catalyst remains on the substrate).

either (Figure 3.44). With the catalyst–substrate interaction being stronger and the particles being bigger, the latter are not pushed up, but remain at the bottom of the emerging tubes. After detachment from the catalyst, the nanotubes produced this way are virtually free of metallic impurities.

In both cases, the decomposition of the carbon source (methane, acetylene, etc.) takes place directly on the catalyst surface. The released carbon partly diffuses into the metal to form the actual catalytically active species. Upon saturation of the metal, further carbon is segregated in the shape of nanotubes.

In an overall view, many different parameters markedly influence nanotube growth in chemical vapor deposition. They include, among others, the size and shape of the catalyst particles, the metal's ability to form carbides, or the question of whether the carbon diffuses on the surface or through the bulk phase of the catalyst particle. Only a deeper understanding of the interplay between these factors will enable the reproducible production of a specific type of nanotube.

3.4
Physical Properties

3.4.1
General Considerations

Extraordinary properties of the carbon nanotubes had been postulated soon after their discovery already, and they are still a major motivation to study these by now wide-spread materials.

The degree to which a certain characteristic is expressed depends considerably on the structure of the nanotube in question. The following chapters will discuss the differences between single- and multiwalled species as well as the influence of the individual tubes' geometry on their electronic and spectroscopic properties.

According to Section 3.2, the nanotube structure can entirely be deduced from the two-dimensional graphite by symmetry considerations and the application of geometrical operations. The same is true now for the physical properties of the tubes. This finding supports the aforementioned hypothesis that, strictly spoken, nanotubes are no modification of carbon of their own. The properties and values observed, however, differ from those of graphite to an extent that justifies the separate examination of their electronic, mechanical, and chemical behavior nevertheless.

Some extraordinary features have been observed in these studies. Single tubes and bundles, for example, differ significantly in their electronic properties, but also different types of SWNTs among themselves show rather individual characteristics. For instance, there are semiconducting and metallic nanotubes, different bands are observed in Raman spectra depending on the tubes' diameter, etc. This chapter will deal with these and a number of further properties.

3.4.2
Solubility and Debundling of Carbon Nanotubes

As with the fullerenes, the solubility of carbon nanotubes is an essential prerequisite for studying their properties. Numerous efforts have thus been made to progress in this field. Yet the size and structure of the nanotubes are a severe impediment to solubilization both in aqueous and organic media. Firstly, they are very big objects already, with their length ranging in the micrometer scale. Consequently, the formation of a classic solution is rather unlikely, and any actual dissolution should better be considered a colloidal system or a dispersion. However, there are other effects besides their length that contribute to their insolubility.

Individual nanotubes approaching one another may establish $\pi-\pi$-interactions. These are very pronounced as the π-electrons in the bent graphene layers are easily polarized. The exchange can take place over the whole length of the tubes. Hence the single, small energetic contributions may sum up to become a large factor favoring bundle formation. The interactions are all the stronger the more uniform the tubes in a sample are. SWNT with equal diameters can easily form a two-dimensional trigonal packing that causes all tubes of the bundle to take a parallel alignment over considerable distance. The interaction forces become maximal this way. The phenomenon becomes obvious in electron micrographs (Figure 3.14b). Hence, single-walled tubes are much harder to debundle completely and to be brought in solution or dispersion than MWNTs because the latter usually have a wider distribution of diameters. The existence of amorphous carbon in the sample further promotes bundle formation. The unordered material normally adheres to the outer wall of single tubes. If these are the outer individuals of a bundle, the structure is shielded by the amorphous coating, thus increasing the resistance to a chemical attack for tube separation. In addition, the hydrophobic character of nanotubes takes effect especially in polar, hydrophilic solvents. The stability of the bundles is even increased here as the interaction with the hydrophilic environment would be much less favorable from an energetic point of view.

There are several strategies nevertheless to obtain solutions or at least stable suspensions of single- and multiwalled nanotubes. The most important ones are presented below.

Pure carbon nanotubes do not solve in any of the common organic or inorganic solvents, only the preparation of fairly stable suspensions succeeds by the application of ultrasound. This method may also cause a first partial debundling. Solvents with amide or amino functions, like DMF or N-methyl-2-pyrrolidone (NMP) proved their worth for the dispersion with ultrasound. The free electron pairs of the nitrogen can interact with the π-system of the tubes. This enables an at least partial dissolution of the bundles, which increases dispersibility. A true solution is, however, not achieved – slow sedimentation sets in after ending the ultrasound treatment. Still this simple procedure is sufficient for many investigations or for a homogeneous mingling of nanotubes with, for example, polymers. The action principle of ultrasound is based on the supply with mechanical energy. The reactive positions of the tubes, that is, their defects, are attacked first. To some extent, a shortening of the tubes occurs (Section 3.3.6.3), and the bundles are partially separated. Moreover, the amorphous carbon present in the sample is dissolved by the ultrasound treatment. It is less stable due to its smaller particle size and its uneven, partially unsaturated edges. Hence, it is destroyed and dispersed faster than the carbon nanotubes themselves. Similar observations are made for the reaction with oxidizing substances: Here as well the distinctly less stable unordered material is attacked first.

The simplest, and at the same time the most effective way to achieve dispersion in water or aqueous media is the addition of amphiphilic molecules: carbon nanotubes without any functional groups are hydrophobic. This characteristic may be employed to enclose them in micelles. The surfactant molecules arrange around the single tubes with their hydrophilic head directed outward and their hydrophobic tail oriented toward the nanotube in the center. In this manner, each individual tube is surrounded by an envelop that enables a dispersion, for example, in water. Yet complete solubilization requires considerable amounts of the surfactant. "Solutions" obtained this way may in fact contain up to 80% of detergent and only 20% of carbon nanotubes. Surfactants suitable to enclose nanotubes in micelles include, for example, sodium dodecylsulfate (SDS), Triton X-100, or octadecyltrimethylammonium bromide (OTAB).

Choosing a block-copolymer with both hydrophilic and hydrophobic domains as the amphiphilic substance, the application of suitable linking methods can lead to the formation of an irreversibly closed capsule around every single nanotube. The reaction with a diamine linker may for instance initiate a crosslinking in the polyacrylic acid domains of an amphiphilic polystyrene/polyacrylic acid copolymer. Upon addition of water to a solution in DMF the latter form micelles with the nanotubes in the center. Contrary to the simple micelles mentioned in the previous paragraph, the crosslinked species can be dried and redispersed.

Another strategy is pursued in the direct functionalization of nanotubes. The solubility of the carbon material in different solvents can be controlled by the attachment of various functional groups. In Section 3.5, numerous examples for a solubility enhancement of single- and multiwalled carbon nanotubes are described. A rather simple functionalization usually occurs already in the first steps of purification. The reaction with concentrated mineral acids does not only remove amorphous carbon and catalyst particles, but it also generates functional groups on the tubes' surface. After the acid treatment, especially polar groups containing oxygen are found. They increase the solubility or dispersibility in polar solvents. A higher concentration of functionalities intensifies the interaction with the solvent and consequently increases solubility. Therefore, an attachment of the highest possible number of functional groups should be aimed for. Conducting this process with concentrated nitric acid does markedly increase this number, yet the reaction may also do damage to the sidewall of the nanotubes, especially upon prolonged duration of action. However, these defects can be cured by thermal annealing at about 1000 °C. This is clearly indicated by a decreasing intensity of the Raman D-band (Section 3.4.5.1), which shows more dominantly after the acid treatment. Besides nitric acid, there is a series of further oxidizing agents that suit to the attachment of groups containing oxygen to nanotube walls. The variety of applicable system may be outlined just exemplarily by a mixture of sulfuric acid, $(NH_4)_2S_2O_8$ and P_2O_5. The elemental composition of the sample after reaction (C:O:H = 2.7:1.0:1.2) proves a significant increase in the oxygen portion. With the surface concentration of carboxyl groups being very high, nanotubes of this kind can form hydrogels in aqueous solutions concentrated as low as 0.3%. The gel network is held together by hydrogen bonds between the functional groups of different tubes. However, a certain part of the tubes is attacked severely in the oxidation. Their tubular structure is destroyed and unordered material is generated that normally deposits on the surface of the remaining nanotubes.

The solubility in aqueous media may be further increased by linking additional polar chains to the primary functional groups, especially to –COOH. Suitable reagents may be ethylene glycols of various chain lengths or alkyl chains with terminal carboxyl groups, but also peptides and proteins that enhance solubility especially in physiological media. Concanavalin A, although not linked covalently, is an example of a peptide that can adhere to nanotubes. It features a hydrophilic exterior and a hydrophobic pocket, and obviously it is able to arrange on the nanotubes so the latter interact with the pouch, while the exterior is in exchange with the solvent. Concanavalin A itself forms a tetrameric structure on the nanotubes' surface.

Another strategy suggests itself if the solubility in rather unpolar, organic solvents is to be increased. Long alkyl chains or aromatic compounds are suitable here to promote solvent interaction. The attachment of solubility-enhancing substituents is achieved, for example, by amide-bonding to the carboxyl groups of acid-treated nanotubes. The fluorination of the nanotube surface (Section 3.5.4.1) increases the solubility in organic media to a similar extent as unpolar alkyl chains do. The direct linking of structures bearing alkyl chains is possible as well. Examples include the 1,3-cycloaddition of azomethine ylides to nanotube sidewalls (Section 3.5.4.1).

The noncovalent attachment of solubility-enhancing groups is another alternative to obtain stable suspensions or solutions of carbon nanotubes. The wrapping with various polymeric substances serves especially well to this purpose. The formation of noncovalent conjugates with carbon nanotubes is discussed in detail in Section 3.5.4.2.

Quite often the polymer Nafion (a perfluorosulfonated material) is employed to disperse single- or multiwalled nanotubes. Usually less than 5% (w/w) of polymer in a solution suffice to obtain a stable solution. However, a debundling is not effected, but the bundles are enveloped and solubilized as a whole.

It would surely be attractive for a variety of examinations and applications to avoid bundle formation immediately at production. As for the time being this has not succeeded, a postprocessing is required in all cases. This is aimed at removing impurities from production (e.g., catalyst) as well as at dissolving the bundles of nanotubes. These efforts result in stable solutions of carbon nanotubes. Still the obtainable concentrations are not very high and range about a few milligrams per milliliter.

The flow chart of a typical purification and debundling procedure of carbon nanotubes is shown in Figure 3.37. It is evident there that only combining several of the methods presented here can lead to success.

The question remains of how to control the success of debundling a nanotube sample. The suitability of electron microscopic methods is limited both for the reasons mentioned in the section on purity testing and due to the way of sample preparation. Alternatives like Raman- or XPS-spectroscopy do not yield unambiguous data either. It is only measurements of fluorescence that allow for a reliable assessment of sample agglomeration. The method takes advantage of the fact that only isolated tubes show bandgap fluorescence. It is observed in the infrared region at about 1060 nm when exciting the sample with red or infrared light (e.g., 785 nm for 5,6-nanotubes). Any eventual bundles effectively quench fluorescence because radiationless deactivation occurs in interactions between neighboring tubes. A fluorescence of the solution therefore indicates that at least a certain portion of the nanotubes is isolated.

3.4.3
Mechanical Properties of Carbon Nanotubes

An early motivation to study carbon nanotubes has been the expectance of extraordinary mechanical properties for both single- and multiwalled species. The perspective to obtain highly strain-resistant composites or fibrous materials liberated large resources in nanotube exploration. At the peak of nanotube euphoria, the catchphrase of the so-called sky elevator created a sensation in the media and indicated what great expectations were set in the newly discovered material. The concept was based on the fact that the tensile strength of carbon nanotubes is much higher than required to carry their own weight. Contrary to a steel rope of equal diameter, a fiber made from nanotubes would therefore not collapse under its own weight once a certain length was exceeded. This imaginary experiment is an exaggeration of course, and the actual setup would fail for several practical reasons, but the remarkable strength of carbon nanotubes stands true nevertheless.

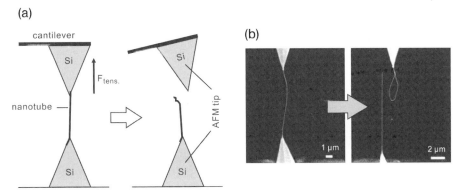

Figure 3.45 Measuring the mechanical properties of a single nanotube by means of two atomic force microscopy tips; scheme (a) and microscope image (b). The torn and markedly elongated nanotube is clearly to be seen (© ACS 2000).

When the first suitable samples became available, measurements revealed stupendous values for Young's modulus and tensile strength indeed. A "superelastic" nanotube has been reported just recently. It stood an elongation by 280% without the tubular structure collapsing.

Exact measurements turned out difficult, however, as the manipulation of such tiny objects led to experimental problems. The performance of tensile tests, for example, requires a device to clamp objects just a few nanometers thick. The employment of two atomic force microscopy (AFM) tips movable against each other proved its worth here (Figure 3.45).

Measurements of the Young modulus (modulus of elasticity) of single-walled carbon nanotubes yielded variable values between 0.4 and 4.15 TPa, which is indicative of the difficulties to obtain exact results. The majority of values, however, has been found to be something about 1.0–1.25 TPa, which is the highest figure ever reported for any material. This strength correlates to the $C_{||}$-constant of elasticity in parallel to the basal plane of graphite. The results for the modulus of elasticity depend on the diameter; 1.4 TPa, for instance, were measured for a tube 1 nm thick, whereas it has been 0.7 TPa for a 2 nm tube. Furthermore, the *Young* modulus was found to depend strongly on the quality of the nanotubes under examination. SWNTs from the catalytic decomposition of acetylene, for example, bear many defects and actually exhibit values of about 50 GPa only.

The modulus of elasticity has been determined for multiwalled carbon nanotubes as well. The results range about 1.0–1.3 TPa, which is even slightly above those for single-walled tubes. This strength originates from the strongest SWNT within the respective MWNT and from a small additional contribution of the van der Waals interaction between the individual tubes. However, this is valid only for measurements on single MWNTs that are clamped on both ends.

The mechanical properties of nanotube composites are easier to determine because these can be studied in larger dimensions. Sufficient adherence of composite matrix (e.g., a polymer) and nanotube provided, a modulus of elasticity of 45 GPa may be attained. For an MWNT, not only the wetting of the tube with the polymer, but also the additional possibility of its constituent tubes gliding telescope-like one inside another must be considered. The latter effect reduces the material's rigidity. In SWNT, on the other hand, the stress acts directly on the C–C-bond that collapses under much larger forces only. Hence the matrix interaction is decisive here for composite stability (Section 3.5.5).

Numerous calculations on the mechanical properties of single-walled nanotubes have been performed as well. For a tube with a constant bond length of 1.466 Å, that is, for a completely delocalized π-electrons, the modulus of elasticity was calculated to be 0.764 TPa. Likewise a Poisson's ratio of 0.32 has been obtained. These values show that in comparison to other materials, carbon nanotubes are very sturdy. This stability is based on the absence of a conventional crystal lattice, so contrary to crystalline materials, there are no preferred cleavage planes and all C–C-bonds are stressed to the same extent. It is worth mentioning in this context that due to their structure, carbon nanotubes possess a kind of dual character: They are molecules with clearly defined dimensions on the one hand, but on the other they exhibit translational symmetry over a distance that is sufficient for them to show crystal-like properties in parallel to the tubular axis. Clearly distinct properties regarding different directions in space are characteristic for these structures with their large ratio of length-to-diameter. Tensile strain acting along the tube's axis is absorbed by the elasticity of the tube up to the point when the breaking of C–C-bonds leads to the collapse of the structure. Forces acting in perpendicular to the nanotube cause bending stress, especially when working close to the tube's ends. An overload will cause a collapse by the formation of kink defects, which is a totally different mechanism of stress relief. Multiwalled nanotubes are normally more rigid than SWNT, but strength decreases with growing diameters. The mechanism of deformation under excessive strain changes with increasing load. For larger MWNT, the unordered bending at the site of highest strain turns into a system of wave-shaped defects on the concave side of the curve (Figure 3.46a).

A peculiarity of nanotubes under tensile stress shows especially at very high loads. At first, in the Hookeian range, the answer of the system is perfectly elastic and reversible. Exceeding this range, however, will lead to irreversible changes to the tube's structure. The diameter is decreased by the migration of C-atoms or defects, which lastingly affects the chirality and the electronic properties of the tube under examination. Metallic nanotubes may become semiconducting, and a tube structure with varying diameters is formed (Figure 3.46b). It may bear domains of different helicity. The whole effect results from the formation of pentagon–heptagon defects that tend to occur pair-wise (Stone–Wales defects). This leads either to immediate breaking of the bonds under highest strain, or they "glide" apart on the nanotube structure to relieve strain this way (Figure 3.46b). The latter alternative causes a difference of tube diameters to both sides of the defect, so finally there will be two adjacent sections with different helicity. The result of this deformation depends largely on the chirality of the starting material. Then it is only upon extreme loads that the tubular structure actually collapses. In

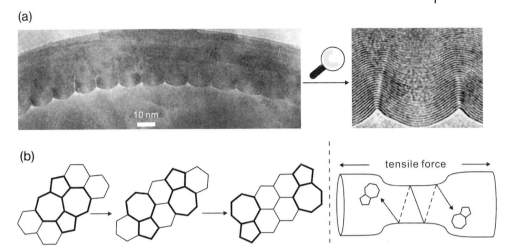

Figure 3.46 The effect of mechanical strain on carbon nanotubes: (a) evolution of undulated defects on the concave side of a bent MWNT, (b) migration of Stone–Wales defects upon tensile stress (© AAAS 1999).

Figure 3.47 Carbon nanotubes with very large diameters flatten on a substrate, while smaller tubes largely keep their circular cross-section.

a transition state before ultimate rupture, one or more threads of carbon atoms are observed between the separating units.

Carbon nanotubes exposed to mechanical stress are generally observed to give an electronic response in the manner of altered conductivity and density of states. This feature may be employed, for example, to construct electromechanical systems that can be switched by mechanical stress. The changes under tensile or bending stress, respectively, have quite different causes. In the case of tensile strain, changes of the structural parameters n and m and of the associated diameter and conductivity are measured (refer to Section 3.4.4 on the electric conductivity of different nanotubes). Bending stress, on the other hand, leads to an increasing rehybridization from sp^2 to sp^3 at the newly arising defect sites. The consequently reduced delocalization of π-electrons then decreases conductivity.

Being hollow objects, carbon nanotubes possess yet another parameter of structural stability: the tubes become increasingly softer the larger they get in diameter. The effect shows in the interaction with a substrate – wide tubes flatten and their contact area with the supporting material grows (Figure 3.47). The effect is less pronounced for MWNT as the van der Waals interactions between the constituent tubes provide additional stabilization of the cylindrical shape.

3.4.4
Electronic Properties of Carbon Nanotubes

3.4.4.1 Band Structure and Density of States of Carbon Nanotubes

It is essential to give a correct and conclusive description of the electronic properties of carbon nanotubes to get an understanding of their broad potential for applications. Chemists and physicists have developed two fundamentally different concepts of the matter. They are either based on considering electrons in molecular orbitals, especially in frontier orbitals, and on examining the π-system, or they pursue a solid-state physical approach that employs density-of-state functions and the band structure in the Brillouin zone of a two-dimensional graphene.

Both concepts have their pros and cons. On the whole the "chemical" approach as outlined by E. Joselevich shall be presented here, followed by a short discussion of the solid-state physical way that is given in more detail in many special monographs on the physical properties of carbon nanotubes.

First experiments and calculations revealed the electronic properties of carbon nanotubes to be in parts rather extraordinary. The small diameter, for instance, causes the occurrence of quantum effects. The tubes behave like a quasi-one-dimensional molecular wire, which is very useful for some electronic applications. However, the electronic properties of nanotubes are also related to those of the two-dimensional graphene as the first formally result from the rolling up of the latter. Changes and unexpected phenomena then arise, for example, from the curvature of the graphene lattice.

Measurements indicated that a part of the nanotubes is of metallic character, whereas the rest is semiconducting. Now a suitable model for the description of electronic properties must be able to explain the different electric conductivity of individual tubes, while at the same time it must correctly reflect the quantum effects. Moreover, it should allow for the prediction of the electronic properties of any chosen nanotube.

The organic molecule polyacetylene, for a start, may be considered a kind of low-molecular homolog of carbon nanotubes with regard to some characteristics. It consists of a chain of alternating single and double bonds that constitute a molecular wire with a completely conjugated π-system. Making some simplifying assumptions on the actual conditions, theoretical predictions claim electric conductivity for this molecule. In reality, polyacetylene is a semiconductor with a bandgap of 0.93 eV for the all-*cis* configuration or 0.56 eV for the all-*trans* species. The band structure of polyacetylene is shown in Figure 3.48a. It is derived from the Hückel description of the π-system, which is based on the assumption that only π-electrons contribute to electric conductivity, whereas the skeleton of σ-bonds has no significant influence on the density of states and on the occupation close to the Fermi level. In first approximation, the resulting band structure represents a continuum of allowed states that crosses the Fermi energy (the energy constituting the border between occupied and vacant states at $T = 0\,\text{K}$).

The distribution of state densities features occupied states up to the Fermi level and a symmetrical distribution of allowed, but vacant states beyond (dashed curve

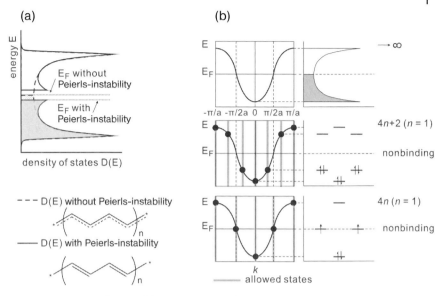

Figure 3.48 (a) Distribution of the density of states in a polyacetylene with and without Peierls distortion, (b) allowed states for annulenes with $4n + 2$ or with $4n$ π-electrons (indicated by the gray lines) and the resulting energy levels. The representation is made in the wavenumber space (k-space) (according to E. Joselevich © Wiley-VCH 2004).

in Figure 3.48a). This is the typical behavior of a metallic compound. These conditions, however, are not stable, but they are subject to the so-called Peierls distortion (solid curve). This is a characteristic phenomenon in one-dimensional molecular conductors. It is caused by structural distortions and by a coupling of electronic wavefunctions to modes of lattice vibration, and it leads to a splitting of the highest partly occupied band in such a one-dimensional metallic conductor. In the case of polyacetylene, it results in the opening of a small bandgap and thus in semiconducting properties.

Considering now finite sections of an ideal polyacetylene, a compound with cyclic conjugation may be conceived by ring closure or "one-dimensional rolling up." These so-called annulenes are quasi-zero-dimensional for being approximately point-shaped objects. The smallest species in this row are cyclobutadiene, benzene and cyclooctatetraene (COT). They are valuable model compounds for the frontier orbital concept as well as in considerations on aromaticity. In the case at hand, they represent zero-dimensional analogs of carbon nanotubes. The number of π-electrons (four in cyclobutadiene, six in benzene) serves as structural parameter n. It should be mentioned that the planarity of n-annulenes is assumed for further considerations, although in reality this is not the case for COT and higher homologues.

A restriction is forced upon the system by this cyclization. In a specific annulene, not every electron wave fits in to form a stationary wave. Hence, all "nonfitting" wavenumbers are forbidden. This results in a quantization. It can be described in dependence on the perimeter (which in its turn is related to the number of π-electrons involved). Actually, this is an extension of the "particle in a box" concept that describes the effect of confining a single electron to a defined space. The quantization condition for the transformation of polyacetylene into annulenes reads (Eq. 3.10):

$$C_h k = 2\pi q \tag{3.10}$$

with C_h being the circumference of the annulene, k the one-dimensional wave vector, and q an integer. Representing now the band structure of the polyacetylene in the reciprocal k-space (see below), only those waves and energy levels related to the wavenumbers defined by Eq. (3.10) are still allowed. This means that instead of a continuous curve, only single points remain in the band structure diagram $E(k)$. Plotting the permitted electronic states then leads to the well-known molecular orbital diagram in Figure 3.48b. The more π-electrons an annulene bears, the more permitted states exist (that move ever closer in the $E(k)$-representation). Very large annulenes thus approximate again the electronic structure of polyacetylene. There are now two types of annulenes to be distinguished that differ in the number of π-electrons. The classical aromatic compounds according to the Hückel theory like benzene possess $4n + 2$ π-electrons that can occupy $4n + 2$ permitted electronic states. Only half of these, however, lie below the Fermi energy, so they are occupied completely, whereas those states above the Fermi level remain vacant. Consequently, all π-electrons are situated in bonding orbitals. Nonbonding orbitals do not exist here as there are no permitted states on the Fermi level. Compounds of the cyclobutadiene type, on the other hand, possess $4n$ π-electrons that may be filled into just as many permitted states. In this case, nonbonding orbitals are situated at the Fermi level. They are all half-occupied in the [$4n$]-annulenes (Figure 3.48b). In both cases $q = n$ is true.

[$4n + 2$]-annulenes feature low-lying occupied orbitals and a comparatively large HOMO–LUMO gap, which corresponds to a large bandgap in higher molecular compounds. Hence the removal or addition of electrons is unfavorable from an energetic point of view, while in [$4n$]-annulenes it is possible without much effort because the semioccupied, nonbonding orbitals are situated on the Fermi level. It is true that, due to a Jahn–Teller distortion, the orbitals in actual structures are no longer degenerate, which might result in a narrow HOMO–LUMO gap. It is, however, infinitely small compared to those in aromatic compounds. The annulenes with $4n$ π-electrons thus constitute the analog of metallic nanotubes, whereas those with $4n + 2$ π-electrons can serve as model compound for semiconducting tubes.

This conclusion by analogy shall be explained in the following. Instead of the one-dimensional polyacetylene, a structure completely conjugated in two dimensions, that is, a graphene layer, will be considered. Due to this dimensional extension the calculated electronic band structure can no longer be drawn in dependence on just one parameter (the wavenumber k), but it has to be plotted over the two-dimensional Brillouin zone (Figure 3.49a). The contour lines represent different

3.4 Physical Properties | 197

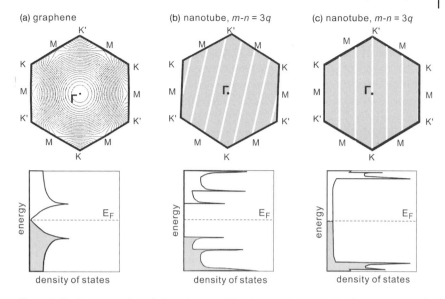

Figure 3.49 Representation of allowed states within the two-dimensional Brillouin zone (a) of graphene, (b) of semiconducting, and (c) of metallic nanotubes (top) with the corresponding distributions of densities of states (bottom) (according to E. Joselevich © Wiley-VCH 2004).

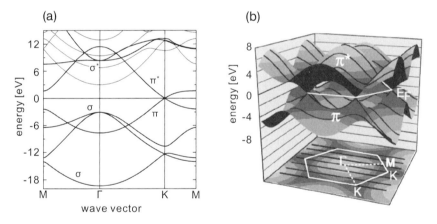

Figure 3.50 Band structure along a basal plane of graphene (a) (© Wiley-VCH 2004) and representation of the π- and π^*-orbitals above the Brillouin zone of graphene (b) (© ACS 2002).

energies here. The Fermi level is reached at the points designated with **K**. The density of state distribution obtained from this band structure is depicted in Figure 3.49. A representation of the band structure along a basal plane of graphene is given in Figure 3.50. Those bands closest to the Fermi level can be seen to touch at point **K**.

The plain, two-dimensional graphene layer is a conductor with a zero bandgap. Its density of states exhibits just one maximum each above and below the Fermi level, respectively. These correspond to the energy minima at the saddle points **M** in the band structure representation. The point Γ designates the energetic minimum that corresponds to the lowest bonding orbital in a molecular-orbital consideration – that is, the bonding combination of all p_z-orbitals across the entire graphene layer (LCAO-concept: *l*inear *c*ombination of *a*tomic *o*rbitals). There is also a related antibonding orbital which as well represents a linear combination of atomic orbitals at the point Γ. The points **K** correspond to nonbonding orbitals on the Fermi level. In the end, it is the symmetry of these orbitals that determines the properties of the nanotubes resulting from a rolling up of the graphene. There is an infinite number of such orbitals in an ideal graphene layer as the latter consists of an infinite number of carbon atoms.

Now, like with the polyacetylene, a part of the graphene sheet is rolled up to be a tube. From a chemical point of view, this is an electrocyclic reaction along the "seam" of the resulting nanotube. There is an infinity of possible directions for rolling up because contrary to polyacetylene the graphene is a two-dimensional structure. Hence, the two structural parameters n and m must be considered for the tubes obtained from graphene. A periodic boundary condition is forced upon this system nevertheless, and only selected wavefunctions are still solutions to the Schrödinger equation after the rolling-up. The quantization can be described in analogy to Eq. (3.10), with $\bar{C}_h$ now being the perimeter vector of the respective nanotube, and k representing the two-dimensional wave vector. The band structure thus falls into several smaller subbands that lie the closer together, the longer $|\bar{C}_h|$ (the circumference) gets. This effect corresponds to the molecular orbitals moving closer upon increasing diameters in the annulenes considered above.

In real space, that is, in a chemical view, electrocyclic reactions are governed by the Woodward–Hoffmann rules of orbital symmetry conservation. According to Fukui only the frontier orbitals are regarded here to decide whether a given reaction is permitted or not. The highest occupied orbital must be examined for a thermal process, which applies to the imaginary formation of a carbon nanotube from a graphene layer. (The first unoccupied orbital would have to be considered for a photochemical process – which is not the case here.)

The highest occupied orbitals are those on the Fermi level. Two distinct scenarios can be envisaged for the roll- up procedure. One of them consists in the encounter of two carbon atoms with equal LCAO-coefficient, for example, + and +. The orbitals would thus retain the same periodicity in the nanotube as they had before. According to Woodward–Hoffmann there are permitted states then on the Fermi level, which results in metallic behavior (Figure 3.51a). In the other scenario, it is just carbon atoms with different LCAO-coefficients, that is, + and –, that meet. Periodicity, and thus orbital symmetry, cannot be maintained here. Hence the molecular orbitals formally formed in this process are forbidden according to the orbital symmetry conservation law. So there are no permitted states on the Fermi level and a bandgap arises. Consequently the resulting nanotubes are semiconductors (Figure 3.51b).

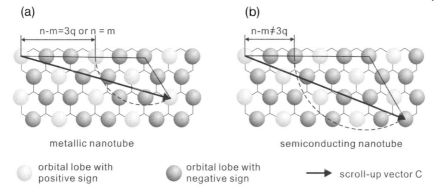

Figure 3.51 LCAO-representations of the rolling up of (a) metallic or (b) semiconducting nanotubes, respectively. In the case of metallic tubes, two orbital lobes with identical sign meet, whereas for semiconducting tubes they do not (© Wiley-VCH 2004).

After having explained which nanotubes are metallic or semiconducting, respectively, the question remains whether there is an inherent law that enables a prediction of which tube will exhibit which behavior. Another look on the nonbonding orbital of graphene on the Fermi level as depicted in Figure 3.51 can serve to this purpose. The pattern of positive and negative LCAO-coefficients repeats every three length units along both unit vectors of the two-dimensional elementary cell. Hence the condition for an encounter of identical LCAO-coefficients – and thus for metallic behavior – reads that the difference of the structural parameters n and m must be divisible by three. In all other cases, atoms with different LCAO-coefficients meet, and the nanotube resulting from rolling up is a semiconductor. Therefore Eq. (3.11) is valid (with q being an integer):

Metallic nanotubes: $n - m = 3q$ (3.11a)

Semiconducting nanotubes: $n - m \neq 3q$ (3.11b)

Equivalent results are obtained considering the band structure in reciprocal space and subsequently determining the density function of electronic states. The latter is obtained from the multiplication of the energy range between E and $E + dE$ by the density of wavenumbers (depending on the shape of the Brillouin zone) in the reciprocal k-space. Only allowed wavenumbers k are considered here. Now, the periodic boundary condition (Eq. 3.10) is introduced so again only selected solutions of the Schrödinger equation are still acceptable. The quantization of the graphene's band structure as shown in Figure 3.50 gives rise to different arrangements of bands for the resulting nanotubes depending on the direction of roll-up. It is an effect of this quantization that permitted states in the hexagonal first Brillouin zone of graphene exist now only along the white lines drawn in Figure 3.49. Their orientation depends on the structural parameters n and m. In some cases, the permitted lines intersect at the points K (i.e., the Fermi level) of the Brillouin zone. The respective nanotubes are metallic because valence and conduction band are in touch here. This statement applies to all tubes where $n = m$ or $n - m = 3q$. A closer inspection of the latter, however, reveals that actually there is a very small bandgap that is caused by curvature effects. Yet it decreases with growing diameters of the tubes, and at room temperature the respective nanotubes can be considered metallic for most practical applications. If, on the other hand, the point K is not intersected by any of the lines representing permitted wavenumbers, a significant bandgap and semiconducting properties are observed. It decreases as well for tubes with larger diameters, but even then it is wide enough to maintain semiconducting behavior. Summing up, it can be said that there are three distinct types of carbon nanotubes:

- Metallic carbon nanotubes where $n = m$.
- Semiconducting nanotubes with very small bandgap where $n - m = 3q$ $(q \neq 0)$ that can be considered metallic under realistic conditions.
- Semiconducting nanotubes with wide bandgap where $n - m \neq 3q$.

A first look on the density function of states of different carbon nanotubes reveals a number of singularities (Figure 3.52). These arise from a multitude of

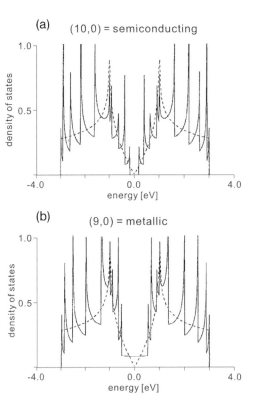

Figure 3.52 Densities of states for a metallic (b) and a semiconducting carbon nanotube (a). The sharp maxima of the density function are called van Hove singularities. (The dashed lines indicate the density of states for graphene. © AIP 1992).

states concentrating in very narrow intervals of energy in the vicinity of each energetic minimum of the individual subbands. Calculating the density of states consequently results in sharp peaks for these energies. They are called *van Hove* singularities and constitute a distinct characteristic of the electronic structure of carbon nanotubes. The first van Hove singularities on both sides of the Fermi level in semiconducting nanotubes correspond to the HOMO and LUMO of the respective tube. The energy difference between them represents the distance between valence and conduction band (the bandgap). As for the aforementioned model compounds, that is, for the [$4n + 2$]-annulenes, there is also a large HOMO–LUMO gap between a low, occupied bonding and a high, vacant antibonding orbital. In the band structure of metallic nanotubes, on the other hand, HOMO and LUMO are situated on the Fermi level and cannot be distinguished. This is equivalent to a zero bandgap and corresponds to the [$4n$]-annulenes from the zero-dimensional analogy. The first van Hove singularities in the density of states diagram of metallic nanotubes are found only a bit farther from the Fermi level.

It can be seen from the above that the electronic behavior of carbon nanotubes can be described correctly on a qualitative level using either the chemical or the solid-state physical approach. In doing so, the "chemical" method has the clear advantage of being quite illustrative because it does not abandon real for reciprocal space. Still the quantitative consideration of densities of states and band structures remains exclusive to the stringent approach that starts from solving the Schrödinger equation. Here the transition into reciprocal space markedly facilitates calculations. Exhaustive descriptions of the method and its implications are given in the references cited at the end of the book.

Up to now only the electronic properties of single carbon nanotubes have been regarded. In reality, however, frequent use is being made of multiwalled tubes or of bundles of single-walled species. Their electronic properties differ to some extent from those of isolated SWNT. Yet the growing complexity of the systems increasingly complicates the analytical description of the density of states and band structure. Still some recent publications on the matter discuss the phenomena observed upon the assembly of several tubes.

In MWNT, the individual tubes are in such close proximity that electronic interactions come to bear. Indeed the behavior experimentally observed for macroscopic samples resembles that of the semimetallic graphite. When examining individual MWNT by means of scanning tunneling spectroscopy (STS), however, both metallic and semiconducting species were found, although it is not settled if maybe just the properties of the outermost single tube have been measured.

Band coupling is observed for nanotubes with suitable structure and arrangement. The orientation of the constituent tubes relative to each other is the decisive parameter here, yet this may vary over time due to the mobility of tubes (Section 3.4.3) against each other. Combinations like (9,0)@(18,0) or (5,5)@(10,10) exhibit suitable geometries, yet the properties change with the angle of rotation. A (5,5)@(10,10)-nanotube with a mirror plane, for instance, is metallic, whereas a small bandgap opens upon a rotation of the inner tube. MWNTs made up from less compatible single-walled species, on the other hand, show little interaction between the individual tubes for any relative orientation. Consequently, there is no effective

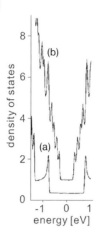

Figure 3.53 The density of states for multiwalled nanotubes; (a) density function of a (6,6)@(11,11)-nanotube, (b) density function of a (6,6)@(11,11)@(16,16)@(21,21)@(26,26)-MWNT (© World Scientific 1994).

band coupling, and the density of states represents more or less the sum of its constituent densities (Figure 3.53).

Intertubular interactions occur as well in bundles of SWNTs. They reduce the symmetry of the system and thus influence the band structure. For metallic tubes this results in the opening of a bandgap because the valence and conduction bands for wave vectors in parallel with the tube's axis do no longer touch on the Fermi level. Still the bands overlap in another manner due to their dispersion rectangular with respect to the tube axes, so actually this is a pseudo-bandgap.

3.4.4.2 The Mechanism of Electric Conduction in Carbon Nanotubes

The electrons in a conventional electric conductor move toward the positive pole under the influence of an external electric field. In doing so they experience a resistance due to scatter on lattice defects and phonons (lattice vibrations). Finally, a stationary state of constant current is established that is described by the Fermi function. The conductivity of the material decreases with rising temperature because the scatter on phonons becomes more efficient due to thermal excitation. It is true that the electrons scatter on lattice defects, too, but for being temperature-independent, these play just a minor role at elevated temperatures. However, the effect becomes important at low temperatures because phonon scatter ceases under these conditions, and the specific residual resistance almost exclusively arises from scatter on lattice defects. Hence, the residual resistance is a measure for material's purity: it lessens with increasing purity and defect density of a sample.

Ohm's law holds for macroscopic samples of an electric conductor. It states that voltage and current are proportional (with the ohmic resistance as proportionality factor). The electric resistance itself for a macroscopic wire depends on length and cross-sectional area of the object. Yet these conditions are no longer effective for a nanoscale conductor. The resistance of such a nanowire is independent of its length because charge transport is achieved by so-called conduction channels.

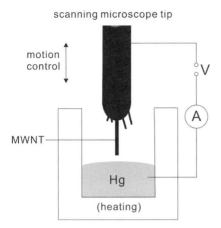

Figure 3.54 Experimental setup proving the quantization of electrical properties in carbon nanotubes (© AAAS 1994).

There is a very large number of them, and each of these exhibits a resistance of ~13 kΩ (in consideration of spin degrees of freedom, otherwise ~6.5 kΩ). Strictly speaking, however, this statement is true only if the respective one-dimensional wire is absolutely free of defects. In reality, this is not very likely to be the case, and so in actual nanowires there are, for example, interactions with the electrons of the atomic lattice.

To understand the electric properties of carbon nanotubes, it is important to be aware that the conductance is quantized. This means that raising the external electric field applied does not lead to a continuous, but to a stepwise increase in the current intensity. An elucidating experiment has been reported by W. de Heer and co-workers in 1998. It consists in dipping a carbon nanotube into mercury and drawing it out again with a voltage applied (Figure 3.54). The resulting current is measured in relation to the position of the tube. A classical conductor would show a continuous variation of conductivity, but for the carbon nanotube, a series of steps was observed. Furthermore, the tubes were found not at all to be damaged in spite of high voltages applied (up to 6 V). Assuming a conventional mechanism of conduction, the lattice interactions would have heated up the system until damages by superheating would have been inevitable. So there is strong evidence that the electrons travel long distances through the nanotube without interacting with the carbon framework.

This kind of electron movement is also termed *ballistic transport* because the electrons do not experience any resistance on their free paths. In a particle model they might be considered as objects flying freely before interacting again with the material at the end of a free path. It is due to this mechanism that the conductor does not heat much as there is no interaction with phonons and consequently the lattice is not excited to perform stronger vibration. This is of special interest for the development of efficient electronic devices. Conventional materials are limited in their tolerance to current density because too much heat is developed above a certain current density.

According to calculations, the free path Λ in a defect-free carbon nanotube is about 1 μm, while values of about 100 nm have been determined experimentally. This is an extremely large value in comparison to classical electric conductors (e.g., lithium: 110 Å, copper: 430 Å, silver: 560 Å).

Measurements on single nanotubes fixed between two contacts revealed that the transport of electrons through the contact sites is achieved by tunneling and that the electronic wavefunction extends from one contact to the other. Hence, at the given temperature of the experiment (mK range), a nanotube of this kind represents a coherent quantum wire.

The electric conductivity of carbon nanotubes is largely influenced by the presence of defects. Even effects as modest as axial strain with bond expansion change the band structure. Stone–Wales defects and other imperfections diminish the electric conductivity as well. This effect is especially pronounced for defects with two adjacent vacancies. The resistance of a 400 nm long SWNT, for example, increases by a factor of 1000 if the tube bears as little as 0.03% of these double vacancies. Single vacancies, on the other hand, do not cause such dramatic changes. In any case, however, the free path of the electrons is reduced considerably by the defects (in parts down to a few nanometers). Still, due to the multitude of existing conduction channels, this has no large influence on the overall conductivity.

Impurities of the sample may have significant influence on the electric conductivity, too. Upon cooling to very low temperatures, for instance, the resistance of carbon nanotubes has been observed not to decrease continually until approximating the specific residual value. There has rather been an increase in resistance at extremely low temperatures; so the $R(T)$ curve exhibits a minimum. This has been explained with the magnetic properties of residual catalyst particles that cause a Kondo effect. The latter is based on an exchange interaction between the magnetic moments of the contaminant atoms and the electrons, and it leads to an increase in resistance. In the meantime, however, there have been other publications that attribute the resistance minimum to an intrinsic phenomenon, namely to the so-called contact tunneling.

3.4.4.3 Field Emission from Carbon Nanotubes

One feature of carbon nanotubes is of special interest for a variety of applications: the field emission. This is the ability to emit electrons upon the application of an electric field. Field emission is no phenomenon exclusive to carbon nanotubes. It has been known for long that electrons can be extracted from the surface of conductive materials. Still it takes in parts extremely high field intensities of up to several kilovolts per micrometer to enable a tunneling through the energy barrier constituted by the surface. Normally such values are not practicable. Hence other approaches are required to facilitate the extraction of electrons. It is a known effect that an electric field is locally intensified at the pointed ends of a sample because the flux lines are more concentrated at highly curved sites. A suitable material must further exhibit a low work function, which means that little energy is required for the removal of an electron from the solid to the outside.

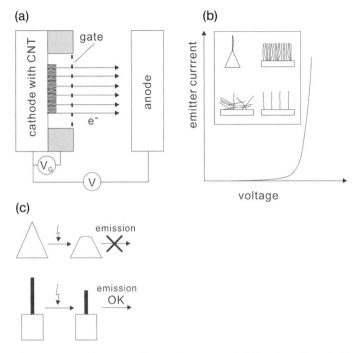

Figure 3.55 Field emission from carbon nanotubes: (a) scheme of an emitter array consisting of carbon nanotubes, (b) emission curve and possible geometries of CNT-emitters (insert), (c) advantages of nanotubes over other emitter tips: emission ability is conserved even after damage.

Carbon nanotubes are virtually ideal to fit these requirements. Especially the fact of the tubes' tips being extremely thin and bent causes an extraordinary local strength of the electric field. The tubes emit electrons already at field intensities below $1\,V\,\mu m^{-1}$, while the resulting current density can amount to more than $3\,mA\,cm^{-2}$ (Figure 3.55). Contrary to the emission of electrons from thermionic sources (e.g., tungsten filament), the source is not heated to more than $1000\,°C$ here, and less energy is lost by radiation. Another essential advantage of nanotubes over tips made by mechanical processing is illustrated in Figure 3.55c: The high current densities as well as the bombardment with ions and radiation generated by the electric field can easily damage the emitter tip. This might bear fatal consequences for a device made by mechanical means: The site of highest field intensity will be destroyed if it breaks, and in the worst case the emission of electrons would completely cease. For carbon nanotubes, on the other hand, the geometry will hardly change if the uppermost part is destroyed by wear effects. The diameter of the tube is largely constant, and emission proceeds continuously. It has even been found that opened nanotubes are better emitters than those still closed with a cap. This may be due to the chains of carbon atoms present at the open ends. These constitute extremely pointed and thus even more efficient tips.

All known types of carbon nanotubes are potential field emitters (Figure 3.55b). Single and bundled SWNT as well as multiwalled species show this behavior. Individual tubes may be attached to a conductive support, but they might just as well be employed in the shape of ordered or unordered films or as structured arrays (Section 3.3.5). The respective products can be selected depending on the desired application. The choice of single- or multiwalled species influences emission properties, too: SWNTs feature a low work function, whereas MWNTs better suit continuous use because they are clearly more stress resistant.

3.4.5
Spectroscopic Properties of Carbon Nanotubes

Not only the electronic and magnetic characteristics, but also the spectroscopic properties of carbon nanotubes are of special interest. On one hand, the utilization of these features raises hopes for various applications, but, for being quasi-one-dimensional objects, they might also serve to a deeper understanding of numerous spectroscopic phenomena.

The model of permitted wave vectors and the concept of reciprocal space (Section 3.4.4) that both proved their worth in the discussion of electronic properties are as well suitable to describe a number of spectroscopic features. Especially those phenomena that are based on the behavior of phonons can be treated this way.

The essential spectroscopic properties will be presented in the following. In most cases, a stringent mathematical and physical deduction will be omitted as there is a multitude of monographs and reviews on that topic, and it would be far beyond the scope of this introductory text.

3.4.5.1 Raman and Infrared Spectroscopy of Carbon Nanotubes

From their first measurement and theoretical discussion on, the Raman spectra of carbon nanotubes provided valuable information on the shape and composition of the structure. They can also serve to prove the existence or nonexistence of nanotubes in a sample, which is assessed, among others, by considering a distinct Raman band. It is called *radial breathing mode* (RBM) band and corresponds to a synchronous radial vibration of all carbon atoms in perpendicular to the tube's axis. This band is the characteristic of the nanotube structure, and for the time being, nothing like this has ever been observed for any other carbon material studied.

While the electronic wavefunction has been considered to describe electronic properties, the discussion will now focus on the phonons, that is, the quantized normal modes of the lattice. Again the zone-folding model of a graphene layer being rolled up to become a nanotube proved to be a useful approach. In comparison to graphene, the resulting density of phonon states exhibits many sharp peaks similar to the van Hove singularities in the density of electronic states (Section 3.4.4). Graphite itself shows only one single Raman band at 1582 cm^{-1} resulting from E_{2g}-phonons. Carbon nanotubes with diameters below 2 nm show high intensities of the Raman bands due to the existence of the van Hove singularities. Therefore, even measurements on single nanotubes are possible here. This is of special interest when studying the differences between metallic and semiconducting species. For symmetry reasons, different types of carbon nanotubes posses distinct sets of Raman and IR-active bands, respectively. For mixtures of different tubes, the interpretation of spectra with regard to the identification of individual types of tubes is rather complicated.

Some essential modes in the phonon band structure are worth mentioning in detail. These are the highly energetic longitudinal acoustic modes (vibrations in parallel with the tube's axis), the transversal acoustic modes (vibrations in perpendicular to the axis), and two acoustic twist modes that describe rotations about the tubular axis. The latter two bands are meaningful especially for heat conductance and for the scatter of charge carriers on the lattice. The low-lying optical modes in the center of the Brillouin zone are important for the coupling of the electrons to the lattice. They are also responsible for the typical IR- and Raman signals of carbon nanotubes. These include, among others, a mode with E_1-symmetry at about $118\,cm^{-1}$, an E_2-mode (so-called *squash mode*) at ca. $17\,cm^{-1}$ and an A_{2g}-mode at about $165\,cm^{-1}$. This so-called RBM is situated in a region of the spectrum otherwise "silent" for carbon materials. It further shows an inversely proportional relation of its wavenumber to the tube's diameter d_{SWNT} (Eq. 3.12, value for a (10,10)-nanotube: $165\,cm^{-1}$). This mode corresponds to a "breathing" of the molecule, that is, a synchronous "inward and outward" vibration of all carbon atoms of the tube. It originates from to the "out of plane" translation of a basic graphene layer before rolling up.

$$\omega = \frac{234\,\text{nm cm}^{-1}}{d_{SWNT}(\text{nm})} \tag{3.12}$$

In a mixture of nanotubes with similar diameters, then the so-called R-band results from combining the RBM of the individual tubes. Besides this breathing mode, there are two further important bands in the Raman spectrum of SWNT, namely the G- and the D-band (Figure 3.56). The G-band is found at ca. $1580\,cm^{-1}$. It is a superposition of several individual signals with the two major components being termed G^+- and G^--band. Still other components exist which are caused by phonons with E_1- and E_2-symmetry. The G^+-band at about $1593\,cm^{-1}$ corresponds to a vibration in parallel with the tube's axis. Its frequency is shifted upward or downward by the presence of electron donors or acceptors, respectively, which shows a sensibility to charge transfer. The signal position of the G^+-band is not observed to change significantly depending on tube diameter or chirality angle, which is plausible with regard to the underlying vibration.

The G-band, on the other hand, varies with the tube's diameter, while it does not react much either to changes of the chirality angle. The underlying vibration may be described as tangential along the perimeter, and the signals show different shapes depending on the conductivity of the tube: Semiconducting nanotubes give rise to Lorentz-shaped bands, whereas metallic species exhibit bands of the Breit-Wigner-Fano type.

The D-band is observed at ca. $1347\,cm^{-1}$. It results from disorder defects in the nanotube lattice and originates from phonons close to the point K of the Brillouin zone. The signal is strongly dependent on the irradiated laser energy. Changing,

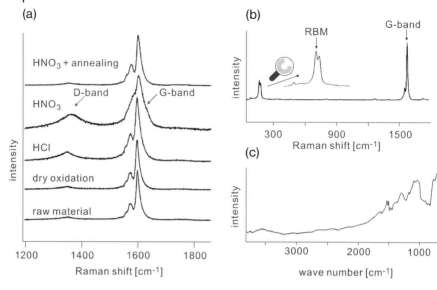

Figure 3.56 (a) Raman spectra of carbon nanotubes after different kinds of treatment (© ACS 2004), (b) apart from the D and G band also the so-called radial breathing mode (RBM) is observed at low wavenumbers in the Raman spectrum (© RSC 2005). (c) The IR-spectrum of a nanotube sample (© Royal Soc. 2004).

for example, the excitation energy by 1 eV causes a shift of 50 cm^{-1}. For metallic nanotubes there is another signal at 1540 cm^{-1} in addition to those bands described so far. It can be used to identify this type of tubes, yet for the time being, the exact origin of this band has not been fully elucidated.

Experimental studies on nanotubes revealed that the energy of the exciting radiation has a general influence on the Raman spectrum. This is indicative of resonant processes with the resonance resulting from the accordance of the excitation energy E_{exc} and the transition energy E_{ii} between the van Hove singularities of valence and conduction band.

In comparison to SWNTs, MWNTs give rise to a much more graphite-like Raman spectrum with a dominant signal at ca. 1575 cm^{-1} and a weaker one at about 868 cm^{-1}. The determination of structural properties from the Raman spectra of MWNTs is anything from complicated to impossible due to the variety of constituent tubes and their respective interactions.

For single-walled carbon nanotubes, it is also possible to predict the structure of their infrared spectra. It turns out that depending on the structure of the tube under consideration, a different number of bands should be IR-active. Zig-zag nanotubes possess IR-active A_{2u}- and two E_{1u}-modes; for armchair tubes there are three E_{1u}-modes, and chiral species exhibit one A_2- and five E_1-modes. The signals can be expected to lie chiefly in one of two spectral regions about 870 and 1590 cm^{-1}. The first of these is always *Raman*-inactive, whereas for the latter this is only true for achiral tubes. The band positions observed are close to those found for graphite (shifted toward higher wavenumbers by only 6–10 cm^{-1}), which is why the contamination of a sample with graphitic particles can lead to the erroneous identification of nanotubes in the IR-spectrum. What is more, the observation is complicated as only weak signals are obtained due to the strong inherent absorption of carbon nanotubes. Figure 3.56c shows the IR-spectrum of a pristine nanotube sample.

3.4.5.2 Absorption and Emission Spectroscopy of Carbon Nanotubes

Compared to the vibrational spectroscopy of carbon nanotubes, their absorption and luminescence spectroscopy kind of lives in the shadows. This is, however, not due to a lack of information these methods could provide to the understanding of the nanotubes' electronic structure. There are rather experimental complications that arise from the inhomogeneity of the available materials.

These methods could especially elucidate the structure of bands farther distant from the Fermi level. Due to the one-dimensionality of nanotubes, markedly structured absorption and luminescence spectra can be predicted from theoretical considerations. The extinction should depend on the structure of the respective tube. Especially in the lower energy range (infrared) most of the transitions between π-bands are observed. Figure 3.57 shows the diagram of the calculated density of states for a metallic and a semiconducting nanotube with the dipole-permitted transitions drawn in. These are easily recognized to exist between mirror-symmetrical van Hove singularities with a difference E_{ii} of energies corresponding to the absorption energy. The position of absorption maxima consequently depends on the diameter as well as on the chirality angle of the tube under consideration as both parameters influence the distance between singularities in the density of electronic states (Section 3.4.4).

At first, however, these structured spectra could not be confirmed by experiment due to the poor solubility of nanotubes: No satisfactory samples for absorption

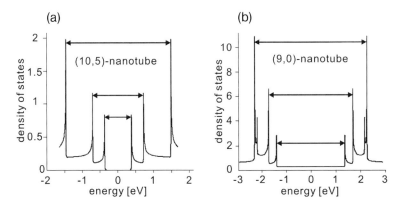

Figure 3.57 Allowed electronic transitions in a (a) semiconducting and in a (b) metallic nanotube, respectively (© Wiley-VCH 2005).

spectroscopy were available. Moreover, examining bundles of different tubes hardly provides any information as the assignment of individual signals is virtually impossible. A strong intertubular electronic coupling causes a significant mixing of electronic states, and so the fine structure of the spectra does not allow for data evaluation with regard to individual tubes. Studying bundles of nanotubes by photoluminescence spectroscopy did not turn out too well either. Here the emission of light is efficiently quenched due to fast transfer processes between metallic and semiconducting tubes. Therefore, the fluorescence of nanotubes can at least be taken as an indicator for successful debundling of SWNT: Bundled samples do not show infrared fluorescence, whereas it is easily measured for separated nanotubes.

Some groups then succeeded in preparing solutions of shortened nanotubes by ultrasonication and acid treatment, while the production of films made by airbrush techniques or spin-coating was developed as an alternative. Spectra could also be obtained for debundled nanotubes, ensuring in the experiments that the signals observed did not arise from the surfactant (usually sodiumdodecyl sulfate). A debundling by surface functionalization is unsuitable here because firstly, the functional groups might severely disturb the electronic structure of the nanotubes, and secondly, they would contribute their own absorption or emission bands, respectively, thus rendering the spectra unnecessarily complex.

After all, the absorption spectra of carbon nanotubes below ca. 2 eV are more or less independent from contaminants or defects in the sample. For nanotubes separated by surfactants or other techniques, absorption or emission spectra with a distinct structure and coinciding signal positions are obtained. The characteristic bands are situated at 0.68 eV (1823 nm), 1.2 eV (1033 nm), and 1.7 eV (729 nm). They correspond to the first and second transition in semiconducting tubes and to the first permitted transition in metallic species. Furthermore, there is the broad signal at 4.5 eV (275 nm) that represents a π-plasmon band (Figure 3.58).

The fluorescence of nanotubes is not very pronounced. The quantum yield ranges about 10^{-4} to 10^{-3} only. The lifetime of the excited state is below 2 ns, and the shift between absorption and emission is rather low, which is indicative of little geometric differences between ground state and excited state. These values correspond well to what might be expected for a spin-permitted process arising from a singlet exciton. An emission spectrum as obtained with an excitation energy of 2.3 eV (532 nm) is shown in Figure 3.58b. The initial excitation leads to the population of higher states (e.g., E_{22}). The charge carriers thus relax into the first excited state before recombination, and only then E_{11}-emission occurs. Consequently, the fluorescence on the band edge is observed. In Figure 3.58c the excitation spectrum of the bandgap fluorescence at 875 nm is shown. The band at 581 nm is assigned to the absorption of the second van Hove singularity. Accordingly the examination of fluorescence also provides information on states in a certain distance from the Fermi level.

However, the experimental data discussed so far still constitute superimposed spectra of many different carbon nanotubes. Measurements on individual tubes would mean an improvement here. This requires some experimental effort, but by

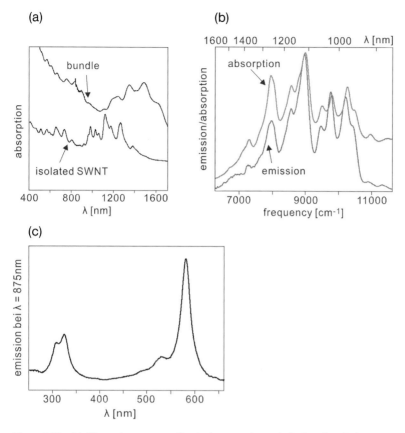

Figure 3.58 (a) Absorption spectra of a single nanotube and of a bundle of tubes, (b) emission spectrum of a CNT-sample at $\lambda_{exc.} = 532$ nm in comparison to the absorption spectrum, (c) excitation spectrum of the same sample at $\lambda_{exc.} = 875$ nm (© AAAS 2002).

now it has been accomplished by means of fluorescence microscopy. Some remarkable results have been obtained this way: The directed excitation of individual tubes succeeded expectedly as due to their various band structure they absorb different wavelengths from the exciting radiation. The signals measured show a Lorentz-shape and do not possess any fine structure. Yet even for nanotubes with identical structure parameters n and m varying fluorescent wavelengths are observed depending on the position of the measurement, which is explained by contributions of defects and fluctuations that cause a disturbance of the band structure.

Combining it with the observation of the radial RBM, fluorescence microscopy can be employed to determine the structural parameters n and m of the tubes under examination. Table 3.2 collects experimental and theoretical values of carbon nanotubes characterized this way.

Another feature of the fluorescence of SWNT turns them into interesting candidates for quantum optical applications: Contrary to all semiconductor quantum

Table 3.2 Determination of structural parameters m and n from Raman and fluorescence data.

Calculated values		Experimental data		Determined structural parameters
ν_{RBM} (cm^{-1})	$h\,\nu_{em}$ (eV)	ν_{RBM} (cm^{-1})	$h\,\nu_{em}$ (eV) (λ_{em} (nm))	
281.9	1.212	282	1.212 (1023)	(7,5)
307.4	1.272	308	1.270 (976)	(6,5)
298.1	1.302	293	1.298 (955)	(8,3)
307.4	1.359	309	1.355 (915)	(9,1)
335.2	1.420	337	1.407 (881)	(6,4)

Acc. to A. Hartschuh, H. N. Pedrosa, J. Peterson, L. Huang, P. Anger, H. Qiang, A. J. Meixner, A. Steiner, L. Novotny, T. D. Krauss, *Chem. Phys. Chem.* **2005**, *6*, 577–582.

dots (e.g., CdS) their fluorescence does not exhibit the otherwise characteristic "blinking" behavior. This means nanotube fluorescence is stable on the time scale examined and does not appear or vanish coincidently. Consequently, carbon nanotubes are suitable as stable single photon sources. More particular measurements revealed, however, that actually blinking can be detected for a part of the nanotubes at low temperatures. For the time being, it is an open point whether the effect only sets in in the cold, or if the blinking of nanotubes occurs on a much faster, hitherto unresolved time scale at room temperature.

3.4.5.3 ESR-Spectroscopic Properties of Carbon Nanotubes

Electron spin resonance (ESR), which is also termed as electron paramagnetic resonance (EPR), is another spectroscopic method to provide valuable information about the electronic structure of carbon nanotubes.

The spectrum is a plot of the absorption intensity over the magnetic field strength. Both the g-value of the sample and the shape of the signals are employed for the interpretation of results. The ability of the applied magnetic field to induce local currents in the sample varies. Depending on the extent of this induction, the g-value actually measured deviates more or less from $g_e = 2.0023$, the value for a free electron. Metallic conductors accordingly feature the largest deviations from g_e.

In graphite, an ESR-signal can only arise from a resonance of the free conduction electrons. As expected for a material that is conductive only in directions x and y, g is found to be 2.0026 ($\sim g_e$) upon application of the field in parallel with the lattice planes. Orienting the field in perpendicular to the graphene layers, the value is higher by 0.047. The signal shape obtained (Figure 3.59) is called dysonic and is typical of metallic materials. The less ordered carbon materials give rise to distinctly more complex ESR-spectra due to the additional unpaired electrons localized in defects.

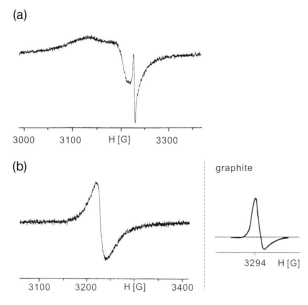

Figure 3.59 ESR spectra (a) of an pristine nanotube sample at 10 K (© Elsevier 1994), (b) of a purified sample at 4 K (© Elsevier 1995) and of graphite (right, © APS 1960).

ESR-experiments with MWNT yielded signals much resembling those of graphite. Hence one may conclude that a large portion of the sample has metallic properties or at least exhibits a very small bandgap. ESR-measurements on bundles of single-walled nanotubes indicated that these must be considered at least semi-metallic according to the signals' shape and position.

The purification of the nanotubes and the removal of lattice defects play a crucial role in sample preparation for this kind of measurements. The ESR signal of an ideal nanotube would be determined by the conduction electrons alone. Like with graphite, anisotropy of g-values in parallel with and perpendicular to the tube's axis can be expected. In detail, the value in axial direction should as well be close to g_e. The deviation measured at right angles to the axis, on the other hand, should be less pronounced than in measurements perpendicular to graphite layers because unlike the latter, the induced orbital currents are not completely closed in nanotubes. If the nanotube samples contain defects or impurities like catalyst particles, further ESR signals are observed that cannot be assigned to conduction electrons. These signals fade when, for instance, the samples are heated strongly as residual catalyst is eliminated this way just as well as defects are (Figure 3.59b).

3.4.5.4 Further Spectroscopic Properties of Carbon Nanotubes

Besides those spectroscopic properties presented so far, there is a variety of further phenomena that have been studied on carbon nanotubes, either in experiment or in theoretical consideration. The methods employed include, among others, NMR-spectroscopy and electron energy loss spectroscopy (EELS).

^{13}C-NMR-Spectroscopy ^{13}C-NMR-spectroscopy can also provide valuable information on the structure and the bond type of the carbon nanotubes. Yet the full potential of the method cannot be exploited due to experimental problems. Most of all, the heterogeneity of the samples, their poor solubility in common NMR solvents and the presence of ferromagnetic impurities (catalyst particles) complicate the recording of instructive NMR spectra.

By now, it has been possible to overcome at least some of these problems and to obtain spectra of carbon nanotubes both in solution and in solid state. Especially the removal of catalyst residues by magnetic separation led to a clear improvement of spectrum quality.

Solid-state NMR-spectroscopy shows a broad band extending from 120 to 130 ppm which can be assigned to the carbon atoms of the nanotubes (Figure 3.60a). This shift range corresponds to the values expected for sp^2-hybridized carbon atoms in conjugated systems (e.g., benzene: 128 ppm). In the spectrum obtained from solution, there is also a band with a maximum at 132 ppm. The respective samples had been dissolved in D$_2$O by functionalization with ethylene glycol, which is why there is also a sharp peak at about 70 ppm besides the band due to nanotubes (Figure 3.60b). Upon thorough investigation, at least two overlapping peaks have been found to constitute the ^{13}C-band of carbon nanotubes. They could be separated by deconvolution to yield two signals with maxima at 128 and 144 ppm, respectively. Supposedly these can be assigned to semiconducting (high-field shift) and metallic (low-field shift) species. The high-field shift

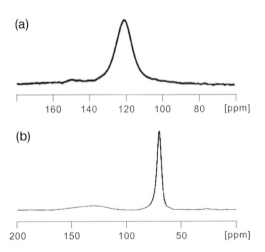

Figure 3.60 (a) Solid state ^{13}C-NMR spectrum of a nanotube sample, (b) ^{13}C-NMR-spectrum of a sample solubilized by functionalization with polyethylene glycol (© ACS 2005).

of semiconducting tubes is attributed to the existence of localized ring currents (Section 3.2.2).

Electron Energy Loss Spectroscopy (EELS) The EELS is a method that suits well to the examination of light elements like carbon. Among others, it provides information on the oxidation state and the kind of bonding in the material under examination. The dielectric properties of the substance can be determined as well as the degree of hybridization of its constituent atoms. sp^2- and sp^3-hybridized carbon, for instance, give rise to different typical signals. The spectrum records the energy loss of the irradiated electrons after passage through the sample. It can be divided into two regions that correspond to different types of transitions. The so-called low-loss region results from inelastic electron scatter and comprises signals of interband transitions and collective excitations, for example, the bulk plasmons of the material. The so-called core-loss region represents transitions from inner electronic levels into vacant states, which are caused by the inelastic scatter of high-energy electrons (~200 kV). Therefore, even information about the band structure beyond the Fermi level becomes available here. At small scattering angles, only dipole excited transitions are permitted.

For the element carbon, both regions of the loss spectrum provide their share of information (Figure 3.61). In the high-energy range (core loss region) between 260 and 320 eV, electrons from the orbital closest to the nucleus (1s-orbital) are excited. The energetic differences measured and the shape of signals allow for statements on the ratio of sp^2- and sp^3-carbon. Actually the loss spectra of nanotubes are mostly similar to that of graphite in this region, and so a very small portion of sp^3-hybridization can be assumed.

In the low loss region from 0 to 40 eV, plasmon excitation occurs. At ca. 5.2 eV, a signal of π-plasmons is observed that corresponds to a collective oscillation of π-electrons along the tube's axis. At about 21.5 eV, the loss spectrum shows another signal that is attributed to (π + σ)-plasmons. The signals in the low-energy region

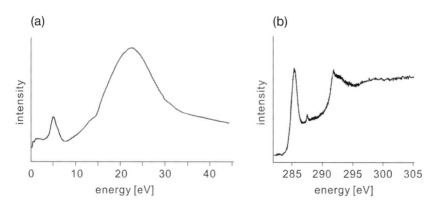

Figure 3.61 EELS analysis of a carbon nanotube sample: (a) low-loss region of the spectrum (© Oyo Butsuri Gakkai 1994) and (b) core-loss region (© Elsevier 1999).

of the loss spectrum depend on the diameter of the respective tube. The smaller it is, the lower the signal is observed. This is explained, among others, by a reduced delocalization of the π-electrons in smaller tubes. Consequently, these electrons contribute less to excitation.

3.4.6
Thermal Properties of Carbon Nanotubes

Besides the electronic and spectroscopic properties of carbon nanotubes, their thermal behavior is of interest, too. Most of all a distinct anisotropy should be observed. The thermal behavior is predominantly governed by the phonon properties of the sample. With the density of phonon states being available from theoretical considerations, it is possible to compare these expectation values directly to the experimental results. Again, however, it is true that exact assignment and explanation of phenomena require measurements of single nanotubes whose structure has been determined. This is some challenge for the time being.

3.4.6.1 Specific Heat Capacity of Carbon Nanotubes

The thermal characteristics of carbon nanotubes are related to those of the two-dimensional graphene, yet the rolling up induces a quantization of these properties. It shows most in nanotubes less than 2 nm in diameter.

The specific heat capacity of a substance comprises an electronic contribution C_{e^-} and a contribution C_{Ph} of the phonons. The latter is dominant in carbon nanotubes, regardless of their structure. C_{Ph} is obtained from integration over the density function of phonon states and subsequent multiplication by a factor that considers the energy and the population of individual phonon levels.

Below the Debye temperature, only the acoustic modes contribute to heat capacity. It turns out that within a plane there is a quadratic correlation to the temperature, whereas linear behavior is observed for a perpendicular orientation. These assumptions hold for graphite, which indeed exhibits two acoustic modes within its layers and one at right angles to them. In carbon nanotubes, on the other hand, there are four acoustic modes, and they consequently differ from graphite in their thermal properties. Still at room temperature enough phonon levels are occupied for the specific heat capacity to resemble that of graphite. Only at very low temperatures the quantized phonon structure makes itself felt and a linear correlation of the specific heat capacity to the temperature is observed. This is true up to about 8 K, but above this value, the heat capacity exhibits a faster-than-linear increase as the first quantized subbands make their contribution in addition to the acoustic modes.

3.4.6.2 Heat Conductivity of Carbon Nanotubes

The transport of heat through a solid is mainly effected by low frequency phonons. Hence the heat conductivity κ_{zz} along the axis of a carbon nanotube can be described as the sum of all phonon states and their respective heat capacity C_{Ph}:

$$\kappa_{zz} = \sum C_{Ph} v_z^2 \tau \qquad (3.13)$$

with v_z being the group velocity and τ being the relaxation time of the individual phonon state. It is evident that phonons with a high band velocity or a long free path, respectively, contribute the most to heat conductivity. Therefore, carbon nanotubes along their axis feature the highest heat conductivity ever measured for any material. The value at right angles to the tubular axis, on the other hand, is only about a hundredth of κ_{zz}. Multiwalled carbon nanotubes and bundles of single-walled species should thus be expected to possess about the same heat conductivity like their single components, whereas samples with an unordered arrangement of individual tubes should show lower values due to intertubular coupling. These assumptions have been confirmed by experimental results. Heat conductivities of up to $3000\,W\,m^{-1}\,K^{-1}$ have been measured for single MWNT, whereas for mats of SWNT, they only range from 35 to 200 $W\,m^{-1}\,K^{-1}$ depending on the degree of parallelism. The temperature dependence of heat conductivity has a maximum at about 310 K that is due to phonon-phonon scatter occurring above this value. However, the effect is not as pronounced as in graphite and the maximum is shifted toward higher temperatures because due to the one-dimensionality of nanotubes there are less phonon states available to scatter to.

3.5 Chemical Properties

3.5.1 General Considerations on the Reactivity of Carbon Nanotubes

For carbon nanotubes just as well as for the fullerenes discussed in Chapter 2, there is a direct correlation between their structure and their respective reactivity. Further important conclusions can be drawn from the relation to graphite. Considering a nanotube quickly reveals that there are three clearly distinct sites available for a chemical reaction (Figure 3.62): the most reactive tips, the outer side

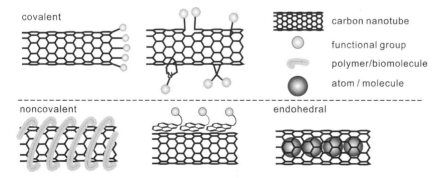

Figure 3.62 Possible ways of carbon nanotube functionalization.

wall, and the cylindrical inner surface of the tube. In addition, there is the marked reactivity of defect sites that will also be discussed in this chapter.

With the carbon nanotubes formally being a graphene sheet rolled up around the z-axis, one might expect their chemical behavior to resemble that of graphite. Yet this is only true in parts. Carbon nanotubes, as mentioned in Section 3.2.2 on structure and aromaticity, possess an extended network of delocalized π-electrons like graphite. Yet contrary to a graphene layer, the cylindrical shape of the tube enforces a divergent orientation upon the π-orbitals. According to the aforementioned considerations on aromaticity, this induces a certain mixture of σ- and π-orbitals. The latter then come to protrude radially from the tube's surface, which causes a less effective overlap of orbitals toward the outside. This is all the more true for nanotubes with small diameters because they are particularly curved. Very large tubes like the outer walls of MWNT mark the other extreme of the range. Their carbon network is much less curved and the respective interactions remain more efficient. In the limiting case of an infinite diameter a curvature of zero is achieved. All π-orbitals are in parallel with each other and the structure represents a graphene sheet. Hence, the diameter plays a decisive role for the reactivity of nanotubes. The thinner a tube gets, the more easily it actually reacts. The hybridization of carbon atoms is shifted more toward sp^3 with a stronger curvature of the tubular surface. This facilitates a complete transition into sp^3-hybridization in the course of any reaction.

Comparing fullerenes to nanotubes, it is obvious that the first are curved with regard to all three directions in space, whereas no curvature at all is observed in the axial direction of nanotubes. The fullerenes further exhibit a rich double bond chemistry as the incorporation of five-membered rings affects the distribution of double bonds in the six-membered rings to the effect of a reduced delocalization. From these considerations, nanotubes can be assumed to be less reactive than fullerenes, which is indeed confirmed by actual observation. It is true that nanotubes enter into a number of reactions known for fullerenes, yet normally more drastic conditions or longer reaction times, respectively, are required to achieve similar results. What is more, samples of carbon nanotubes never consist of just one type of structure with exactly one pair of descriptors n and m. Usually it is rather all three types of tubes with various helicities being present. In addition, there is always a distribution over a certain range of diameters and lengths. Hence unlike with the fullerenes, no uniform compound is obtained from the functionalization of carbon nanotubes, but inevitably a mixture of substances with various amounts and arrangements of functional groups is produced.

Another aspect of the structure must by all means be regarded when discussing the reactivity of real nanotubes: Contrary to the "ideal" tubes considered so far, the ones actually produced by the methods described in Section 3.3 contain defects, sometimes even many of them. These defects may be, for example, holes in the side wall bearing sp^3-hybridized carbon atoms around their rim. The latter may either be saturated with functional groups or they may exist as "dangling bonds." The reactivity of the respective nanotube is markedly increased at these positions.

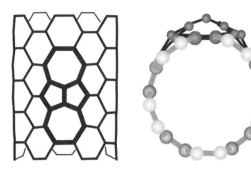

Figure 3.63 The *7-5-5-7* defect as site of increased reactivity. It is attacked more easily due in particular to the stronger curvature.

There is a number of further defects that also cause a local increase in chemical reactivity. These include the so-called Stone–Wales defects that result from a Stone–Wales type rearrangement of two six-membered rings into a structure with annealed five- and seven-membered rings. Two of these defects in direct contact give rise to the *7-5-5-7*-arrangement shown in Figure 3.63. Obviously the tube exhibits an increased curvature at this site, and the respective π-orbitals possess a higher degree of sp^3-hybridization. Therefore, the carbon atoms situated around such a defect are more easily attacked by potential bonding partners. Still not all bonds of a defect site feature increased reactivity. The bond connecting the two seven-membered rings, for instance, is less easily attacked than a normal 6,6-bond of a nanotube. The positions of higher pyramidalization and consequently increased reactivity are the contacts between seven- and six-membered rings as well as those between five- and six-membered ones. After all, it seems that the decisive parameter is not the kind of rings in contact with each other, but the arrangement of the respective bond with regard to the tube's axis: bonds oriented in perimetral direction are more easily attacked than approximately axial ones. Altogether, Stone–Wales defects mean a disturbance to the electronic and geometric structure and thus increase the susceptibility of nanotube side walls to chemical attack.

Electron micrographs show yet another kind of defect that causes a local rise of reactivity. They are positions along a nanotube where a bend is generated by a five- and a seven-membered ring, which are positioned on opposite sides along the perimeter, respectively (Figure 3.64). Especially the convex side of the kink, where the five-membered ring is situated, is much more reactive than the usual side wall. The position of the seven-membered ring at the concave side of the kink is less prone to attack due to the steric requirements. However, these kinks are over all rather susceptible to oxidative reactions which, in many cases, cause the tubes to break apart. The reaction results in markedly shortened segments that usually exhibit no kinks any more.

In principle, the aforementioned considerations hold true for both single- and multiwalled nanotubes. The reactivity of both types differs mostly due to the

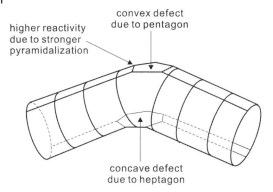

Figure 3.64 Scheme showing a kink defect. The reactivity of carbon atoms is increased especially on the convex side.

curvature of the outermost wall which is normally less pronounced in MWNT. Hence compared to SWNT, the state of the respective π-orbitals is more alike the one in graphite here.

Furthermore, there are interactions between neighboring walls within a multiwalled tube that cause an additional stabilization. Still it is no serious problem to derivatize multiwalled carbon nanotubes as well. In doing so, one takes advantage of the increased reactivity at the tubes' ends and of the vulnerability of defect sites in the outer wall of the MWNT.

The electronic structure of nanotubes naturally influences their chemical behavior, too. Provided electrons close to the Fermi level are involved, the reactivity is expected to differ for all variants: metallic as well as semiconducting, either with small or with large bandgap. The exact position of the Fermi level, however, is largely dependent on the type and position of defects in the carbon lattice, so there is no simple correlation to be observed experimentally. Yet it has turned out in the course of time that certain reactions exhibit remarkable selectivity for specific types of nanotubes. These include, among others, the reaction with diazonium salts and the photochemical osmylation.

The subsequent sections will now deal with the different methods of nanotube functionalization. The reactivity of MWNT will be mentioned separately only where significant differences exist between the behavior of single- and multiwalled tubes.

3.5.2
Redox Chemistry of Carbon Nanotubes

Carbon nanotubes can act both as electron donor or acceptor and consequently exhibit an abundant redox chemistry. Still oxidizing reactions mainly take place at the tubes' ends and at defects. The direct oxidation of carbon atoms in an intact side wall, on the other hand, rarely ever occurs. It would require much

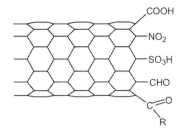

Figure 3.65 Functional groups potentially present after an acid treatment of carbon nanotubes.

harsher reaction conditions that normally cause the nanotube structure as a whole to break up, resulting in smaller fragments with functional groups containing oxygen on their rims. A very important feature of oxidations on nanotubes is the opening of the tubes' tips. Firstly, this improves sample homogeneity, but secondly it also opens access to the cavity inside the tube (Section 3.5.6). Actually, the functionalization of terminal carbon atoms enables the multifarious chemistry of nanotube ends (Section 3.5.3). Oxidizing methods further suit to change the length of crude nanotubes, this is termed as the "oxidative cutting" of CNT (Section 3.3.6.3).

The reaction with hot, concentrated oxidizing mineral acids like nitric or sulfuric acid introduces carboxyl groups to the ends of the tubes and to defects of the side wall. Yet this reaction is not too selective, and a variety of further oxidized structures is formed besides the carboxyl functions. They include keto and hydroxmethyl groups, anhydrides, and sulfonic acids (Figure 3.65). Nitration can also occur at the six-membered rings of the nanotubes.

Carbon nanotubes can also react with other oxidants besides nitric or sulfuric acid. These may be hydrogen peroxide, chromosulfuric or perchloric acid, $KMnO_4$, K_2IrCl_6, nitrating acid and molecular oxygen, or air at elevated temperatures, to name a few. Oxygen dissolved in water can also serve as oxidant. The latter reaction simply requires to work at acidic pH to let the redox process take its course. From a thermodynamic perspective, this reaction is possible with tubes of any size as the potential for the half-reaction of oxygen ($4H^+ + 4e^- + O_2 \rightarrow 2H_2O$; 840 mV at pH 7) is higher than the oxidation potential of even the smallest carbon nanotube.

$$4\,SWNT + O_2 + 4\,H^+ - 4\,SWNT^+ + 2\,H_2O \qquad (3.14)$$

The oxidation of nanotubes can be reversed by the reaction with reductants like $NaBH_4$ or $Na_2S_2O_4$. Provided the samples are more or less homogeneous, the redox reaction can be monitored by absorption spectroscopy, which enables the performance of redox titrations.

The reactivity of some oxidizing agents, for example, sulfuric acid, can even be enhanced by simultaneous ultrasonication. This technique reduces the reaction times required and facilitates the cutting of the tubes.

Compared to small graphene fragments or amorphous material, the nanotubes stand the oxidation mostly undamaged. The only considerable exceptions are the opening of the tips and the occasional generation of defects, while the small components of the sample like amorphous carbon are completely oxidized. These methods consequently suit for the purification of carbon nanotubes. The respective products are always shortened tubes featuring terminal or defect functionalization. For MWNT, the intensity of the reaction may as well be adjusted to decide whether or not outer walls of the tubes are oxidatively removed. This allows for a variation not only of the length, but even of the diameter of the tubes.

Some of these reactions are conducted in suspension, while others are not. The latter variant, if applied to bundles of nanotubes, has the general drawback of accessing only the outer surface of these tightly bound aggregates. This results in a quite heterogeneous mixture of tubes, their degree of functionalization ranging from high via mediocre down to "not at all." This problem is most pronounced in reactions of nanotube powders with gases. Therefore, it is crucial to employ samples with the highest possible degree of dispersion and a weak bundling.

Reacting, for instance, bundles of MWNT with oxygen at 700 °C gives satisfying results only if a sufficient debundling has been achieved before, for example, by mechanical means. It has to be considered as well that the tubes' ends are opened in the course of the reaction, thus generating unsaturated bonding sites. Before further use of the partly oxidized nanotube material these have to be healed in an annealing step leading to a toroidal structure at the ends. It interconnects several individual walls of the very same MWNT (Figure 3.10).

Ozone is another reagent to attack carbon nanotubes. In a first step the respective ozonides are formed by addition to the double bonds, and after work up, carboxyl, keto, or hydroxyl functions are obtained. A further oxidation route can be taken in the photochemical decomposition of ozone with concomitant generation of oxygen radicals. This method also results in a partial destruction and, consequently, in a shortening of the tubes. Especially further heating after ozonization causes the elimination of CO_2 and CO from the functional groups bearing oxygen, which gives rise to defects that may, for example, significantly alter the absorption characteristics of the tubes. The electronic properties are affected as well. Metallic tubes may for instance be transformed into semiconductors by introducing a multitude of such oxidized sited.

Supercritical water as well may bear oxidizing effect under suitable conditions. The reaction can also serve to purify nanotube samples as these are less reactive than the impurities present. The removal of catalyst particles and amorphous carbon material is achieved this way.

There is also a large repertoire of reactions available for the reduction of carbon nanotubes (Figure 3.66). The most effective method is the reaction with alkali metals in liquid ammonia with subsequent methanolysis. At first, the alkali metal (lithium in most cases) is oxidized to the respective monovalent cation, while at the same time a solvated electron is generated. The highly reactive species $[e^-(NH_3)_x]$ attacks the carbon atoms of the nanotubes and reduces them to carbanions. The ensuing reaction with methanol leads to the hydrogenation of the carbon atom and to the production of one equivalent of methanolate. This process introduces an sp^3-center into the framework of the tube. This bears strongly on

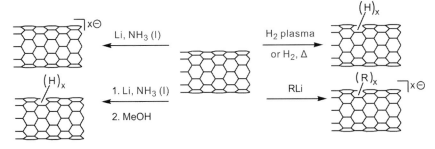

Figure 3.66 Methods of carbon nanotube reduction.

the electronic conditions because the conjugation of π-electrons is interrupted. Furthermore, a wavy, partly unordered structure is formed especially for the outer walls of MWNT, which is due to the new spatial requirements of the hydrogenated carbon atoms. A similar picture is obtained for the reduction of single-walled tubes. The structure of the nanotube surface is disarranged by the sp^3-hybridized centers and exhibits an unordered shape. The composition of the reduced nanotubes roughly corresponds to $C_{11}H$, which means a considerable degree of functionalization especially for the outer wall of MWNT.

The reaction of single-walled carbon nanotubes with lithium in liquid ammonia generates multiply reduced SWNT anions surrounded by lithium cations. In aprotic media, these negatively charged single-walled tubes may then be reacted with alkyl halogenides to yield alkylated nanotubes (Figure 3.66). Using long-chained alkyls like dodecyl residues ($-C_{12}H_{25}$) leads to a markedly improved solubility of the respectively modified nanotubes in organic solvents.

The lithium salts of reduced SWNT themselves have been studied as well (Figure 3.66a). They behave like a polyelectrolyte and feature a relatively good solubility in polar, aprotic solvents even without the application of ultrasound, surfactants or the like. Solutions of up to $2.0\,mg\,ml^{-1}$ in DMSO or $4.2\,mg\,ml^{-1}$ in sulfolane can be obtained. The nanotube salts solve as well in DMF and NMP. The concentration of charge reaches roughly one negative charge per ten carbon atoms.

The direct reaction of nanotubes with organolithium compounds yields reduced tubes as well. Normally, these are saturated immediately with the organic residue, which is why they carry comparatively few charges. Species of this kind can nevertheless be employed to start polymerizations that produce interesting composite materials (Section 3.5.5). Electrochemistry is another means of reducing nanotubes, for example, by using the reductive coupling of aryl residues generated from anilines with the tube. In doing so, electrons are transferred onto the nanotube. Further reactions, for example in supramolecular structures, also include electron transfer processes (like an SET) changing the redox state of the respective carbon nanotube (Sections 3.5.4.2 and 3.5.7). Nanotubes can both take up or set free electrons depending on the kind of interaction at hand.

Some more reactions comprising the oxidation or reduction of nanotubes are described in the sections on side-wall functionalization. They include, for example, the ozonization, the electrochemical reaction with diazonium salts, and the hydrogenation with various reagents. Furthermore nanotubes are either reduced or oxidized, depending on the bonding partner, in the formation of charge transfer complexes. Examples can be found in Sections 3.5.4.2 and 3.5.6 on intercalation compounds and noncovalently functionalized nanotubes, respectively.

3.5.3
Functionalization of the Caps or Open Ends of Carbon Nanotubes

A first covalent functionalization of the tubular structure is already achieved upon purification of the nanotubes obtained from different methods of production. Normally, the reaction consists in an attack of the in general oxidizing reactants to the ends of the individual tubes, be it single- or multiwalled species. The attack to the tubes' tips can lead to different results depending on the reaction time and the extent of reactivity. A complete removal of the cap would be the extreme. This would mean the formation of open nanotubes carrying functional groups (carboxyls groups for the most part) at their rims (Figure 3.65). Under the drastic conditions given, however, it is not only the tubes' ends reacting, but defects of the side wall like holes or Stone–Wales defects (Section 3.5.1) exhibit an increased reactivity too and get attacked accordingly. Yet provided a more or less perfect surface, the reaction at this stage occurs only at the nanotube tips.

Carboxyl derivatives obtained from the oxidative opening of nanotubes can be further modified with the classical methods of organic chemistry (Figure 3.67). The respective products from the attachment of long alkyl chains exhibit markedly increased solubility in organic solvents. The debundling of single-walled carbon nanotubes can as well be promoted by the opening and functionalizing of tubes' ends as the intertubular van der Waals exchange decreases due to the modification. Normally, the overall length of the tubes is reduced to a few hundreds of nanometers in the course of such reactions as well.

The acid groups can be coupled with other organic compounds. Most easily this is achieved by esterification or amide formation (Figure 3.67). There is a multitude of structures that have been bound to nanotubes in this manner. Examples include even biologically active substances that have been attached to the tips of SWNT or MWNT, thus enabling their use in sensor applications. Reacting the terminal carboxyl groups with N-hydroxysuccinimide yields a nanotube derivative that is easily connected to peptides carrying nucleobases. These so-called peptide nucleic acids (PNA) may pair with single stranded DNA. Adding DNA with "sticky ends" (a piece of DNA presenting a short sequence of unpaired single strand at its end) to a suspension of PNA-modified nanotubes leads to a pairing with the complementary bases of the PNA. The DNA is then bound to the nanotube. This allows for self-assembly or a detection by molecular recognition.

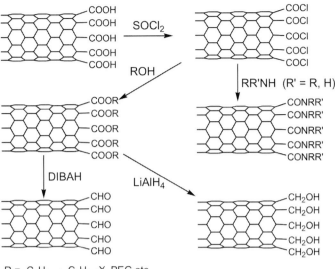

R = -C_nH_{2n+1}, C_nH_{2n}-X, PEG etc.
(X = functional group, e.g. -NH_2, -OH, -halogen, -COOH, -CH=CH_2 etc.)

Figure 3.67 Further reactions of carboxylated carbon nanotubes with carboxylated tips.

Water soluble carbon nanotubes can be produced by attaching polyethylene glycol units to the carboxyl groups (Figure 3.67). In this way, it has been possible to dissolve several hundreds of milligrams in one milliliter of water. Useful derivatives can be prepared by reacting the carboxyl units at the ends of the tubes. One example are tips for a chemosensitive AFM (Section 3.6.1.1).

The asymmetric modification of the ends is a particular challenge. Normally both ends are functionalized the same way. This is detrimental, however, for some applications that require a specific alignment of the tubes. Therefore methods of deliberately modifying both ends in different ways have been sought. Indeed one has succeeded by now in positioning a nanotube film in such a way that one end of the tubes is immersed into the solution of a photoactive reagent while the other side protrudes into air. This is achieved by keeping the tubes afloat on a solvent either due to the latter featuring a high specific density (e.g., 1,1,2,2-tetrachloroethane) or due to repulsion for lack of wettability (e.g., water or ethanol). The tubes in such a film assume a parallel alignment. This setup allows for a selective photochemical functionalization of one side of the tubes, while the other side can be reacted with a different partner only after turning over the film. The nanotubes obtained may, for instance, carry perfluorooctyl groups on one side and 3′-azido-2′-desoxy-thymidine on the other (Figure 3.68a). In general, it is a very promising technique to employ the low density and poor solubility of nanotubes for asymmetric functionalization. A further approach has been reported for SWNT. These have been bound on one end to a gold surface to allow for an asymmetric modification (Figure 3.68b).

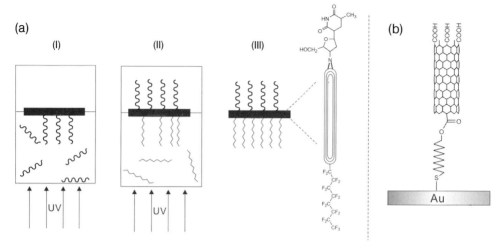

Figure 3.68 Asymmetric functionalization of carbon nanotubes (a) by reaction in a liquid–liquid phase boundary, and (b) by immobilization on a solid substrate.

3.5.4
Side Wall Functionalization of Carbon Nanotubes

3.5.4.1 Covalent Attachment of the Functional Groups

As for the fullerenes, the development of methods for nanotube functionalization began very soon after their discovery. After first successes in opening the tubes and attaching functional groups to their ends, the next attempts were made in applying the common reactions of fullerene modification to the side-wall functionalization of the structurally related carbon nanotubes. Many reactions performed on fullerenes can indeed be applied to nanotubes as expected. However, the latter are generally observed to be less reactive, which has already been discussed in Section 3.5.1.

Suitable reactions for side-wall functionalization are mainly those that attack the π-system of the nanotube. Primarily, these are transformations like the addition or cycloaddition known from the chemistry of double bonds. These transformations shall be discussed in the following. The direct attack to the π-system of the tube has yet another interesting aspect: Unlike the functionalization of the ends it enables a control over the electronic structure of the entire tube. A suitable modification thus allows for the construction of complex electronic systems based on carbon nanotubes. At first, however, the more simple side-wall functionalizations will be discussed.

Hydrogenation The simplest modification conceivable for the side wall of a carbon nanotube is the hydrogenation. Taking it to the extreme, it would yield a tubular, annealed hydrocarbon. The character of a nanotube, however, would most likely be destroyed upon an exhaustive hydrogenation. Actually, experiments like

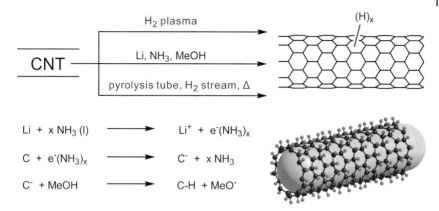

Figure 3.69 Hydrogenation of carbon nanotubes. The theoretical structure of a completely hydrogenated nanotube is shown in the lower right corner.

the reaction with molecular hydrogen in a plasma or with alkali metals (sodium, lithium) in liquid ammonia under modified Birch conditions gave partially hydrogenated carbon nanotubes. High-temperature reactions in a tube furnace under an atmosphere of hydrogen led to similar results (Figure 3.69). A completely hydrogenated nanotube would exclusively consist of sp^3-hybridzed carbon atoms. Calculations revealed that such a species should be stable up to a tube's diameter of 12.5 Å (roughly corresponding to an 8,8-nanotube). Above this value, the incorporation of sp^3-centers succeeds no longer completely due to the reduced curvature.

In analogy to the chemistry of fullerenes, the reaction of carbon nanotubes with boranes could be assumed to provide hydroborated products that might be converted into a number of derivatives, for example, into partially hydrogenated structures by treatment with carbonic acids. It turned out, however, that the reactivity of double bonds in nanotubes does not suffice for a hydroboration. According to calculations, the process should be thermodynamically neutral when performed on a typical SWNT, which means it is anything but a preferred reaction.

Halogenation The fluorination of carbon nanotubes is an important primary functionalization because it may be conducted even as heterogeneous process between gaseous and solid phase. The reaction of fluorine with SWNT, for example, can be performed in a tube furnace at about 150 °C. It yields perfluorinated nanotubes with a degree of fluorination of up to 100%.

DWNTs and MWNTs can be fluorinated, too. In DWNTs, only the outer wall is attacked as the inner tube is protected by its exterior counterpart. The resulting compound has an overall composition of ca. $CF_{0.3}$, which corresponds to a very high degree of fluorination for the outer wall.

The reverse reaction of fluorinated species being converted into unmodified nanotubes can be achieved using hydrazine. The method may also be employed to remove just a part of the fluorine atoms, thus allowing for an adjustment of the degree of fluorination.

$$4\,C_nF + N_2H_4 \rightarrow 4\,C_n + 4\,HF + N_2 \tag{3.15}$$

The mechanism of the fluorination has not yet been fully elucidated. However, radical intermediates are presumed. Normally, the fluorine atoms are found in close proximity, and the addition preferably advances not along the axis, but along the circumference of the tube. This results in domains with a high degree of fluorination existing side by side with regions that are at best scarcely attacked by fluorine. A possible explanation for this effect is based on the assumption of a 1,4- instead of a 1,2-addition to the conjugated six-membered rings. This hypothesis is supported by calculations that indicate the 1,2-addition to be energetically favorable along the tube's axis, while the same is true for a 1,4-addition along the perimeter. However, the question on the thermodynamic stability of the aforementioned products must remain open for the time being. Different calculations predict the stability of either the 1,2- or the 1,4-adduct. Actually, the energetic differences in all cases are determined to be a few kilojoules per mole and fluorine atom only. Hence in reality, both addition pathways can be assumed to exist in parallel.

There is yet another phenomenon that points to a 1,4-addition of fluorine atoms occurring at least partially: The products obtained from the reaction of single-walled carbon nanotubes with fluorine are definite nonconductors (resistance >20 MΩ), while the nanotubes employed feature a resistance of 10–15 Ω only. Considering the structures that result from a consecutive 1,2- or 1,4-addition of fluorine, respectively, reveals that an electric current via conjugated π-bonds would still be possible in the 1,2-adduct (Figure 3.70). In the 1,4-adduct, on the other

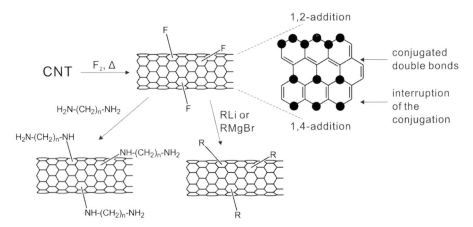

Figure 3.70 Halogenation of carbon nanotubes. The reaction at least partly takes place as 1,4-addition. Fluorinated nanotubes are good starting materials for further functionalization (bottom).

hand, the conjugation is interrupted and the product comes to be an insulator. This calls for the 1,4-addition to take place at least at some positions, which prevents the fluorinated tube from being an electric conductor. Fluorinated nanotubes differ from the unmodified species not only in electric conductivity. There are rather more characteristics being changed. Fluorinated tubes dissolve, for instance, in some organic solvents like DMF, THF, and different alcohols. Most of all, the solubility in 2-propanol and 2-butanol is increased. Hydrogen bonds between the protons of the hydroxy groups and the fluorine atoms of the nanotubes are assumed to cause an effective solvation.

Although the fluorination is a well-established standard method for the primary functionalization of carbon nanotubes, little has been published on the halogenation with chlorine and bromine. Possible reasons for this may include fluorine being by far the most reactive of halogens and consequently being the most efficient in an attack to rather unreactive species. Furthermore, it turned out that the heavier halogens can also enter into noncovalent exchanges that interfere with the actual carbon–halogen bond formation.

Different charge transfer complexes have been described to result from the reaction of SWNTs or MWNTs with bromine. They feature a shift of electron density from the nanotubes toward the halogen atoms. Yet the formation of covalent bonds has not been observed in any bromination of carbon nanotubes. Still the bromine-nanotube adducts obtained exhibit a marked change of electronic properties in comparison to the unmodified tubes. Especially the concentration of charge carriers is increased by more than one order of magnitude during the reaction, which is attributed to the charge transfer complexation. In a simplifying model, a Br_2^- species is generated by the acceptance of delocalized electrons from the nanotube, whereby the latter undergoes p-doping.

Compared to the bromination, the chlorination of carbon nanotubes should be easier due to the higher reactivity of chlorine. This effect is observed indeed. However, the chlorination chiefly occurs at the ends of the tubes or at defects of the side wall. The direct reaction of the side wall itself has not yet been described as a preparative method. Still the chlorination at defect sites provides materials with modified characteristics as well.

Reactions of Fluorinated Carbon Nanotubes The fluorine atoms attached to the side wall of functionalized carbon nanotubes are easily substituted by alkyl residues, which, among other reasons, is due to the C–F-bond being weakened by ecliptic interactions. Organometallic substances like lithium organyls or Grignard compounds are employed in the reaction (Figure 3.70). This way, a direct C–C-junction of functional groups with the nanotube framework is achieved. There are further methods of direct C–C-bonding to be discussed later. Still the pathway via fluoro-intermediates and subsequent alkylation or acylation leads to a wider variety of possible products and, in comparison to other methods, to a higher degree of functionalization.

The reaction of fluorinated carbon nanotubes with alcoholates yields ether-bridged structures. The analogous reaction with primary amines provides species

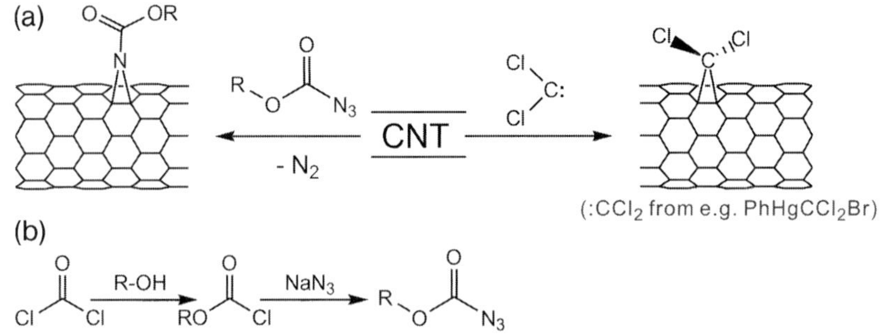

Figure 3.71 Cycloaddition of nitrenes (a) and carbenes (b) to individual double bonds of a carbon nanotube.

with a nitrogen bridge between tube and residue. Employing terminal diamines with a sufficient length of the connecting alkyl chain allows for an interconnection of tubes, thus establishing a covalently bound network. With the methods presented here, modified nanotubes with about every seventh carbon atom carrying a functional group can be produced.

Addition of Carbenes and Nitrenes The addition of highly reactive electron-deficient reagents like carbenes or nitrenes enables the preparation of nanotubes modified with cyclopropane- or azacyclopropane rings, respectively. A large variety of functionalized nanotubes can be obtained using appropriately substituted reagents here. Especially the photochemical or thermal *in situ* generation of oxycarbonyl nitrenes from the respective azidocarbonates allows for a wide variation of functionalities as the respective azides are easily available (Figure 3.71a). For example, sugars can be attached to the nanotube in this manner. These might serve as scaffold for the subsequent synthesis of DNA-sequences directly on the nanotube. In detail, the chlorocarbonates are obtained from reacting an alcohol ROH with phosgene. The intermediate is converted into the azidocarbonate by reaction with sodium azide (Figure 3.71b). Alkylazides, on the other hand, do not suit to this kind of transformation. They would preferrably enter into [3 + 2]-cycloadditions without nitrogen cleavage according to calculations, and in practice even this latter reaction failed.

The addition of dichlorocarbenes also allows for further conversions as the existing substituents can be replaced. Normally, $PhHgCCl_2Br$ is used as carbene source in this reaction, but it succeeds with other carbenes or carbene sources as well. The addition of dichlorocarbene markedly changes the electronic structure of the functionalized nanotube. Metallic SWNT, for instance, turn into semiconductors, which is due to altered electronic transitions close to the Fermi level. What is more, the intensity of interband transitions decreases for both the original metallic and the semiconducting species because the extended π-network is severely disturbed by the functionalization and the concomitant introduction of sp^3-carbon atoms. However, a completely different picture of the carbene or nitrene addition is

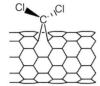

Figure 3.72 Alternative reaction of nanotubes with dichlorocarbene resulting in the opening of the side wall.

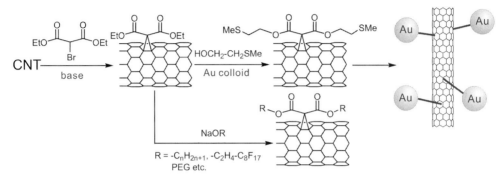

Figure 3.73 The Bingel–Hirsch reaction on carbon nanotubes and examples of subsequent reactions on the respective products.

obtained from calculations. They indicate that no three-membered rings are formed, but that an opening of the nanotube's side wall occurs (Figure 3.72). This way the sp²-structure is conserved and the electronic conditions close to the Fermi level are just slightly affected at least at low degrees of functionalization. It is only at higher coverages (about 20%) that the substituents distinctly change the band structure of the respective tube. The same calculations revealed that, due to the electronic conditions, metallic tubes react more easily than semiconducting ones. But then, at higher degrees of substitution (15–20%, depending on the tube's diameter) the conversion from metallic into semiconducting nanotubes sets in anyway. The method of carbene or nitrene addition might be used for a tailored modification of the bandgap of carbon nanotubes, yet the progresses on this field are not exactly numerous.

The Bingel Reaction Also on carbon nanotubes, the Bingel reaction is performed quite easily. Yet in comparison to C_{60}, a diminished reactivity is observed. Here as well a bromomalonate is deprotonated by a strong base, and the resulting carbon nucleophile attacks on a double bond. The subsequent cleavage of a bromide anion leads to the formation of a three-membered carbon ring carrying two ester groups at its apex. These groups then provide the opportunity of further functionalizations. The Bingel reaction thus constitutes the starting point for the preparation of numerous nanotubes modified in different ways (Figure 3.73). Saponifying the ester groups yields the free acid – another example for the attachment of a

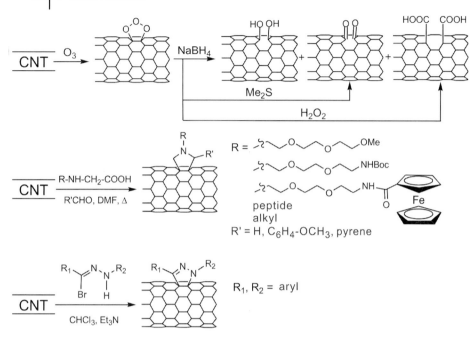

Figure 3.74 Examples of [3+2]-cycloadditions and of the multifarious derivatives obtained. Some of these reactions are of interest for biological applications in particular as they tolerate a multitude of functionalities.

carboxyl group without direct connection to the nanotube's framework. Transterifications or the reactions of the free acid enable the bonding of various functional moieties.

Reactions with Ozone The reaction with ozone proceeds according to the classical Criegee mechanism that starts with the 1,3-dipolar cycloaddition of an ozone to the respective double bond. The immediate product is the primary ozonide which, depending on the way of work up, can be converted into a number of functional groups. Reactions with hydrogen peroxide, dimethyl sulfide, or with sodium borohydride have been reported in the literature. Depending on the reagent used, they allow for the generation of carboxyl, keto or hydroxyl groups, respectively, instead of the original double bond (Figure 3.74a). It turned out that to some extent the ozonization of SWNTs is diameter-selective with thinner tubes reacting much more easily than larger ones. A future method for the size-selective separation of carbon nanotubes by functionalization might be envisaged here. The oxidative as well as the reductive work up of the ozonides break bonds in the nanotube, so the ozonization introduces not only functional groups, but also defects into the tube's side wall. This causes distinct changes of the electronic properties.

1,3-Dipolar Cycloadditions In analogy to the chemistry of fullerenes, 1,3-dipolar or [3+2]-cycloadditions succeed with carbon nanotubes as well. Like in most other reactions, too, the tubes are less reactive due to their weaker curvature, so normally more drastic conditions have to be applied.

The reaction with azomethine ylides is a typical [3+2]-cycloaddition. It leads to the products shown in Figure 3.74b. The reaction is particularly suitable to couple biologically active moieties to carbon nanotubes, which is of great relevance, for example, for the preparation of nanotube–peptide composites. The reaction is normally performed on a nanotube suspension in DMF that is treated with an N-substituted glycine and paraformaldehyde or with an aldehyde carrying a further residue to be coupled. The azomethine ylide is then formed *in situ*; it constitutes the actual reagent. The method is applicable to single- and multiwalled nanotubes. The degree of functionalization achieved is about $0.4\,\mathrm{mmol\,g^{-1}}$ for SWNT and about $0.7\,\mathrm{mmol\,g^{-1}}$ for MWNT. Bundles of single-walled tubes are not completely broken in the course of the reaction, while multiwalled species are completely separated afterward.

The solubility of the resulting nanotube derivatives can be tuned by employing different kinds of residues attached to the nitrogen atom of the pyrrolidine ring. Those patterns of functionalization that increase the solubility in physiological media are of special interest for biological applications. Oligo- and polyethylene glycols (PEG) covalently inserted between the attachment site and the residue's essential functionality proved their worth here (Figure 3.74). Many different groups and molecules may be bound to the end group of the spacer attached to the nanotube. They include, among others, amino acids, peptides, DNA, and peptide-based DNA.

Redox-active substances can be bound to nanotubes this way, too. The first photo-induced charge transfer from a nanotube to a ferrocene has been observed on a tube connected with the ferrocene in this manner (Figure 3.74). The longevity of the charge-separated state renders the material interesting for electronic applications.

Another [3+2]-cycloaddition on carbon nanotubes may be performed with nitrile imines. In doing so, a pyrazole ring is formed on the tube's surface (Figure 3.74c). Upon choosing an appropriate substituent, an electron transfer from the electron-rich residue to the nanotube can occur.

The ozonization of carbon nanotubes formally constitutes a [3+2]-cycloaddition, too. Nevertheless, it has been discussed separately here because only the products after work-up are relevant for the further functionalization and examination of nanotube derivatives.

[4+2]-Cycloadditions The Diels–Alder reaction on the side wall of carbon nanotubes had been predicted theoretically rather early, still the success in actually performing it was long in coming. Finally, in 2004, the conversion of SWNT functionalized on their end-caps into the respective product of a cycloaddition succeeded by the microwave-assisted reaction with o-quinodimethane (Figure 3.75). The latter is generated *in situ* from 4,5-benzo-1,2-oxathiine-2-oxide and constitutes the diene component, while the double bond of the nanotube functions as dienophile. Applying pressure and adding $Cr(CO)_6$ also enables the Diels–Alder reaction of electron-rich dienes on unfunctionalized nanotubes. Here again, the

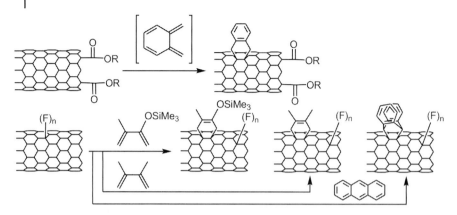

Figure 3.75 Examples of the Diels–Alder reaction on the side wall of carbon nanotubes.

analogy to fullerene chemistry becomes obvious as those always act as dienophiles, too. It is conceivable, though, that the less curved outer walls of multiwalled tubes may also react with strong dienophiles.

The Diels–Alder reaction succeeds as well with fluorinated nanotubes. Different dienes can be connected this way to the double bonds of the nanotube surface (Figure 3.75). A 5% degree of functionalization is achieved in doing so. The fluorination enhances the reactivity of the remaining double bonds of the carbon nanotube as, firstly, the tension is increased by adjacent sp^3-centers and, secondly, due to the electron-withdrawing effect of fluorine atoms. Once more the double bonds of the nanotube always function as dienophile.

Altogether the electrophilic character of nanotubes is less pronounced than that of fullerenes. Regarding that in addition their reactivity is diminished by the weaker curvature, one would not expect too strong an affinity to typical dienes. Still further reports on examples of the Diels–Alder reaction on single- and multiwalled nanotubes may be foreseen for the future.

Photochemical and Radical Reactions Most photoreactions in the field of nanotube chemistry serve to the generation of reactive intermediates that attack the nanotube afterward. The functionalizing step itself is rarely photochemical. Examples of such a preparatory step prior to functionalization include the conversion of azides into nitrenes or the radical generation from acyl peroxides, iodoalkanes, etc. In the photochemical reaction of nanotubes with osmium tetroxide, on the other hand, the essential step occurs only under irradiation.

Comparing fullerenes to carbon nanotubes, one would expect for the latter a pronounced reactivity in photochemical [2+2]-cycloadditions, too. For the time being, however, this has not been confirmed experimentally. The reactivity of nanotubes obviously does not suffice to enter into this transformation. The dimerization by irradiation in solid phase as known from the fullerenes has not yet been realized either.

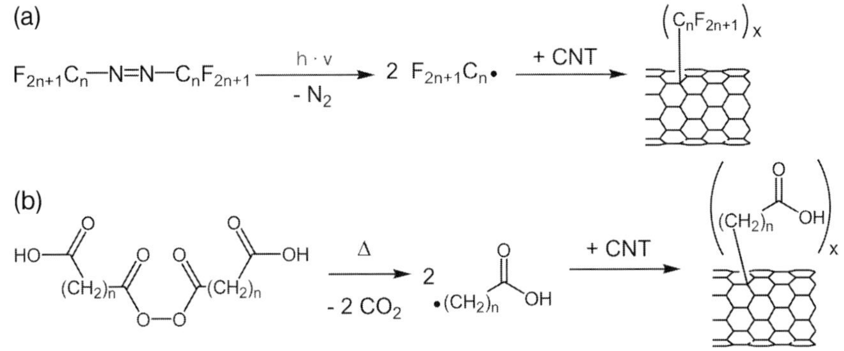

Figure 3.76 Examples of radical reactions on carbon nanotubes with photochemical (a) and thermal (b) initiation.

Radical reactions on carbon nanotubes can be performed in close analogy to those on classical double bonds. It is possible, for instance, to connect them with perfluorinated alkyl residues obtained from the photolysis of the respective perfluoroazoalkanes (Figure 3.76, top). The reaction with alkyl iodides and lithium yields alkylated nanotubes as well (also refer to Sections 3.5.4.1 and 3.5.4.1.3). The formation of lithium iodide is the driving force in this process. The direct reaction with alkyl halogenides has been reported, too. For example, SWNT may be converted directly into their alkylated derivatives by milling them in an atmosphere of trichloromethane, trifluoromethane, hexafluoropropane, or tetrachloroethylene. The reactive sites on the nanotube are generated by mechanical energy here.

The reaction with diacyl peroxides under the action of heat constitutes another way of directly attaching alkyl residues to carbon nanotubes. In an initial step, alkyl radicals are generated by the cleavage of CO_2. These attack directly on the double bonds of the tube (Figure 3.76a). A variety of diacyl peroxides, for example, such as lauryl or phenyl residues, are suitable for this reaction. An analogous process is observed when using succinic acid peroxide or glutaric acid peroxide. These compounds also cleave carbon dioxide, thus generating an alkyl radical. The products obtained are nanotubes carrying alkyl chains with terminal carboxyl groups. Accordingly, these are not situated at defects, but they are connected to the tube's wall by a spacer of one or three carbon atoms, respectively.

From an electronic perspective, the radical addition to nanotubes induces so-called impurity bands close to the Fermi level if the addends are positioned in a sufficient distance one from another. Upon addition to neighboring carbon atoms, on the other hand, the defects thus introduced interact strongly with each other. As a consequence the band structure changes, resulting in an intersection at the Fermi level. This means that in spite of the newly formed sp^3-defects, the band structure remains largely unaffected, and that, for example, the metallic character of such modified nanotubes is conserved.

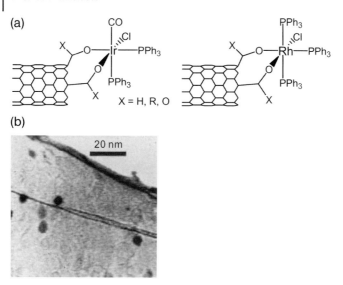

Figure 3.77 (a) Examples of transition metal complexes bound to the ends of carbon nanotubes. (b) HRTEM image of an oxidized nanotube with attached nanoparticles of CdSe which in their turn are functionalized with mercaptopropionic acid (© Wiley-VCH 2003).

Coordination Compounds with Transition Metals Coordination compounds formally do not count among the covalently bound structures. Still they shall be discussed in this chapter as, compared to other noncovalent variants, they are rather related to the covalently modified nanotubes. One can presume that nanotubes like fullerenes form coordination compounds with transition metals. Yet the tendency to do so is much less pronounced which is, among other reasons, due to the lack of five-membered rings. Hence, firstly, the delocalization of double bonds is less disturbed and, secondly, the energy gap between HOMO and LUMO is wider. The latter effect hinders an efficient backbonding and thus decreases the affinity of the nanotubes toward electron-rich metal systems. The cyclopentadienyl units as existent in fullerenes can stabilize the complexes once they come into being, whereas their absence in nanotubes hampers the formation of η^2-complexes.

In recent years a variety of examples with single-walled tubes has been demonstrated employing the Wilkinson complex [RhCl(PPh$_3$)$_3$] and the Vaska complex [Ir(CO)Cl[PPh$_3$)$_2$] (Figure 3.77a). In many cases, the nanotube under examination already carries a number of functional groups containing oxygen. A coordination to these is definitely favored over the formation of an η^2-complex. This allows for a selective bonding of metal species to the tips and possible defects of the nanotubes, which may also be employed for an attachment of small metal clusters of, for example, platinum, mercury, or rhodium to carbon nanotubes. Quantum dots of CdTe or CdSe can as well be connected to nanotubes in this manner (Figure 3.77b).

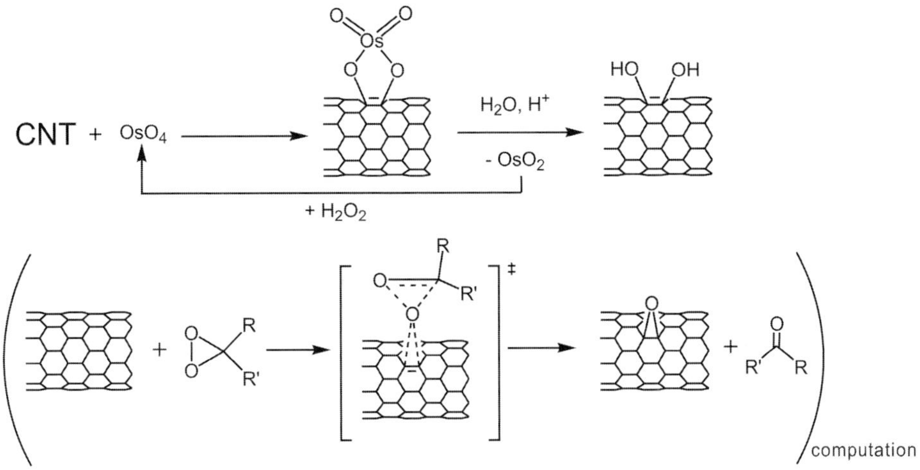

Figure 3.78 Osmylation and epoxidation of carbon nanotubes. The epoxidation has not yet been realized in experiment.

Osmylation and Epoxidation With osmium tetroxide, carbon nanotubes react as expected for a compound containing double bonds. The osmylation adduct with the respective double bond being replaced by two C–O-bonds is formed as shown in Figure 3.78. However, the process is normally conducted in a photochemical way here. The intermediates thus obtained can be transformed into hydroxylated nanotubes by hydrolysis. In doing so, it is advisable to effect a reoxidation of the resultant osmium(VI) by hydrogen peroxide in order to minimize the consumption of osmium. The osmylation of carbon nanotubes is reversible; so the process may also be employed for purification or separation steps. Contrary to an ozonolysis with subsequent reductive work-up, the osmylation does not give rise to holes in the side wall. Hence the electronic structure is less affected.

It turned out that the osmylation in organic solvents (as compared to a reaction from the gas phase) preferably takes place on metallic nanotubes with higher electron density close to the Fermi level. This finding is possibly applicable to a selective isolation of nanotubes with certain electronic properties. The effect has been noticed by the uneven decrease of the Raman resonance at different wavelengths.

The side-wall epoxidation of single-walled carbon nanotubes has been identified by calculations as feasible from an energetic point of view, but an experimental confirmation has not yet been achieved. Again the deviating behavior might be due to the decreased reactivity as compared to fullerenes. The calculations indicated that the reaction with dioxiranes should yield epoxidated nanotubes (Figure 3.78) that would be attractive intermediates for further transformations. For the time being, it is unclear whether other epoxidizing agents like peracids or HOF that have proven their worth in epoxidations of olefins or fullerenes can act respectively on carbon nanotubes. Once the reaction is successfully established, the

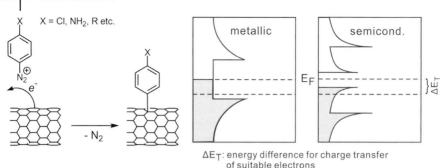

Figure 3.79 Reaction with aromatic diazonium salts. The extent of the ability to donate electrons to the substituent depends on the density of states in a certain region around the Fermi level (ΔE_T) (© AAAS 2003).

well-known chemistry of organic epoxides should open the way to numerous new derivatives of nanotubes.

Reactions with Diazonium Salts The reaction with diazonium salts of aromatic compounds yields directly arylated derivatives of carbon nanotubes (Figure 3.79). It has first been reported for HiPCo tubes as these are very reactive due to their small diameters and the related high curvature. Other nanotubes are lager in diameter and, consequently, less reactive. Hence, the reaction occurs only under more drastic conditions, but still good yields may be achieved.

The arylated carbon nanotubes are formed by the diazonium salt cleaving molecular nitrogen in an electrochemical reaction and the resulting aryl radical attacking on the π-system of the nanotube (Figure 3.79). Apparently electrons are extracted from the tube and employed for bond formation. When performed in solution, the functionalization features a remarkable selectivity for metallic nanotubes that is explicable by the high electron affinity close to the Fermi level (Figure 3.79). A charge transfer complex is formed from reactant and nanotube that provides electrons for the stabilization of the transition state. Finally the aryl radical is generated in a one-electron reduction and enters into a C–C-bond with the nanotube. The disturbance of the tube's electronic structure subsequently also enhances the reactivity of adjacent carbon atoms, which leads to a strong functionalization of the metallic species. Up to 5% of the carbon atoms might be arylated. The required diazonium salts can be generated, among others, by the treatment of aniline with nitrosonium tetrafluoroborate or isoamyl nitrite, respectively. In the latter case, the reaction is conducted in substance, that is, without a solvent. The input of mechanical energy by a magnetic stirring bar is sufficient here to at least partly break up the nanotube bundles. The arylated nanotubes may then be reacted with a variety of substances via functional groups existing on the aryl groups. This approach, for example, even allows for the attachment of polymers to carbon nanotubes. In the latter case the connection may be effected over the whole length of both tube and polymer strand. Using a suitable spacer, biomolecules can be connected as well. *p*-Aminoaryl groups are employed here for the initial functionalization.

Derivatization of Carboxyl Groups on the Side Wall One of the most frequent functional groups on the side wall of nanotubes is the carboxyl group. It might be formed during oxidative purification of the tubes as well as by other reactions like ozonolysis with oxidative work-up. The acid group suits perfectly well to a facile connection of the nanotubes with molecules and functional units via ester or amide bonds. Some examples have already been given in the discussion of end-cap functionalization (Section 3.5.3). Further ways of derivatizing carboxyl groups on the side wall shall be presented now.

Reacting nanotubes bearing carboxyl groups on their side walls or end-caps with octadecyl amine (ODA) yields a product soluble in many organic solvents provided the content of ODA in the material exceeds a threshold value of about 28%. Below this degree of functionalization, the dispersibility remains insufficient and a preparation of stable solutions fails.

The attachment of DNA and related structures is achieved in the same manner as described for end-cap functionalization. Contrary to the products obtained there, however, the presence of carboxyl groups on the nanotube side walls leads to a more or less even distribution of the attached structures over the full length of the tube instead of a concentration at the ends. Besides DNA, there is a variety of other biologically active substances, like peptides, which can be covalently linked to carbon nanotubes this way. This is of interest for a number of applications in biosensing and enzyme sciences. The amide bond is sufficiently stable here to stand physiological conditions, but still labile enough to be cleavable by acidic hydrolysis when required. The functionalization also succeeds with complex structures like the flavine-adenine cofactor (FAD cofactor) that may be employed for the attachment of glucose oxidases to nanotubes. The covalent linkage of DNA oligonucleotides with films of SWNT further enables the construction of systems for biosensing.

Here as well, another attractive way of derivatization consists in the connection with bifunctional units, which allows for a multitude of further possible reactions. Examples include amines with ω-thiol functions. They act as a kind of spacer between the carboxyl group and the thiol (Figure 3.80). By way of the sulfur groups

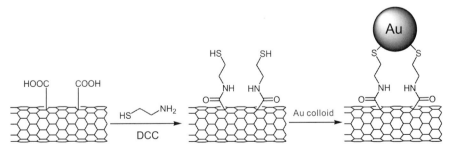

Figure 3.80 Derivatization of carboxylated carbon nanotubes. The bonding may be effected by amide or ester bridges.

then the nanotubes may be deposited on or selectively bound to thiophilic substances like gold nanoparticles or respective surfaces. These self-organized structures are very important in the production of sensor devices.

Using a diamine with sufficient chain length as reagent even allows for the connection of two nanotubes via an alkyl chain. This method may be employed to assemble covalent networks of nanotubes.

An attachment of structures bearing isocyanate groups by amide bonding is feasible, too. However, a diisocyanate is employed here with one of its functional groups being converted into an amide with the carboxyl group. Carbon nanotubes are thus modified suitable for the preparation of nanotube–polymer composites, especially of the polyurethane type.

Not least of all, the reduction of carboxyl derivatives on the ends or side walls of carbon nanotubes is rather easy to achieve. The reduction of carboxylic acid amides (e.g., of the didecyl amide) on the side wall of MWNT with lithium aluminum hydride yields aminoalkylated nanotubes with the tubular structure not being affected by the reaction.

3.5.4.2 Noncovalent Attachment of Functional Units

Besides the covalent attachment to functional groups situated at the ends, side walls or defects of carbon nanotubes, one may also make use of their tendency to enter into strong intermolecular exchanges. As mentioned in the chapter on purification (Section 3.3.6), nanotubes tend toward forming solid bundles. These are held together by noncovalent interactions alone, so replacing this exchange with neighboring tubes by another one of at least the same strength would mean a noncovalent attachment of molecules to the nanotube surface.

Wrapping with Long-Chain Molecules A very simple, but extremely efficient method for carbon nanotube functionalization consists in establishing an interaction with long-chain molecules that wind around individual tubes. These are not covalently bound to the nanotube, but only by van der Waals interactions. Usually, they are polymeric compounds featuring an ever recurring pattern of functionalities that enables a regular wrapping around the nanotube. There are different kinds of structures capable of entering into this type of interaction. These include biopolymers like amylose or synthetic polymers like poly-m-phenylenevinylene as well as compounds of the tetraalkylammonium salt type or pyrenes (Figure 3.81).

All these substances have in common their way of functionalization: The comparatively strong attraction forces that cause bundle formation especially with single-walled tubes are replaced by an interaction with the reagent. To this purpose at least a part of the structure must be rather hydrophobic in character so it can enter into a sufficiently strong interaction with the nonpolar nanotubes. Provided the polymer additionally contains aromatic rings, like for instance the polyphenylenevinylenes do, the polymer and the nanotubes are also bound by π-stacking. The single contributions of individual benzene rings may only be small, but the large number of aromatic units leads to a significant stabilization of the exchange between nanotube and polymer nevertheless.

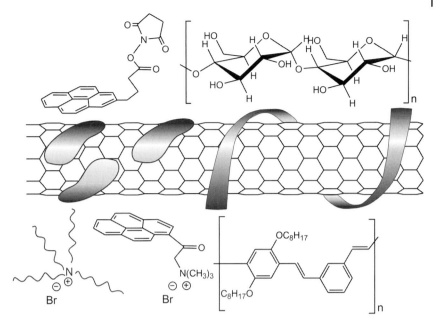

Figure 3.81 Possible ways of noncovalent functionalization carbon nanotubes. Depending on the structure of the complexing agent, planar, or helical interactions result.

The interactions are effective over large distances along the tubular axis due to the usually long-chained nature of the compounds employed to this end. Thus the formation of new bundles is efficiently suppressed (Figure 3.81), which significantly enhances the solubility of such nanotube derivatives. By varying the substances employed, it is also possible to take an influence on the type of solvent that suits best to solubilization. Nanotubes derivatized with biopolymers like amylose, certain sugars, peptides, or with polar synthetic polymers (e.g., polyvinylpyrrolidone, polystyrenesulfonate), for example, are soluble in water and physiological media.

Polymers featuring a helical shape are especially apt to wind around nanotubes. Hence peptides with an α-helical structure are frequently used to this purpose. One even succeeded in the directed synthesis of respective amphiphilic peptides that coil around nanotubes by simply reacting a peptide solution with SWNT under ultrasonication. Single-walled tubes thus derivatized dissolve in water. Nanotubes wrapped in unpolar synthetic polymers like poly-*m*-phenylenevinylene (PmPV), on the other hand, are soluble in organic agents such as THF, chloroform, or hexane. Yet a side chain functionalization with polar groups may again give rise to a solubility in aqueous media. The products of wrapping in PmPV feature completely new properties that render them attractive materials. The strong luminescence of PmPV, for instance, is combined with the electric conductivity of nanotubes (about eightfold increase compared to the neat polymer), while at the same time a marked mechanical stabilization of the composite is achieved.

A derivatization with tetraalkylammonium salts such as tetraoctylammonium bromide leads to a noncovalent functionalization of the carbon nanotubes. In these cases, the long alkyl chains of the ammonium salt interact with the tube (Figure 3.81). This kind of modification is also suitable to stabilize charges on the nanotubes, which are induced by an electric field. This allows for the formation of parallely aligned adducts protruding rectangularly from the electrode. This arrangement lasts only as long as the field is applied, so a way of switchable SWNT agglomeration has been realized here.

Interactions with Larger Aromatic Systems As mentioned before (Section 3.2.2), carbon nanotubes possess an extended system of delocalized π-electrons. Aromatic compounds may hence interact with this π-system by way of a so-called π-stacking. However, as might be seen from the insolubility of nanotubes in aromatic solvents like benzene or toluene, the interaction with single benzene rings is not sufficient here. It rather takes larger π-systems to achieve a sufficient effect. The phenomenon has first been described for the exchange between nanotubes and the π-system of pyrene, which could be proven by a shift of the proton signal of the latter in the ^{1}H-NMR-spectrum.

A suitable derivatization of the pyrenes allows for a directed influence on the solution behavior in different solvents. To this end, a respective residue with an end group causing, for example, solubility in water is attached to the pyrene. Common substituents exhibit terminal ammonium groups – among others, trimethyl-(2-oxo-2-pyrene-1-yl-ethyl)-ammonium bromide is frequently used (Figure 3.82a). Long-chained alkyl substituents of the pyrene, on the other hand, convey solubility in organic solvents. SWNTs derivatized with DomP (1-docosyloxymethylpyrene), for instance, are soluble in THF and may be examined by spectroscopy. These studies revealed that the electronic interaction not only affects the π-system of the pyrene, but also gives rise to changes of the bandgap of semiconducting nanotubes (Figure 3.83a). Furthermore, the formation of a noncovalent composite can be achieved by reacting carbon nanotubes with polymers carrying pyrene units in their side chains. For instance, a connection of multiwalled nanotubes with suitable polymethacrylates may be realized this way (Figure 3.82b). The latter are obtained by copolymerizing methyl methacrylate with the 1-pyrene-methyl ester of 2-methyl-2-propenic acid. Further functionalization of the pyrene's side chains is possible, too. Succinimidyl esters, for example, can be reacted with proteins or DNA after cleavage of the succinimidyl group. The resulting products are of interest for biological applications of nanotubes (Figure 3.82c).

The interaction between the π-systems of the aromatic compound and the nanotube is based on an overlap of the respective orbitals. Bent graphene layers, as already stated in the chapter on fullerenes, exhibit a curved π-system. On the one hand, this has an influence on the degree of electron delocalization, while on the other hand it sets a limit to the overlap with planar aromatic compounds. Contrary to fullerenes, however, the nanotubes are not curved in all three directions in

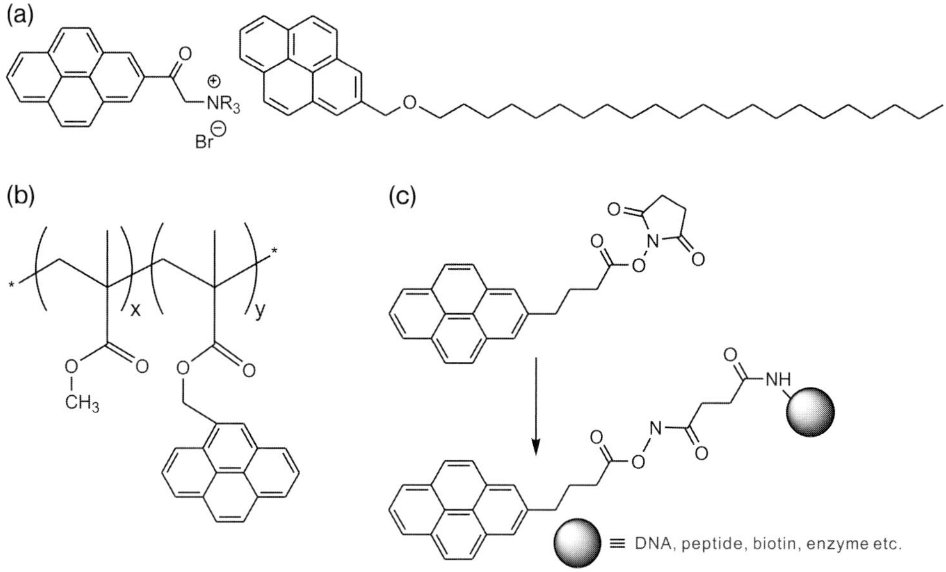

3.5 Chemical Properties | 243

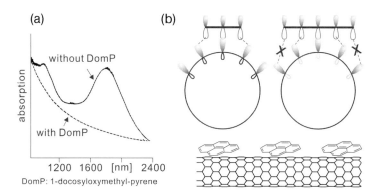

Figure 3.82 Exemplary derivatives of pyrene for a reaction with CNT. (a) The adducts' solubility may be controlled by the substituents present; (b) a noncovalent incorporation into polymers succeeds if the latter is carrying pyrene units attached to its side-chains; (c) noncovalent adducts with various biologically active substances may be formed via substitution with an activated ester.

Figure 3.83 (a) The influence of a noncovalent interaction with pyrene on the bandgap also reflects in a significant change of the absorption spectrum (© ACS 2004); (b) the strength of interaction depends on the size of the complexing agent's aromatic system just up to a certain extent. Oblong aromatic systems like pyrene suit best to this task.

space, but the π-orbitals are aligned in parallel at least along their axis due to the cylindrical shape.

An efficient exchange between a planar aromatic system and a carbon nanotube consequently depends on a largest possible number of overlaps existing along the tubular axis. Still the interaction becomes weaker with increasing distance from an assumed center line (Figure 3.83b) because the distance between atoms grows correspondingly. This is why oblong aromatic molecules like pyrene are most suitable here: they exhibit a preferred direction that can be arranged in parallel with the tube's axis. Possible interactions are thus maximized. A relatively large diameter of the functionalized nanotubes further favors a strong exchange as the curvature of the π-system is less pronounced in these species. Hence, the peripheral distance between aromatic system and nanotube does not increase that much and the delocalization within the nanotube is enhanced, too.

The reaction with pyrene is largely established as fundamental example for this kind of noncovalent functionalization for several reasons. Firstly, it allows for a wide scope of derivatization, secondly, it can be aligned in parallel with the tube's axis, and finally, the solubility of pyrene compounds ensures an easy handling. In fact, larger aromatic systems already give rise to similar solubility problems like fullerenes or nanotubes themselves.

An interesting idea would thus consist in a functionalization with aromatic compounds bent around the z-axis. These should interact especially well with the π-system of the nanotubes, possibly even featuring certain selectivity for nanotubes with different diameters. A first attempt of this kind has been presented just recently. A belt-shaped aromatic molecule has been put over a fullerene here (refer to Section 2.5.6). Such a procedure should in principle be applicable to nanotubes as well. A success in performing this reaction would mean a big step toward a size-selective functionalization and possibly even toward a directed separation of individual species. Here an interaction with aromatic compounds immobilized within tubular templates, for example, within zeolites, would be another conceivable strategy.

Further Noncovalent Derivatizations of Carbon Nanotubes By now there are numerous different approaches to derivatizing single- and multiwalled carbon nanotubes by way of noncovalent interactions. A multitude of further organic or inorganic substances can be bound to nanotubes besides those mentioned above.

The functionalization with porphyrines is an interesting variant here, especially with regard to electronic applications. Van der Waals interactions and π-stacking again play the major roles in the binding. Porphyrines are known to be adsorbed as well on graphite surfaces. The attachment to carbon nanotubes is presumably based on similar effects. It does not take more than the simple ultrasonication of SWNT in solutions of the respective porphyrine in DMF to obtain deeply colored solutions that are stable for weeks. The solubility amounts to about $20\,\mu g\,ml^{-1}$, which roughly corresponds to the value for pyrene-modified nanotubes. A marked attenuation in the porphyrine's characteristic fluorescence is observed for the

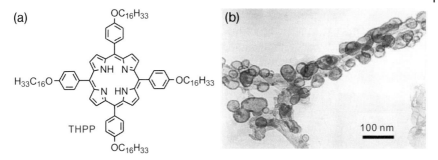

Figure 3.84 (a) Example of a porphyrin employed for the noncovalent functionalization of nanotubes, (b) HRTEM image of a carbon nanotube surrounded by hollow spheres of MgO (© RSC 2004).

adducts. It can be attributed to an energy transfer emerging from the porphyrine to the nanotube.

Another important discovery has been reported in 2004. A typical mixture of metallic and semiconducting SWNT was treated with a derivative of porphyrine (Figure 3.84a). Actually one succeeded in selectively functionalizing the semiconducting species and consequently transferring them into solution, while the metallic tubes did not react and thus remained in the sediment. This procedure might once enable the separation of both types of nanotubes on a larger scale. This would be of eminent importance for the construction of electronic devices from nanotubes.

Inorganic substances like small metal clusters, metal oxides etc. may also be deposited on the surface of single or bundled carbon nanotubes. It is possible, for instance, to "decorate" multiwalled nanotubes with nanoparticulate oxides of zinc or magnesium. To this end, a surfactant-mediated dispersion of MWNT is solubilized in cyclohexane with Triton X-114 as a surface-active agent to obtain a water/oil emulsion. Aqueous solutions of the respective metal acetates are added afterward and are then found in the aqueous portion of the emulsion. Subsequently increasing the pH value to about 9.5 causes a precipitation of the metal hydroxides that deposit on the nanotube surface in the shape of hollow spheres. Final calcination at 450 °C transforms these hollow particles into the crystalline metal oxides. Particles measuring about 5 nm or 30–40 nm have been observed for ZnO or MgO, respectively (Figure 3.84b).

Pure metals as well can be deposited on the surface of carbon nanotubes. The reductive precipitation of gold nanoparticles on MWNT coated with citrate ions may be given just as one example. Besides dispersing the MWNT, the citric acid is also responsible for the reductive generation of the gold particles from $HAuCl_4$. Other metals like platinum, palladium, titanium, and iron can be deposited on the nanotube surface, too.

A completely different kind of interaction can be observed with alkali metals, some metal halogenides and with tungsten hexafluoride: In close analogy to the behavior of graphite, multiwalled nanotubes are able to incorporate atoms or molecules into the spaces between their constituent tubes. These intercalation compounds are distinguished by an extended distance between tubes. They are of interest with regard to the production of materials for lithium ion batteries or hydrogen storage devices. Normally, the intercalates penetrate the tubes by diffusion from the gas phase or from a melt. Defects must exist in the individual tubes for the additional species to reach interstitial spaces situated farther to the inside.

3.5.5
Composite Materials with Carbon Nanotubes

The diameter of carbon nanotubes is rather small in comparison to their length. This feature predestines them to be used in the production of fiber reinforced composite materials. Like for classical carbon reinforced plastics, a wider range of applications is envisaged for the respective polymers after the addition of nanotubes as the latter should enhance the mechanical, chemical, and electronic properties. The homogeneous distribution of fibers or nanotubes, respectively, in the polymer constitutes a general problem in the preparation of such composites. This is due to a marked tendency toward aggregate formation, which may cause large differences in concentration and thus interfere with a homogeneous profile regarding the properties of interest. Interactions on the contact surface between polymer and filler (carbon fiber or nanotube) as well play an important role for the generation of a stable composite: a separation can only be avoided by sufficient wetting or bonding. For classical carbon fibers, the largest problems arise from this very interfacial exchange as, compared to length and circumference, only few functional groups are generated. Hence an efficient covalent connection is hard to achieve. Carbon nanotubes, on the contrary, are more easily dispersed in the polymer due to their comparatively low size, and the attachment of numerous functional groups is established as standard procedure. However, nanotubes strongly tend toward the formation of bundles (Section 3.4.2) that impede a homogeneous distribution in the polymer. Efficient methods of nanotube deaggregation are thus required for a successful preparation of homogeneous composites.

Carbon nanotubes should altogether suit well to the production of fiber-reinforced composite materials because their large outer surface allows for a strong interfacial exchange. With their inherent properties like conductivity and mechanical endurance they open up a wide range of attractive applications that cause significant effects at low concentrations already. The large surface is also influential regarding the transfer of mechanical stress from the polymer to the reinforcing filler. The covalent bonding further enhances the adhesion of the polymer to the nanotube and, consequently, the resistance of the composite.

The mechanical properties of isolated SWNT in particular (Section 3.4.3) turn them into nothing short of a filler *par excellence* for polymer composites. However, the embedding of individual, completely debundled carbon nanotubes in the polymer matrix still poses large problems. This is why, for the time being, the materials produced do by far not reach the envisaged improvement of characteristics: Bundles of nanotubes feature much worse mechanical properties as there is a pronounced tendency for shearing away single tubes, and what is more, the transfer of strain onto individual nanotubes is less efficient than for isolated species. It becomes obvious again that a successful strategy must aim at generating a composite material that bears separated carbon nanotubes with a stable connection to the matrix.

The loading capacity of a nanotube composite depends, among others, on the arrangement of the individual tubes within the polymer matrix. For a material that has to carry tensile or compressive strain in a preferred direction, an alignment of the tubes most possibly in line with the strain proposes itself. Each individual tube may then be stressed up to its maximum capacity, which is highest in parallel to the tubular axis (Section 3.4.3). To obtain such composites, it is possible, for instance, to employ polymer materials moldable by melt extrusion. These are oriented by pressing the bulk composite through a nozzle. In doing so, the nanotubes align in parallel to the direction of extrusion. Another method consists in applying an electromagnetic field during polymerization. Good experiences have been made here with polyaniline composites regarding an arrangement of the nanotubes in parallel to the field lines.

An interesting question arises considering composites with multiwalled carbon nanotubes. In such materials there are two possible ways of stress relief upon mechanical strain: On one hand, like in other fiber-reinforced composites, the interfacial interaction may be overcome by the mechanical forces so the polymer peels from the nanotubes. On the other hand, however, the individual walls of the MWNT may slide one inside another like a sword in a sheath (so-called interwall sliding, Figure 3.85). Both effects may occur to varying extents depending on how strongly the multiwalled tubes are bound to the matrix.

Composites of multiwalled tubes are distinct from those with SWNT filler in other respects too. MWNT composites normally exhibit a different resistance regarding compression and expansion. This may be explained by the compression acting on all individual walls of the MWNT, whereas only the outer shell is exposed to tensile forces. Especially in the latter case the interwall sliding mentioned above becomes relevant.

In general there are two ways of embedding the nanotubes. Firstly, such composites with the interaction between polymer and nanotube exclusively basing on noncovalent forces may be prepared. These can be obtained by simple admixture.

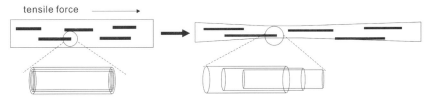

Figure 3.85 The effect of tensile strain on a composite bearing MWNT. The nanotubes may delaminate from the polymer and/or a sliding of individual tubes one inside another may occur (interwall sliding).

Secondly, however, the carbon nanotube may be covalently functionalized with monomer units, initiator molecules or crosslinking groups, which leads to a significantly better interaction with the matrix.

3.5.5.1 Composites with Covalent Bonding of the Polymer

A carbon nanotube must be chemically functionalized to enable the covalent attachment of suitable polymer molecules. This may be achieved, for example, by the methods described in Section 3.5.4.1. There are various procedures to prepare covalently bound composites. These shall be discussed in detail below. The methods can altogether be classified with respect to their type of connection and to the moment of composite formation.

Bonding During the Polymerization Process and to "Living" Polymers In this case the incorporation of the carbon nanotube is realized by scavenging reactive positions at the ends of the growing polymer chains. The resulting polymer strands thus cannot be bound to the nanotube at more than two positions. The diffusion approach of additional polymer strands and their reaction with the nanotube is further complicated by a loss of conformational entropy. As a consequence only low degrees of grafting are achieved (<10%) and a considerable portion of the polymer is not at all connected to the nanotubes. The method is mainly of historical interest because the first nanotube composite materials ever have been prepared via esterification and amide formation.

Polymerization Originating from the Nanotube A polymerization that originates from the carbon nanotubes is realized as follows: initiator molecules are bound to the nanotube surface by chemical functionalization. The polymerization then starts from these upon the addition of suitable reagents (Figure 3.86). This proce-

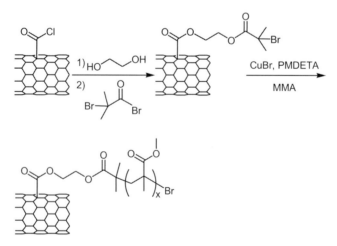

Figure 3.86 Attaching an initiator molecule like bromoisobutyric acid has proven a suitable strategy for the atom-transfer radical polymerization (ATRP) starting from carbon nanotubes.

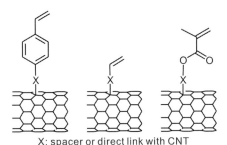

Figure 3.87 Examples of carbon nanotubes decorated with monomer units. The latter may be incorporated into a polymer matrix by copolymerization with the respective free monomers.

X: spacer or direct link with CNT

dure allows for high degrees of polymer bonding to the tubes. It is not trivial, however, to control either this number of polymer molecules attached or their chain length. Only a little influence can be exerted by the number of existing initiator sites.

The generation of dendritic structures on a nanotube surface represents a special case of a polymerization that originates from nanotubes. The method is of interest for the production of composites with a large number of terminal functional groups.

Copolymerization The carbon nanotubes to be incorporated in the composite might carry functional groups that can act as monomer unit of a polymer. In such a case a composite material can be obtained by classical copolymerization (Figure 3.87). The method particularly enables a rather exact adjustment of product stoichiometry as well as the preparation of block copolymers. A certain drawback exists, however, in a sometimes intricate handling of the functionalized nanotubes. Upon imprudent storage or the like, these may already react with each other so they are no longer completely available to the copolymerization. It is advantageous, on the other hand, that generally the materials properties can easily be adjusted by the degree of polymerization and by the number of attachment sites.

Chemical Attachment of the Nanotubes to Already Existing Polymer Chains Provided the polymer chains carry reactive groups (like amino or hydroxyl groups) on their ends, these may be bound to suitably functionalized carbon nanotubes (e.g., with acid chloride groups). In the composite materials thus obtained, the polymer strands again are connected to the nanotubes on a few positions only. Still this problem can be circumvented by using polymers that carry the functional groups in their side chains. These can react with the nanotube over its whole length and consequently establish a strong attachment.

3.5.5.2 Composites with Noncovalent Attachment of the Polymer

It has already been mentioned in Section 3.5.4.2 that a functionalization of carbon nanotubes can also be achieved by noncovalent intermolecular exchange. This concept can as well be applied to the production of composite materials. Here a

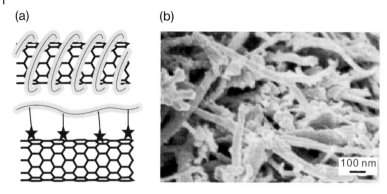

Figure 3.88 (a) Possible modes of a polymer interacting noncovalently with carbon nanotubes, (b) SEM image of a nanotube–polyaniline composite (© Wiley-VCH 2003).

distinction should be made between an interaction of isolated functional units attached to the side chain of the polymer on the one hand and a direct wrapping of the nanotube by the polymer main strand on the other (Figure 3.88a).

There is a drawback, however, to noncovalently bound nanotube composites. The samples usually contain an excess of polymer that is not attached to the tubes. This leads to inhomegoneities inside the material, and problems may arise concerning the wetting or segregation of components. Hence the materials' properties are hard to correlate to their content of carbon nanotubes. The homogeneous distribution of individual tubes poses another difficulty in the production of noncovalently bound nanotube composites. Due to their strong tendency toward bundle formation it normally turns out no simple task really to disperse single tubes in the matrix. This problem can be overcome by adding a surface-active substance like SDS to the mixture of nanotubes and polymer. It leads to a debundling in the beginning, and subsequently causes a homogeneous distribution of the nanotubes in the solution.

The preparation of noncovalently bound nanotube composites proves most simple. The desired products are obtained by mixing the constituent substances (in solution or in a melt) or by polymerizing the monomer in the presence of the nanotubes. The melt extrusion of a nanotube–polymer mixture may yield such materials as well. In the latter case, the nanotubes are already aligned in parallel to some extent, which has a beneficial effect on the product's rigidity. Figure 3.88b shows an example of a noncovalent composite.

3.5.5.3 Nanotube Composites with Different Polymers

In the following sections some examples of covalently and noncovalently bound carbon nanotube composites are presented. They are ordered by classes of substances as in most cases composites with both modes of bonding are known.

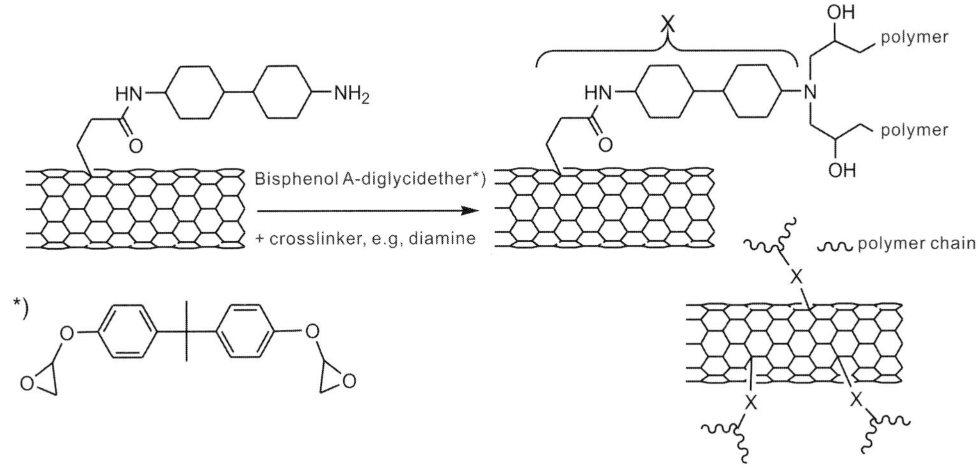

Figure 3.89 Preparation of epoxy resins cured by carbon nanotubes decorated with amino functions. The nanotubes are incorporated directly into the resin this way.

Epoxy Resins   Epoxy resins constitute a large group among the composite materials of nanotubes examined so far. There are covalently as well as noncovalently bound variants. Provided single-walled nanotubes can be dispersed to be a metastable suspension and subsequently curing of the epoxy resin is initiated (e.g., by diamines or phthalic acid anhydride), the SWNT can be embedded in the polymer in a very homogeneous and largely debundled way. The dispersion in dimethyl formamide under ultrasonication, for instance, has successfully been employed to generate noncovalently bound composites of nanotubes with epoxy resins.

With the curing agents often carrying terminal amino groups, such nanotubes with amino functions can also be covalently embedded immediately during the hardening process (Figure 3.89).

Polymethacrylates   Polymethacrylates have found a broad range of applications as durable, transparent plastics (e.g., plexiglass), which rose considerable interest in PMMA–nanotube composite materials. Once more covalent as well as noncovalent composites are known with the first variant noticeably gaining ground from the latter due to its distinctly improved properties. What is more, the interfacial exchange with noncovalently bound PMMA molecules is complicated by the hydrophobic character of the nanotube surface. The desire to conserve the transparency of the polymer matrix is another aspect here. It can be fulfilled by using sufficiently small nanotubes whose size is markedly below the wavelength of visible light.

The procedure normally employed in the production of PMMA–nanotube composites is the surface initiated polymerization (SIP). The initiator molecule may be tethered to the nanotube via various linkers. Bromoisobutyric acid is frequently used with the polymerization being started by the addition of copper(I) salts and a suitable ligand (Figure 3.86). This type of polymerization is termed atom transfer radical polymerization (ATRP) and represents one possible variant of SIP. The linkers tying the initiator to the nanotube may be either polar or rather nonpolar units, depending on the desired solubility. With ethylene glycols as linking unit, the functionalized nanotubes obtained are even soluble in water. Initiator molecules can also be attached to carbon nanotubes by a 1,3-dipolar cycloaddition of azomethine ylides (Section 3.5.4.1).

The resulting PMMA composite materials are insoluble in both organic or aqueous media. Hence alternative acrylates soluble even in their polymerized state have been sought for. Especially poly-*tert*-butyl acrylates suit well to this purpose. Furthermore, the *tert*-butyl residues are easy to remove from these composites, so water soluble polyacrylic acids with additional carbon nanotubes are obtained.

Polyanilines Polyanilines (PANI) are another class of substances that have been studied early as potential partners to carbon nanotubes in a composite (Figure 3.88b). Materials with noncovalent bonding have normally been produced in doing so. They feature interesting mechanical and electronic characteristics. Meanwhile, however, covalently bound PANI–nanotube composites have been reported, too.

Polyanilines themselves already feature electric conductivity and interesting redox characteristics, and what is more, they are stable both in air and in water. These properties turn them into an ideal candidate partner for nanotubes in a composite material. Their structure (Figure 3.90) reveals that an intermolecular exchange with the tubes should be favorable as there is an opportunity of π–π-interactions. The addition of nanotubes to PANI causes a drastic change of materials properties. The electric conductivity, for example, increases by one order of magnitude at a nanotube content of no more than 10%.

Another species of polyanilines, the poly-(*m*-aminobenzenesulfonic acid) (PABS) forms nanotube composites as well. Up to 5 mg ml^{-1} of these dissolve in water due to the existing sulfonic acid groups. The first internal doping in a nanotube composite ever has been observed for these materials. It has been detected by the signal position in the IR-spectrum that is indicative of an electronic hybrid structure with states situated between those of the composites in their ground state.

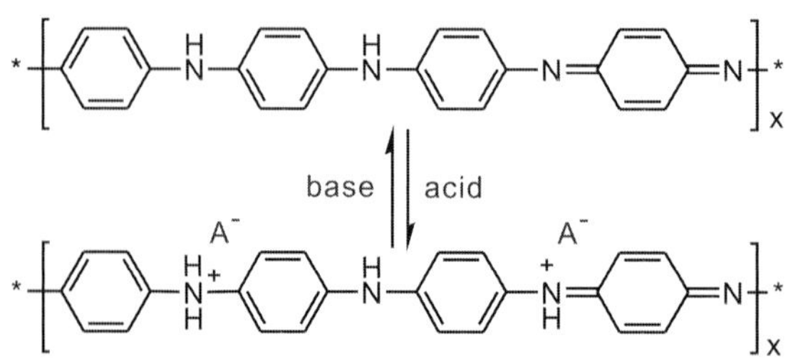

Figure 3.90 The structure of polyaniline (PANI).

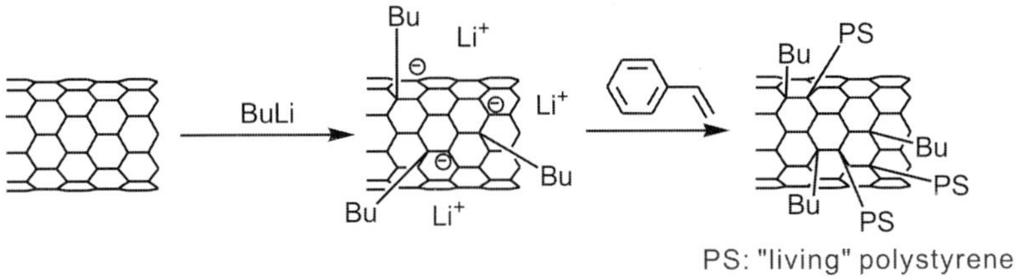

Figure 3.91 Polystyrene–nanotube composites may be obtained for example, by anionic polymerization.

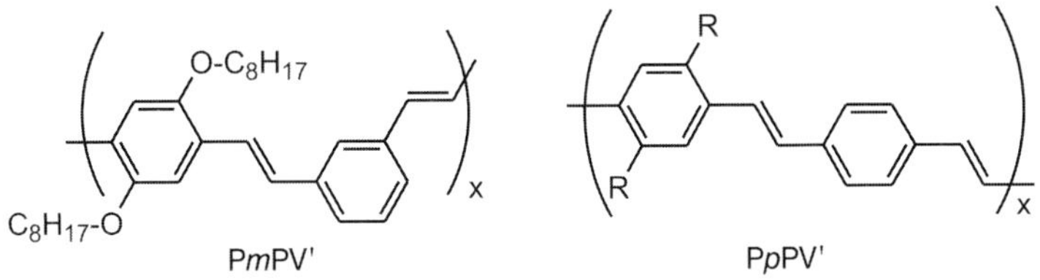

Figure 3.92 Exemplary polyphenylvinylidenes.

Polystyrene For polystyrene, both covalently and noncovalently bound composite materials with carbon nanotubes are known. Polystyrene, like the methacrylates, can be generated by surface-initiated radical polymerization on nanotubes functionalized with initiator molecules (in analogy to Figure 3.86). Suitable substances then actually include analogous compounds like in the case of polymethacrylates.

Another approach consists in the activation of carbon nanotubes with lithium organic compounds like butyl lithium and subsequent reaction with styrene in the sense of an anionic polymerization (Figure 3.91). In the course of the reaction, it is even possible to break up the bundles of tubes, thus obtaining separated, polymer coated nanotubes.

Polyphenylenevinylidenes and Related Polymers *meta*-Polyphenylenevinylidenes (PmPV) are rather frequently used for the production of nanotube composite materials. Normally, such derivatives with functional groups ensuring a better wetting are employed (e.g., PmPV'). An exemplary substance commonly used is depicted in Figure 3.92a. The adhesion of the polymer to the nanotube is mainly based on π–π-interactions with the phenyl rings oriented in parallel to the surface of the tube. The long side chains of substituted PmPV additionally contribute to the interfacial exchange.

Figure 3.92 indicates that there are different types of connectivity between the phenylenes and the vinylidene units. Experimental examples of nanotube composites with interesting features are known with both *meta*- and *para*-bound polyphenylenevinylidenes.

The composite preparation is achieved by dispersing the SWNT in a solution of the polymer (e.g., in chloroform) and subsequently removing the solvent *in vacuo*.

The electric conductivity was found to be increased to about the 10-fold, whereas the luminescence characteristics remain largely unaffected.

Composites with Other Polymers   Besides the composite materials presented so far, a multitude of further polymer composites with carbon nanotubes has been prepared and studied regarding their properties. After all, any given polymer is suitable to some extent to interact with different carbon nanotubes (pristine or functionalized). The number and range of possible combinations surpass the scope of this text, so the examples mentioned below inevitably have to remain incomplete.

Even simple polymers like *polyethylene* or *polypropylene* can form composites with carbon nanotubes. The embedding, however, is limited to noncovalent interaction due to the complete lack of functional groups. Yet on the other hand, the hydrophobic nanotube surface hardly poses any problems in a wetting by the nonpolar polymer chains or in the formation of noncovalently bound composites.

Besides binary composites consisting of only one polymer component and the carbon nanotubes, the use of ternary or even quaternary mixtures was also found useful in some cases. There is, for instance, a composite of polymethacrylate with carbon nanotubes previously treated with polyvinylidene fluoride (PVDF). It is obtained by melt blending. There may be no covalent bonding between nanotubes and methacrylate, the coating with the significantly less polar PVDF causes a good wettability by the matrix polymer. The PVDF obviously serves as composite mediator without which a separation of the other components might occur.

Nanotubes may also be modified with *gum arabic* so that they can form composites with rather polar polymers. In aqueous emulsion, for example, it is possible to prepare a composite of polyvinyl acetate (PVAc) with carbon nanotubes pretreated this way. Upon removal of water, the nanotubes accumulate in the interstitial space between the PVAc emulsion droplets, so after drying and molding into a compact workpiece a network of carbon nanotubes in the polymer matrix is obtained. Even very low concentrations (~0.04% w/w) already give rise to a significant electric conductivity. This so-called percolation threshold – that is, the critical concentration at which electric conductance in the material can just be detected – could markedly be lowered by using this emulsion polymerization. Thus polymers with lowest contents of carbon nanotubes are available. (For comparison, the analogous system with carbon black as filler has a percolation threshold of 4% w/w.) This is relevant not only with regard to the constantly high price of nanotubes, but also to the optical properties of the respective polymers: the lower the concentration of nanotubes and the more homogeneous their distribution (individual tubes should possibly exist), the less the color and transparency of the material are affected.

3.5.5.4 Nanotube Composites with Other Materials

Besides combinations with polymers as mentioned above, there are also some composites of carbon nanotubes with other, and especially with inorganic materials. These are, in most cases, metals or ceramic compounds like aluminum oxide.

The intention is again to use the properties of the nanotubes for an improvement of the materials' characteristics. The main focus in doing so is on mechanical reinforcement.

Common processes in the production of ceramic or metal composites employ sinter or hot-pressing methods to blend the respective partners. However, these procedures have some serious drawbacks once it comes to the preparation of carbon nanotube composites. Firstly, a wetting of the nanotubes by the composite partner often is complicated, if not impossible, in the solid state, and secondly, it still poses considerable problems to incorporate the tubes into the matrix in a homogeneous and, where possible, individually dispersed way.

The classical methods used for the production of ceramic or metal composites normally employ their starting materials as powders. Hence a sufficient dispersion cannot be generally achieved. Yet a procedure called "molecular mixing" is suitable to prepare composites with a portion of nanotubes from solutions of inorganic salts. Solutions of nanotubes with different functions may, for instance, be transformed into metallic composites by adding the respective metal salt, evaporating the homogeneous mixture to dryness, calcinating the resulting product and, finally, converting the metal oxide into the element under reductive conditions (hydrogen atmosphere). An exemplary composite of this kind consists of acid-modified nanotubes and elemental copper that is obtained from Cu(II)-salts via an intermediate CuO. The material exhibits remarkable properties. Its strength, for instance, is higher than that of copper alone by a factor of three, and the Young modulus still experiences a doubling. These values markedly surpass the reinforcing effect of other fillers, which once more underlines the outstanding position of carbon nanotubes in materials sciences. The improvement of mechanical properties is based on the strength and multitude of interfacial exchanges that are yet increased by a good dispersion in the matrix. Not only copper, but also silver as metallic or aluminum oxide as ceramic component may be employed in composite production. For the ceramic compounds, the reduction of the oxidic powder can be omitted as the metal oxides constitute a major part of the final product anyway. Ceramic materials filled with carbon nanotubes feature a markedly increased mechanical resistance as well.

3.5.6
Intercalation Compounds and Endohedral Functionalization of Carbon Nanotubes

Within the bundles of single- or multiwalled carbon nanotubes, cavities exist that may be filled with atoms or molecules. It has already been mentioned in the introductory chapter on the reactivity of carbon nanotubes that different sites are available for this intercalation. In detail, these are the channels in between the tubes, the interstice between individual layers in a multiwalled nanotube, and the central cavity of the tube. Examples of all three types have been proven by experiment (Figure 3.93).

The intercalation into the interlayer space of MWNT has already been discussed in Section 3.5.4.2. The intertubular incorporation could especially be observed for

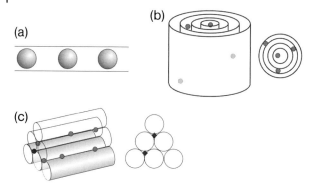

Figure 3.93 Possible sites for the intercalation of small particles or atoms inlude the tube's inner void (a) as well as the interstitial space in multiwalled tubes (b) and the hollow space within bundles of nanotubes (c).

alkali metals. The space between the constituents of nanotube bundles can also be filled with alkali metals or with a variety of further substances. On this occasion, normally a charge transfer occurs from the host (nanotube bundle) to the guest (intercalate).

As mentioned in Section 3.5.4.1, the covalent bromination of carbon nanotubes does not succeed. A noncovalent aggregate with the bromine atoms situated between the tubes of a bundle is obtained instead. Br_2^--units arise from the charge transfer, and the electric resistance decreases by a factor of 15. The stoichiometry of the compound can be adjusted by varying the temperature applied during preparation. The higher it is, the less bromine is incorporated from the gas phase. Similar intercalation compounds can be produced from potassium vapor and bundles of SWNT. In this case electrons are transferred onto the nanotube, which formally corresponds to n-doping. In analogy to graphite, a composition of C_8K is obtained once the maximum amount of potassium has been received. The benzoide hexagon network of graphene layers obviously has only a limited capacity to accept further electrons, and these bounds are reached at a composition of C_8M. One of the reasons for this is the bond dilatation caused by the electron uptake. For the time being, there are no detailed structural models of the arrangement of intercalates inside the tubes, yet in any case a certain disorder can be assumed to exist because the nanotube samples are always inhomogeneous.

It has already been mentioned in the introduction on general chemical properties that the central cavity of carbon nanotubes is another conceivable site for chemical modification. Normally, however, covalent interactions with potential bonding partners cannot evolve inside of a nanotube as due to the surface curvature the bonding orbital lobes are only little in size and intensity (Figure 3.94). Still this effect is less pronounced than in the fullerenes, and there is indication, for instance, of nanotubes with large diameters being at least partially coated also on the inside upon reaction with for example, ammonia (see below).

For being largely inert, the central cavity of carbon nanotubes may be employed as a kind of container or reaction vessel. Calculations on reactions that comprise

Figure 3.94 The orbital lobes protruding into the tube's core are comparatively small and an interaction with intercalated compounds is not very effective.

a charge separation step revealed that the environment (i.e., the SWNT cavity) acts like a solvent with a low dielectric constant.

In principle, the incorporation should succeed with any atom or molecule whose diameter is compatible with the cross-section of the nanotube in question. It must be assured, however, that an interaction with the outer surface of the tube is not markedly favored for energetic reasons. Otherwise an intercalation between the tubes would occur instead of endohedral complexation. The incorporation can take its course by diffusion from either the gaseous or the liquid phase, in the latter case being assisted by capillary effects. To this end, however, the closed tips of pristine nanotubes have to be removed to allow for the entry of guest substances or solutions containing them. Their respective surface tension must not exceed 100–200 mN m^{-1} as otherwise a wetting of the nanotubes would not take place. For tip opening, the methods discussed in Section 3.3.6 are employed. Subsequently various guest atoms or molecules can be filled into the open tubes. Atoms of noble gases (e.g., Kr, Xe) may for instance be absorbed to the inside of chemically opened, single-walled nanotubes, which has been proven by measuring the absorption isotherms of krypton and xenon with closed and opened SWNTs. A comparison of the resulting values revealed that already at low relative pressures of the respective gas an additional absorption occurred with the open tubes. Small molecular or atomic gases like hydrogen, argon, or nitrogen can be incorporated as well. They also enter into the inner cavities by diffusion. The incorporation of hydrogen is of special interest regarding a possible application of nanotubes as hydrogen storage in fuel cells.

Metals can also be deposited inside of carbon nanotubes. This kind of endohedral functionalization could in principle already be considered feasible from the examination of nanotubes prepared using metal catalysts: they contain catalyst particles included in the tubes' tips. Yet a later introduction of metal clusters is possible, too – for example, gold clusters 1–2 nm in size may be generated by the reductive deposition of $HAuCl_4$. Beforehand, the multiwalled tubes are opened by heating them in a stream of ammonia and thus providing them with amino groups. It has not yet been fully elucidated if in this process amino groups are really formed on the inner surface of the MWNT as well (Figure 3.95).

Besides gold there is a variety of metals that can be intercalated into carbon nanotubes to form nanowires. Examples include the low-melting metals lead,

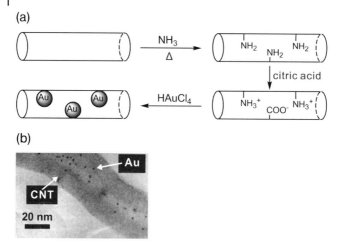

Figure 3.95 (a) Reaction with ammonia most likely allows for a covering even of the inner surface with NH$_2$-groups. Species like gold nanoparticles may subsequently be immobilized there; (b) HRTEM image of gold particles inside an MWNT (© Elsevier 2003).

bismuth, copper, sodium, and silver. In the latter case, silver nitrate is employed as low-melting salt. It is only after incorporation then that elemental silver is released by heating. The diameter of the resulting wires corresponds to that of the hosting tubes and their length is up to 120 nm. The endohedral functionalization with metal nitrates is valuable also for other elements, especially for those with a high melting point, as generally the nitrates melt low, and they are easily converted into the metals or metal oxides. Cobalt oxide is an example of a metal oxide incorporated this way. Once more, nanofilaments are obtained that could possibly be employed as one-dimensional conductors (Figure 3.96).

Furthermore, the production of magnetic carbon nanotubes is possible by filling suitable nanoparticles into the tubes. Nanotubes generated by CVD in a template matrix with defined pore diameter can for instance incorporate nanoparticles of iron oxide from ferrofluids. Upon removal of the solvent, the magnetic Fe$_3$O$_4$-particles remain inside the tubes and give rise to a respective behavior like, for example, aligning in a magnetic field (Figure 3.96). Nanotubes bearing fluorescent nanoparticles (e.g., of modified polystyrene) are obtained in a similar manner.

Examining the inclusion of crystalline compounds like salts or metal oxides, it is a frequent observation that the two-dimensional confinement of the space of crystallization causes the formation of structures that differ sometimes completely from those found in the bulk phase. Antimony(III) oxide and potassium iodide may serve as examples here. For the Sb$_2$O$_3$-filaments inside nanotubes a *valentinite* structure with a lattice distortion inflicted by the outer restriction is observed (Figure 3.79d). Normally, this phase is found at high pressures. At standard conditions, the Sb$_2$O$_3$ adopts as cubic *senarmonite* structure with discrete Sb$_4$O$_6$-units.

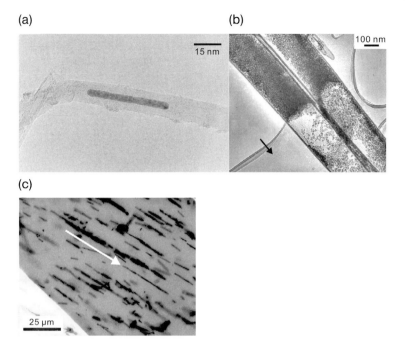

Figure 3.96 Metals or oxides may be intercalated into the void of nanotubes. These confer new electronic and magnetic properties to the respective composite; (a) silver wire inside an MWNT (© Springer 1998); (b) particles of iron oxide inside a large MWNT (the arrow designates the film support, © ACS 2005); and (c) nanotubes filled with iron oxide align in a magnetic field (with the arrow indicating the field direction, © ACS 2005).

With the outer confinement given, potassium iodide as a binary halogenide forms so-called Feynman-crystals. Their number of atomic layers is exactly determined by the inner cavity. For example, in an SWNT with a diameter of 1.4 nm a 2 × 2-lattice is observed, whereas a 3 × 3-lattice is formed in a tube measuring 1.6 nm across (Figure 3.97). Besides these structural limitations, there is also a change to the accustomed coordination numbers. In the 2 × 2-type, for instance, the coordination is reduced from 6:6 to 4:4 with an additional lattice expansion taking place along the tubular axis. For the 3 × 3 arrangement, on the other hand, 6:6-, 5:5- and even 4:4-coordination exist simultaneously with phase transitions being evened out by lattice distortions.

A very attractive way of filling the inner cavity in single-walled carbon nanotubes consists in using a material that in spite of complexation leads to a product that is still made of carbon alone. The compound incorporated is the fullerene C_{60}. These products are called "peapods" as the single fullerene molecules are actually arranged like peas inside their pod (Figure 3.98). The preparation of peapods does not pose larger problems. The purified and opened single-walled tubes are reacted

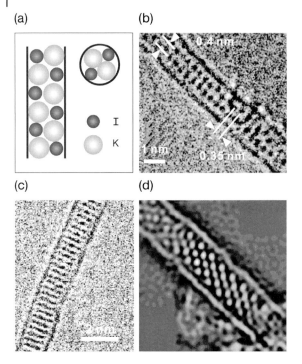

Figure 3.97 The spatial confinement inside a nanotube gives rise to crystal structure completely different from those observed for the bulk phase: (a) scheme showing the structure of a potassium iodide crystal inside a nanotube; HRTEM-images (b) of KI@SWNT with $d = 1.4$ nm; (c) of KI@SWNT with $d = 1.6$ nm, and (d) of Sb_2O_3@SWNT (© ACS 2002).

directly with solid C_{60} for about a day at 400 °C in a closed ampoule. The complexation takes place by diffusion of the C_{60}-molecules. The success of the reaction is checked by electron microscopy. Furthermore, peapods can be produced in solution as well. In this case the fullerene molecules enter the tubes through defects in the side wall.

Apart from C_{60}, also larger fullerenes like C_{70} may be employed in this reaction. The sole prerequisite is the use of nanotubes with a sufficient inner diameter. Especially the intercalation of endofullerenes has been studied extensively. The resultant products are doubly nested objects as a metal atom is contained inside each of the fullerene molecules incorporated in the tube (Figure 3.98). These substances may be considered M@C_n@SWNT (M = Gd, Dy, La, Sm). The endofullerenes included are not necessarily isolated from each other. Upon heating to 1200 °C or irradiation of electrons, there is rather observed a coalescence of neighboring fullerenes and, to some extent, an oxidation of the metal atoms (Figure 3.98). In the latter process the nanotubes act as electron acceptor. Yet it is not only mononuclear endofullerenes, but also dimetallofullerenes that can be intercalated. Among others, these may be $La_2@C_{80}$, $Gd_2@C_{92}$ and $Ti_2@C_{80}$.

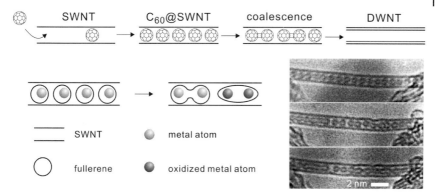

Figure 3.98 Generation of peapods by the diffusion of fullerene molecules into an SWNT. Upon electron bombardment and heating DWNTs are formed (top). Endofullerenes are incorporated, too. Upon coalescence, the metal atoms are partly oxidized (bottom left). The HRTEM-image shows $Sm@C_{82}@SWNT$ with coalescence of fullerene units increasing from top to bottom (© ACS 2001).

For the C_{60}-peapods, a conversion into double-walled nanotubes can be observed as well. The fusion of adjacent fullerene molecules occurs, for example, upon heating the peapods to 1270 K *in vacuo* or under electron bombardment in a transmission electron microscope. Maintaining the reaction conditions for long enough leads to the formation of a new carbon nanotube inside the peapod (Figure 3.98a), which means a selective generation of the double-walled nanotube. The mechanism of fullerene fusion has been subject to vivid discussion. It is quite probable that Stone–Wales rearrangements as well as dimerizations (via cyclobutane rings or single covalent bonds) are involved here. The actual mechanism, however, has not yet conclusively been proven, but it is rather certain that initially neighboring fullerenes fuse to be C_{120}-units that then continue to react with further fullerenes or dimers. The interaction between individual fullerenes and the tube's wall is increased by the small distance of 0.3 nm only. This value roughly corresponds to the van der Waals distance between two carbon atoms.

Recently, perylene-3,4,9,10-tetracarboxyl dianhydride (PTCDA) has been reported to be incorporated into an SWNT. The filled nanotube could then be converted into a DWNT by heating to about 1000 °C. PTCDA is known to form graphene layers upon heating to 2800 °C. Now given the spatial confinement inside the tube, it is forced to adopt a tubular geometry.

The C_{60}-molecules intercalated to an SWNT themselves may also enter into reactions. Heating, for instance, the complex $C_{60}O@SWNT$ to about 260 °C leads to the formation of polymeric chains from the $C_{60}O$-molecules. The outer nanotube serves as a template to this reaction that actually consists in an opening of the epoxide to give a furanoid structure connecting two adjacent fullerene cages. Hence the distance between the fullerene molecules is larger than in dimers connected via C–C-single bonds or cyclobutane rings, which is confirmed by HRTEM-exposures. Contrary to the three-dimensional polymerization of $C_{60}O$ without a

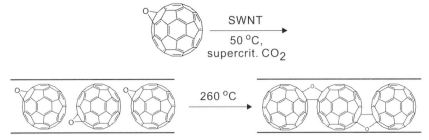

Figure 3.99 Owing to the spatial confinement, unusual reactions become possible inside SWNT. Shown here is the reaction of epoxidized C_{60} being linked to its neighboring molecule via a furan bridge (© RSC 2005).

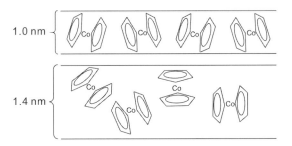

Figure 3.100 The diameter determines the way of incorporating smaller compounds. At too large a diameter the molecules may escape again from the tube.

nanotube, a totally different linear and unbranched product is obtained here (Figure 3.99).

A direct attack of oxygen on the included fullerenes, on the other hand, is much more complicated as it would have to happen through the tube's wall. Thermogravimetric measurements did show that the oxidation only sets in once the outer tube is oxidatively destroyed as well. Normally C_{60} is oxidized at lower temperatures, but in case of the peapods the reaction is suppressed by the impossibility of oxygen molecules entering into the filled tube.

Besides the substances mentioned so far, functionalized fullerenes like the simple Bingel adduct can be intercalated into nanotubes as well (Section 2.5.5.2). The formation of peapods has further been described for metallocenes (e.g., ferrocene), porphyrines (e.g., erbium phthalocyanine complex) and small fragments of nanotubes. The most important prerequisite for the feasibility of inclusion is always a suitable proportion of sizes of both the tube and the structure to be embedded. For example, this effect can be observed for the intercalation of different cobaltocene derivatives into SWNT. The endohedral functionalization only takes place at an internal diameter of 0.92 nm or above (Figure 3.100). But there is also an upper limit to successful incorporation. When the diameter of the nanotube is too large, the embedded species can easily diffuse away again from the host. Only few molecules are consequently found inside such a wide tube.

Purely organic compounds can be incorporated into nanotubes as well. One example consists in filling the inner cavities with ethylene glycol which, in electron micrographs, is depicted as liquid situated in the hollow. It is even possible then to perform a polymerization inside the tube to obtain a wire-shaped polymer molecule constricted by the nanotube. Further organic compounds include other solvents (e.g., ^{i}PrOH) as well as larger entities like single- or double-stranded DNA with up to 2000 base pairs. A common characteristic to all these experiments is the application of high pressures. The filling with nucleic acids, for instance, is performed at 100 °C and 3 bar/43.5 psi.

3.5.7
Supramolecular Chemistry of Carbon Nanotubes

Single- and multiwalled carbon nanotubes represent attractive components for supramolecular structures. Their rod shape predestines them as guest molecules to be included inside host compounds that are designed like rings or hollow cylinders. From an arrangement with the host wrapped helically around the nanotube, another type of supramolecular compound arises. The tubes can further serve as scaffold to supramolecular structures that form by a complexation of single tubes or bundles. Several examples of supramolecular patterns have already been discussed in Section 3.5.4.2 on the noncovalent functionalization of nanotubes. Hence only some other specimens will be presented here.

It takes no more than mixing carbon nanotubes with aqueous solutions of surfactants in a suitable range of concentrations to achieve an arrangement of the surfactant molecules on the nanotube surface. For amphiphilic compounds like SDS or OTAB, electron micrographs show periodical structures that correspond to the formation of half cylinders of surfactant molecules arranged on the surface of individual nanotubes (Figure 3.101). These half cylinders are never aligned in

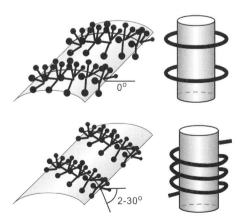

Figure 3.101 Some surfactants like SDS or OTAB form regular structures on the surface of carbon nanotubes. The specific pattern presumably depends on the tube's chirality angle.

parallel to the tubular axis. They are rather oriented at right angles to it, or at an angle of 2–30° with respect to a cross-sectional plane. Hence it follows that the resulting structure of surfactants on the nanotube surface is annular in case of the perpendicular arrangement or helical in case of the tilted. The pattern presumably aligns with the orientation of the individual tubes' graphene network. However, this assumption has not yet been proven as, due to the complete covering, a statement on the structure of the underlying nanotubes cannot be made. It would require samples with a uniform structure of tubes, which are unavailable for the time being.

Yet there are more factors playing a role in the self-arrangement of surfactant molecules. The ability to form micelles as well as the existence of long alkyl chains obviously contribute to the effect. Triton X-100, for instance, is a nonionic detergent based on polyethylene glycol. It does not tend toward patterning at all, but entirely covers the individual nanotubes with an unordered layer. Reagents with two long alkyl chains that are insoluble in water do not show structured deposition on nanotubes either. Still they may be arranged on nanotubes in an ordered manner by employing an adjuvant that might form micelles (e.g., SDS) and subsequently removing it by dialysis.

Interesting supramolecular structures can furthermore be obtained by the controlled crystallization of polymers like polyethylene or nylon-6,6 on carbon nanotubes. The latter serve both as template and as nucleus of crystallization. There are in principle three modes of how a polymer crystal might grow on a nanotube surface: (a) Phase separation may occur in the course of the crystallization so the initially dispersed carbon nanotubes reagglomerate and precipitate. (b) The polymer coils around the individual tubes whose solubility increases in the sequel. (c) Epitaxial crystallization takes place on the nanotube surface. In fact, variant (c) is observed. The crystallized polymers exhibit morphologies similar to those obtained in shear and elongation flow fields. Yet as no such forces are exerted during crystal formation, the structure is most likely induced by the MWNT present. A characteristic feature is the occurrence of disk-shaped crystallites situated in periodic distances of 50–70 nm. All these disks measure 60–80 nm across (Figure 3.102).

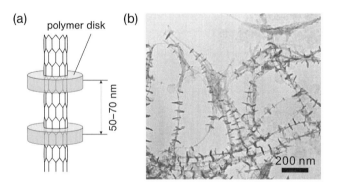

Figure 3.102 (a) The crystallization of polymers (polyethylene in this case) leads to regularly spaced disk-like structures; (b) HRTEM-image (© ACS 2006).

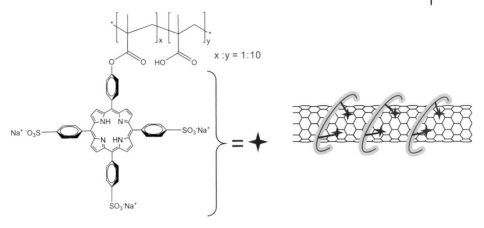

Figure 3.103 Helical entanglement of CNT with chains of PMMA functionalized with phthalocyanines.

Carbon nanotubes are also suitable templates to the generation of nanowires from phthalocyanines (Pc). The crystallization of $HErPc_2$, an erbium phthalocyanine, on SWNT has been reported, to name just one example. From a certain critical concentration on, a continuous layer is formed on the nanotube surface which then can act as a nanowire. A photoinduced electron transfer from the HOMO of the electron donor (the phthalocyanine in this case) to the nanotube has been observed for this system. A long-lived, charge-separated state evolves that is of some interest regarding electronic applications. Phthalocyanines attached to polymer chains are as well able to form photoinduced charge transfer complexes with nanotubes. In doing so, the polymer (e.g., PMMA) winds around the single tubes, while these are in direct electronic exchange with the phthalocyanine moieties present in the side chains. Likewise may metal-porphyrines that are bound to a polymer interact with SWNT, which results in the formation of a donor–acceptor system (Figure 3.103).

An important way of generating supramolecular structures consists in the recognition of attached DNA-strands: Carbon nanotubes functionalized with oligodesoxyribonucleotides may be connected to complementary nucleic acids or to DNA-modified nanoparticles by way of base-pairing. This fact offers a wide field of applications to biosciences (biosensing, markers, targeted drug delivery etc.), or it may be employed for the construction of complex, self-assembled structures for electronic applications. To this end, however, the site of functionalization on the nanotube surface must be exactly controlled. In Section 3.5.3, some methods have already been presented that allow for a directed modification of either apical positions (by amide bonding to acid groups at the tubes' tips) or of the side wall (functionalized pyrenes).

For long time a way was sought to achieve a size-selective complexation of carbon nanotubes by generating supramolecular structures. By now it has turned out that at least a partial separation can be achieved by the reaction with cyclodex-

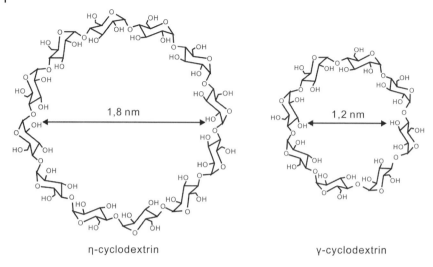

Figure 3.104 Examples of cyclodextrines able to form complexes with carbon nanotubes.

trines (cyclic polysaccharides) of different diameters. γ-Cyclodextrines, for example, lead to a specific complexation, and thus solubilization, of nanotubes measuring 0.7 nm across. The markedly larger η-cyclodextrines with their diameter of 1.8 nm, on the other hand, prefer nanotubes with a diameter of ca. 1.2 nm (Figure 3.104a). However, not only tubes with a defined pair (n,m) but all species with a fitting diameter are complexed in this process. These include, in the latter case, (9,9)-, (15,0)-, (15,1)-, (14,2)-, (13,4)-, (12,5)-, (11,6)-, and (10,8)-nanotubes. Obviously, the method is suitable to the separation of certain sizes, but not of a certain helicity. The selected nanotubes can subsequently be isolated by thermal decomposition of the respective cyclodextrine at about 300 °C as SWNTs do not suffer any damages in this range of temperatures.

Other polysaccharides as well suit the formation of supramolecular structures with carbon nanotubes. For instance, a complex may be formed with schizophyllan. This β-1,3-glucane exists as triple helix and winds around the SWNT accordingly. Functionalizing the termini of the glucane, for example, with lactosides, provides a compound that is soluble in water. Furthermore, it may confer to the product an increased affinity, for example, to lectine.

3.6
Applications and Perspectives

Rather soon after the discovery of the carbon nanotubes it became obvious that their extraordinary properties render them interesting materials for a multitude of applications. By now some fields have crystallized where significant technological progress might be achieved by the employment of nanotubes. This chapter will present some of them.

3.6.1
Electronic Applications of Carbon Nanotubes

3.6.1.1 Nanotubes as Tips in Atomic Force Microscopy

Atomic force microscopy is a powerful method for surface characterization. It is based on an interaction between a tip mounted to a cantilever and the substrate. The latter is systematically scanned to obtain a three-dimensional picture of its surface (Figure 3.105). Contrary to other methods in high-resolution microscopy, the samples can be examined at ambient conditions, and even nonconducting materials do not require coating with a metallic conductor, so the effort for sample preparation is markedly reduced.

The efficiency of AFM depends on several parameters pertaining to apparatus. Essential prerequisites for a good z-resolution include the flexibility of the cantilever and a sensitive detection of its movement. An accurate positioning and the sharpness of the microscopic tip, on the other hand, are substantial contributions to the x,y-resolution. Upon optimal adjustment of all parameters, the method is able to provide pictures with atomic resolution. The size and the shape of the microscopic tip constitute the limiting factor to the resolution of small objects, and especially of narrow, deep slits. Normally, tips made of silicon are used in atomic force microscopy. They are pyramidal in shape and exhibit a diameter of ca. 10 nm on their apex. These tips are unsuitable to examining deep grooves and, due to their brittleness, their mechanical resistance leaves much to be desired (Figure 3.106). Moreover, the resolution decreases when the tip breaks. Carbon

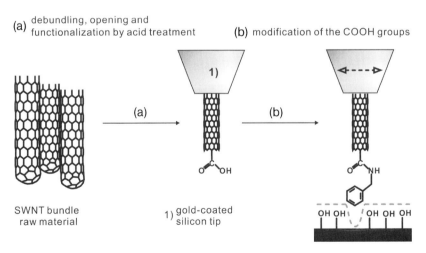

Figure 3.105 Scheme showing a CNT tip for atomic force microscopy. For chemoselective applications it might be even functionalized with suitable groups.

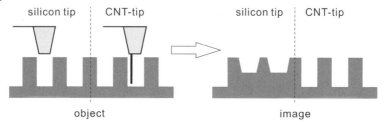

Figure 3.106 One of the advantages of carbon nanotubes is their large length : diameter ratio. Thus even deeply trenched structures may be correctly imaged.

nanotubes, on the other hand, offer a number of favorable characteristics predestining them for a use as microscopic tip. Firstly, they exhibit a suitable shape due to the large proportion of length to diameter, and secondly, they feature large mechanical resistance against the bending strain effective in AFM.

Both single- or multiwalled nanotubes may be used to prepare tips for atomic force microscopy. Several examples in the literature show that in comparison to conventional tips made of silicon or silicon nitride, a better resolution can be achieved. The first report on a nanotube tip describes a multiwalled tube stuck to a pyramid of silicon and trimmed to the desired length by a current impulse. It is crucial that exactly one MWNT protrudes at the tip. The resolution attainable with this setup is about 100 nm for deep slits and ca. 10 nm in lateral direction.

Single-walled tubes are promising too as due to their smaller diameter they should allow for a further increase in resolution. The values obtained to date are about 5 nm. SWNTs feature yet another advantage: they may be grown directly on the silicon substrate after preparing the latter with a suitable catalyst. In principle, this would enable a mass production of such tips. However, it is still a problem that not only one tube is generated, and that the nanotubes do not necessarily protrude at right angles from the tip.

Moreover, the tips may be chemically modified so in addition to a high resolution, also a chemical selectivity is achieved. Thus, it is possible to visualize not only the topology, but also the chemical structure of the respective surface. Scanning, for instance, a hydroxylated surface with an AFM tip that is functionalized with an acid amide allows for detecting the distribution of hydroxyl groups on this surface (Figure 3.105b).

3.6.1.2 Field Emission

It has been mentioned before (Section 3.4.4.3) that with an electric field applied, carbon nanotubes are able to emit electrons from their tips. This effect is termed "field emission" and may be employed to various applications. The production of field emission displays is one of the most attractive among them. These devices can be designed much brighter and more efficient in energy consumption when using carbon nanotubes. Just recently, a fully functional picture screen has been presented (Figure 3.107). Its production became feasible thanks to a new, ink-jet related method of generating structured nanotube patterns. Other techniques

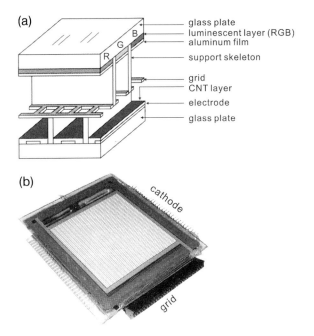

Figure 3.107 Scheme of a CNT-based field emission display (a) and a complete display (b) (© Springer 2002).

useful for this application employ a deposition from the gas phase on prestructured catalyst patterns. Thus a selective decoration of certain, addressable positions on the substrate surface with nanotubes can be achieved.

Contrary to other field emission displays that consist of arrays of tips generated by mechanical means, carbon nanotubes are stable up to much higher field intensities, while they emit already at a rather low field of $<1\,V\,\mu m^{-1}$. They can also stand high current densities of more than $1\,A\,cm^{-2}$. Both SWNTs and MWNTs are suitable for the production of field emission devices. The actual absence of defects is a more decisive parameter here. It correlates well with the efficiency of emission. MWNTs are more robust on long-term use, whereas for SWNTs, the small diameter of individual tubes is an attractive feature.

Besides the application of their field emission characteristics in display technology, other possible uses of carbon nanotube have been published, too. For example, the development of so-called cold cathodes containing nanotubes will become at least one element in advanced tube technology. Nanotube-based materials for field emission may further be employed in X-ray generation or as a microwave amplifier (Figure 3.108).

3.6.1.3 Field Effect Transistors

Upon heating, semiconductors exhibit a typical drop of electric resistance as consequently an increased number of charge carriers can cross the bandgap. The

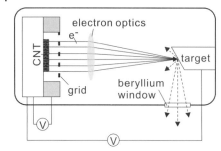

Figure 3.108 Scheme showing a "cold" cathode. The nanotubes are placed directly on the cathode.

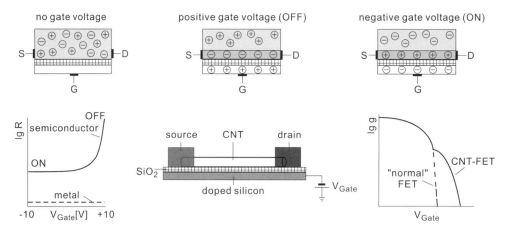

Figure 3.109 Construction scheme and operation of field effect transistor (g: channel conductance, © ACS 1999).

electric conductivity of semiconducting nanotubes may be influenced not only by this effect, but also by the application of an external field. This phenomenon is termed field effect. The electric induction of the acting field reduces or augments the number of charge carriers in the respective zones of the material. This feature may be employed for the construction of field effect transistors (FET) containing nanotubes (Figure 3.109).

Semiconducting carbon nanotubes deposited on a substrate normally show the characteristics of a p-doped conductor, which is apparent from a larger number of holes as compared to electrons. Hence the application of a negative gate voltage leads to a higher concentration of charge carriers and, consequently, to a marked increase of electric conductivity. The resistance of the nanotube can be modulated over several orders of magnitude this way (Figure 3.109). Metallic nanotubes, on the other hand, do not respond to a command voltage. The kind of contacts to the remaining circuit is another essential parameter influencing the $g(V_g)$-characteristic (dependence of channel conductivity from gate voltage) of a carbon nanotube FET. On the interface of nanotube and metallic contact, a Schottky barrier evolves. It gives rise to a kink of the $g(V_g)$-curve as the charge carriers have to tunnel through this barrier (Figure 3.109). Oxidation exerts a massive influence on the Schottky barrier, which actually proves its existence in nanotube FET.

3.6.2
Sensor Applications of Carbon Nanotubes

Many properties of nanotubes are highly dependent on the conditions given. While this may pose problems in the characterization, it can still be employed to study those very environmetal conditions. The nanotubes are thus used as probe or sensor. One of these applications, namely that as a tip for atomic force microscopy, has already been described in Section 3.6.1.1 as after all, these nanotube tips, too, act as sensors that detect variations of the surface structure on a substrate. The sensor applications hitherto presented in the literature may be divided into two groups: physical sensors on the one hand, and chemical ones on the other. These differ in the kind of properties that they serve to examine.

3.6.2.1 Physical Sensors
Physical sensors are sensitive to external parameters like temperature, pressure, mechanical strain etc. In the process, the sensor generates a signal that can be measured and assigned to a certain value. Carbon nanotubes can be employed to measure a multitude of quantities. Just a few examples shall illustrate this versatility here.

Carbon nanotubes respond to mechanical strain by a signal shift in the Raman spectrum. Especially the change of the G-band is proportional to the pressure and can be used to measure the latter. In a similar manner, nanotubes may serve as an indicator for strain in polymers. This is of particular interest with regard to nanotube–polymer composites.

Carbon nanotubes are also suitable as flow meters. The signal measured here is the induced current, which is proportional to the flow rate of the respective fluid. A reverse setup, that is, the action of an electric signal and a response from the nanotube by way of motion, is being tested for the construction of small actuators. These are molecular objects set in motion by an (electric) driving power. A charge injection into bundles of nanotubes, if performed in a suitable environment, can cause a considerable flexion of the bundles. Devices mimicking the function of a muscle may hence be envisaged all the more as the mechanical power of such nanotube actuators is much larger than the potential of magnetic analogs.

3.6.2.2 Chemical Sensors
While physical sensors are used to examine environmental conditions, the chemical ones serve to detect the kind and concentration of substances in this environment. Carbon nanotubes suit this task very well especially because all their atoms are found on the surface of the structure. The analytes can often be detected at

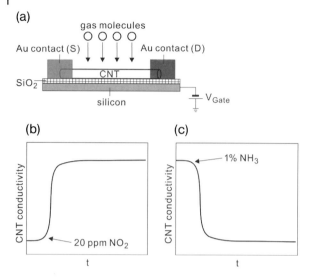

Figure 3.110 Application of carbon nanotubes as sensors. A scheme is shown in (a). The electric response upon addition of NO_2 or NH_3 is clearly dependent on the concentration ((b) and (c)). The arrows indicate the time of analyte addition (© AAAS 2000).

concentrations down to the ppt range. At the same time, nanotube-based sensors are already operative at room temperature, whereas the common semiconductor devices work only above 200 °C. The absorption of the analyte to the nanotube and their interaction by partial charge transfer alter the concentration of charge carriers or the potential barrier between the tube and its contacts. Usually, the nanotube's conductivity, a change of the resonance frequency or the electric resistance are measured in dependence from the concentration of analyte present. The ability to act as either electron donor or acceptor is an essential feature of suitable analytes. Nitrogen dioxide NO_2 (acceptor) or ammonia (donor), for example, are easily determined by means of nanotubes (Figure 3.110). Detection can take place both in the gas phase or, at least partially, in liquids. Analytes without or with less pronounced donor/acceptor characteristic can only be detected with sufficient sensitivity if the nanotubes have been surface modified before. A sensitive measurement of molecular hydrogen, for instance, is achieved only after the electrochemical deposition of palladium particles on the nanotube. The hydrogen then is cleaved into atoms by the palladium and subsequently diffuses into the metal–nanotube interface. Here, as an effect, a dipole layer evolves that acts like a microscopic gate electrode. The setup thus works in the manner of a field effect transistor (Section 3.6.1.3). To remove the hydrogen, however, the electrode must occasionally be regenerated in air or oxygen.

Carbon nanotubes functionalized with organic residues can act as sensors, too. For example, the limit of detection of NO_2 drops below 100 ppt using nanotubes modified with polyethylene imine. Strongly electron-withdrawing molecules as well can be detected this way. Sensors consisting of nanotubes may further be integrated into superstructures so the sensor molecule can be employed in electronic circuitry.

One example for the application of electrochemical sensors in biological systems is found in the measurement of the content of glucose by way of glucoseperoxidase (Gox) immobilized on a semiconducting carbon nanotube. The actual detection takes place by amperometry. The system of nanotube and enzyme could be demonstrated to be a rather efficient electrochemical sensor for glucose when combined with a source of electrons (ferrocenyl carboxylic acid in this case). However, the entire setup becomes operative only by the connection of nanotube and Gox. Apart from the latter, also further enzymes and bioactive substances have been immobilized on and activated by direct electron transfer from the nanotubes. They include horse radish peroxidase (HRP), cytochrome c and myoglobin. The structure of the tubes with their large ratio of length-to-diameter renders them ideal 1D-electron channels that conduct the electrons directly to the redox center.

3.6.3
Biological Applications of Carbon Nanotubes

Besides those carbon nanotubes modified with biologically active compounds to be sensing devices as described in Section 3.6.2.2, there are further perspectives for a use of nanotubes in biology and medicine. They have been proposed as carrier for antigenes, as blocker for ion channels, biocatalyst, bioseparator, and for directed cell growth, for example, of neurons, and promising first results have already been presented indeed.

Yet for the time being, it is still unclear in how far functionalized carbon nanotubes are toxic. The pristine tubes possess a nonnegligible cytotoxicity, which is, among others, due to their insolubility and the residual catalyst contained. First experiments with functionalized derivatives indicate, however, that at least this toxicity is markedly decreased once the surface has been modified. Anyhow, further studies are required here, and real applications in the medical sector are still a vision for the future.

3.6.3.1 Recognition of DNA Sequences

It has already been mentioned in Section 3.5.3 that carbon nanotubes are easily functionalized with PNA. Such tubes decorated with single PNA strands may be employed to enter into specific interactions with single-stranded DNA ("sticky ends"), realizing thus a molecular recognition of DNA fragments. To this purpose, a single-stranded DNA might as well be attached to the nanotube, but PNA offers some advantages like solubility in DMF or stability against enzymatic decomposition. Furthermore the uncharged PNA backbone stabilizes the PNA–DNA conjugates by causing less electrostatic repulsion. Nanotubes functionalized with PNA can be employed not only to the recognition of DNA fragments, but also to the generation of supramolecular structures. Nanotubes with suitable, complementary functionalities may combine in a supra-network, which is of interest for the construction of larger integrated circuits.

3.6.3.2 Delivery of Drugs and Vaccines; Gene Therapy

An essential prerequisite to the application of carbon nanotubes in a biological context is their solubility in physiological media. Unfunctionalized nanotubes do not meet this requirement. Hence, in a first step a sufficient solubility must be achieved by a suitable modification of the surface. One method consists in employing a nanotube functionalized with pyrrolidine rings by [3+2]-cycloaddition. To these may be attached triethylene glycol units with terminal Boc-protected amino groups (Section 3.5.4.1). Subsequent deprotection provides the terminal amino groups (Figure 3.111). The solubility of nanotubes modified this way amounts to no less than 200 mg ml^{-1} at a surface covering of about 0.4 mmol of ammonium groups per gram of material.

Well-established coupling strategies can now be applied to covalently link the functional groups of the nanotubes with peptides, maleimide linkers (for subsequent attachment of proteins via cysteine units), or the like. It is possible, for instance, to immobilize viral proteins (e.g., of the virus causing foot-and-mouth desease) while conserving their immunological properties (Figure 3.111). Thus the nanotube acts as a carrier for antigenes and might also be employed as a vehicle for vaccines. Nanotubes possess an important advantage over the carrier proteins commonly used for this purpose: It is true that the correct immune response to the antigene is evoked in both cases, and also that in a mouse model an immunity against the actual virus is generated, but to the nanotubes themselves, there is no immune response at all. Carrier proteins, on the other hand, give rise to such a reaction by inducing the formation of unspecific antibodies.

Conjugates of nanotubes and peptides also feature some beneficial properties for diagnostics. For example, peptides thus immobilized are bound more easily to ELISA plates, and they are better accessible for the recognition process. Carbon

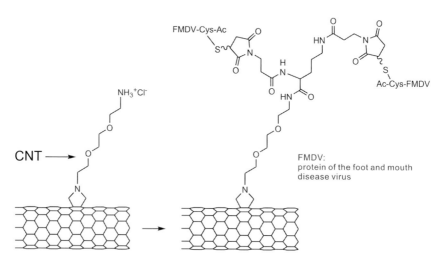

Figure 3.111 The immobilization of enzymes, antibodies and other biologically active substances is achieved by way of a pyrrolidine spacer.

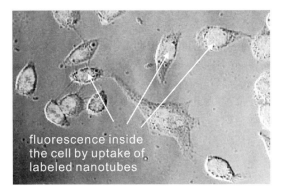

Figure 3.112 Suitably functionalized carbon nanotubes may be channeled into living cells and detected there directly, for example, by fluorescence labeling (© RSC 2005).

nanotubes can further present several epitopes (regions on the antigene surface determining its valency) at the same time, which leads to a more accurate detection.

Furthermore, carbon nanotubes offer some interesting prospects regarding the uptake and release of drugs in cells. Such carrier systems help to minimize or even overcome some typical problems in targeted drug delivery like low solubility or dispersion of the active, bad selectivity, or side effects to healthy tissue. It has already been demonstrated that nanotube peptide conjugates equipped with a fluorescent marker of fluorescein isothiocyanate (FITC) or nanotube streptavidin conjugates can be channeled through cell membranes without destroying it (Figure 3.112). However, the details of the incorporation into the cell by either endocytosis or other mechanisms have not yet been fully clarified.

In gene therapy, suitable nanotube conjugates offer similar advantages like in drug delivery. Most viral vectors commonly employed today cause severe side effects, ranging from unwanted immune responses via inflammatory processes to the generation of cancer cells, and this is where the nonviral nanotubes offer themselves as an alternative means of transport for gene-coding DNA or RNA. First experiments on their suitability as gene-transfer vector, performed with a noncovalently bound plasmid DNA, took a promising course. The DNA nanotube complex infiltrated the cell, and the coded enzyme (β-galactosidase in this case) was expressed many times more strongly. It is true that, for the time being, the efficiency as compared to other vector systems is still too low, but these preliminary studies give reason to expect a considerable enhancement for the future.

3.6.4
Materials with Carbon Nanotubes

Due to their fibrous structure and remarkable properties, carbon nanotubes are suitable to the production of polymer composites. Their extreme tensile strength

(>1 TPa) and also the somewhat elevated compression stability predestine nanotubes to be fillers in heavy-duty materials. The bonding between polymer matrix and nanotubes is quite durable due to the strong van der Waals interactions. Yet for the same reason, only rather nonpolar plastics can be reinforced with nanotubes as otherwise the attractive forces would not suffice to incorporate the carbon material into the matrix. In many cases, bundled nanotubes are used to prepare such composites. The advantage in doing so is to avoid the strenuous debundling, but at the same time the homogeneous distribution of tubes in and the strength of the interaction with the matrix are impaired.

A variety of substances fits a use as polymer components in these materials. Examples described in the literature include polystyrene, polypropylene, PMMA, polyaniline, PmPV, epoxy resins etc. (Section 3.5.5). Polymers with conjugated π-bonds suit best to the incorporation of carbon nanotubes because the additional π–π-interaction strengthens the bonding and leads to a wrapping of the (single or bundled) nanotubes in polymer molecules.

In any case, the mechanical loading capacity of the material increases upon the addition of carbon nanotubes. The composite's properties are considerably enhanced, especially with regard to tensile or bending stress. In workpieces charged with pressure, on the other hand, nanotubes do not convey much benefit as their resistance to compression along the axis is limited.

Depending on the method of production, the nanotubes in the polymer material obtained may be arranged either at random or in parallel. If no influence on the alignment is exercised during polymerization, a material with isotropic properties will be obtained. Methods like melt extrusion or the application of an electric or magnetic field during polymerizing proved their worth on the other hand, to achieve an alignment. To spin suspensions or solutions of nanotubes in the presence of the solved polymer or its precursor yields a particular degree of alignment, and thus highly anisotropic properties of the resulting material. In the *in situ* polymerization of caprolactame to nylon 6, for example, a very homogeneous distribution of nanotubes in the emerging fiber is achieved by adding carboxylated tubes to the monomer. The first dissolve in the latter, and then, during polymerization, the nanotubes are evenly incorporated into the fiber, thus enhancing its mechanical properties.

In a similar approach, it is not the monomer, but a solution of the prefabricated polymer (polyacrylonitrile in this case) in DMF that is being used. Herein the SWNTs are very finely dispersed. The product then also contains nanotubes aligned in the fiber's longitudinal direction. Another procedure resembles the method of producing carbon fibers from PAN (Section 1.2.3). Here the composite fibers are carbonized to yield a material of nanotube-reinforced carbon fibers. At a nanotube portion of as little as 3%, it already exhibits markedly improved mechanical properties.

The direct production of fibers exclusively consisting of carbon nanotubes succeeded as well. The method makes use of the deposition from the gas phase on a finely subdivided iron catalyst that is generated *in situ*. Ethanol added with <2%

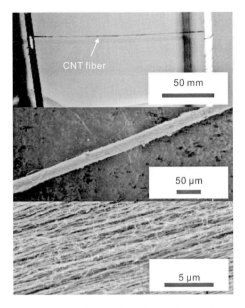

Figure 3.113 Fibers exclusively consisting of carbon nanotubes may be obtained by deposition from the gas phase. In this process, the resulting aerogel is reeled up directly at the reactor's end. Inside the fiber there are well-aligned nanotubes (© AAAS 2004).

of ferrocene and <4% of thiophene serves as carbon source. The liquid mixture is injected into a stream of hydrogen carrier gas that is led through an oven held at about 1100 °C. Here the nanotubes grow on the catalyst particles. Depending on the conditions chosen, either SWNTs or MWNTs are obtained. They form an aerogel that can be reeled up at the cool end of the oven to become a nanotube yarn this way. The tubes in this fiber exhibit a high degree of alignment (Figure 3.113).

Still for an industrial scale application of carbon nanotubes as filler in polymer materials, their price remains to be markedly reduced. Hence, for the time being, nanotube composite materials are only available for special examinations and applications.

3.6.5
Further Applications of Carbon Nanotubes

3.6.5.1 Heterogeneous Catalysis

For other carbon materials, and especially for the activated carbons, it has been known for long that they are serviceable supports for heterogeneous catalysts. Counting among the reasons for this is their large specific surface. Carbon nanotubes are suitable catalyst supports as well. Apart from a better control over

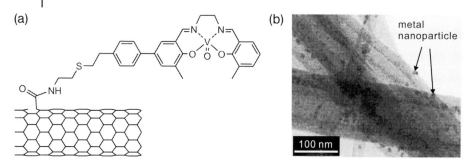

Figure 3.114 (a) Immobilization of a homogeneous catalyst complex, (b) MWNT with Pt–Ru particles deposited there by reductive precipitation (© AIP 2004).

morphology and chemical composition, they also offer the advantage that the respective catalysts can covalently be bound to the carbon matrix. Even originally homogeneous catalysts can be immobilized on the nanotube surface this way. This serves to an easier separability and thus to a recovery of the catalyst. The fixation of the organovanadium complex shown in Figure 3.114a is just one example here. This species can be employed in the cyanosilylation of aldehydes.

Nevertheless, classical heterogeneous catalysts like particulate noble metals may be immobilized on the nanotube surface as well. Nanoparticles of platinum or rhodium, for instance, can be deposited on cup-stacked carbon nanotubes by reductive precipitation (Figure 3.114b). The catalysts obtained this way suit an application in fuel cells run on methanol. Electrodes made from the nanotube material exhibit twice the efficiency as compared to the classical material XC-72-carbon. The particles of noble metal on the nanotube surface catalyze the direct conversion of methanol into CO_2 ($MeOH + H_2O \rightarrow CO_2 + 6H^+ + 6e^-$). A material to be employed in such fuel cells has to meet some essential requirements, including a maximal specific surface, a defined porosity and a high degree of crystallinity. Carbon nanotubes are endowed with exactly these characteristics, which is why they are the most suitable material for electrodes. Their high price, however, is still prohibitive to an industrial scale application.

3.6.5.2 Hydrogen Storage in Carbon Nanotubes

Increasing efforts are taken to explore new sources of energy that might substitute for the fossil carburants hitherto employed once the current supplies of mineral oil and gas are exhausted. A promising strategy, especially for the automotive sector, consists in power-harnessing from hydrogen. In doing so, the latter reacts in an environmental-friendly way in a controlled oxyhydrogen reaction to give water. For the time being, however, a safe and efficient storage of the gas constitutes the one problem insufficiently solved in the overall concept. A storage device assumed suitable for routine application should be able to bear more than 6.5% of its own weight or $62\,kg\,m^{-3}$, respectively, of hydrogen.

One method of storing gases in a safe way is their adsorption to solids, and so it has been for obvious reasons that quite early in the development of hydrogen fuel cells activated carbons were tested as storage media. But although their large specific surface seemed favorable at first glance, they turned out unsuitable because their pores are too wide, and only weak interactions occur between the hydrogen molecules and the carbon surface.

Carbon nanotubes, on the other hand, present pores of adequate size in the shape of their central cavity. Hence they are attractive candidate materials for efficient hydrogen storage. The values actually measured already range up to more than 7% of their own mass, which is about the minimal required storage capacity. What is more, nanotubes can already collect these amounts of hydrogen at room temperature, whereas other materials require significantly higher temperatures. Bundles of CNT show good storage capacities as well. Here the incorporation of hydrogen is assumed to take place also in the interstices between individual tubes.

From a mechanistic point of view, there are at least two different sites for hydrogen adsorption. The desorption from the positions binding less strongly sets in at about 400 °C already, while from those bound more tightly, it only occurs above 600 °C. The latter sites presumably are those inside the tubes, whereas the weaker binding is related to positions on the outer wall. Existing defects should also contribute to hydrogen adsorption, yet chemisorption should be considered in this case.

Other tubular structures like carbon nanohorns (SWNH, Section 3.3.4) are able to store large amounts of hydrogen as well. Only the availability of sufficient amounts rendered the use of these materials practicable, and much effort is taken to obtain operational hydrogen storage devices on nanotube basis as soon as possible. Once more, however, it is true that the material is still too expensive for industrial-scale use, and only a mass production of nanotubes would enable such widespread application. Moreover, the reproducibility of results turned out to be problematic. The ability to store hydrogen largely depends on the quality of the nanotubes employed. Their pretreatment and possible defects consequently affect this capacity. Hence it is crucial to establish protocols for a reproducible quality of samples.

3.6.5.3 Carbon Nanotubes as Material in Electrical Engineering

There is also some benefit in employing carbon nanotubes in lithium ion batteries. They are a suitable additive to the anode material for several reasons. Firstly, their small diameter allows for an even distribution in the electrode. Secondly, their electric conductivity substantially contributes to that of the electrode material, and finally, they are able to absorb the stress arising from lithium intercalation. On the addition of a few percent of carbon nanotubes, the cycle efficiency remains close to 100% even after many charging cycles, whereas untreated electrodes show a decline here.

Apart from the use in batteries, carbon nanotubes may also be added to materials employed in electric double layer capacitors. In this case, the nanotubes ensure a high capacity at markedly higher current densities than might be achieved using carbon black.

3.7
Summary

Carbon nanotubes are one of the most important classes of "new" carbon materials. Distinctions are made between single- and multiwalled as well as between zig-zag, armchair, and chiral nanotubes. The structure is characterized by the descriptors n and m. These structural parameters allow for a prediction of the electric conductivity. Only armchair nanotubes (n,n) and such species with $m-n = 3q$ are electric conductors. Any other nanotube is semiconducting. These statements have been established from symmetry considerations and from determining the band structure by way of the zone-folding method. There are different approaches to the production of single- and multiwalled nanotubes. Important methods of preparation are:

- chemical deposition from the gas phase (CVD methods);
- arc discharge between graphite electrodes;
- laser ablation;
- HiPCo process.

Box 3.1 Structure and electronic properties of carbon nanotubes.

- armchair nanotubes (n,n): electric conductors;
- zig-zag nanotubes $(n,0)$: semiconductors if $n \neq 3q$; electric conductor at room temperature if $n = 3q$;
- chiral nanotubes $(n,m$ with $m \neq n$, $n > m)$: semiconductor if $n - m \neq 3q$; at room temperature electric conductor if $n - m = 3q$;
- In the density of state function there are characteristic van Hove singularities.

If a catalytic process has been applied, residual catalyst must be removed from the crude nanotubes obtained. Further impurities include amorphous carbon and, in parts, fullerenes. A customary purification comprises an acid treatment which, at the same time, removes the end caps of the formerly closed tubes.

Major difficulties in the study and application of nanotubes are their tendency toward bundle formation and their poor solubility. Hence numerous methods have been developed to overcome these obstacles by a functionalization of the nanotubes.

> **Box 3.2 Chemical modification of carbon nanotubes.**
>
> - *Covalent:* Bingel reaction, [3+2]-addition of azomethine ylides, halogenation, oxidative opening of the nanotube tips
> - *Noncovalent:* functionalization with derivatives of pyrene, formation of micelles with various surfactants, wrapping in polymers (including starch, peptides)
> - *Endohedral:* peapod formation, intercalation of metal atoms, storage of hydrogen

The chemistry of nanotubes is closely related to that of the fullerenes. Their reactivity, however, is less pronounced due to the curvature in just one direction. There are three kinds of functionalization: covalent or noncovalent, respectively, on the outer wall as well as a filling of the central cavity. Important types of reactions are given in Box 3.2.

> **Box 3.3 Applications of and perspectives for carbon nanotubes.**
>
> - field emission displays with high luminosity and low energy consumption;
> - tips for atomic force microscopy;
> - field effect transistors;
> - sensors for small molecules;
> - composite materials bearing large mechanical charges due to high tensile strength of the nanotubes;
> - carrier material for systems in targeted drug delivery to living cells.

Carbon nanotubes possess a large potential for a use in electronics, sensing, and medicine. The applications below have either been realized already, or they are in an advanced state of development:

The results obtained so far impressively demonstrate the potential of these tubular carbon structures. For a continually successful development, however, the directed preparation of certain types of structures is eminently important. This task will consequently be a major challenge to further research. Moreover, efforts must be taken to produce budget-priced carbon nanotubes, and large amounts of them.

4
Carbon Onions and Related Materials

After having made the acquaintance of fullerenes and of single- and multiwalled carbon nanotubes, the question arises on the existence of multiwalled fullerenes. Such carbon cages concentrically arranged one inside another are also called carbon onions. In comparison to other new carbon materials, they have by far been studied less. Chiefly this is because only small amounts of those are available. Still they represent an interesting structural variant of carbon. This chapter deals with their structure, different methods of preparation, and first results regarding their properties.

4.1
Introduction

As early as 1980, even before the discovery of the fullerenes, that is, S. Iijima reported on the preparation of multilayered, spherical particles of graphitic character. He conceived them to be an sp^2/sp^3-hybrid material, and his results went largely unnoticed. The structures described then were first interpreted as carbon onions only after the determination of the fullerenes' structure and after D. Ugarte's finding that particles of fullerene soot may be transformed into multilayered fullerenes by electron irradiation (Figure 4.1).

The carbon onions present a link between the fullerenes and the multiwalled nanotubes. From one point of view, they may be considered fullerene cages concentrically contained one inside another and constantly increasing in size toward the outside. On the other hand, they might just as well be interpreted as multiwalled nanotubes with length zero so only the caps of the constituent tubes remain. With irregular caps, carbon nanoparticles of uneven shape result.

284 | *4 Carbon Onions and Related Materials*

Figure 4.1 Frontispiece of the October 22, 1992 issue of *Nature* (© Nature Publishing Group 1992).

4.2
Structure and Occurrence

4.2.1
Structure of Carbon Onions

There are two types of multilayered graphitic objects: the real carbon onions exhibiting a concentric structure of spherical shells and the onion-like, graphitic nanoparticles with a strongly faceted shape and a markedly larger inner cavity. Furthermore, a strong agglomeration is frequently observed for both types. In some cases, individual shells may even contain several smaller objects, thus connecting them inseparably (Figure 4.2).

All of these species are closely related, and they are in parts easily convertible one into another (Section 4.3.5). The unifying principle for all of these structures is the existence of closed carbon cages that are put together like the individual puppets of a Russian doll (Figure 4.2). The resulting structure is related to graphite in that covalent bonds exist only within the single layers, whereas the interaction between neighboring shells is limited to van der Waals forces. At first now, the completely filled, spherical carbon onions will be discussed, whereas the faceted nanoparticles will be detailed in Section 4.2.2.

Carbon onions, even if obtained from different procedures (Section 4.3), normally exhibit a similar structure. High-resolution TEM exposures usually show concentric circles with a distance between layers of 0.34 nm. This observation suggests a largely perfect spherical shape and a graphite-like interaction between the constituent shells (Figure 4.3). Just seldom, slightly faceted onions are found. At a suitable viewing angle, their projections exhibit symmetrical polygons (e.g., a decagon with a rotational symmetry, Figure 4.3c). Now then, the question arises of how these observations can be brought in line with theoretical considerations on the structure of the onions.

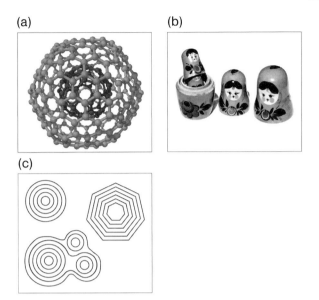

Figure 4.2 Carbon onions consist of fullerenes contained one inside another (a), thus resembling in a way the structural principle of a Matryoshka (b). Apart from perfectly spherical onions, however, there are also faceted species and such with several onions connected in one particle (c).

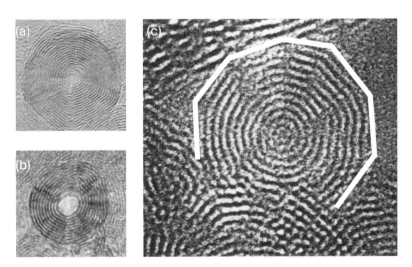

Figure 4.3 Different types of carbon onions. (a), (b) largely spherical onions, (c) carbon onions with faceted projection, also highlighted by the white line (© Elsevier 1996).

It has already been discussed in the chapter on fullerenes that the formation of a closed, cage-like structure from a planar, hexagonal lattice requires the presence of at least 12 pentagons. This rule is deduced directly from Euler's Theorem (Eq. 4.1 that correlates the numbers of faces (F), corners (C), and edges (E) in a polyhedron:

$$F - E + C = \chi \tag{4.1}$$

For two objects belonging to the same topological class, χ is a constant, characteristic number. For the class of polyhedra, χ takes on a value of 2. This is why at least 12 pentagons are required to close a hollow carbon cage. Assuming icosahedral structures, and obeying further the other rules for fullerene stability discussed in Section 2.2.3, the possible shells of a carbon onion are as follows: $C_{60}@C_{240}@C_{540}@C_{960}@C_{1500}$. ... This sequence of different fullerenes can also be inferred from simple geometrical considerations. With a predetermined number of 12 five-membered rings per shell and H six-membered rings, the number of atoms N in the respective shell is obtained from Eq. (4.2):

$$N = 20 + 2H \tag{4.2}$$

Assuming now the atoms to be distributed in a way to give a sphere, it is obvious that its surface has to equal that of its constituent deformed five- and six-membered rings (Eq. 4.3):

$$4\pi r^2 = 12A_5 + HA_6 = (12A_5 - 10A_6) + \tfrac{1}{2}NA_6 \tag{4.3}$$

One may further assume that due to the constant bond length of all shells, the density of atoms per unit surface is equal for all shells of a carbon onion. Hence follows $12A_5 = 10A_6$. With the surface of a hexagon of $A_6 = a^2(\sqrt{3}/2)$, Eq. (4.4) is obtained:

$$4\pi \cdot r^2 = (\sqrt{3}/4) \cdot a^2 \cdot N \tag{4.4}$$

This allows for calculating the difference of numbers of atoms between neighboring shells (Eq. 4.5):

$$\Delta N = N_d \cdot (1 + 2\sqrt{N/N_d}) \text{ with } N_d = (16\pi/\sqrt{3}) \cdot (d/a)^2 \tag{4.5}$$

Bearing in mind the length $|\bar{a}| = 0.246$ nm of the unit vector of a graphene cell and the constant distance of 0.34 nm between shells, the following results for a carbon onion consisting of $i = 1, 2, 3...$ shells and with $N_1 = N_d$ The number of atoms per shell with $\Delta N = N_1 \cdot (1 + 2i)$ is obtained from Eq. (4.6):

$$N_i = N_1 i^2 \tag{4.6}$$

So, if the innermost shell is a C_{60}-molecule, the abovementioned sequence of Goldberg polyhedra results: $C_{60}@C_{240}@C_{540}@$. ...

However, fullerenes with an I_h-symmetry are increasingly faceted with growing size and exhibit a distinctly icosahedral structure with 12 tips (Figure 4.4). Likewise, the projections of the Goldberg polyhedra show pronounced facets from any direction. Viewing along the C_5-axis, for instance, reveals a regular decagon. Thus,

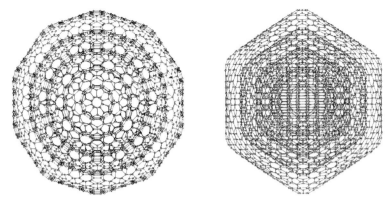

Figure 4.4 Projections in different lines of the sight of an ideal five-layered carbon onion consisting of Goldberg polyhedra. The faceted structure is clearly to be seen (© Elsevier 1996).

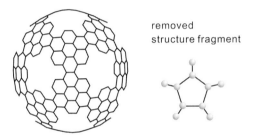

removed
structure fragment

Figure 4.5 A possible way of explaining the spherical shape of most carbon onions is the removal of fragments containing five-membered rings from the icosahedral apices. This would result, however, in a large number of dangling bonds.

it may be explained that the faceted structures are occasionally found in HRTEM pictures, but it does not account for the concentric circles observed in the projection of most onions.

There are now different ways of developing the fullerene approach; so it might explain the spherical structure as well. One variant consists in hypothetically cutting off the 12 tips of the model and considering the resultant orifices as defects of the shell (Figure 4.5). Such an approach would lead to the structure with a multitude of unsaturated bonds along the rims of the holes. A strong ESR signal should consequently be observed. It is true now that about 10 spins per carbon onion are measured (Section 4.4.1.5); still these do not suffice to prove a structure defective to such an extent. One possible reason could be the saturation of these bonding sites with functional groups. The spectra of existing samples, however, do not indicate anything like that. Under standard conditions, such highly energetic, defective structures most likely would collapse for their instability so they would exist as intermediates at best during the formation of onions, for example,

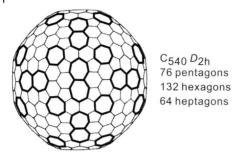

C₅₄₀ D_{2h}
76 pentagons
132 hexagons
64 heptagons

Figure 4.6 Another, much more realistic proposal for the structure of spherical giant fullerenes and onions is based on the presence of five- and seven-membered rings. The seven-membered rings (bold lines) in direct proximity to the five-membered ones evenly distribute the curvature over the entire surface (© APS 1998).

in the electron beam of an HRTEM (Section 4.3.5.2). Hence, this model does probably not suffice to describe in all detail the structure of spherical carbon onions.

An alternative approach is based on defective fullerenes, too, but the cage-like structure that it postulates is still closed. This concept endeavors to disperse the curvature of the onion shells as evenly as possible over the entire surface. This may be achieved by the incorporation of seven- or eight-membered rings causing a concave curvature of the surface. If, at the same time, an equal number of five-membered rings is inserted, the overall curvature will not be affected, but it will be spread over a larger area (Figure 4.6). In doing so, the five-membered rings balance the concave curvature of a heptagonal defect, while an eight-membered ring requires two pentagons for compensation. Eight-membered rings, as compared to seven-membered ones, are less likely to occur as they cannot arise from a single rearrangement.

A pentagon–heptagon pair, on the other hand, may be formed from two neighboring six-membered rings in the Stone–Wales rearrangement mentioned before (Figure 3.63) couples of defects normally result from this transformation. These are in direct contact with each other and originate from four six-membered rings arranged in the manner of a pyrene. Incorporating a sufficient number of such defects confers an almost perfectly spherical shape to the once icosahedral fullerene cages (Figure 4.6). What is more, one is no longer limited to the sizes of the Goldberg polyhedra due to the defects present. The spherical fullerenes, positioned at a mutual distance of 0.34 nm, take a random orientation among themselves. A graphitic interaction between adjacent shells will consequently form in very small regions at best. Obviously, this model is apt to support the electron microscopic observations also from a theoretical perspective.

The question remains on the size of the innermost shell. An estimate can be made by measuring the diameter in projections of entirely or partially filled carbon onions. The values obtained this way frequently correspond to the diameter of C_{60}. It is not possible, however, to exclude fullerenes of similar size like C_{50}, with certainty; all the more as the low stability observed for these species in their isolated state might be compensated by the pressure prevailing inside the onions (Section 4.5.2). Only nonspherical structures do not exist in all probability.

In HRTEM pictures of some samples of carbon onions, spiral patterns are observed that correspond to three-dimensional, nautilus-like spiral objects. These are especially found as an intermediate in the generation of concentric nano-onions from other forms of carbon. A more detailed discussion of these structures and their relevance is found in Section 4.3 on the mechanisms of onion formation.

4.2.2
Structure of Faceted Carbon Nanoparticles

It has been mentioned before (Section 4.2.1) that apart from the spherical carbon onions, there are also markedly faceted structures. Some of these exhibit a large central cavity (Figure 4.7a). They may be generated, for instance, by heating spherical nano-onions, or by direct methods like arc discharge or others. Structures of this kind are furthermore observed as faceted shells of metal nanoparticles that fill the void within the carbon structure.

Electron micrographs of these carbon particles clearly show parallel arrangements of graphitic layers that exhibit the typical distance of 0.34 nm between them. Some of these nanoparticles resemble short multiwalled carbon nanotubes. At the facets' contact points, the individual layers are heavily bent, which is to be attributed to a comparatively larger number of pentagon defects concentrated in a limited area (Figure 4.7b). In these regions, the structure in parts completely deviates from that of graphite. The structural features present here can also be observed in the caps of multiwalled carbon nanotubes.

Due to this largely graphitic character, the faceted carbon nanoparticles are more stable than their spherical analogs. In the latter, the graphitic interaction between layers is much less pronounced due to the multitude of defects and the random arrangement of shells. In the parallel domains of the nanoparticles, on the other

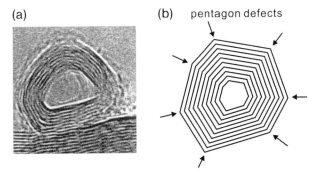

Figure 4.7 (a) HRTEM image of a faceted carbon nanoparticle (© Royal Soc. 1996); (b) scheme of such a particle. The positions of pentagon defects are designated by arrows.

hand, considerable attractive forces may act. The increased stability of the faceted particles is also evident from the fact that they can be produced from the spherical species, for example, by tempering at high temperatures (Section 4.3.5).

4.2.3
Occurrence of Carbon Onions and Nanoparticles

The carbon onions are an artificially generated modification of carbon. For the time being, no terrestrial occurrence is known. Just some highly defective structures similar to carbon onions have been observed in classical soot (Section 1.2.3), yet these do not normally have a closed shell and consequently cannot be considered carbon onions in the true sense of the meaning.

In the cosmos, however, carbon onions possibly exist. Soon after their discovery they have already been discussed as a potential reason for an up to then not interpretable absorption at 217.5 nm in the spectrum of interstellar space. The absorption spectrum of carbon onions closely resembles that of interstellar dust indeed. A red shift is observed on the occasion, yet this may be explained by the measurements being made in different media (water or vacuum, respectively; Section 4.4.1.3).

The generation of carbon onions in space has not yet been fully elucidated. However, it seems reasonable to assume that they originate from nanoscopic diamond particles. These may be converted into carbon onions upon heating, electron bombardment, or intensive irradiation (Section 4.3.5.4). The existence of nanodiamonds in extraterrestrial material could be confirmed by analyses on different meteorites. Especially the Allende meteorite contains significant amounts of tiny diamond particles (Section 5.1.2).

Actually, the first signs of carbon onions existing in such objects have been found in this very meteorite. Electron microscopic pictures (Figure 4.8) show

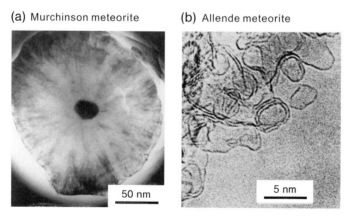

Figure 4.8 Spherical (a) and onion-like (b) carbon structures have been observed in chondritic meteorites (© Amer. Astron. Soc. 1996, Elsevier 2000).

circular structures with an interlayer distance of 0.34 nm. This is a clear indication of onion-like structures and supports the thesis on the cosmic occurrence of carbon onions.

4.3
Preparation and Mechanisms of Formation

In principle, two ways of preparing carbon onions are to be distinguished: On the one hand, they may be obtained from the transformation of other forms of carbon, which requires a considerable amount of energy in the shape of heat, highly energetic particles, or electromagnetic radiation. Another kind of preparation is based on the segregation of carbon from condensed phases that, due to its low solubility, are supersaturated with regard to carbon.

4.3.1
Arc Discharge Methods

The arc discharge in vacuum or inert gases at low pressure is one of the most common methods of fullerene or nanotube production. A similar procedure should hence be feasible for the carbon onions closely related to those structures, especially as all of them consist of the same structural elements.

The fact of the nano-onions formally being multiwalled nanotubes with a tubular length of zero is of particular relevance here. The growth of carbon species takes place inside a plasma zone. This has to be designed in a way to prevent (or at least disfavor) linear growth of occasionally emerging nanotubes and thus to promote the formation of spherical structures instead. One possible method consists in raising the pressure inside the reaction chamber to ensure a quick dissipation of heat.

First experiments with an arc discharge between graphite electrodes in helium at about 300 Torr/5.8 psi yielded spherical objects indeed, but these were still rather amorphous in character. Actually, this is due to the originally desired distribution of heat as now the emerging objects do not have enough time to graphitize properly before finally cooling down. However, most of the product obtained from attempts to overcome this problem by decreasing the pressure consisted of larger graphitic structures. Hence another way of restricting the reaction zone has to be sought which is not based on a higher gas pressure. This may be achieved, for instance, by confining the gas phase to bubbles with the surrounding medium simultaneously acting as a coolant. In this setup, the carbon vapor included in the bubbles can condense and the resulting carbon structures are collected at the gas–liquid interface. So essentially, the concept is to conduct the arc discharge in a suitable liquid, which would have the additional advantage of doing without expensive vacuum technology.

Especially water turned out an appropriate medium to generate a stable arc between graphite or other carbon electrodes (Figure 4.9a). In the course of the

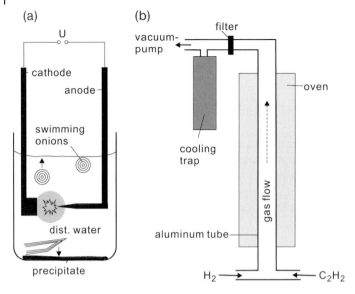

Figure 4.9 Methods of carbon onion preparation. (a) Arc discharge in a coolant, and (b) low pressure CVD-method with the resulting onions being carried from the reactor with the gas current.

reaction, a bottom sediment grows in the vessel that consists of larger graphitic fragments, nanoparticles, and nanotubes. On the water surface, on the other hand, a film containing a large portion of carbon onions is formed. No deposit is found on the cathode. The carbon demand of the reaction is solely covered by the anode, which is continually being consumed by the arc. The diameter of the latter electrode is only half that of the cathode, which serves to maintain a stable arc. The resultant carbon onions measure 5–40 nm in diameter (mainly 25–30 nm) at a density of $1.64\,g\,cm^{-3}$. On first sight then, it seems strange that the onions float on water in spite of their higher density – normally they should sink to the bottom of the reaction vessel. However, they feature an extremely hydrophobic surface which renders impossible the wetting and thus sinking of particles that are small. Additionally, the formation of agglomerates on the water's surface causes a further segregation of the nano-onions from the other carbon species formed. When taken from the water and dried, the material exhibits an extremely large specific surface of more than $980\,m^2\,g^{-1}$. This is, on the one hand, due to the small particle size, and on the other hand, the individual onions bear defects that give rise to a coarse surface.

Onion formation is also observed when using liquid nitrogen, but the product obtained contains more defects, and its concentration decreases. In theory, the higher cooling capacity of liquid nitrogen should prevent the formation of the undesired nanotubes and nanoparticles. In reality, however, this is not the case as, due to its low boiling point, the medium evaporates vigorously. As a consequence, the bubbles formed and the arc itself are rather unstable. Furthermore, a higher inhomogeneity of pressure and density inside the bubbles can be assumed. Hence, liquid water is the prevailing medium today. Its comparatively high boiling temperature, its nontoxicity, and the general availability render it an ideal solvent for a synthesis on larger scale.

4.3.2
CVD-Methods

Currently, there are few reports on the preparation of carbon onions and nanoparticles by means of chemical vapor deposition. It is obvious that this method just exceptionally suits the production of multishell fullerenes.

First experiments on making carbon onions by CVD were performed in the mid-1990s. They employed acetylene as a starting material which, in a process called "low pressure CVD" (LPCVD), was decomposed without catalyst at high temperature (Figure 4.9b). To this end, a laminar flow of a carrier gas (hydrogen in this case) mixed with the acetylene is passed through the reactor that is run at a temperature of 1150–1250 °C and, usually, at a pressure of 200 mbar/2.9 psi. Strictly speaking, this method is no deposition process as the resulting carbon structures are not condensed on a substrate, but carried from the reaction oven with the gas current and only collected in a cooling trap at the end of the apparatus.

The reaction yields nano-onions with an average diameter of 50 nm. With increasing gas pressure, larger onions are observed. The distance between individual layers is widened to about 0.35 nm, which indicates the existence of defects. With 1.9 g cm^{-3}, the density of the material obtained is significantly lower than that of graphite. Moreover, a strong agglomeration of the primary particles is observed with some of them even sharing one or more graphene layers. There are also regions within the onions that are not perfectly concentric. Summing up, the structure can be said in many aspects to resemble the carbon black obtained from acetylene at considerably higher temperatures (Section 1.3.1). Finally, there are tarry depositions of highly condensed aromatic compounds on the surface of the individual onions, yet these may be removed by extraction, for example, with toluene.

Another kind of onion-like carbon objects could be obtained from plasma-enhanced CVD. These are spherical graphitic particles with a diameter of ca. 1 μm. Their formation takes place by depositing on a substrate the decomposition products of methane (concentration 0.5%) in a stream of argon. The resulting carbon film consists of aggregated spherical particles (Figure 4.10). Furthermore, irregular carbon onions with diameters below 30 nm are generated, but they bear in parts large cavities and a multitude of defects in single shells. The method is based on the UHF microwave plasma CVD. Again no catalyst is employed as otherwise growth of multiwalled nanotubes would occur. The formation of onions, on the other hand, does not require linear growth and can take place immediately on the substrate without admixing transition metals.

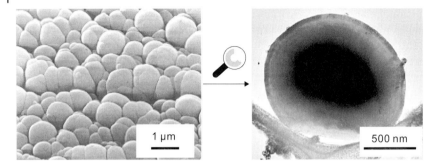

Figure 4.10 SEM and TEM images of spherical graphitic particles generated by plasma-enhanced chemical vapor deposition (© IOP Publishing 2003).

Figure 4.11 Carbon onions generated by ion implantation into a silver substrate (© Elsevier 1996).

4.3.3
Preparation of Carbon Onions by Ion Bombardment

In attempts to generate diamond films by the implantation of carbon ions into copper substrates, T. Cabioc'h and co-workers made a remarkable discovery: instead of the expected diamond crystallites, they found large, spherical carbon particles with graphitic structure. In further experiments, they could demonstrate the technique to be applicable to other substrates.

In the initial work, so-called giant onions with diameters from 50 up to 1000 nm were obtained (Figure 4.11). High-resolution electron micrographs reveal the material's graphitic character as well as a certain portion of amorphous carbon. EELS examinations further confirm the atoms' sp^2-hybridization. The position of the signals also points to amorphous and turbostratic portions within the material. The carbon onions are situated on the surface of the metal substrate with some of them being embedded in the latter or in a film of amorphous or turbostratic carbon.

Suitable substrates include silver and copper besides further metals like gold, tin, or lead. An essential feature of all relevant materials is a poor solubility of carbon in them. Consequently, segregation occurs upon carbon atoms penetrating

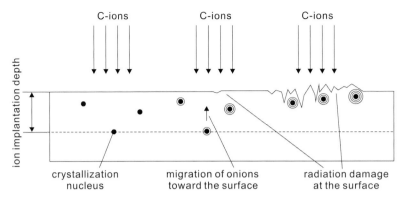

Figure 4.12 Mechanism of the formation of carbon onions in a metal substrate.

the metal matrix. It can take place either right on the surface or, at deeper penetration, in the shape of inclusions inside the bulk material. Bombarding the metal substrate (at 500–750 °C, depending on the experiment) with carbon ions will therefore generate nuclei of crystallization. Under the evenly distributed pressure of the surrounding matrix and under the influence of the emerging interface, these evolve into spherical objects inside the metal. Normally C^+-ions with an energy of 120 keV are employed (flow rate $\sim 1 \times 10^{13}\,\mathrm{cm^{-2}\,s^{-1}}$). The depth of penetration achieved is about 150 nm, depending on the metal, and actually this is where the carbon onions grow. Subsequently, they migrate to the surface through damages the material suffers from the irradiation (Figure 4.12).

Apart from the kind of metal, it is also the structure of the substrate affecting the size of the resulting carbon onions. Bombarding silver, for instance, results in a markedly decreased particle size of 10–20 nm, which can be reduced even further down to 5–10 nm by using a 300 nm silver film on a silica support. Furthermore, there is hardly any amorphous or turbostratic carbon observed for the implantation into silver. The presumable reason for the formation of smaller onions may be either or both of the following: firstly, the diffusion coefficient of carbon is higher in silver, which leads to an increased generation of nucleation centers, and secondly, the irradiation causes a larger number of defects. The additional reduction in particle size in the irradiation of the silver film mentioned above is due to the different grain sizes of the substrates used. The silver crystallites in the vapor deposited film are significantly smaller, so with nucleation taking place at the grain boundaries and with the current of carbon ions being constant, more and hence smaller nano-onions are inevitable to arise.

The carbon onions can be isolated from the metal substrate by thermal treatment. The silver or copper evaporate *in vacuo* at temperatures above 850 °C to release the onions. Especially when using a thin film of silver on a silica support, the nano-onions can easily be collected from the silica gel left after heating as there are virtually no interactions between them (Figure 4.12). If, on the other hand, silicon or steel is employed as a support for the silver layer, a strong binding of the carbon onions is observed after evaporation of the substrate metal. Most likely this is due to a local formation of carbide structures. Some 10 mg of carbon onions as a powder can easily be obtained in the manner described above, yet for production on a larger scale the method is too cumbersome.

4.3.4
Chemical Methods

It has already been mentioned in Section 2.3.5 that a rational synthesis of even the simplest fullerene makes high demands on the preparation of a suitable precursor, which has to be closed to the carbon cage in a subsequent pyrolytic reaction. This problem becomes all the more salient in attempts on designing a strategy for the rational synthesis of multishell fullerenes: it is not only the curvature of the carbon skeleton that must be considered here, but also the supramolecular pattern of concentric fullerenes of suitable sizes. Hence, for the time being, a directed synthesis of these multilayered species has not yet succeeded.

Besides a directed formation of covalent bonds at the correct positions, an approach to onion synthesis has to consider the arrangement of five-membered rings in neighboring layers. The central C_{60}-molecule could possibly exert a kind of template effect here that gives rise to the required π–π-interactions. A model application of this method has recently been reported. The authors succeeded in generating a complex consisting of a C_{60}-molecule and two carbon nanorings encircling it (Figure 4.13). The exchange between the π-systems is strong enough to render the complex stable even in nonpolar solvents. However, measurements of the complexation constant revealed that the interaction of the two outer rings is already diminished due to the lesser curvature. To transpose this model to the production of real carbon onions, similar interactions would have to be established

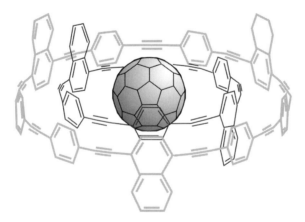

Figure 4.13 Fullerene molecule in a complex with two carbon nanorings. This is a first step to a rational synthesis of multilayered carbon objects.

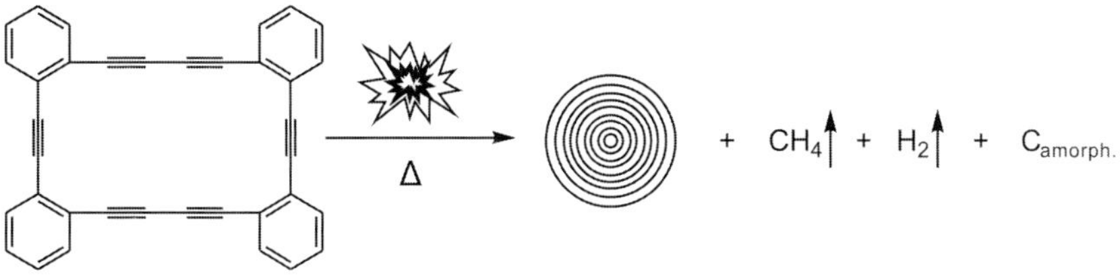

Figure 4.14 Thermolysis of benzodehydroannulenes yields a carbon material containing onions.

between hemispherical aromatic molecules of suitable size. Still the poor solubility of such compounds in common organic solvents is very likely to pose problems in doing so.

Nevertheless there is, like in fullerene synthesis, a variety of methods that start from – in parts synthetic – precursors and, in a final step, convert them into elemental carbon. This may, among others, exist as carbon onions. The closest thing to a "synthesis" surely is the thermal decomposition of benzodehydroannulenes that disintegrate into hydrogen, methane, and elemental carbon in an explosive reaction at about 245 °C *in vacuo* (Figure 4.14). The resulting mixture contains different graphitic species with carbon onions constituting a certain portion of them. The formation presumably takes its course by a stepwise crosslinking process, which would account for the multitude of further allotropes (especially graphite, amorphous carbon, and nanotubes). Fullerenes, on the other hand, are not generated. Larger carbon onions with a diameter of ca. 60–90 nm can be prepared by reducing glycerin with magnesium at about 650 °C in an autoclave. The magnesium oxide obtained as by-product is soluble in diluted hydrochloric acid and may thus be removed from the product mixture. Apart from glycerin, ethanol and butanol can be employed as carbon source as well.

Furthermore succeeds, with the support of hydrogen, the reductive transformation of supercritical carbon dioxide into nano-onions ($CO_2 + 2H_2 \rightarrow C + 2H_2O$). A platinum catalyst $[Pt(\eta^2\text{-}C,S\text{-}C_{12}H_8)(PEt_3)_2]$ is employed here which presumably releases $[Pt(PEt_3)_2]$ as reactive species. The resultant carbon onions exhibit a partially spiral structure, or they form aggregates enclosed in a common shell.

The chlorination of titanium carbide can as well lead to the formation of onion-like carbon structures. The resulting titanium(IV) carbide is removed from the reaction mixture by distillation. The carbon blend obtained has a large specific surface of about $1400 \, m^2 g^{-1}$ and contains, among others, carbon onions with a diameter of 15–35 nm. Further products include considerable amounts of amorphous carbon and small quantities of nanotubes.

Another method to generate onion-like structures (besides nanotubes and irregular carbon particles) is the ultrasonication of chloroform, dichloromethane, or similar solvents containing halogens. The decomposition of the solvent takes place on hydrogen-terminated silicon nanowires which, at the same time, also serve as a template to the formation of tubular structures.

Under suitable conditions, the reaction of perfluorinated hydrocarbons like teflon (PTFE) or perfluoronaphthalene with alkali amalgam yields carbon nanostructures as well. The reductive dehalogenation generates intermediate polyynes that subsequently condense to give the respective graphene structures. Normally, a mixture of carbon nanotubes and onions is obtained.

The decomposition of methane on a catalyst (with rare earth or transition metal oxides as suitable components) at 1100 °C provides spherical carbon structures with a rather uniform diameter of ca. 210 nm. These are, however, not classical carbon onions as the individual shells are not closed. They are rather objects made from graphitic units stacked up one over another that, to some extent, resemble classical soot particles.

4.3.5
Transformations of Other Carbon Species

Different modifications of carbon may be interconverted under adequate environmental conditions. This phenomenon has already been described in Section 1.3.2 regarding the production of artificial diamonds, which was formed from graphite on the application of enormous pressure and heat.

Extreme conditions in general enable a phase transition. These may not only be just high pressure alone, but also extreme temperatures, bombardment with electrons or other particles, or the application of energy-rich electromagnetic radiation. The crucial step is to remove a carbon atom from its equilibrium position to enable the redeposition in the shape of another modification. In such a process, the product formed is not necessarily the one thermodynamically most stable as kinetic effects may influence the outcome.

Usually, the vaporization of the entire carbon sample prior to conversion is not required in this type of reaction. It will do if a sufficiently large number of carbon atoms are mobilized and settle at a new equilibrium position. This may be achieved by thermal vibrations at elevated temperatures, but the so-called knock-on effects at particle bombardment or irradiation may serve to the same end. In the sections below, the products of thermal treatment and of various irradiations will be presented starting from different forms of carbon. All these procedures have in common that normally a heterogeneous mixture of products is obtained. Still these may often be prepared in macroscopic amounts, thus presenting the opportunity to study the physical and chemical properties of onion-like carbon materials.

4.3.5.1 Thermal Transformations of Soot-Like Structures

Carbon soot, as might be seen in Figure 1.10, already possesses a structure very much alike that of nano-onions. Only the roof tile arrangement of the graphene platelets in soot differs from the concentric pattern of intercalated fullerenes in the onions. Hence, it is self-suggesting to try preparing carbon onions from diverse soot materials.

Depending on the applied temperature, products of the most different kinds are obtained from heating arc furnace black in metal tubes. The structures formed at

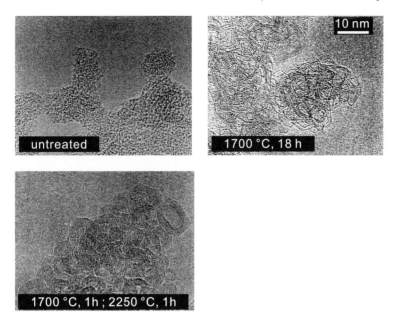

Figure 4.15 Thermal conversion of soot into onion-like material (© Elsevier 1993).

1700 °C are already onion like, yet they strongly resemble glassy carbon and feature entangled ribbons of carbon as well (Figure 4.15b). It is only upon heating to markedly more than 2000 °C that the formation of real carbon onions commences. Depending on the actual temperature of the second heating step, different numbers of onion shells are obtained as well as varying amounts and kinds of by-products. At 2100 °C, the onions possess three to four shells and considerable amounts of glassy and amorphous carbon are found. The portion of these by-products rapidly decreases at 2250 °C, and the onions exhibit up to eight layers (Figure 4.15c). This trend, however, does not hold at higher temperatures, and at 2400 °C, the number of onion shells is down to three or four again, which presumably is due to vaporization of the outer layers. In addition, larger structures like multiwalled nanotubes are observed. Examining the single carbon onions reveals some common structural features although they are not perfect spheres. In general, they are oval objects about 3 nm thick and 4–10 nm long with a central cavity. The dimensions of the latter roughly correspond to the two innermost fullerene shells which, due to the strong curvature, seem to be unstable under the prevailing conditions. Larger structures about the size of C_{60} may well be formed at first, but if so, they are consumed again in further course of onion growth. A finding in support of this thesis is the observation of single-walled fullerene-like objects measuring less than 1 nm across inside some of the products obtained.

A close electron microscopic inspection of carbon black shows the single soot particles to contain one or more nucleation centers that, in part, exhibit an onion-

Figure 4.16 Scheme of a soot particle with more than one nucleation center. In the respective subdomains' cores, there may be either concentric or spiral structures.

like structure. Both concentric and spiraly patterns have been described (Figure 4.16). Furthermore, there are single-walled fullerenes with diameters roughly corresponding to C_{60} or C_{70}. Summing up, there is strong indication that fullerene- and onion-like structures are involved in the growth mechanism of soot particles in incomplete combustion like, for instance, the production of furnace black. Hence, the question arises whether an early, sudden cooling of the emerging soot particles might stop the growth at the stage of carbon onions and whether a subsequent annealing of the product is feasible.

The structure of products from sooting benzene/oxygen flames has also been studied in this context. It turned out that the sooty film precipitating on the cool parts of the apparatus consists of tubular and onion-like carbon species. Still the product obtained is rather inhomogeneous, so for the time being, combustion methods are no effective means of generating carbon onions.

4.3.5.2 Irradiation of Soot-Like and Other sp^2-Hybridized Carbons

In irradiation, like in the thermal conversion of sp^2-hybridized carbons, it is crucial to achieve some structural fluidity to enable a transformation of the starting material. The absorption of energy from electron irradiation, for example in an HRTEM, also causes a local heating of the sample and a consequent breaking of C–C-bonds by excitation of electrons. In addition, high-energy particles like the electrons in the scanning beam of an HRTEM may transfer their momentum onto atom nuclei of the sample and shift them from their initial equilibrium position to an interstitial site. Normally, a flow of electrons 10 to 20 times higher (200–400 A cm^{-1}) than that employed in pure electron microscopic examination is applied to this end.

First results on the irradiation of arc furnace black have been reported by D. Ugarte as early as 1992. In these experiments, soot consisting of tubular structures, nanoparticles, and amorphous carbon were completely converted into largely spherical carbon onions within ca. 30 min (Figure 4.17). Graphite itself also changes into onions on electron bombardment. Depending on the duration of irradiation, the nano-onions may be anything up to 50 nm, and other authors even described the generation of "giant carbon onions" measuring up to 80 nm across and bearing 115 shells (Figure 4.18). Apart from the increase in diameter, a longer electron irradiation is also associated with a slight change of the onions' shape – actually their outer shells become increasingly faceted then. Onions of this

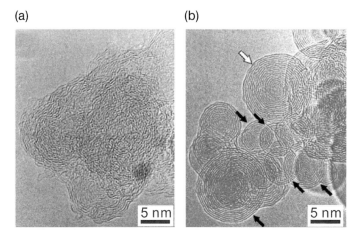

Figure 4.17 The conversion of soot (a) into carbon onions (b) may also be affected by the electron beam under an electron microscope (black and white arrows indicate onions with concentric or with spiral core, respectively, © ACS 2002).

Figure 4.18 Detail from an HRTEM image of a giant carbon onion (© Taylor & Francis 1995).

kind, which were subject to such an annealing process, are afterward stable in air for months.

It is striking that in the end, all samples examined so far could be transformed into multishell fullerenes under electron bombardment and that the products are stable even in air after suitable post-treatment. Hence, it seems reasonable to assume that all sp^2-carbons change into nano-onions, provided that the energy input is sufficient. These would consequently represent a metastable high-energy modification of carbon. Although they are not the thermodynamically favored species at room temperature, the structure of the onions kinetically stabilizes itself due to the complete lack of dangling bonds. A breaking of the bonds in the closed cages would require considerable amounts of energy. The structure of defective carbon onions, on the other hand, is easily destroyed, and they quickly transform into unordered graphitic material both *in vacuo* and under standard conditions. The defects constitute points of attack here, and a sufficient number of them too, and so the conversion can set in easily. Therefore, the respective samples have to be irradiated with electrons for long enough to cure all structural defects before storage.

Other forms of carbon can also be converted into nano-onions upon electron bombardment. Nanoparticles bearing a metal cluster in their center are just one example here. These objects are obtained from an arc discharge between suitably pretreated graphite electrodes in a manner similar to the preparation of endohedral fullerenes. The electrodes may, for instance, contain lanthanum oxide or gold inside a central bore. For lanthanum, the encapsulation in fullerenes or carbon nanoparticles has been known for long. Gold, on the other hand, does not form endofullerenes, and its wettability with carbon is poor, but in this manner, an intercalation into multilayered nanoparticles can be achieved nevertheless. A subsequent irradiation of this metal-carbon capsule causes an outward migration of the metal cluster from the center of the particle. The remaining void is immediately filled with carbon material, and a largely perfect carbon onion is obtained this way (Figure 4.19). Obviously, the electron irradiation creates inside the capsule a pressure large enough for the metal cluster to open a hole in the particle's wall and to migrate through it as a whole. The shape of the metal particles often change in doing so, which may be another indication of elevated pressure. Apart from

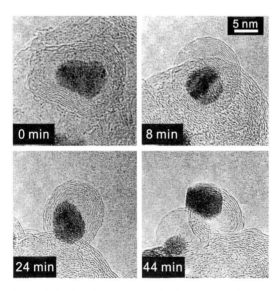

Figure 4.19 A carbon nanoparticle filled with metal transforms into an onion under electron bombardment while the metal particle emerges from the onion intact. Shown here is the migration of a gold particle from the center of an evolving carbon onion (© Elsevier 1993).

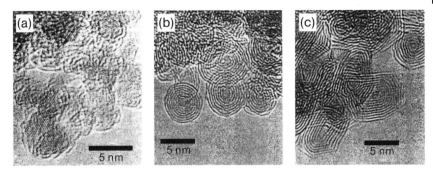

Figure 4.20 Thermal conversion of nanodiamond into carbon onions: (a) starting material, (b) treatment at 1700 °C, and (c) at 2000 °C (© Elsevier 1999).

lanthanum and gold, aluminum is another element to form such nanocapsules suitable to the preparation of carbon onions by electron bombardment. However, the encapsulated metal particles here are generated *in situ* under the electron microscope and may be transformed by subsequent irradiation.

4.3.5.3 Thermal Transformation of Diamond

From the detonation of carbon-rich explosives, diamonds with primary particles measuring about 5 nm can be obtained. Electron microscopic examination also reveals graphitic portions of the material that partly exist as onion-like structures or as multilayered graphitic shells on the diamond particles (Section 5.2.2). Hence, it seems reasonable to assume that a complete graphitization of detonation diamond leads to onion-like particles.

In experiments with nanodiamonds being heated *in vacuo* to 1000–1500 °C, various onion-like structures are obtained indeed. There are, on one hand, the desired spherical carbon onions, and oval onions as well as faceted nanoparticles on the other (Figure 4.20). It turned out that small nanodiamonds tend toward forming real onion structures whereas the faceted particles arise from larger diamonds. Still the latter product may further be converted into carbon onions by heating to 1700 °C. The minimal temperature for the graphitization of individual nanodiamonds to set in has been determined by experiment. It is considerably lower than that for the conversion of soot which is peculiar as the latter already exhibits a structure much alike the desired product. However, the (111)-planes of the diamond have many structural features in common with the (001)-plane of graphite. The transformation thus preferably takes place on these planes. Moreover, the tendency toward minimizing the surface energy leads to a reduction in structural strain and further favors onion formation (Section 4.3.7).

The resulting carbon onions normally consist of five to ten shells and agglomerate to be grape-like structures. The individual nano-onions are spherical in parts only, while others are oval in shape (Figure 4.20). In the center of the particles, there is often a shell corresponding in diameter to a C_{60}-molecule. At very high conversion temperatures, however (>1900 °C in general), the carbon onions initially formed alter their shape and become faceted. This is in analogy to the irradiation of soot particles in an electron microscope where upon continued exposure to heating, faceted particles are readily formed as well after an initial formation of spherical onions.

The particle size of the diamonds employed influences the conversion temperature, too. The larger the diamond particle, the higher is the temperature required for graphitization to begin. Nanodiamond with particles measuring ~5 nm starts graphitizing already at about 940 °C, while it is 1100–1500 °C for samples with particle sizes on the two-digit and lower three-digit nanometer scale. In bulk material, finally, the transformation into sp^2-carbon sets in only at 1800 °C. This points to an activation barrier that becomes increasingly higher with growing particle dimensions and which must be overcome before the conversion into graphitic material begins.

Altogether the thermal transformation of nanodiamond turned out a suitable method to prepare macroscopic amounts of onion-like carbon. It is true that the products obtained are inhomogeneous to some extent and that the resulting onions show various deficiencies (defects, deviations from spherical shape), but still the heating of diamond *in vacuo* constitutes the best method to date to generate larger amounts of carbon onions and study in principle their physical and chemical properties.

4.3.5.4 Irradiation of Diamond Materials

Small diamond particles, as indicated in the previous section, suggest themselves as precursors for carbon onions for the existence of sp^2/sp^3-transition regions on the particle surface. Besides the thermal conversion described above, nano-onions can also be generated by irradiation with electrons. Usually, the method is performed with a focused electron beam in an HRTEM. The flow rates applied amount to 10^7 electrons per square nanometer, which corresponds to a current density of $150\,A\,m^{-2}$. The duration of irradiation depends on the properties of the starting materials. Intervals normally reported range from 15 to 30 min. Upon prolonged exposure to the radiation, a curing of defects is observed for the prepared nano-onions. Contrary to soot, however, the size of the onions is conserved even over longer periods of irradiation. The product consists of largely spherical objects with 10 to 15 shells.

Nanoscale (~5 nm) materials as well as larger diamonds with plate-like morphology and particle sizes of several micrometers have successfully been transformed into onion-like structures. The constitution of the particle surface plays an important role in this process: on larger diamonds, carbon onions will only grow if there is a sufficient occurrence of surface imperfections. On a smooth surface free of defects, on the other hand, a plane, even layer of graphite is formed (Figure 4.21). Nanoscale diamond, as the other extreme, is directly converted into exactly one carbon onion per diamond particle. Each of these onions exhibits a larger volume than the diamond particle it arises from, which is explained by the lower density of the graphitic material. Upon electron bombardment on diamond particles that are markedly larger than nanodiamonds (i.e., >>5 nm; Section 5.5.3), the forma-

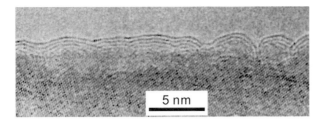

Figure 4.21 A sufficiently smooth surface provided, electron irradiation can cause the formation of a graphitic layer on diamond. Irregularities, on the other hand, give rise to onion-like structures (© Elsevier 1996).

tion of onion-like structures is observed too, yet these coalesce to become bigger, grape-like agglomerates before the outer shell of the respective objects is closed.

The action principle of electron bombardment corresponds to that of the irradiation of graphitic materials: it is chiefly the knock-on effect that causes the displacement of single carbon atoms. A rough estimate of the required energy is 50–80 eV. Thus, the electron radiation generated inside the microscope (at about 200 kV acceleration voltage) is just sufficient for a momentum transfer onto the carbon atoms' nuclei. The probability of a knock-on event taking place is significantly increased for a surface atom as a lower energy is required here to shift their equilibrium position. In addition to this displacement reaction, electronic effects are conceivable, too. In this case, the excitation of electrons to higher orbitals would lead to a bond breaking with subsequent reorientation of the atoms.

There is a major drawback, however, to the preparation of carbon onions by electron bombardment: the amounts obtained are extremely small and thus render the examination of bulk properties virtually impossible. Only high-energy electron sources outside an HRTEM would enable the production of macroscopic amounts. Still a further development of this method is of considerable interest as the carbon onions made from diamond are very uniform in quality.

4.3.6
Further Methods to Produce Carbon Onions

Apart from the methods described so far, there are several other procedures mainly or partly leading to the formation of nano-onions or onion-like structures. Some of these experiments will be presented below.

One way of producing carbon nanomaterial in the condensed phase is the electrolytic conversion of graphite electrodes in a melt of alkali halogenides. Depending on the reaction conditions, various carbon nanostructures are formed especially in molten lithium chloride and bromide. In the apparatus shown in Figure 4.22, the depth of immersion and the amperage are the leading parameters to control whether MWNT, carbon onions, or amorphous material are the main product. Apart from that, the temperature is another important factor as carbon nanostructures will only be formed above 500 °C. The carbon onions observed in

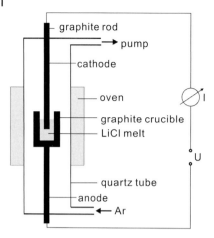

Figure 4.22 Scheme showing an installation for the preparation of carbon onions in molten alkali halogenide. The kind of carbon material obtained depends on the immersion depth of the graphite electrode.

the resultant mixtures measure 20 nm on average and exhibit numerous defects. In some cases, the internal structure suggests a spiral growth. At least in the present form, the method does not suit to generating macroscopic amounts of neat carbon onions. Still it serves to demonstrate that the nucleation of these species does not necessarily have to take place in the gas phase, but can also occur in condensed matter.

Provided that a suitable substrate has been chosen, the irradiation with high-energy laser light can also cause the formation of onion-like carbon structures. Irradiating, for example, acetylene in an oxyhydrogen flame with a CO_2 infrared laser yields soot mainly consisting of onion-like particles. Injecting the acetylene into such a flame without simultaneous irradiation only provides flat graphene layers and particles corresponding to normal acetylene black. A subsequent irradiation of the acetylene soot does not furnish carbon onions either. Obviously the action of radiation is required for the nucleation of onions to take place.

Furthermore, carbon onions can be obtained from irradiating silicon carbide with a pulsed UV laser (KF-laser, 248 nm, 25 ns) at a substrate temperature of ca. 600 °C. The action principle here is a selective vaporization of the silicon whose vapor pressure at the conditions prevailing is much higher than that of elemental carbon or of the silicon carbide itself. In this transformation, the onions grow both on the surface of the substrate and inside the SiC-phase, so two growth mechanisms obviously exist. On the surface, the selective vaporization of the silicon portion is followed by a thermal conversion of the remaining carbon into onion structures. This process should be similar to the formation of nano-onions from soot. Inside the SiC-phase, on the other hand, the poor solubility of carbon in silicon carbide becomes effective. The depletion of silicon causes a partial segregation of carbon from the surrounding phase, and spherical structures evolve under the prevailing pressure (Section 4.3.3). Once more, for the time being, the method is only suitable to provide small amounts of product with additional problems arising from the removal of residual silicon carbide.

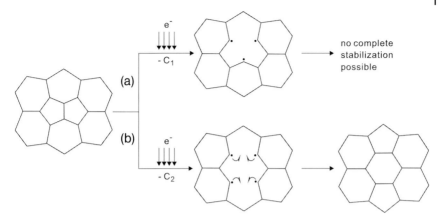

Figure 4.23 Electron bombardment causes single (a) and double defects (b) in graphitic materials. The latter lead to closed structures again.

4.3.7
Growth Mechanisms of Carbon Onions

The formation of carbon onions and related materials follows different mechanisms depending on the type of starting material and the method of preparation. A special distinction must be made between those procedures achieving the formation of onions from other carbon structures by bombarding them with high-energy particles and, on the other hand, such methods employing the energy input to generate small C_x carbon fragments that subsequently assemble to be carbon onions.

4.3.7.1 Growth Mechanisms of Carbon Onions Obtained by Electron Irradiation

The action of energy-rich electrons and other particles on a material may cause structural modifications that are characterized by single atoms or groups thereof being translocated to new positions. The acting particles require a minimum energy to do so. Its value depends on several factors like the particles' own mass and the activation barrier of displacement in the respective material being irradiated. For graphitic carbon, for example, soot, the energy required to translocate a single atom amounts to 15 eV.

Therefore, due to their small mass, the electrons used must convey a minimum energy of 100 keV to transfer in an elastic collision the momentum required for the displacement.

The electron irradiation gives rise to different types of defects in graphitic materials. Frequently a so-called single defect is observed. It is caused by knocking a single atom out of its position that leaves a vacancy behind and itself occupies an interstice (Figure 4.23a). Then, at a high concentration of such interstitial atoms, cluster formation from several carbon atoms may occur. The hole in the graphene structure cannot be compensated for by rearrangement of adjacent atoms and thus

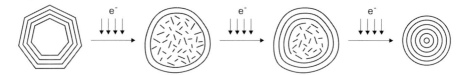

Figure 4.24 Mechanism of the formation of carbon onions from faceted nanoparticles.

will be stable until it is filled by another single migrating atom. Depending on the temperature, such defects may therefore have a remarkable lifetime. At elevated temperatures, however, the mobility of interstitial carbon atoms increases and the single defects are faster to recuperate. On the other hand, there are "double defects" generated by removing two carbon atoms at a time. These are unstable regardless of the actual temperature (Figure 4.23b) as they may pass into a closed structure by a sort of Stone–Wales rearrangement (cf. Figure 3.63). In doing so, two five-membered rings are formed at the concomitant loss of a C_2-unit. Consequently, the overall structure shrinks and experiences an additional curving. The finding of onion formation from soot or other graphitic materials being largely independent of the temperature suggests a mechanism with considerable contribution from double defects. However, spiral intermediates are observed as well in experiments on the electron irradiation of soots. The mechanism underlying here will be discussed in Section 4.3.7.2.

For the conversion of faceted carbon nanoparticles, an increasing order of the structure is found with the outer shell of the onion being formed first, while the inside initially suffers increasing disorder (Figure 4.24). Upon further irradiation, onion shells are also formed on inside by connecting the available graphene fragments. In the course of the reaction, the tendency to avoid dangling bonds also leads to the formation of five-membered rings that give rise to the required curvature.

The electron irradiation of mainly sp^3-hybridized carbon materials like nanodiamond (Section 5.5.3) yields carbon onions as well. In the case of bulk phase diamond, an energy of up to 80 eV is necessary to displace a single carbon atom, so electrons with an energy of not less than 330 keV would be required. Actually, however, the transformation of nanodiamond into carbon onions also takes place with electrons of much lower energy of, for example, 200 keV and at flow rates of 10^9–10^{10} electrons per nm^2. It is impossible to knock atoms out of the diamond lattice under these conditions. Still on the particles' surface there are carbon atoms that may be translocated at a markedly lower energy input. The same is true for defect sites. Actually, the conversion of nanodiamond particles into graphitic material under electron irradiation is observed to progress inward from the particle surface, and the resulting graphitic structures feed on the diamond material underneath (Figure 4.25). In this process, there will be also at first defects which are characterized by interstitial atoms that are quite mobile at room temperature already. These defect atoms constitute a transition between sp^2- and sp^3-hybridization. Therefore, they can act as bridging atoms between layers during the forma-

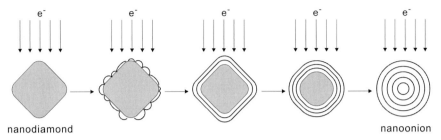

Figure 4.25 Conversion of a diamond particle into a carbon onion. Here as well graphitization begins on the particle surface.

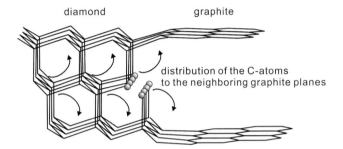

Figure 4.26 The mechanism of distributing the carbon atoms of a (111)-plane in diamond among the emerging graphite layers (© AIP 1999).

tion of individual shells because they contribute to the saturation of bonding sites and thus help to stabilize the emerging structure. On the diamond surface, cap-like graphitic structures arise that resemble the nuclei in the formation of multiwalled nanotubes (Section 3.3.7). These nanocaps keep growing to become onion shells. The latter are incomplete at first, but close in the further course of reaction with defects being healed by the diffusion of carbon atoms. It turned out that three (111)-planes of diamond are transformed into two (001)-planes of graphite (Section 4.3.7.2) as the respective lattice spacing shows only minor differences. The sp^3-carbon atoms situated amidst the new sp^2-layers are distributed between them (Figure 4.26).

4.3.7.2 Growth Mechanisms of Carbon Onions Obtained by Thermal Treatment

A thermal treatment of different carbon forms can lead to the formation of onion-like species as well. As for diamond particles, their surface structure plays an important role for the actual outcome of the process. If it is covered with functional groups, the bonding sites are saturated which renders a graphitization more difficult. From dangling bonds, on the other hand, graphitized domains will arise that can serve as nucleation center to the formation of carbon onions. In this process, a suitable orientation of lattice planes as well as a small particle size that entails a larger portion of partly graphitized structures favor the generation of onion-like material. Bulk diamond is hard to be converted into carbon onions, whereas nanoscale diamond particles already change into – initially unordered – sp^2-structures at about 1170 °C. The conversion expectedly starts on the surface. The heating serves, among others, to the removal of functional groups from the surface and to the concomitant partial graphitization.

At higher temperatures then a faster generation of sp^2-structures sets in and a particle with an onion-like shell on a diamond core is formed as an intermediate (Figure 4.25). For the close proximity of sp^2-shell and sp^3-core (the distance is never more than 0.35 nm), one can well assume a chemical bonding between them. Just like in the process induced by electron radiation, the transformation preferably starts from the diamond's (111)-planes as due to a reconstructive mechanism, they are prone to form graphitic surface structures (Section 6.2.2). Again the (111)-planes of diamond are converted into the (001)-planes of the curved graphitic material (Section 4.3.7.1). Depending on the structure of the starting material, either an exfoliation of graphene layers from the diamond surface or a transformation of (111)-planes confined on the edges is observed (Figure 4.26). Sometimes onions with a three-dimensionally spiral structure are found.

The formation of spiral intermediates is observed for the conversion of sp^2-carbons (induced both by electron bombardment or thermally) and, in parts, of nanodiamond as well. The resulting objects represent the three-dimensional equivalent to the rolled up graphene layers that have been discussed as a hypothetical structure of multiwalled nanotubes (Section 3.2.3). This growth pattern has been termed "spiral" or "snow accreting mechanism." It is based on the assumption of the onion growing outward from the center, which is supported by electron microscopic observation. Bowl-shaped aromatic compounds constitute the initial nuclei in this process. They are available especially in soot-like starting materials. The continued addition of further carbon atoms results in an ever larger and increasingly spherical object which ideally is closed to a cage. In most cases, however, the open edges do not meet due to an unsuitable curvature, so an overlap is formed instead. The basic unit of a three-dimensional spiral has come into being. From this moment on, the van der Waals forces constantly affect a sufficiently low distance between the outermost edge and the layer below (Figure 4.27).

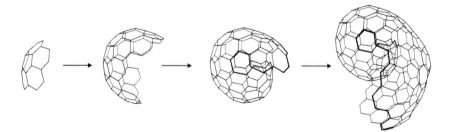

Figure 4.27 Formation of onion-like carbon by the so-called snow accreting mechanism that leads to the onions growing outward from the core (© Nature Publ. Group 1988).

Figure 4.28 The nautilus is an example of a naturally occurring three-dimensional spiroid structure. The distance between the turns increases from the inside to the outside.

Further growth then always takes place on this outer edge with the mechanism resembling the accretion of a snowball. Due to this interaction with graphene layers situated farther inside, the addition may be considered some kind of "epitaxial" process, which ensures a constant interlayer distance. In macroscopic spiroids like the nautilus (Figure 4.28), on the other hand, this distance normally increases toward the outside.

The spiroids formed are always in equilibrium with partly concentric and really onion-like structures. Due to the strong curvature and the high concentration of dangling bonds, the inner edge of a carbon spiroid is rather reactive. Hence, it is able to attack on the neighboring outer layer (Figure 2.29). In doing so, a closed fullerene cage is completed in the center and a new edge of dangling bonds is formed on the adjacent shell. Likewise, the transformation proceeds toward the outside. Provided the electron radiation or heating are kept up, a completely concentric carbon onion is formed in the end. The processes and species involved here include both radical reactions and aryne structures (Figure 4.29).

It is clearly to be seen from electron micrographs that the conversion of carbon spiroids into nano-onions really proceeds from the core to the periphery. They show entirely spiral objects at first. These are transformed into completely concentric onions via the hybrid form of an onion core with a spiral shell (Figure 4.30). It is presumably even just a part of the spiroid structures actually present in the sample that are detected in these examinations as their projections appear onion-like at unfavorable orientation.

At very high temperatures like in the plasma zone of an arc discharge apparatus, a vaporization of small carbon clusters occurs, and even atomic carbon may be generated. In this case, the carbon onions do not grow by a displacement of atoms inside an already existing structure, but they rather assemble stepwise, starting from a nucleation center. Among the species observed in the gas phase of such a process, it is especially C_2-clusters that are found in high concentration. At first,

4 Carbon Onions and Related Materials

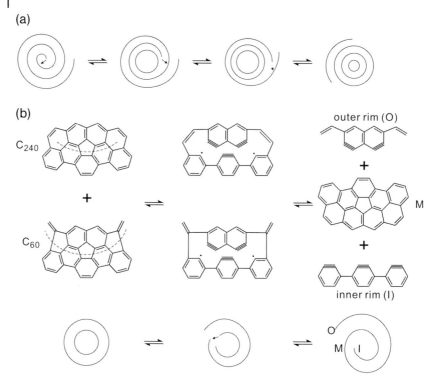

Figure 4.29 (a) Mechanism proposed for the formation of carbon onions via spiroid intermediates. At any time there is a subtle equilibrium between concentric, closed species and spiroid, open forms (b).

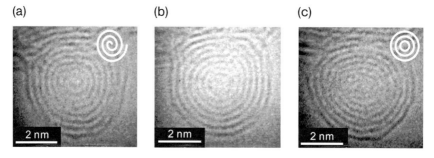

Figure 4.30 Conversion of a spiroid carbon particle (a), obtained from soot into a concentric onion (c). The transformation is induced by the electron bombardment under an electron microscope with the rearrangement proceeding from the outside to the inside (b), ©ACS 2002).

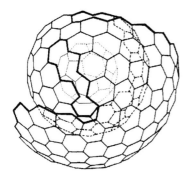

Figure 4.31 The accretion of curved graphitic structures and of annealed aromatic compounds also leads to a formation of carbon onions. The inner shells are not always completely closed, which explains for defects of the onion structure (© ACS 1986).

polycyclic aromatic structures emerge. Due to the tendency toward bond saturation, they will soon form the five-membered rings that also induce curvature.

This structure then keeps growing by the addition of further carbon units. The size and number of defects generated in doing so and the proportion between them give rise to different degrees of curvature which, in their turn, result in closed cages (Section 2.2.3) or in the aforementioned spiroids (Figure 4.27). The conversion into carbon onions then takes place as described above for the spiral mechanism. Still the growth of onion-like structures could also be explained by already closed cages adding further atoms and clusters. Especially in a plasma zone that is strongly confined by cooling, existing cages serve as nucleation centers indeed. Onion-growth occurs shell-wise according to this model, and inner layers do not necessarily have to be completed before another cluster can add to the surface (Figure 4.31). This hypothesis would also account for defects that are often observed inside of carbon onions.

Altogether it is reasonable to assume that different mechanisms, and predominantly among them a spiral and a shell-wise growth, proceed in parallel, regardless of starting material and method of preparation.

4.4 Physical Properties

4.4.1 Spectroscopic Properties

Numerous spectroscopic methods have been applied to examine the physical properties and to elucidate the structure of carbon onions. They include IR- and Raman spectroscopy, X-ray diffraction, electron energy loss spectroscopy (EELS), absorption, and photoluminescence spectroscopy and NMR-spectroscopy. Each of these methods gives account of certain aspects of the geometric and electronic structure, so altogether quite a detailed picture is obtained of the situation in carbon onions and related materials. There is, however, a strong dependency on

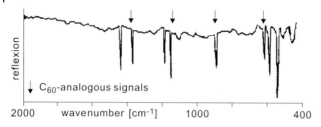

Figure 4.32 IR-spectrum of an onion sample (© Elsevier 1998).

the quality of the samples under test, and the data consequently show considerable variation. In the sections below, some of the most important methods of examination and the results of these experiments will be discussed.

4.4.1.1 IR- and Raman Spectroscopy

The structure of the IR-spectra available so far from various samples of carbon onions obtained by ion bombardment or arc discharge experiments is simple in a way (Figure 4.32). Up to nine different signals have been identified. Especially in transmission spectra, they are rather broad and comprise several vibrational modes. The relationship to the spectra of C_{60} is striking. The whole four signals of the smallest stable fullerene are found as well with the nano-onions examined, which means C_{60} frequently to be the innermost shell. The remaining signals could be identified as belonging to the larger shells. Computational simulations of the IR-spectrum of C_{240} indicate, for example, that at least a part of these signals correspond to vibrations of the second shell. Furthermore, there are additional bands above $1430\,\text{cm}^{-1}$ that are assigned to the even larger shells of the onion.

Deviations from the computed values result, on one hand, from the calculation method itself, and on the other hand should the actual structure of the IR-spectrum be influenced by the interaction between individual shells of the onion that alter the vibrational behavior.

The Raman spectrum of carbon onions is rather simple, too. Essentially, there are two bands situated in the regions about 1350 and $1580\,\text{cm}^{-1}$. In principle, a Raman signal should be expected for every single shell. These signals are, however, presumed to lie in close proximity and thus a cumulative spectrum with broad signals is obtained.

In Figure 4.33, the signal at $1350\,\text{cm}^{-1}$ is clearly to be seen, which is an induced band, only visible due to loosening of the conservation law of crystal momentum. This signal is characteristic of unordered graphitic material. The band situated above $1570\,\text{cm}^{-1}$ consists of several components. There are contributing signals at 1584 and $1572\,\text{cm}^{-1}$ (Figure 4.33). The G-band corresponds to phonons at point Γ of the Brillouin zone. For bulk graphite, it lies at about $1600\,\text{cm}^{-1}$. The shift toward smaller wavenumbers observed here ($1584\,\text{cm}^{-1}$) results from the bond bending and the associated strain brought about by the curvature.

The additional band at $1572\,\text{cm}^{-1}$ can be assigned to the onion structure. It originates from the in-plane vibration of the six-membered rings (E_{2g}-mode) of the

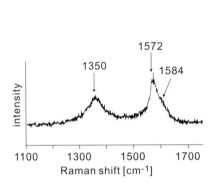
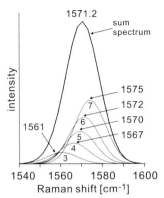

Figure 4.33 (a) The Raman spectrum of carbon onion shows typical bands in two ranges: the signal at ~1350 cm⁻¹ and the band at above 1570 cm⁻¹. The latter can be assigned to the in-plane vibration of six-membered rings in the onion shells. It may be deconvoluted for the contributions of fullerenes with different dimensions (right, the figures 3–7 correspond to the individual shell numbers, © Elsevier 1998).

onion shells. Computer simulations revealed that, depending on the size of the shell, different wavenumbers should be observed for this oscillation. The actual signal, however, can easily be interpreted as a summation of these individual vibrations (Figure 4.33). The half-width of this peak is an indicator for the number of defects in the nano-onion at hand. The sharper the peak, the less defects the material bears.

The Raman spectrum of carbon onions exhibits further, less intensive signals at 250, 450, 700, 861, and 1200 cm⁻¹. These as well become visible due to a breach of the selection rules that is associated with the bent graphene layers.

4.4.1.2 X-Ray Diffraction

X-ray diffraction can provide valuable information on the presence of different crystalline variants of carbon in a material. Hence, it suits well to the determination of sample purity after the preparation of carbon onions. Upon incomplete transformation, there are still signals of the starting material (e.g., of diamond) to be found.

Figure 4.34 shows the XRD-spectra of a pure sample of onion-like carbon and of a material that still bears residual diamond. The signals associated with the diamond lattice are clearly visible in Figure 4.34b. The pronounced line broadening can be explained by the small particle size (Section 5.4.1.3). Moreover, the spectrum features broad signals that are typical of unordered graphitic material. The spectrum in Figure 4.34a, on the other hand, shows the signals of carbon onions alone. The main peaks observed are the (002)-, (100)-, (004)-, and the (110)-reflexes of graphitic material. Still these signals exhibit strong broadening as well, which allows for an estimate on the average particle size being about 3 nm. In comparison to the value obtained from HRTEM, this diameter is a bit smaller,

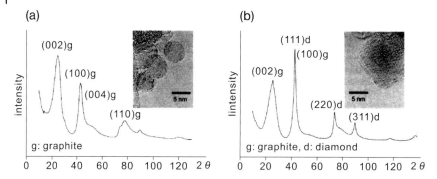

Figure 4.34 X-ray diffractograms of carbon onions without (a) and with (b) contaminations of diamond. Hence, an evaluation is possible of whether a thermal conversion of nanodiamond into carbon onions has been complete (© Elsevier 2002).

which is due to the fact that an XRD-spectrum represents an average of the entire sample including very small particles, while these latter components are normally underrated or disregarded in electron microscopy.

Actually, X-ray diffraction in general is a useful supplement to high-resolution electron microscopy because it measures the mean characteristics of the whole sample, whereas the HRTEM can only cover extremely small sections of it. The XRD results confirm the sp^2-character of carbon atoms in onions, low particle size as well as a certain deviation from perfect crystallinity. This imperfection results from the nonordered packing of individual shells, the different sizes of individual onions, and the defects present in the fullerene shells.

4.4.1.3 Absorption Spectra of Carbon Onions and Related Materials

Soon after the discovery of carbon onions, their absorption behavior has been studied both in visible and UV light. One motivation to do so was the assumption that they might be the long-sought source of an absorption band at 217.5 nm (4.60 μm^{-1}) in the spectrum of interstellar dust.

In the examination of nano-onions dispersed in water, really there is observed an absorption band that strongly resembles the shape of the interstellar spectrum (Figure 4.35). Furthermore, the band does not experiences a significant shift, but rather a broadening upon a variation of particle (or shell) size, which is also in good agreement with the behavior of the astronomic absorption. The signal is a π-plasmon band as expected for a multilayer fullerene. Only the position of the peak maximum at 264 nm (3.78 μm^{-1}) differs from that observed in the interstellar spectrum. The reasons for this bathochromic shift include, among others, the different media from which the spectra were obtained. The laboratory measurements were performed in aqueous suspension, while in space, vacuum prevails. However, the dielectric function of the surrounding medium has a considerable impact on the position of the band, and that is why the presence of water causes a marked red shift of the absorption. What is more, the carbon onions are partly agglomerated in these experiments, which also leads to a

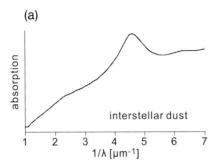
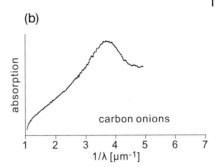

Figure 4.35 The absorption spectra of interstellar dust (a) and of carbon onions (b). It is true that the position of the absorption maximum is slightly shifted, but it still seems plausible that onion-like carbon structures might be responsible for the cosmic spectrum (© Elsevier 1993).

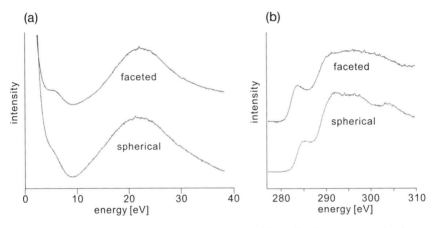

Figure 4.36 Electron energy loss spectra of spherical and faceted carbon onions: (a) low-loss region and (b) core-loss region (© Elsevier 1999).

bathochromic shift as compared to the measurement *in vacuo* where isolated onions should exist.

4.4.1.4 EEL-Spectra of Carbon Onions and Related Materials

The electron energy loss spectroscopy is an important means to determine the portions of sp^2- and sp^3-hybridized carbon in a given material. Both the low loss and the core loss regions show signals giving information about the structural features of carbon onions. In the range from 0 to 40 eV, there are two characteristic bands at 5.7 and 24 eV (Figure 4.36a). The signal at 5.7 eV results from the plasmon excitation of the π-electrons (π-plasmon peak), whereas the signal at 24 eV originates from a collective excitation of π- and σ-electrons (σ + π-plasmon peak). In the same spectral region, diamond or bulk graphite would give rise to signals at 33 eV or 6.6 and 26 eV, respectively. So, it is obvious from the plasmon peaks already that the dielectric properties of the onions differ from those of bulk graphite.

The signal close to the edge in the core loss region gives further important indication of the structure of carbon onions (Figure 4.36b). Most of all, it is the shape of the spectrum which provides information on unoccupied states of the conduction band and thus about the closest neighbors of covalently bound atoms. Besides a characteristic peak at 285.4 eV resulting from 1s→π*-transitions that is observed for graphitic materials in general, the spectrum shows an increase at 290 eV (1s→σ*-transition), a shoulder at 297 eV, and another signal at 303 eV. The fine structure above the 1s→σ*-transition is indicative of the spherical onions also containing a small portion of sp^3-hybridized carbon. A perfect nano-onion, however, should consist of sp^2-hybridized atoms alone, so where do these signals of sp^3-carbon originate from? There are several ways to explain this: firstly, if prepared from nanodiamond, the onions might still harbor a tiny diamond core that may no longer be visible in HRTEM, but well enough give rise to a signal in the very sensitive EELS. Eventual defects of the onion shell, on the other hand, could contain sp^3-hybridized carbons as well. These might, for example, be bridging atoms that connect neighboring shells. This hypothesis is supported by the loss spectrum of faceted carbon particles that are formed by further heating of spherical onions (Section 4.3.5). In the core loss region of the spectrum, there is no more indication of sp^3-carbon while the signals of graphitic structures are much more pronounced.

4.4.1.5 Further Spectroscopic Properties of Carbon Onions and Related Materials

ESR Spectroscopy In the ESR-spectrum of carbon onions made from nanodiamond, a sharp signal with a g-value of 2.002 is observed. It does not originate from an amorphous surface layer, but can really be assigned to the nano-onions themselves (Figure 4.37b). Keeping in mind that carbon onions are known from other measurements (e.g., EELS) to contain a certain portion of sp^3-atoms, an intermediate state between sp^2- and sp^3-hybridization can be assumed. It is characterized by the sp^3-atoms being situated at the edges of incomplete sp^2-layers, so dangling bonds are present here. These localized free electron spins are the reason for the ESR-signal of carbon onions. An estimate on the spin density yields a number of ca. 3.9×10^{19} spins per gram, corresponding to about 10 free valencies per nano-onion. These are obviously far from being perfect, but then, considering a number of about 12 000 carbon atoms per onion (at a diameter of ca. 5 nm), the concentration of defects is not very high either.

A signal of delocalized π-electrons, on the other hand, is not observed in spherical nano-onions. This means that the dimensions of conjugated sp^2-domains are rather limited and that most π-electrons are localized instead. An additional broad signal arises, however, for the better graphitized, polyhedral nanoparticles obtained

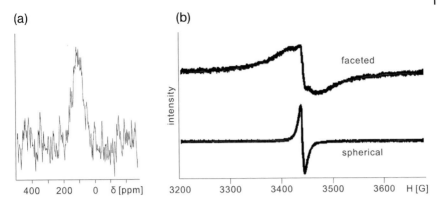

Figure 4.37 (a) Solid-state ^{13}C-NMR-spectrum of carbon onions (© Elsevier 2000); (b) ESR-spectra of spherical and faceted carbon nanoparticles (© AIP 2001).

from stronger heating of the spherical onions. The higher the degree of graphitization, the more pronounced this signal gets (Figure 4.37b). It can be assigned to the electrically conducting π-electrons as they become increasingly delocalized with a decreasing number of defects. Still even for these carbon particles with the faceted onion structure, a signal of spins from unsaturated bonding sites is observed. However, with ca. 7.6×10^{18} spins per gram there are obviously less defects.

^{13}C-NMR-Spectroscopy The solid-state ^{13}C-NMR-spectrum of carbon onions shows a signal at 100 ppm with a half-width of 117 ppm (Figure 4.37a) that can be assigned to sp^2-hybridized carbon atoms. In comparison to graphite (179 ± 10 ppm) and C$_{60}$ (143 ppm, sharp signal), it experiences a significant high-field shift and a strong broadening. This signal position arises from the action of the ring current of aromatic structures on neighboring shells, which causes a pronounced anisotropic effect. The considerable broadening that is wide even for a solid-state measurement results from the presence of many similar, but not equivalent carbon atoms and from the shell structure. However, a signal at about 40 ppm is not observed, so presumably the portion of sp^3-carbon in nano-onions, if any, is rather small. Bulk diamond, for instance, shows a ^{13}C-NMR-signal at 38 ppm, and nanoscale diamond resonates in the same range, too.

4.4.2
Thermodynamic Properties

The question of the thermodynamic stability of carbon onions has been subject to controversial discussion ever since their discovery. Some groups postulate them to be the thermodynamically most stable phase for this size of particles, while others describe them as metastable and existent under certain conditions only. Some of the arguments brought forth will be discussed in the following.

The size of the particles under consideration is essential in this discussion. Especially in the low nanometer range, the number of atoms is decisive for the stability of a given type of structure. Quantum chemical calculations reveal, for example, that up to a diameter of 1.9 nm, fullerenes are the most stable modification of carbon, while graphite is the most stable structure at diameters of more than 5.2 nm. For the range in between, it turned out that nanodiamond should be more stable than the respective sp^2-phases (Section 5.2.3). Carbon onions have some advantages over other nano-scale forms of carbon due to their optimal surface-to-volume ratio and their closed shells. For example, they normally do not bear dangling bonds, which considerably lowers the total energy of the system. Still the formation of nano-onions is obviously hindered kinetically at least as for the time being, no spontaneous conversion into onion structures has been observed under standard conditions. Sufficient energy to induce transformation will only be available under extreme conditions. The onions obtained, however, are rather stable and can be stored without serious damage even in air. Only the outermost shell is subject to a slight decay resulting in amorphous carbon.

Other authors, on the contrary, postulate the spherical carbon onions to be extremely unstable at least upon observation (not irradiation!) in the HRTEM, so they would retain their ball shape only at sufficient electron intensity. They decompose into unordered graphitic material inward from the outer shell. The instability is attributed to the system being far from thermodynamic equilibrium. The electron radiation maintains a steady state of onion formation and decomposition balancing each other. When it stops, the structure will tend toward equilibrium. Thus, carbon onions are considered a high-energy modification of carbon in this case.

For spherical onions obtained from heating carbon materials such as nanodiamond, a conversion into faceted particles is observed upon heating to even higher temperatures. The latter species possibly constitute the real thermodynamically stable modification of carbon in this range of dimensions. However, the phase diagram in the region up to ca. 10 000 carbon atoms is rather complex, so one should keep in mind that the slightest variations to particle size or conditions prevailing might lead to considerable changes of the stability situation. Therefore, a statement on the thermodynamically most stable modification of nano-scale carbon is hard to give.

4.4.3
Electronic Properties

Appreciable amounts of perfectly spherical carbon onions are hard to obtain, and so only few experimental data on their electronic properties are available. For the irregular onion-like carbon that may for instance be prepared from nanodiamond, on the other hand, the conductivity and other parameters have been studied much more extensively.

Figure 4.38 Model explaining the electrical conductivity of a carbon material consisting of carbon onions and related structures. Conductive channels develop with increasing graphitization of the material (© MRS 2002).

It turned out, for example, that upon heating nanodiamond, the specific resistance decreases from $10^9 \Omega$ cm to below 0.3Ω cm. This is actually due to the formation of carbon onions and the concomitant increase of graphitization. The onion-like structures in these materials exist as tightly bound agglomerates that are held together by defective graphene layers and C–C-bonds connecting several onions. Owing to these covalent and noncovalent links, conductive channels evolve inside the onion agglomerates, and a significant increase in conductivity is found (Figure 4.38). In this process, the structural difference between the almost spherical onions formed at first and the faceted species obtained by further annealing at higher temperatures becomes obvious: the initial free path in the onion structures is about 1.2 nm long, which roughly corresponds to the size of the graphitic fragments in this defective material. For faceted onions, on the other hand, a value of 1.8 nm reflecting the expansion of graphitic domains is found.

4.5
Chemical Properties

4.5.1
Reactivity and Functionalization of Carbon Onions and Carbon Nanoparticles

For the time being, little is known about the reactivity of carbon onions. This is, on the one hand, because only small amounts of the starting material are available and, on the other hand, for reasons that have already been discussed for the fullerenes and the nanotubes (uniformity, solubility, etc.).

In the broadest sense, the structure of the individual shells (including the outermost one) corresponds to that known from fullerenes. Hence, their chemical behavior can be expected to resemble that of large fullerenes. Contrary to the latter, however, carbon onions normally bear many defects. These may be five- and seven-membered rings as well as holes or connections to inner shells. Accordingly, an

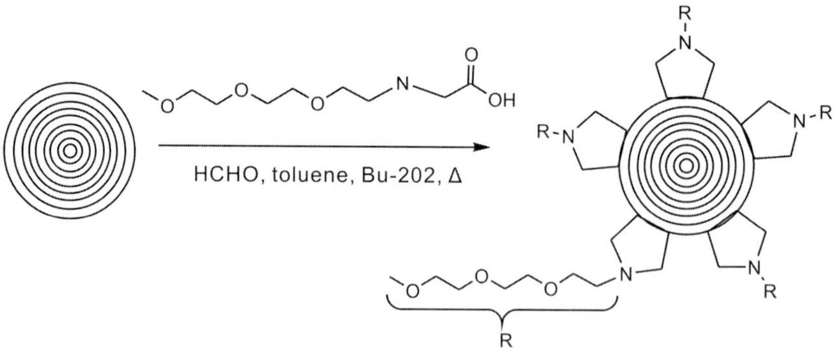

Figure 4.39 Reaction of carbon onions with azomethine ylides.

increased reactivity should be observed for some sites as the π-system would severely be disturbed there. The interaction with bonding partners in noncovalent functionalizations should be weaker for the same reason, and actually no such example has been reported so far.

In onions with large diameter, the reduced curvature leads to a decrease in reactivity. This phenomenon has already been discussed in detail in Sections 2.5 and 3.5, so it may be repeated just as catch words that the energy gain from strain relief in curved objects is less at larger diameters.

Multiwalled nanotubes suggest themselves as a comparison structure for onions the same as the fullerenes do. Like with the MWNT, the poor solubility of carbon onions poses problems in chemical conversions. Further complications arise from the inhomogeneities regarding diameter and number of shells as instead of defined products, there will always be a mixture that is much harder to characterize.

One example for the functionalization of carbon onions (measuring 50–300 nm across in this case) was presented in 2003. A reaction commonly performed on fullerenes or nanotubes was employed here: the addition of azomethine ylides (Figure 4.39). Using this reaction, some degree of solubility (5–7 mg per 100 ml) in different organic solvents could be achieved by derivatizing the reactant with triethyleneglycol units.

The reaction is conducted the same way as with MWNT with the sole difference of employing toluene instead of DMF as a solvent. This is because the nanotubes present as impurity in the sample are insoluble in toluene, whereas in DMF, they would be dispersed too, and participate in the reaction with the azomethine ylides. In toluene, on the other hand, only fullerenes and onions are sufficiently dispersed. The obtained samples feature interesting nonlinear optics (NLO) properties. In the meantime, one has also succeeded in generating on the onions' surface carboxyl groups subsequently to be functionalized by reaction with amines (Figure 4.40).

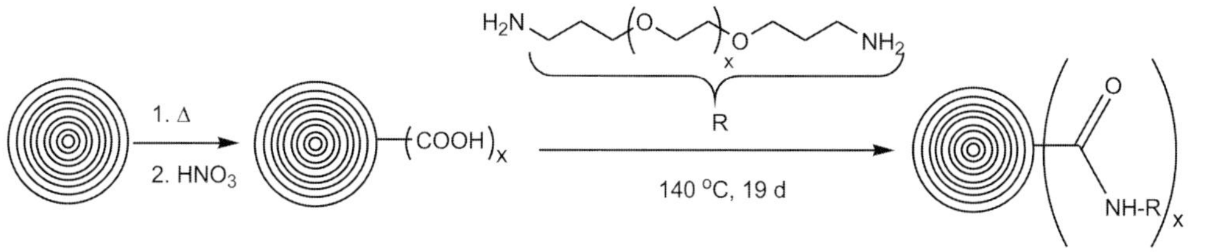

Figure 4.40 The attachment of organic groups to carbon onions via an amide bond succeeds by oxidation and subsequent reaction with an amine.

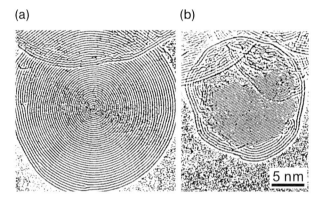

Figure 4.41 (a) In very large carbon onions, the interlayer distance decreases markedly from the outside to the center, which is indicative of an increased internal pressure. (b) Heating and concomitant electron bombardment in a TEM convert these onions into nanoscopic diamond particles (© Elsevier 1998).

4.5.2
Conversion into Other Forms of Carbon

Owing to their curved and defective structure, carbon onions are quite easily converted into other forms of carbon. The transformation of spherical particles into faceted nanoparticles by heating to at least 1900 °C has already been described in Section 4.3.5.3 on the thermal production of nano-onions from diamond particles.

It is strange to say then that carbon onions of sufficient size can as well be transformed into particles with diamond structure by electron irradiation. The phenomenon has been observed for large onions with 30–50 shells. Upon bombardment with energy-rich electrons at high flow rates (100 A cm^{-2}), they change into diamond from the core outward, provided that the sample is held at a temperature of about 600 °C during irradiation (Figure 4.41). Normally, a single graphitic shell is found on the surface of the converted particle. Obviously, it serves as a trap for unsaturated bonding sites.

A remarkable observation is made on close inspection of the large carbon onions that can undergo this transformation: the distance between individual layers is drastically reduced from the outer shells (0.34 nm) toward the core (0.22 nm; Figure 4.41). Pressures of more than 100 GPa would be required if one was to compress graphite to this extent! Very high pressure can thus be assumed to prevail in the heart of big carbon onions, although it is hardly possible to determine an exact value by experiment. This pressure in combination with the high temperature and the energy input from the electrons cause many defects in the nano-onions that lead to a large number of interstitial atoms. These are mobile in between the layers and give rise to bridges between them in the course of the experiment. So over time, the nucleation of a diamond core is achieved this way. The transition from sp^2- to sp^3-carbon is further facilitated by the curvature being strongest in the center of the onion. Altogether, the onions act as a kind of nanoscale pressure reactor for the nucleation of diamond.

On first sight, one would expect that the growth comes to a halt once the immense pressure is relieved and that a particle with graphitic shell, but with a diamond core is formed. Actually, however, the diamond formation proceeds outward unto the last but one shell of the former onion (Figure 4.41). There are different approaches toward explaining this effect. Possibly the further growth takes place by a similar mechanism such as the deposition of diamond from the gas phase (Section 6.3.2). The only difference would be the carbon atoms originating from the remaining onion shells which are knocked out by the radiation. These defects are much easier to generate in graphitic material as the energy required for a displacement of single carbon atoms is only 15 eV as compared to >50 eV in diamond. This is also why the diamond core, once generated, is not destroyed by the continuous irradiation. Another theory claims that the thermodynamic stability of the various carbon phases alters during irradiation as the electron bombardment and the high temperatures shift the system far out of thermodynamic equilibrium. This model assumes diamond to be more stable than the defective graphitic phase at sufficient electron radiation and in a certain range of temperature. Consequently, the atoms would have to settle as sp^3-carbon after displacement.

Apart from the irradiation with high-energy electrons, the conversion of carbon onions into diamond also succeeds by bombardment with ions like Ne^+. The latter are 36 000 times heavier than the rather light-weight electrons. Consequently, they require far less velocity and thus smaller accelerator voltages to bear the same effect. Diamond-like structures can further be generated by thermal treatment in air at 500 °C or by irradiation with a CO_2-laser.

4.6
Applications and Perspectives

So far, there are no commercial applications of carbon onions due to the small amounts of material produced as well as for the problems in assuring a homogeneous and reproducible sample quality. The attempts presented below should therefore be regarded as first perspectives of which direction the development might take.

4.6.1
Tribological Applications

Owing to their mechanical properties, carbon onions and related materials are suitable to tribological applications (reduction of friction between moving parts) or, in other words, they may be employed as a kind of lubricant. The nano-onions are predestined to this application due to their low size of about 10 nm, their spherical shape, and the weak van der Waals interactions between their individual shells that enable an easy ablation of layers. In the case of an outer shell free of defects, there is also little exchange with the surrounding material, which is of importance for good lubricating properties as well. Moreover, the carbon onions might be applied as a film, which means a further improvement of mechanical characteristics. The nano-onions may be used dry as well as in mixtures, for example, with conventional lubricating fats or oils. Possible conditions of use range from vacuum applications via air at room temperature to humid surroundings. To normal lubricants, these are extreme conditions hard to handle. Especially in humid air, the addition of carbon onions significantly prolongs the running life of moving parts. A similar effect is observed for graphite used as a lubricating agent. The onions act as a kind of back-up that comes to full effect when the basic lubricant fails. On the components, a graphitic film is formed on this occasion.

The small size and the spherical shape give rise to another mode of lubricating action. Contrary to a conventional fat or oil, the movement of the nano-onions across the surface is rather a rolling than a gliding (Figure 4.42). A similar action principle has been presented for onion-like structures of molybdenum sulfide. The onions are "sacrificed" in the reduction in friction as under mechanical stress they are destroyed by the prevailing shear forces just before the lubrication fails (Figure 4.42). However, the resulting film of graphitic and amorphous fragments and remaining intact onions still exhibits good tribological properties, especially as a carbon onion deprived of its outer shell is still an onion.

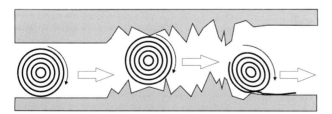

Figure 4.42 Operation principle of carbon onions in tribological applications. The lubricating effect is conserved even upon destruction of the onion structure.

4.6.2
Applications in Catalysis

A wide variety of carbon materials is known for their applicability as a catalyst or support of it. Especially the activated carbons with their large specific surface can be employed. But also carbon nanotubes and, to some degree, the fullerenes may be used to the same end. Besides, carbon onions and onion-like carbons are promising for catalytic applications, too. They feature a considerable specific surface, bear little structural defects (at suitable preparation), and they are stable over a wide range of temperatures.

The catalytic production of styrene by dehydrogenation of ethylbenzene constitutes an example of carbon onions being applied in the process of industrial relevance (Figure 4.43). Styrene is a basic chemical that is prepared on a scale of millions of tons. Normally, it is generated by thermal dehydrogenation, which, for being endothermic, requires supply with large amounts of energy. Furthermore, the catalyst (hematite with added potassium) is quickly deactivated, so altogether the process efficiency is limited. Hence, it is worthwhile searching for alternative procedures and catalysts. The oxidative dehydrogenation to styrene, for example, is exothermic and consequently far less demanding with regard to its energy consumption. A series of catalyst materials such as alumina, various phosphates, or metal oxides was found indeed. It turned out, however, that the actual catalytically active substance was a film of carbon that formed on the surface of the respective support. Therefore, it was self-suggesting to directly employ carbon materials themselves.

Still, when using porous carbons, the desorption of newly generated styrene from the catalyst surface was found to be hindered, and so the conversion was limited. The lack of any porosity in carbon onions should clearly be beneficial here, and indeed the conversion and the yields of styrene could be increased when using nano-onions or onion-like material. In samples of catalyst that had already been used, the onion structure was found to be partly destroyed. Actually, the catalyst reaches full activity only in this state, which also accounts for the induction period observed. On the onions' surface, presumably carbonyl groups and quinoid structures constituting the real active sites are formed.

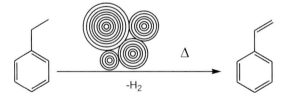

Figure 4.43 The dehydrogenation of ethylbenzene may be catalyzed by the addition of onion-like carbon.

4.7 Summary

Carbon onions consist of carbon cages in a concentric arrangement, the sole interaction between them being of the van der Waals type.

Box 4.1 Structure of carbon onions.

- Carbon onions consist of a multilayered arrangement of individual, closed fullerene shells. Ideally, each contains $N_i = N_1 \cdot i^2$ carbon atoms.

- In general, carbon onions are spherical, which implies a considerable number of defects in every shell as otherwise they would have to be Goldberg fullerenes with a faceted icosahedral symmetry.

- The central cavity is less than 1 nm wide. Hence, C_{60} and other spherical fullerenes about that size are candidate structures for the innermost shell. C_{70}, on the other hand, can be excluded due to its ellipsoid shape.

- Owing to a small number of dangling bonds and a favorable surface-volume ratio, carbon onions possibly represent the most stable form of small carbon clusters under conditions that favor high-energy structures.

Carbon onions can be produced by converting other forms of carbon by electron bombardment or heating. Furthermore, modified arc discharge methods or the implantation of carbon ions into metal substrates can be employed.

The physical properties of carbon onions clearly demonstrate their relationship with fullerenes and multiwalled nanotubes. They exhibit characteristic signals in their Raman and IR-spectrum, and electron energy loss spectroscopy provides important information on the structure of the π-electron system.

Little is known so far about the chemical properties, yet first results suggest a reactivity similar to that of multiwalled carbon nanotubes. Furthermore, a transformation of nano-onions into other forms of carbon can be achieved by heating (equilibration as faceted nanoparticles) or electron bombardment. In large carbon onions, a formation of small diamond clusters due to internal self-compression has been observed. These grow up to be nanoscale diamond particles under complete consumption of the onion structure.

The full range of possible applications has not yet been marked out due to the small amounts available and the inhomogeneity of existing samples. There are, nevertheless, promising perspectives for tribological applications and uses as catalyst material.

5
Nanodiamond

5.1
Introduction

Apart from naturally occurring diamond there is by now a variety of artificial carbon materials that feature diamond structure as well. These include the synthetic diamond generated by high pressure and temperature, but also films, polycrystalline materials resembling the *carbonados* (Section 1.3.2) and the so-called nanodiamond.

Now what kind of a substance is this latter carbon material? Of course nanodiamond is nothing else but diamond to begin with, meaning it to possess the same crystalline structure like the macroscopic material including the existence of cubic and hexagonal variants. Yet in addition, it is characterized by its small particle size that can range from a few to some hundreds of nanometers. In a narrow sense, it is especially particles measuring less than ca. 50 nm across that are called nanodiamond. The individual crystallites of nanocrystalline diamond films, on the other hand, do have diameters in the one-digit nanometer range. They are, however, frequently termed "ultrananocrystalline diamond" (UNCD) and will be described in Chapter 6.

The small particle size gives rise to a number of interesting effects. Especially the large portion of surface atoms contributing to the total mass makes itself felt in the physical and chemical properties.

5.1.1
Historical Background to the Discovery of Nanodiamonds

Nanodiamond as such is not really a new material. Indeed, essential experiments on the production of nanoscale diamond crystallites have already been performed in the late 1950s and early 1960s. No later than 1959, DeCarli and Jamieson found that tiny diamond crystallites could be obtained from the action of a shock wave on graphitic material. The procedure was patented in the 1960s, and from that time on, the DuPont Corp. has been producing about 2 million karat of synthetic diamond per year in this manner (Section 5.3.2).

Carbon Materials and Nanotechnology. Anke Krueger
Copyright © 2010 WILEY-VCH Verlag GmbH & Co. KGaA, Weinheim
ISBN: 978-3-527-31803-2

Figure 5.1 Detonation diamond as powder (a), as unstable suspension in water (center) and as completely deagglomerated dispersion in water (b).

At about the same time like the shock synthesis above, the basic principle of detonation synthesis was discovered. Around 1963, the Russian scientists Volkov and Danilenko found that the soot provided by the detonation of carbonaceous explosives bore diamond (Figure 5.1). Their results, however, remained unpublished due to their military relevance (the starting materials were explosives, after all). It was not before the late 1980s when the detonation method was discovered "anew" by several groups, and still a lot of these results were published only years later for confidentiality reasons (Abadurov 1987, Greiner 1988). The history of nanodiamond synthesis thus represents a revealing example of how the propagation of scientific knowledge may be hampered by political circumstances. Only now, more than 40 years after the first performance of a detonation synthesis, researchers worldwide can gain access to their colleagues' results.

The research on nanodiamonds was additionally stimulated by the discovery of the material in presolar meteorites (Section 5.1.2) and by the detection of nanoscopic diamond particles in interstellar matter. Insight into the formation of diamonds in space was provided especially by studies on their general ways of formation and on their spectroscopic properties. The first plant for an industrial scale production of nanodiamond with an output of several tons per year went operative in the early 1980s as well. This is, by the way, a striking advantage of the material: Contrary to other new carbon materials there is already an established method to prepare large amounts, so the subject of research is available in sufficient amounts at reasonable prices, and the starting material for future applications could easily be provided even on a large scale. Today big plants exist, for instance, in Russia, Belarus, Ukraine, China, Germany, the USA, and Japan.

The research on nanodiamond has by now developed into a multifaceted and very active field with the main interest focusing on the physical and chemical properties. Numerous applications have already been realized, and further dynamic development can be envisaged once it becomes possible to examine and modify individual nanodiamond particles.

5.1.2
Natural Occurrence of Nanodiamond

In terrestrial substances, no nanoscale diamond particles have been detected so far. As for the carbon in interstellar space, however, about 30% of it are assumed to be diamond. Proof of the existence of extraterrestrial diamond material is furnished by meteoric objects containing presolar diamond particles. The best-known among these are the Allende and the Murchinson meteorite both of which belong in the class of primitive chondrites. If at all, the matter in this type of meteorites has only little been affected during the formation of solar system, so their presolar composition has been conserved. This is documented especially in the isotopic ratio of certain elements (see below) that differs significantly from the otherwise uniform distribution within the solar system. The meteoric objects found contain something between 1 and 1400 ppm of nanodiamond with its particle size ranging from 0.2 to 10 nm (the portion decreases the more the surrounding material underwent thermal metamorphosis). The average diameter was found to be 2.7 nm in the Allende and 2.8 nm in the Murchinson meteorite. The diamond particles are isolated by an acid treatment that dissolves the metallic and mineral constituents of the meteor so the diamond portion remains as residue.

The formation of diamond particles in space is still the subject of lively discussion. The one thing undisputed is the fact that the particles have been generated beforehand and not inside the meteorite. There are different possible mechanisms for their formation. On one hand, the diamond may originate from a kind of shock wave synthesis inside a supernova. In this case the material should closely resemble diamond obtained from a shock wave or detonation synthesis. Another possibility consists in a deposition from the gas phase (related to the CVD process). These materials should be similar to the nanodiamond synthesized in the gas phase. A radiation-induced transformation of other, sp^2-hybridized carbon particles has been discussed as well (also refer to Section 4.2.3). A distinction of these processes and a weighting of their importance can be achieved by examining the microstructure of the meteoritic particles and comparing them to synthetic materials. In some of the isolated diamonds, an unusual enrichment of the heavy xenon isotopes ^{134}Xe and ^{136}Xe is found. The latter occurs with a frequency of 4×10^{12} xenon atoms per gram of diamond. The generation of the heavy isotopes can be attributed to a nucleosynthetic process inside the supernovae. At least a certain part of the diamond must consequently be assumed to originate from such a dying star. The basic mechanism here is similar to a shock wave synthesis. For a considerable portion of the meteoric material, however, an isotropic and not-too-fast growth must be considered. It is suggested especially by the existence of nonlinear twinning, the frequent occurrence of star-shaped twins (Figure 5.2) and by the comparatively large number of single crystals. What is more, no dislocations are observed as would be expected for a martensitic process (Section 5.3.2). Hence one may assume homoepitaxial growth here, which is typical of a gas phase deposition. Altogether, the majority of experimental data is indicative of such a gas phase process as it may occur in the expanding outer shell of a Red Giant (a class of stars).

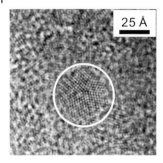

Figure 5.2 Diamond particle of meteoritic origin with star-shaped twinning (© Elsevier 1996).

5.2
Structure of Nanodiamonds

5.2.1
The Lattice Structure of Nanodiamond

The lattice structure of nanodiamond has been examined with various spectroscopic and crystallographic methods. It turned out to correspond to that of bulk diamond in all essential features. Especially in X-ray diffraction experiments, the lattice constant was found to differ just slightly from the value of 2.456 Å determined for the macroscopic material. The only significant difference is a broadening of the signals due to the little particle size (Section 5.4.1.3). Both cubic and hexagonal phases (the latter only in coexistence with the first) have been observed with the cubic variant being much more frequent.

However, the particles' small dimensions give rise to one of the characteristic structural features of nanodiamond: The large portion of surface atoms causes strain within the particles that shows in an altered bonding situation close to the surface. In NMR-examinations, for instance, the carbon atoms on the surface exhibit another chemical shift than those situated in the core of the particle. This is due to the attachment of functional groups or due to sp^2-hybridized atoms arising from a reconstruction of the surface (Section 5.2.2).

Apart from this distinct influence of the surface structure, there are also lattice defects that make themselves particularly felt in such tiny particles. One possible defect found in nanodiamond is the so-called N-V-center (nitrogen vacancy center) consisting of a nitrogen atom incorporated in the lattice and an adjacent vacancy (Figure 5.3). Such a negatively charged defect consequently possesses an additional unpaired electron that can be observed as single spin. Moreover, the N-V-centers are responsible for the fluorescence typical of nanodiamond particles containing nitrogen (Section 5.4.1.4).

Further defects include, for example, the directed doping with boron, nitrogen or nickel. These confer certain electronic or optical properties to the nanodiamond particles (Section 6.2.3). Experimental as well as theoretical results show that only few elements like boron, nitrogen, silicon, oxygen, or phosphorus can be incorpo-

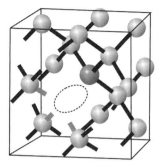

Figure 5.3 N-V-defect in a diamond lattice. Shown here is a detail containing a vacancy (dashed line) and a substitutional nitrogen atom (dark gray).

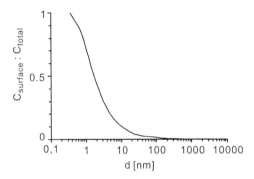

Figure 5.4 The proportion of surface atoms relative to the total atoms for octahedral particles. The smaller the diameter d, the higher the portion of atoms on the surface. For a diamond particle 2 nm in size, for example, the portion of surface atoms (C_{AO}) is already more than 40%.

rated as stable defects into the diamond lattice. Aluminum, arsenic, antimony, sulfur, and many more, on the other hand, are unsuitable because they cannot form stable structures within the diamond lattice.

5.2.2
The Surface Structure of Nanodiamond

For nanodiamond, the same like for any nanoparticle is true: the atoms below the surface exert considerable influence on the material's properties. Therefore, it is very important to know exactly the structure of the particle surface. The ratio of surface atoms per mass increases the smaller a particle gets. Their portion relative to the total number of carbon atoms is drawn in Figure 5.4 for octahedral particles. In the range of diameters relevant for detonation diamond (3–5 nm), there are no less than 20–30% of the atoms situated on the surface. In theory they have to possess at least one unsaturated bonding site each (Figure 5.5). Still these so-called dangling bonds are energetically rather unfavorable and highly reactive, and

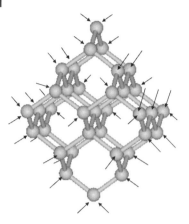

Figure 5.5 Theoretical structure of an octahedral diamond particle with unsaturated bonding sites (indicated by arrows).

different processes take place that serve to saturating these positions. They include reactions with external partners as well as the reconstruction of the surface (also refer to Section 6.2.2) leading to the formation of more or less conjugated π-bonds.

In the case of π-bond formation, a partial graphitization of the particle surface occurs. It is characterized by the sp^2-hybridization of the atoms concerned. In the course of an extensive surface reconstruction, fullerene-like partial structures evolve. Like in the model compound, they adapt to the curvature of the particle surface by forming five-membered rings (Figure 5.27). Sometimes these fullerenoid structures are connected to the diamond core by covalent bonds, but there are also species without any covalent interaction between the sp^3-hybridized core and the sp^2-hybridized shell (so-called bucky diamond). Unsaturated bonding sites are enclosed here between the diamond and the graphite moiety that can persist for being protected by the outer shell. In some cases, the graphitic structure is even observed to "sprout" out of the diamond surface with two graphene layers each originating from three neighboring lattice planes of diamond (Figure 4.26).

Theoretical studies revealed the bucky diamonds to be more stable than the respective hydrogenated diamond particles, and indeed the latter are not observed in the products obtained from detonation or shock wave synthesis. Mostly, however, this is due to the reaction conditions being rather oxidative. It is true that upon water cooling during detonation (Section 5.3.1) the products show signals of a C–H-vibration in the IR-spectrum, but these can presumably be assigned to CH_2–OH-groups. A subsequent hydrogenation of the nanodiamond surface turned out complicated, too. It is true that C–H-bonds may be formed by a reaction with hydrogen at about 900 °C or in hydrogen plasma, but the particles are never hydrogenated in a completely homogeneous manner. This is mostly due to the strong agglomeration within the solid that impedes the access to the inner regions of the aggregates' surface (Section 5.2.4).

Besides the partial graphitization and the saturation with hydrogen, there is also a variety of functional groups existing on the surface that are influential for nanodiamond. The elemental composition (Section 5.3.4) suggests a dense covering with groups containing oxygen. The major part of nitrogen present is situated inside the diamond lattice (but not necessarily on lattice positions), so surface groups containing this element should rather play a minor role. IR-spectra support this assumption by revealing almost no bands associated to amido, amino, nitro, or cyano groups. Just in some cases, especially after treatment with highly concentrated nitric acid at elevated temperature and pressure, signals indicative of amido and nitro groups are observed.

The elemental composition of crude nanodiamond materials, however, does not only result from the presence of functional groups on the particle surface. It is rather a multitude of impurities contained in the sample like metals, metal oxides, in parts material from the reactor walls and, most of all, considerable amounts of water. The latter is easily adsorbed on the large surface of the nanoparticles and accounts for a good part of the oxygen contained in pristine diamond samples. The water can at least partly be removed by heating *in vacuo*, yet a partial graphitization is accepted in doing so if too high a temperature is chosen. Metallic impurities are removed by an acid treatment and subsequent washing with water.

During such an oxidative work up with mineral acids (e.g., in the purification of detonation diamond), further functional groups are generated on the particle surface. Usual samples exhibit pronounced signals of different carbonyl and ether functions in their IR-spectra. Furthermore, OH-vibrations and, in samples obtained by "wet synthesis" (Section 5.4.1.2), also CH-vibrations are observed. The latter signals do not show in samples prepared without water cooling, so a reaction with water, which is a supercritical fluid at the prevailing conditions, is considered to be the cause of this functionalization. Altogether it becomes obvious that there is a complex mixture of various groups rather than a homogeneous covering of the surface.

The carbonyl groups as well as the ether and hydroxyl structures have been identified in several spectroscopic examinations (IR, XPS, NMR, etc., Figure 5.6). Carboxylic acids, lactones, simple carbonyls and, in some cases, amides and esters are found the most. Anhydrides are rarely ever observed as evident from the absence of their broad carbonyl band that would otherwise appear at very high wavenumbers ($\sim 1800\,cm^{-1}$). Furthermore, there are numerous hydroxyl groups situated on the nanodiamond surface, which most of them presumably exist as tertiary alcohols. The formation of ether structures is promoted by the oxidative treatment. Altogether, the surface functionalization is rather inhomogeneous, and actually it contributes to the strong agglomeration of primary particles (Figure 5.7).

There are of course, in addition to the sp^3-carbon atoms on the surface, also the unsaturated ends of superficial graphitic structures that are saturated by the respective functional groups. Furthermore, a reaction with the cooling medium can take place both on the diamond and on eventual graphitic structures. It is known, for instance, that lactones may be formed by the addition of carbon dioxide (coolant in the so-called dry synthesis, Section 5.3.1) to a reactive graphene layer.

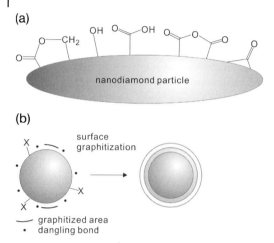

Figure 5.6 (a) The inhomogeneous surface functionalization resulting from the production conditions, (b) graphitization arising from eventual dangling bonds and, amorphous carbon.

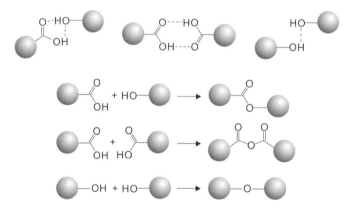

Figure 5.7 The formation of tightly bound agglomerates is promoted by the existence of hydrogen bonds and covalent bonding.

5.2.3
Diamond or Graphite? Stability in the Nanometer Range

The experimental results indicate that in the range of nanometers other rules apply to the stability of different carbon structures than would on a macroscopic scale. This is evident, for example, from the spontaneous formation of various diamond materials with nanoscopic particles in different methods of preparation like CVD or detonation and shock wave synthesis. The products obtained include polycrystalline materials with particle dimension of 1–60 μm that consist of primary particles about 1–50 nm in size. Furthermore there are "free" nanodiamond particles with diameters ranging from 3 to 10 nm that form tightly bound aggregates, and films of so-called ultrananocrystalline diamond consisting of nanoscopically small diamond crystallites. All these species indicate that in the lower range of nanometers there is a region where diamond is more stable than graphitic structures.

At the same time, numerous studies were published dealing with the transformation of carbon materials by various external effects. It turned out that a subtle balance exists between graphitic and diamondoid structures that is governed by the particle dimensions and the temperature prevailing. Nanodiamond particles measuring 2–5 nm, for instance, change into carbon onions when heated to above ~800 °C (Section 4.3.5.3). During this transformation from a pure nanodiamond into a nano-onion, an increasing number of graphene layers is formed starting from the surface of the particle and proceeding inward, so an intermediate sp^3-/sp^2-nanocomposite is obtained. The reverse process can be conducted as well: Upon electron irradiation with simultaneous heating, carbon onions are converted into nanodiamond, completely consuming the initial structure in doing so (Section 4.5.2). Normally, a kind of intermediate may be observed here, too, which can be considered as an sp^3-hybridized core with an sp^2-hybridized shell (Figure 4.41). Altogether, it is obvious that a delicate equilibrium exists between nanodiamond and carbon onions, which is easily influenced by the choice of ambient conditions.

The experimental findings indicate that small variations of these conditions induce in parts drastic changes to the relative stability of sp^2- and sp^3-phases. Theoretical calculations considering different parameters, and especially the cluster size, were performed for this reason. They revealed nanodiamond to be the most stable structure within a narrow window between ca. 2 and 6 nm. Clusters with both bigger and smaller numbers of atoms, on the other hand, took on structures consisting of sp^2-hybridized carbon. Rings are the most stable arrangement in the range of up to 20 atoms, with some major contributions of open chains in the one digit region. Between ca. 20 and 30 carbon atoms there is a kind of "gray area" where rings, bowl-shaped aromatic compounds and fullerenoid structures each may be the most stable species, and in parts the energy differences are calculated to be rather small. At more than ca. 30 carbon atoms then the realm of fullerene stability begins, which extends to about 10^3 atoms. In its upper part, the fullerenes coexist with closed carbon nanotubes and carbon onions.

It is not below a number of about 1500 carbon atoms (or a diameter of ca. 2 nm) that diamond appears as a stable structure. Exactly speaking, it is the bucky diamond on this scale which is more stable than nanodiamond of the same size, but saturated with hydrogen. From about 3 nm on, the stability of nongraphitized nanodiamond increases to pass a maximum at ca. 6 nm and subsequently to decrease down to about 10 nm. From this size on, graphite represents the most stable structure, which actually does not change anymore up to the macroscopic region (Figure 5.8). The "window" of nanodiamond stability thus ranges from ca. 2 to 10 nm. This result may also serve to explain the size of particles found in the detonation and shock wave synthesis (Section 5.3): The short duration of

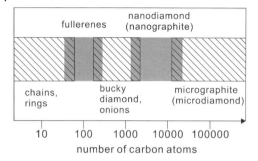

Figure 5.8 Stability ranges of carbon nanomaterials. In the lower range of nanometers there is a subtle equilibrium between different carbon clusters.

the pressure wave allows for particles with exactly these dimensions to evolve and persist during the subsequent drop of pressure and temperature, while smaller as well as larger species cannot survive as diamond and transform into sp^2-hybridized carbon.

In between these individual "islands" of stability, there are regions where several carbon nanostructures coexist. In the range from 1.5 to 2.5 nm, for instance, the simultaneous occurrence of fullerenes (sometimes even onions), bucky diamond and nanodiamond is observed, and between ca. 4.5 and 10 nm, nanodiamond coexists with graphitic structures. In this region, the formation of a specific phase can be favored by a multitude of parameters and their respective interplay. These include the functionalization as well as the energy, stress, and charge of the surface. Taking those structures that experimental data are available for, the theoretical predictions can altogether be said to reflect the actual situation quite well. Hence, one should proceed on the assumption that different carbon structures may (co)exist in the range of up to 10 nm and that the choice of ambient conditions largely influences their formation.

5.2.4
Agglomeration of Nanodiamond

Depending on the method of their preparation, the individual nanodiamond particles do not exist as isolated crystallites, but they form tightly bound agglomerates. Apart from unordered sp^2- and sp^3-hybridized carbon, they may also include other impurities. The latter may originate either from synthesis or purification, for example, finely dispersed material from the reactor walls may contaminate the sample (Section 5.3). This is especially true for material produced by the detonation or shock wave method, whereas hydrogen-terminated diamond nanoparticles do not show this effect.

Agglomerates of nanodiamond feature a hierarchical composition that is characterized by very tightly bound primary aggregates and by "grape-like" secondary

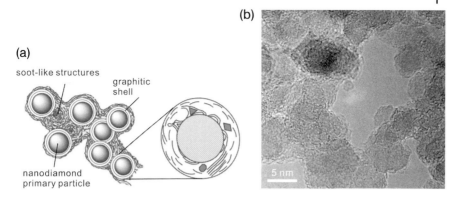

Figure 5.9 Model demonstrating agglomerated nanodiamond (a); the HRTEM-image (b) shows the soot-like structures present on the surface of the diamond particles.

agglomerates that may be up to several micrometers large (Figure 5.9). The latter are held together by electrostatic interactions with the specific functionalization of the nanodiamond particles playing an important role in the process. Owing to their noncovalent bonding, the secondary agglomerates are comparatively easy to destroy. The classical methods of deagglomeration mentioned in Section 5.3.4 may be employed to this end, and also the chemical functionalization may contribute to the reduction of agglomerate size. A markedly improved dispersibility and agglomerate dimensions of less than 200 nm may, for instance, be achieved by increasing the hydrophobic character of the diamond surface by fluorination.

The primary aggregates measure something between 30 nm and a maximum of 100 nm across (200 nm in exceptional cases) with the specific values depending on the methods of preparation and purification. These structures are not held together by electrostatic forces alone, but covalent bonding and a certain portion of soot-like carbon cause a much stronger cohesion of particles (see above). On their surface, groups containing active hydrogen atoms (carboxyl, hydroxyl groups etc.) may enter into hydrogen bonding as well as into the formation of ester, amide, or ether bonds (Figure 5.6). Just a part of them can be cleft by the respective chemical reactions as due to the dense aggregate texture the reagents cannot penetrate the aggregates. The aforementioned soot-like structures interconnect the nanodiamond particles, too. In this case an interaction like in graphitic materials occurs with the participation of sp^2-hybridized domains on the diamond surface (Section 5.2.2). This effect is accountable for the arrangement of primary nanodiamond particles into soot-like primary aggregates. These are very hard to destroy, and there is a strong tendency toward reagglomeration. A complete dispersion of the primary particles (and the formation of a colloid in organic solvents or water) merely succeeds by the application of strong shear forces like in a stirred media mill. These colloidal solutions of nanodiamond are stable even without added surfactants because of the repulsion between individual particles due to their like charge. This holds true, however, just for a narrow range of pH and concentration of background electrolytes. Upon the addition of acid or base to a neutral solution of nanodiamond in water, immediate reagglomeration occurs same as when transferring a sample into the solid state. As a counteraction the surface might be modified in a way to enable complete redispersion after drying. Nonpolar organic groups can serve to this purpose while, at the same time, increasing the solubility in organic solvents.

5.3
Preparation of Nanodiamond

In the phase diagram of carbon there is a region at very high temperatures and pressure where diamond is stable. Hence, normally it is required to sustain according conditions for its production. Only with methods comprising deposition from the gas phase and operating under kinetic control it is possible to produce diamond at low pressure, too. This latter method is described in detail in Section 6.3.1. The generation of high pressures poses some experimental problems. It is true that values in the range of gigapascals can be obtained in suitable presses, yet these apparatus are rather big while only small volumes can be pressurized. Due to the high price of the natural stones, these efforts may be worthwhile for the preparation of macroscopic diamonds. For the production of nanodiamond, on the other hand, there is no need to fall back on this method. Arguments against its use include not only the bulkiness of equipment, the problems in confining suitably small domains (it is always the entire starting material being transformed, so nanoscopic particles of graphite would be required) or the cost, but there are also much more convenient procedures to produce nanodiamond. They include the action of external shock waves or the detonation of explosives in a closed container. Both methods are employed in nanodiamond synthesis.

5.3.1
Detonation Synthesis

The generation of high pressures by means of a detonation in a confined container has been known for long. As early as in the 1960s it has been employed by soviet scientists to prepare nanodiamond, and by now the controlled detonation of explosives is performed even on a large scale.

In principle, there are two procedures that differ in the kind of starting material used. In the first process, the explosive is detonated mixed with a graphitic substance. Two things happen simultaneously then: Firstly, a direct conversion of the already existing elemental carbon, and secondly, a condensation of carbon from the explosive. Together they result in the formation of a polycrystalline diamond product with particle dimensions almost identical to those of the starting material. Hence, an at least partially martensitic process can be assumed for the mechanism of formation. The yield is about 17% relative to the carbon employed or, relative

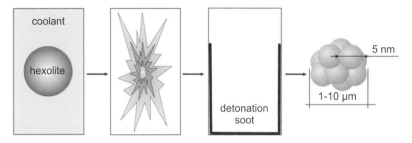

Figure 5.10 Scheme of the detonation synthesis of nanodiamond.

to the explosive, about 3.4%. The product consists of primary particles that sinter during the growth period. Different particles are obtained depending on the ambient conditions. Conducting the reaction in an inert gas leads to the exclusive formation of cubic diamond with the primary particles measuring about 20 nm. When performed in air, on the other hand, the particles are only 8 nm big and contain a certain portion of hexagonal diamond (*Lonsdaleite*).

The second method now is solely based on an explosive or a mixture of different blasting agents (Figure 5.10), so they serve as a source of both energy and carbon. To this end, however, the starting material must have a negative oxygen balance. This is true whenever $z \leq x$ (refer to Eq. (5.1)) as in this case the oxygen provided by the molecule itself does not even suffice to oxidize the entire carbon. A detonation in a closed container will then inevitably generate soot.

$$C_x H_y O_z N_w \rightarrow a\, CO_2 + b\, CO + 0.5w\, N_2 + c\, H_2O + d\, H_2 + e\, O_2 + f\, C \tag{5.1}$$

Sometimes also methane, ammonia, and nitric oxides form in addition to the products given in Eq. (5.1). At the same time the substances present are coupled to each other by several equilibria (Eq. 5.2(a)–(c)).

Water gas equilibrium $\quad CO_2 + H_2 \rightarrow CO + H_2O \tag{5.2a}$

Boudouard equilibrium: $\quad CO_2 + C \rightarrow 2\, CO \tag{5.2b}$

Dissociation of water: $\quad 2\, H_2O \rightarrow 2\, H_2 + O_2 \tag{5.2c}$

A variety of explosives with a negative oxygen balance is available even on an industrial scale. They include such known substances as TNT (2,4,6-trinitrotoluene), hexogen (also called RDX, cyclo-1,3,5-trimethylene-2,4,6-trinitramine), HMX (cyclo-1,3,5,7-tetramethylene-2,4,6,8-tetranitramine) and TATB (triaminonitrobenzene) (Figure 5.11). Particular amounts of detonation soot are formed in the reaction of TNT ($C_7H_5O_6N_3$) that features an exceptionally favorable carbon–oxygen ratio. Pressures of ca. 18 GPa and temperatures about 3500 °C are achieved in the detonation of trinitrotoluene. These conditions thus correspond to the region in the phase diagram where diamond is the thermodynamically most stable modification. Yet the detonation soot obtained from the detonation of pure TNT is found to contain just a small portion (about 15%) of diamond-like carbon. This is because

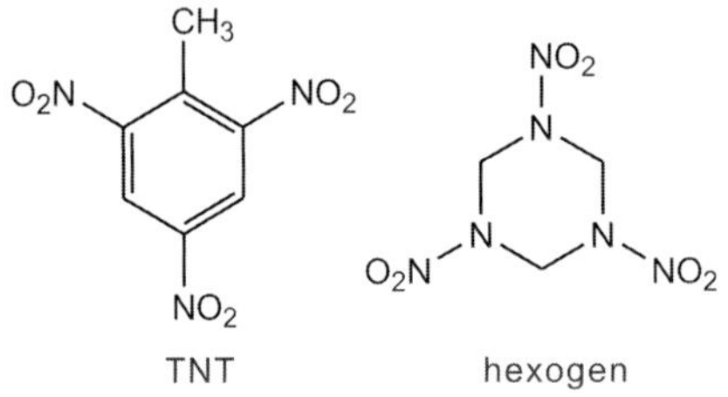

Figure 5.11 Structural formulae of the two explosives most commonly employed for detonation synthesis.

the blasting agent has a comparatively small content of energy, and so the shock wave formed is not sufficiently strong and persistent. Therefore, one has moved on to using the more energetic mixed explosives.

A typical composition is 40% of TNT and 60% of hexogen. This mixture is even commercially available by the name of "hexolite" and finds widespread application for civil and military uses. Hexogen alone has just a slightly negative oxygen balance, so only 1% of its carbon remains as soot after a detonation. Still in the mixture mentioned above, the soot-forming properties of TNT and the high energy content of hexogen complement each another so up to 10% of the available carbon can be isolated as detonation soot. 60–80% of it, depending on the reaction conditions, is nanoscale diamond. The pressure achieved within 10^{-6} s is 20–30 GPa, and the charge heats up to 3000–4000 K for a short period. Further compositions like the binary mixtures HMX-TNT (2:3) or TATB-TNT (1:1) are possible. Normally, however, they bear a smaller portion of diamond in the resulting soot (60% or 10%, respectively, in the examples above).

Furthermore, the detonation is markedly influenced by the size of the bursting charge because, in the first place, a propagating shock wave will only form from a critical diameter $d_{crit.}$ on. Below this limit, the lateral escape of reaction gases causes too big a loss of energy for the decomposition to be self-supporting, and hence no detonation can evolve. $d_{crit.}$ is a characteristic value for each explosive with a dependent relation existing to grain size and density. For TNT with granular dimensions of 1.4 mm, for example, the minimal diameter required is 40 mm. In general, the blasting charge should be at least ten times larger than the length of the reaction zone to ensure a detonation also for compositions with unknown $d_{crit.}$. Finally, the shape of the charge plays a role as well. A spherical design is optimal for a uniform propagation of the detonation. In practice, however, cylindrical charges are employed in most cases.

In a first step, the carbon atoms released from the decomposition of the explosive coalesce to form small clusters. These keep on growing then by diffusion. From a certain particle size on, the growth of diamond particles takes place also at the expense of smaller clusters. It ceases completely, however, once the pressure has dropped. The time-window for diamond formation is very short due to the detonation character of the synthesis, so particles can only be formed up to a certain size in the first place. Considering further that the markedly smaller particles are rather short-lived due to diffusion processes, one may suppose the resulting nanodiamond particles to be quite uniform in size, and indeed the experimental results confirm this assumption.

Contrary to the pressure, however, the temperature achieved by the detonation is rather slow to decline. Hence a continuing formation of soot occurs after the passing of the shock wave. This sp^2-hybridized carbon deposits either as a shell on already existing diamond cores or as loose clusters. Therefore, an efficient cooling is very important to cross the critical range of temperature as fast as possible and thus to obtain satisfying yields. There are in general two ways to achieve this effect: using water or inert gases. Gaseous coolants commonly applied are carbon dioxide, argon, nitrogen or even air ("dry synthesis"). They have the advantage of not causing many side reactions on the surface of the newly formed diamond particles, yet their heat capacity is rather low. With water ("wet synthesis"), on the other hand, the problems do not lie with the heat storing capacity (in comparison the dry synthesis, the portion of graphitic carbon decreases owing to the better cooling ability), but a strong functionalization of the nanodiamonds occurs instead. This is, among others, because water is supercritical at the conditions prevailing in the reactor. As such, it is a highly reactive agent, all the more with traces of metals like, for example, iron being present. Thus various functional groups like hydroxyl or hydroxymethyl units, etc., are formed (Section 5.2.2).

The industrial scale generation of nanodiamond by detonation has been realized quite early. Today there are several suppliers that can produce nanodiamond even by tons. Some of the existing plants alone provide more than ten million karat per year. In general a continuous or semicontinuous process is employed to do so that begins with the reaction inside a large detonation chamber ($20\,m^3$, sometimes even more, Figure 5.12). The amount of explosive needed depends on the reactor dimensions – in a chamber measuring $2\,m^3$, for instance, about 1.2 kg of hexolite are required. The mixture of TNT and hexogen corresponds in composition to the one found optimal in laboratory experiments too. However, it is also possible to use this procedure for the destruction and further processing of other explosives.

Usually, several detonations are performed in a chamber before the soot is removed. Apart from the coolant chosen, the product quality is also influenced by the reactor geometry as it is decisive for the fast heat dissipation. The detonation soot is purified in an autoclave under elevated pressure and at high temperatures

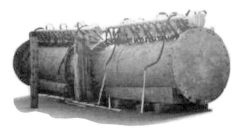

Figure 5.12 Industrial scale installation for the detonation synthesis of diamond at the Alit Corp. in Zhitomir (Ukraine). The reactor's volume is about $100\,m^3$ (© Alit Corp.).

using nitric acid of suitable concentrations. Other methods employ chromosulfuric acid or concentrated perchloric acid as oxidant. The quality of the product obtained varies widely and depends not only on the starting material chosen but also to a considerable extent, on the reaction conditions. Subsequent processing steps like acid treatment or thermal methods then further influence the product characteristics as well.

5.3.2
Shock Syntheses of Nanodiamond

Besides the direct generation of nanodiamond in a detonation, the required pressure can also be achieved by the action of an external shock wave. Usually, the latter is induced by an explosion too and compresses the carbon material that is enclosed in a kind of capsule. A catalyst like, for example, copper, iron, aluminum, nickel, or cobalt is frequently employed in this process. It has already been mentioned in the introduction that nanoscale diamond particles had been prepared quite early by the conversion of other carbon materials in a shock wave. Soon after this discovery, researchers of the DuPont Corp. developed a method also based on shock action that yields very small diamond particles. These are processed by subsequent sintering to give utterly durable cutting and polishing tools.

The process employed by DuPont to obtain the product Mypolex® is based on the transformation of extremely pure synthetic graphite in a closed metal capsule. This is surrounded by a thicker and very stable metal tube, and the whole assembly then is placed in the reactor. The latter is filled with an explosive and the charge is ignited on one end. A circular shock wave runs down the metal tube, which is compressed and, in its turn, transmits the pressure onto the graphite sample inside (Figure 5.13). According to the propagation of the shock wave the carbon material is converted into nanoscale diamond with primary particles measuring about 10–20 nm. The pressure achieved amounts to more than 48 GPa. The com-

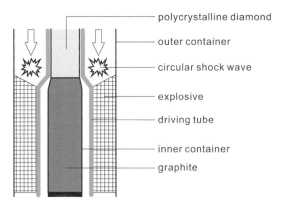

Figure 5.13 Scheme of the diamond production by shock synthesis according to the DuPont method.

paratively long duration of the pressure rise causes the primary particles to agglomerate forming tightly sintered clusters. The temperature prevailing inside the reactor is another parameter influencing product quality. High temperatures promote the reconversion of diamond already formed into graphitic material, which means that a considerable part of the product might be lost again upon insufficient cooling.

The resulting material consists of polycrystalline diamond particles measuring 1–60 µm. They closely resemble the naturally occurring *carbonados* (Section 1.3.2) in appearance and also in their properties. In particular, these diamonds do not feature preferred cleavage planes due to their polycrystalline structure. As a consequence they show enhanced stability and mechanical durability and lead to a longer service life when employed in the manufacture of cutting or polishing tools. More than 2 million karat (>0.4 t) of sintered nanodiamond per year is produced in this manner.

The synthesis of nanodiamond by shock action can also be realized on a laboratory scale (Figure 5.14). The original method presented by DeCarli and Jamieson was based on a shock wave acting on spectroscopically pure synthetic graphite with ca. 30 GPa for about 1 µs. The authors found, however, that the conversion failed upon the use of natural, hexagonal graphite. They attributed this effect to the mechanism of diamond formation, which they claimed to be characterized by a compression of small rhombohedral domains within the artificial graphite along their z-axis. These domains are missing in perfectly hexagonal graphite, and so a shock wave conversion cannot succeed.

Depending on the specific procedure, the duration of the shock action can range from 0.1 to 10 µs. After the fading of the shock wave, the problem arises again that the diamond initially formed is exposed to a temperature that decreases much less quickly. This means massive losses to the final yield as the reconversion into graphitic material is promoted. The solution for this problem consists in the

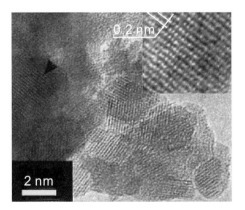

Figure 5.14 HRTEM-image of diamond particles generated by shock synthesis in a laboratory scale reactor (© Elsevier 1998).

addition of cooling elements that heat up less than diamond during the shock, so cooling is achieved owing to a temperature gradient. Values of less than 1800 °C are desirable, but up to 2000 °C can still be accepted. The material used for the coolant must combine a large heat capacity, a high thermal conductivity, and the smallest possible shock impedance to ensure efficient cooling. Possible designs include the use of metal blocks in close contact to the carbon sample or thin metal plates with good dissipating properties. If a catalyst is added to the starting material, it can normally serve as a heat dissipator as well. A mixture of copper and graphite is well-proven here. The properties of the coolant further determine the exact conditions of transformation. A large shock impedance leads to an increase in the required pressure (up to 200 GPa, e.g., at large impedance and with a very dense graphite sample being used), while 20 GPa may already suffice to initiate the conversion at more favorable conditions. The density of the starting material plays an important role, too. Dense samples ($\rho > 1.7\,\mathrm{g\,cm^{-3}}$) change only at higher pressures.

The reaction itself is followed by the isolation of the product. The metal powder and unconverted carbon are removed by the action of concentrated nitric acid. The individual diamond particles are generally covered with a graphitic shell, which is an indication of the processes taking place during particle growth.

Depending on the specific method, the final yield relative to the carbon employed is something between 10% and 25%. However, considerable amounts of the diamond particles initially formed are reconverted into graphitic material before the sample has sufficiently cooled down.

The mechanism of the shock wave-induced conversion has not yet been fully elucidated. In general there are two conceivable ways: A reconstructive process with diffusive contributions and, on the other hand, a martensitic conversion that does not include diffusion. The latter process would, for instance, explain why many samples of diamond prepared by shock synthesis contain significant portions of *Lonsdaleite* (hexagonal diamond). This is actually obtained from hexagonal graphite in a martensitic transformation. On the other hand, however, such a mechanism cannot account for some of the particles generated having a bigger diameter than the crystallites of the starting material. This is only consistent with a lattice reconstruction and a diffusion of carbon atoms. The formation of hexagonal diamond is a kinetically favored process because at less than 2000 K, the rate of *Lonsdaleite* growth is higher than that for cubic diamond. This is why at least a part of the starting material is converted into the hexagonal product. Then at higher temperatures, the *Lonsdaleite* is transformed into cubic diamond.

5.3.3
Further Methods of Nanodiamond Preparation

One other method to generate nanoscale diamond particles has already been presented in Section 4.5.2: In the heart of very large carbon onions, a strong self-compression reduces the interlayer distance from 0.34 nm to less than 0.25 nm

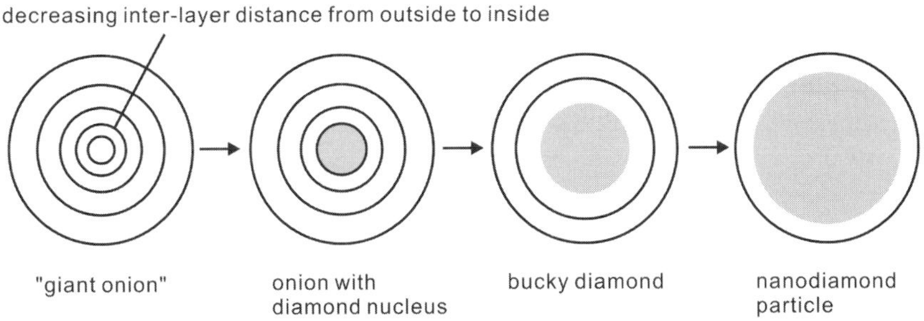

5.3 Preparation of Nanodiamond | 347

decreasing inter-layer distance from outside to inside

"giant onion" → onion with diamond nucleus → bucky diamond → nanodiamond particle

Figure 5.15 Model explaining the formation of nanoscale diamond particles in carbon onions upon simultaneous heating and electron bombardment.

and consequently leads to the formation of initially very small diamond clusters. In this way, the system evades the extreme pressure prevailing there – according to simulations based on the distance of layers found. The diamond nucleus then proceeds to grow up to the surface of the carbon onion even though the pressure there should actually not lead to diamond formation. Still the irradiation with electrons provides enough energy to knock the carbon atoms from their equilibrium positions so they may attach to the already existing nucleus in a kind of epitaxial process (Figure 5.15). Heating the sample to about 600 °C further promotes the conversion, among others by increasing the mobility of the carbon atoms. The method is, however, unsuitable to prepare larger amounts of nanodiamond because it can only be performed by simultaneous heating and electron irradiation, for example, in an electron microscope. It should be possible, nevertheless, to produce at least small amounts by means of other, more intensive electron sources, yet for the time being, no accounts on this subject are known.

Some other methods to prepare diamond particles with diameters in the lower range of nanometers have been described, too. Just some of them shall be presented here as examples. For a comprehensive discussion refer to the specialized literature given in the annexure.

One method related to the high pressure and the shock-wave syntheses consists in exposing graphite or another material containing carbon like fullerenes or polyethylene to high pressure (~13 GPa) before flash-heating it to about 3300 K. From doing so, nanodiamonds with particle sizes between 20 and 50 nm and featuring a cubic lattice are obtained.

From the microwave supported plasma CVD, which is also used to deposit diamond films (Section 6.3.1), a method generating nanodiamond in a flow reactor is derived. The sole modification to obtain single particles is to omit the substrate from the reactor so the diamond particles formed are carried along by the gas current to precipitate only in cooler parts of the apparatus placed downstream after the oven. As carbon source, dichloromethane in a mixture with oxygen is employed. The main product obtained with this method is amorphous, graphitic material. This may, however, quantitatively be removed by a treatment with 70% perchloric acid to leave behind a white, flake-like powder almost exclusively consisting of tiny diamond particles.

It is known, furthermore, that different transformations of carbon materials can be achieved using supercritical water. For example, upon treating glassy carbon at 800–900 °C and 140 MPa, a significant growth is observed for diamond particles added to the starting mixture. Without these seed crystals, however, the synthesis does not succeed. The carbon is deposited as diamond by taking a detour through the fluid phase (e.g., as CH_4), and hydrogen generated from the water presumably plays a role in this reaction, too. The respective mechanism should be related to the one postulated for CVD processes (Section 6.3.2). This is supported by the finding that the conversion does take place inside the golden capsules commonly employed, while in capsules made of platinum (binding hydrogen of course) it does not. Especially the avoidance of graphitic depositions can be attributed to the presence of hydrogen as it keeps the surface from graphitizing by attaching itself to free bonding sites. The reaction may be further promoted by the addition of nickel, platinum, or iron. These transition metals' mode of action has not yet been fully understood. They might, for instance, facilitate the generation of hydrogen from the supercritical water, or the deposition of the diamond phase might actually take place from liquid metal–C_xH_y compounds. In any case, the products of the hydrothermal synthesis are polycrystalline agglomerates of diamond particles measuring 0.1–1 µm.

Tiny diamond particles have also been found in the products obtained from the chlorination of various carbides. The reaction of silicon carbide with gaseous chlorine at more than 900 °C (below this value, SiC and Cl_2 preferably form CCl_4 and Si) yields silicon tetrachloride and elemental carbon. The latter is a mixture of different modifications, among them multiwalled nanotubes and, in parts, amorphous structures, but also the smallest possible diamond crystallites. These do not constitute the major product, however, which is why the synthesis is not suitable to produce larger amounts of the pure material. A method to prepare diamond from a chlorinated carbon compound has been realized just recently. Namely it is the reaction of carbon tetrachloride with sodium at 700 °C in the presence of a nickel or cobalt catalyst. Normally, one would try by all means to avoid these conditions due to an extreme danger of explosion, but in this case the reaction is conducted in a controlled manner inside an autoclave. The yield of diamond may be just 2% relative to the carbon employed, but still the method represents an interesting chemical approach to the synthesis of small diamond particles. Finally, there is a variety of procedures for the generation of nanoscale diamond particles whose mode of action and composition of products is not completely clear in all cases. They include methods that are supposed to yield diamond from the reduction of carbonates or other carbon sources by various metals. Furthermore there are processes using strong magnetic fields or such claiming to induce the conversion by laser irradiation. For detailed information on these methods the original literature should be referred to because in many cases the proof of reproducibility or the demonstration of a benefit for wider application are pending at best.

5.3.4
Deagglomeration and Purification

Two other crucial aspects must be considered for the production of nanodiamonds of reproducible quality: firstly, the impurities must be removed from the sample of course. Secondly, it is important for many further applications to release the primary particles from the agglomerates. Today there are various methods available that meet both requirements.

Depending on the reaction conditions, the detonation soot obtained from the reactor bears 60–80% of diamond, 15–35% of graphitic material, and about 5% of incombustible matter like metals or metal oxides.

At first, the coarse impurities are removed by mechanical means like sieving out large coherent pieces (e.g., material from the reactor wall). In a subsequent step of magnetic separation, nonmagnetic particles (i.e., diamond and other carbons) are divided from magnetic ones (e.g., particles containing iron). This method, comprised in a multiply repeated cycle of sieving, washing and magnetic separation, provides a carbon-enriched raw product that still contains considerable portions of amorphous material nevertheless.

Further purification is achieved by a treatment with concentrated mineral acids at high temperature and, in parts, elevated pressure. One procedure commonly applied uses concentrated nitric acid, which removes the metallic impurities as well as the portion of sp^2-hybridized carbon. The particulate metals or metal oxides are dissolved, while the amorphous carbon is oxidized to be CO_2 in the end. The diamond, on the other hand, is conserved due to its lower reactivity toward mineral acids. Further reagents to be employed besides nitric acid include sulfuric or perchloric acid as well as mixtures of hydrochloric with sulfuric acid or hydrogen peroxide with sulfuric acid. The intention is always to dissolve the impurities and to selectively oxidize the graphitic carbon. Especially with perchloric acid, a significant functionalization of the particle surface with carbonyl and carboxyl groups occurs, whereas with sulfuric or nitric acid, it does not happen to that extent.

The selective oxidation of amorphous graphitic carbon also succeeds with other strong oxidants like lead dioxide or a mixture of KOH and KNO_3. The oxidizing power of the latter for sp^2-carbon is efficient and selective enough to obtain samples that, according to X-ray diffraction, contain no more than trace amounts of graphitic material. These are identified as the graphitic layers bound directly to the diamond surface covering the nanoparticles to at least some extent (Section 5.2).

In some cases, the oxidative purification is followed by a thermal treatment, that is, heating the sample to ca. 700 °C in an atmosphere of argon. It helps annealing surface defects and, partially, causes a thermal removal of functional groups from the surface, too. However, it also leads to a further graphitization of the surface. Hence careful consideration is required whether graphitized or functionalized nanodiamond is more advantageous for a certain experiment or application.

The typical composition of nanodiamond purified this way is given in Table 5.1. Elements like oxygen and hydrogen are mostly situated on the surface of the

Table 5.1 Typical compositions of detonation diamond.

Sample	Carbon	Thereof diamond	Oxygen	Hydrogen	Nitrogen	Residue
Detonation diamond after purification	80–90%	90–97%	0.5–8%	0.5–1.5%	2–3%	0.5–5%

particles, while nitrogen is incorporated into the diamond lattice, which is to contain numerous defects for this reason.

Apart from a wet-chemical work up, the graphitic impurities can also be separated from the diamond particles by sedimentation. To this end, the sample is dispersed by ultrasonication in a suitable solvent (in this case 1,1,2-trichloro-1,2,2-trifluoroethane with its density being $1.56\,\mathrm{g\,cm^{-3}}$). After half the particles have settled, the supernatant is removed with a pipette and the solvent is replenished. The mixture is redispersed, and again the upper part of the liquid is removed after sedimentation. Owing to the large difference between the densities of diamond and amorphous graphitic material, an efficient separation of the latter is achieved this way. It is only the loose graphitic particles, however, that can be removed in this manner, while the graphitic layers directly attached to the surface of the nanodiamond are not affected.

Still many applications and experiments require not only the removal of impurities, but also a breaking up of the very stable agglomerates of detonation diamond. The primary particles measuring 5 nm across had already been mentioned (Section 5.2.4) not to be isolated. They are rather tightly bound one to another both in solid phase and in suspension to the effect that a deagglomeration is very hard to achieve. A considerable input of energy is required in particular because the particles are held together not only by electrostatic forces, but also by chemical bonding. This becomes evident, among others, from the failure of common deagglomeration methods to release the primary nanodiamond particles. The application of, for example, ultrasound (up to 2 kW), concentrated mineral acids or supercritical water, and addition of surfactants or classical ball milling lead to a reduction of agglomerate dimensions to 100 nm at best. A modification of the surface, for example, with fluorine, also provides just functionalized agglomerates with approximate diameters of 150–200 nm. Recent accounts report particle sizes below 20 nm.

Still a complete dispersion of diamond agglomerates is possible in an attrition mill using small milling beads made of zirconia ($\rho = 6\,\mathrm{g\,cm^{-3}}$) with diameters of 30 to 50 μm (instead of the mill a powerful sonicator can be used as well). A suspension of the grinding stock is led through a bed of those beads that is held in motion by means of a stirring gear. The shearing forces generated destroy the diamond agglomerates and liberate the primary particles. These are quite hydrophilic due to their surface covering. Hence, they are stabilized in water, resulting in a colloidal solution of diamond. The deagglomeration can also be achieved in further solvents like DMSO or ethanol. Regardless of the method applied, however, removing the solvent leads to an immediate reformation of the agglomerated structure. The dispersing agent's ability to solvate the primary particles plays an important role, too. Sufficient stability of the diamond colloid is only given in polar solvents, while even in chloroform the system already tends toward quick reagglomeration and sedimentation.

5.4 Physical Properties

5.4.1 Spectroscopic Properties of Nanodiamond

In addition to the electron microscopic examination, valuable information about the material's structure can also be obtained from the evaluation of spectroscopic data. Analytical methods chiefly reflecting the properties of either the surface or the bulk phase will be considered in this section. This distinction is of particular relevance for nanodiamond as surface properties and bulk characteristics differ in parts significantly here.

5.4.1.1 Raman Spectroscopy

As shown before, in the chapters on carbon nanotubes and onions, *Raman* spectroscopy is an excellent means of examining the structure of carbon materials. The same holds true regarding the characterization of nanodiamond. Its *Raman* spectrum is a combination of signals comprising the sp^3-carbon bound in the crystal lattice, the surface structures (usually sp^2-hybridized), functional groups and, where relevant, amorphous carbon (both sp^2- and sp^3-hybridized). The actual structure of the spectrum observed depends on many parameters like the kind of bonding (sp^2 or sp^3), the degree of disorder, the bond lengths, and the degree of cluster formation from primary particles.

However, the sensitivity of sp^2- or sp^3-hybridized carbons varies considerably with the wavelength of the excitation laser. At values in the visible range, the absorption cross section of sp^2-atoms is up to 250 times larger than that of sp^3-atoms, so in these cases the spectrum is dominated by sp^2-signals. It is only at an excitation in the UV-range, for example, at 244 nm, that the cross sections of sp^2- and sp^3-hybridized carbons are roughly the same. The sensitivity of sp^3-detection is enhanced, among others, by an increase in intensity for an excitation at near resonant conditions (the bandgap is about 5.4 eV). Owing to the high-energy radiation, however, the use of UV-lasers in Raman spectroscopy is attended by the risk of inflicting damage to the sample. It is therefore essential to mind working at low power only ($<1\,mW\,cm^{-2}$). Furthermore, the spectrum may show signals resulting from functional groups on the particle surface that absorb in the near UV.

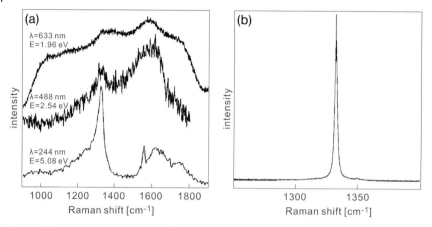

Figure 5.16 Raman spectra of nanodiamond in dependence on the excitation wavelength (a) (© AIP 2005) and Raman spectrum of a macroscopic diamond (b) (© APS 1991).

A typical Raman spectrum of nanodiamond contains a number of signals that can be assigned to either sp^2- or sp^3-portions of the sample (Figure 5.16a). It applies as a rule in doing so that all signals observed above 1360 cm^{-1} are related to the sp^2-portion as this value represents the band limit for sp^3-C–C-vibrations.

For bulk diamond, there is exactly one sharp Raman signal at 1332 cm^{-1} that corresponds to a triply degenerate optical Γ-phonon mode (Figure 5.16b). This finding results from the high symmetry of the diamond lattice (O_h) with two atoms per unit cell. For nanoscale diamond, on the other hand, this band is situated at lower wavenumbers (~1322 cm^{-1} for particles 4–5 nm in diameter), which is due to a size effect and a phonon confinement coupled to it (see below). This effect, when combined with the line broadening observed, may be employed to make an estimate of the diamond particles' size. The figures obtained correspond quite well to those determined by HRTEM. At excitation wavelengths in the visible range, the diamond signal is accompanied by a strong photoluminescence that dominates the background spectrum.

Another signal at about 1100 cm^{-1} that can be assigned to sp^3-hybridized carbon appears upon excitation in the UV. It is termed T-band and results from a resonance amplification of σ-states in amorphous tetrahedral carbon. In many cases, the portion of the latter is not very big, and so the intensity of the respective shoulder is rather low. This band originates from the shell of the individual particles where sp^2- and sp^3-hybridized carbon partially exist side by side.

The sp^2-hybridized constituents of the nanodiamond sample (mostly the outer layers of the particles and, in parts, the material in the interstices of the agglomerates under examination) are characterized by two bands that have already been mentioned in the carbon nanotubes and onions: the G- and the D-band. The exact position of the G-band (observed at about 1580 cm^{-1}) depends on both the excitation wavelength (dispersion) and the particle size. With decreasing wavelength of

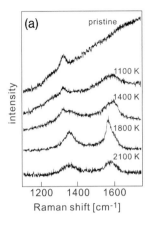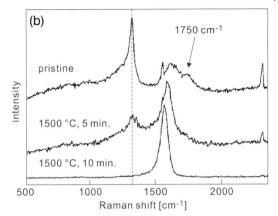

Figure 5.17 (a) The influence of thermal pretreatment and the concomitantly increasing graphitization on the Raman spectrum of nanodiamond (© Elsevier 1998). (b) Apart from the graphitization, functional groups present on the surface give rise to further Raman signals. These disappear upon thermal removal of the groups. Upon vacuum heating, not only the graphitization, but also the appearance of a C=O-band at ~1750 cm^{-1} may be observed (refer to the text, © AIP 2005).

the irradiated light, the wavenumber of this band shifts toward higher values. The effect is the more pronounced the less ordered the material is, with the disorder itself affecting the G-band's position, too.

The model of phonon confinement as mentioned above is frequently applied to explain the shift and broadening of Raman signals as well as changes to other spectroscopic properties as compared to the bulk material. It is based on a consequence of the law of crystal momentum conservation: In an infinite crystal, only phonons from the center of the Brillouin zone may contribute to the Raman signal, whereas contributions from other phonons are forbidden. Now in a very small or defective crystal, the phonons experience a confinement, resulting in an uncertainty of their momentum. The aforementioned restriction is loosened, and phonons from other parts of the Brillouin zone can contribute to the Raman signal, too. At the extremes of a really high density of defects or smallest conceivable particles, phonons originating from the entire Brillouin zone are observed.

The appearance of a G-band provides important information on the surface structure of the nanodiamond particles at hand because it indicates the surface to be at least partially graphitized (Section 5.2.2). In most cases it is already observed for the pristine material. If there is no marked graphitization because the surface is covered with functional groups, a G-band can be produced nevertheless by heating the sample until complete desorption of all functional groups and impurities occurs (>1100 K, Figure 5.17a). The dangling bonds generated by the thermal processing are saturated during graphitization, and a pronounced G-band is observed again for samples treated this way.

In the Raman spectra of surface-graphitized or agglomerated nanodiamond there is, besides the G-band, also another, broad band at about 1350 cm^{-1}. This should not be mistaken for the proper signal of the diamond lattice that appears as a sharp peak at ca. 1332 cm^{-1} (or ~1322 cm^{-1} for nanoscale particles). The signal at 1350 cm^{-1} is termed D-band and results from the disorder of the graphitic material covering the individual diamond particles (also refer to Sections 3.4.5.1 and 4.4.1.1).

UV-Raman spectroscopy has already been mentioned to provide a more balanced sensitivity for sp^2- and sp^3-carbon. Furthermore, it leads to the appearance of new signals. Some samples, for instance, give rise to a comparatively broad signal at about 1750 cm^{-1} which due to its position cannot be assigned to one-photon scattering (Figure 5.17b). Still it may be explained by the existence of functional groups containing oxygen that are situated on the diamond surface. These are as well evident in the IR-spectra of the respective samples. The signal actually corresponds to the C=O-stretch vibration of carbonyl functions. They absorb light in the near UV (the maximum absorption of acetone, for instance, is about 280 nm) as n→π*-transitions can be excited here. This leads to a resonance amplification of the signal upon excitation with a UV-laser, and as a consequence the band at 1750 cm^{-1} becomes visible.

A very informative picture of the respective nanodiamond's structure can be obtained from summarizing the observed signals (diamond peak, G-band, D-band, T-band) and the bands of functional groups. In doing so, however, one must be aware that for particles bearing unordered graphitic material in their shell, the diamond signal may in parts be shielded and, consequently, reduced in intensity. This must be considered in addition to the wavelength-related sp^2-/sp^3-sensitivity when estimating the graphitic portion of a nanodiamond material from Raman data.

5.4.1.2 Infrared Spectroscopy

The IR-spectroscopy yields detailed information on the surface covering of the material under test. Therefore, it constitutes a method complementary to those analytical techniques chiefly characterizing the bulk properties.

A pure diamond phase does not possess IR-active one-phonon modes (first-order absorption processes). Just one second-order band resulting from a two-phonon process is observed at more than 2000 cm^{-1}. In reality, however, nitrogen incorporated in the lattice as well as other defects render the forbidden one-phonon modes allowed. Hence the IR-spectrum of the bulk phase diamond exhibits two broad bands at about 1100 cm^{-1} and between 2000 and 2500 cm^{-1}. Yet this is not true for all types of diamond particles. The second-order band becomes ever weaker with smaller respective particles. At particle dimensions below 5 nm, the signal can no longer be detected at all. This variation is due to the phonon confinement suppressing two-phonon processes. Single-phonon modes as induced by nitrogen or defects, on the other hand, are still allowed, and in parts even several bands are observed at ca. 1360, 1260, and 1180 cm^{-1} (Figure 5.18). However, these signals may always be superimposed by contributions from prevalent functional

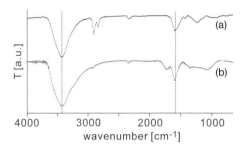

Figure 5.18 IR-spectra of nanodiamond from different sources: (a) detonation synthesis with water as coolant, (b) with CO_2 as coolant. Due to the wet-chemical processing both spectra show strong signals of adsorbed water (dotted lines).

groups, so altogether the IR-spectrum of diamond lattice vibrations does not provide much information in the case of nanoscale particles.

Regarding the surface functionalization of nanoparticles, on the other hand, the IR-signals can be rather elucidating. The kind of functionalization is highly dependent on the way of sample treatment during purification. Upon the use of oxidizing reagents like concentrated sulfuric acid, nitric, or perchloric acid, the spectrum shows very characteristic bands between 1680 and 1780 cm^{-1} with the maximum usually at around 1730 cm^{-1} (Figure 5.18). These signals can be assigned to the carbonyl functions like ketones, aldehydes, and carboxylic groups (esters, lactones, acids) generated by the oxidation of the surface. Only cyclic anhydrides can be excluded in most cases as they would give rise to a band at about 1800 cm^{-1} which, in fact, is missing. The carbonyl band of acid-treated nanodiamond can be removed from the spectrum by hydrogenation at about 800 °C as the respective carbonyl groups are reduced by the hydrogen.

Besides the carbonyl band attributable to the partial oxidation of the diamond surface, the IR-spectrum exhibits a series of further characteristic signals. The intensive band at ca. 1130 cm^{-1} results from the C–O–C-stretch vibration of ether moieties and, in parts, from the bending vibration of OH-groups (Figure 5.18). At higher wavenumbers, with a maximum at about 1330 cm^{-1}, there are several overlapping bands. They are, in detail, the defect-induced single-phonon mode of the diamond lattice and the C–H-bending vibration as well as the C–C–O-stretch vibration of tertiary alcohols (at ~1230 cm^{-1}) and of epoxy groups (~1260 cm^{-1}). A band appearing at 1630 cm^{-1} can be assigned to the OH-bending vibration of adsorbed water (see below). At ca. 2850–3000 cm^{-1} then, the symmetric and asymmetric C–H-stretch vibrations of various alkyl groups (CH, CH_2, CH_3) are observed. They are found especially in materials obtained from wet synthesis as hydrothermal processes occur on the surface of the respective particles. The broad signal at 3430 cm^{-1} normally constitutes the most intensive band in the IR-spectrum. It consists of several individual peaks associated to various O–H-stretch vibrations with the strongest of these subbands resulting from water adsorbed to the surface. The intensity of both the latter and the band at 1630 cm^{-1} corresponding to the

Table 5.2 Signal intensities of X-ray diffraction signals for bulk and for nanodiamond.

Lattice plane	(111)	(220)	(311)	(0002) graphitic
2θ in degrees	43.9	75.3	91.5	~26
Bulk diamond	44%	22%	18%	–
Nanodiamond	85%	14%	0.5%	Variable

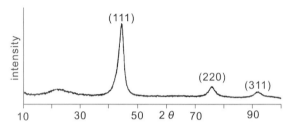

Figure 5.19 XRD-spectrum of detonation nanodiamond.

bending vibration may therefore be reduced by heating the diamond sample. Still these two signals will never disappear completely for two reasons. Firstly, it is rather complicated to remove the adsorbed water in its entirety, and secondly, the water molecules may dissociate upon desorption to generate new hydroxyl groups on the particle surface. Altogether, the IR-data of a nanodiamond sample alone already allow for quite accurate conclusions regarding the situation on the particle surface. The kind of functionalization is revealed, and reactions taking place on existing functional groups may be monitored by IR-spectroscopy.

5.4.1.3 X-Ray Diffraction and EELS

The previous chapters have already illustrated that X-ray diffraction suits very well to analyze crystalline samples. It allows for determining not only the type of lattice, but also the particle size of the crystalline phase. Moreover, the presence of further crystalline or amorphous modifications and of impurities can be detected. The latter two cases become apparent in additional signals or in the existence of a background spectrum.

The X-ray diffractogram of nanodiamond exhibits the signals characteristic for the (111)-, (220)-, and (311)-planes of a diamond type lattice at $2\theta = 43.9°$, $75.3°$, and $91.5°$ (Table 5.2, Figure 5.19). In comparison to bulk diamond, they show another proportion of intensities and a much larger half width. In some cases even the signals for the (400)- and the (331)-plane are observed, with their intensities being rather weak, though.

From the data obtained, a lattice constant of 0.3562 nm has been determined, which is a little less than in the bulk phase (0.3567 nm). With the method of

Selyakov and Scherrer, an average particle size of 4.5 ± 0.5 nm can be determined from the half width of the diffraction signals. This figure is in good accordance with the particle diameters derived from electron microscopy.

The carbon's phase purity may be checked as well by way of X-ray diffraction. The first thing being notable about detonation diamond is that there is no indication of a hexagonal phase (*Lonsdaleite*). The respective signals are found, on the other hand, in the spectra of nanodiamond generated by shock synthesis. The presence of graphitic material can also be proven by X-ray diffraction. In poorly processed samples, the signals of the (0002)-planes of graphite can be observed at $2\theta \sim 26°$. From the signal width, a size of the graphitic domains of ca. 2 nm can be derived (Section 5.2). A treatment with concentrated mineral acids results in a reduction, and finally in a complete disappearance of this band. In the end, just a halo-like signal at $2\theta \sim 17°$ remains, which is due to a scatter on small aromatic structures (sp^2-hybridization). These six-membered rings and condensed patches thereof are situated on the surface of the nanodiamond particles. Depending on the amount and order of these structures the intensity of the halo varies. For nanodiamond prepared in CO_2, for example, it is more intense than for material obtained from wet synthesis. This is because cooling takes longer in carbon dioxide, leaving more time to the graphitic shells of individual particles to reach a certain extension. The halo's intensity increases upon tempering the nanodiamond sample at more than 900 °C because an increasing amount of onion-like structures evolves. Impurities featuring a crystalline structure may be identified by XRD, too. In samples not having been subject to acid treatment, for instance, Fe_3O_4 is found occasionally.

The electron energy loss spectroscopy (EELS) is another suitable means of characterizing diamond particles. The main topic here is the identification of other carbon variants in a sample. The low loss spectrum reveals two signals (Figure 5.20a) with the one at about 34 eV being typical of a crystalline diamond. It is assigned to a collective excitation of the valency electrons with an additional contribution from the excitation of individual electrons. The signal at ca. 22 eV results

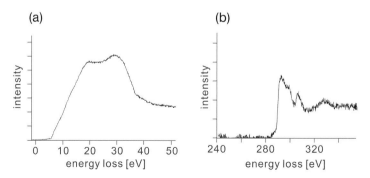

Figure 5.20 Electron energy loss spectra (EELS) of detonation diamond; (a) low-loss region, (b) core-loss region (© Taylor & Francis 1997).

from oscillations of the surface plasmons. For both signals now, the exact position and intensity depends on the particle size. The bands shift toward lower energies the smaller the diamond particles get, and the intensity of the surface plasmon signal increases in the same direction. Significant effects can already be observed at particle diameters of ca. 10 nm. At about 2 nm then, the transition from bulk phase to molecular properties is completed. The spectrum does not show signals at 5–6 eV, so it is reasonable to assume that there are no π-plasmons. In the core loss spectrum, the K-edge typical of diamond-like carbon is observed as well as an additional, weak signal at about 285 eV, which is assignable to the 1s→π*-transition in graphitic structures (Figure 5.20b). Naturally, the signals of sp^2-carbon become more pronounced with more graphitic carbon that a nanodiamond sample contains.

5.4.1.4 Absorption and Photoluminescence Spectroscopy

In the range of wavelengths commonly examined, the UV-spectrum of diamond does not exhibit characteristic bands. This finding corresponds to the transparency observed for bulk diamond from the UV- through to the IR-region. The absence of absorption bands in the visible range of light can be explained by the width of the bandgap of 5.5 eV. For this reason the excitation of electrons from the valence to the conduction band requires considerable amounts of energy which are not provided by visible light. Still the spectrum of diamond features an exponential decay of extinction from the blue region to the red that may be approximated by a function $\varepsilon(\lambda) \sim \lambda^{-2}$ (Figure 5.21). For nanoscale diamond particles, a bandgap in the range of 5.5 eV has been calculated too, and size effects should only be relevant at diameters below 2 nm, so for the nanodiamond materials commonly examined, no influence of the particle dimensions on the bandgap is to be supposed. Hence the spectrum of nanodiamond should largely resemble that of the bulk material.

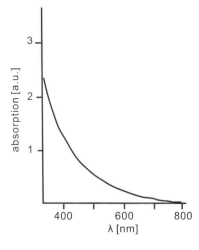

Figure 5.21 UV-Vis-spectrum of a detonation nanodiamond sample dispersed in water.

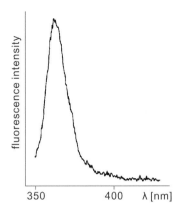

Figure 5.22 Fluorescence spectrum of detonation nanodiamond with the luminescence maximum at 364 nm (© Springer / MAIK Nauka 1997).

The luminescence properties of nanoscale diamond have as well been subject to extensive study. For nanoscopic particles, as compared to the respective bulk material, deviating characteristics are generally expected due to the large portion of surface atoms and a potentially distorted band structure. Yet for diamond, the bandgap is unaffected by particle dimensions (at least in the relevant range), and the luminescence of the nanomaterial has many features in common with that of the bulk phase.

Due to the large bandgap, a direct excitation of electrons from the valence to the conduction band can only be achieved by radiation providing a minimum energy of 5.5 eV, which corresponds to a maximum wavelength of 223 nm. Upon application of suitably shortwave UV-light, a strong photoluminescence is observed that can be assigned to the recombination of electron–hole pairs and to optical transitions. X-radiation may also serve to generate a high concentration of electron–hole pairs by promoting electrons from the valence to the conduction band. Apart from nonradiative deactivation, a pronounced fluorescence occurs then, which results from a recombination of holes and electrons. The first are situated at the upper edge of the valence band, while the latter originate from states within the bandgap that are induced by impurities (Figure 5.22).

Defects and impurities, in general, play a comparably important role for the luminescence properties of nanodiamond like they do for the bulk material. Owing to their existence, there are electronic states situated within the bandgap, which allow for inducing luminescence in nanodiamond samples also with longer wave radiation. Upon excitation with wavelengths between 300 and 365 nm, fluorescence bands are observed at more than 400 nm. They arise from various nitrogen defects. In comparison to bulk diamond, the lifetime of the excited states is rather short, which possibly is due to the effect of surface states and to the increased density of excitons on the surface.

The so-called N-V-centers (nitrogen-vacancy centers) constitute very interesting defects of the diamond lattice. As described in Section 5.2.1, they consist of a nitrogen atom incorporated into the lattice and an adjacent vacancy. Fluorescence in the red to infrared range of the spectrum can be induced by excitation with

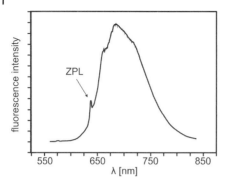

Figure 5.23 Luminescence spectrum of an N-V-defect in the diamond lattice at room temperature (© F. Jelezko).

green light (532 nm, Figure 5.23). The "zero phonon line" (ZPL) is situated at 637 nm (1.945 eV). The corresponding transition takes place as $^3A\rightarrow{}^3E$-transition originating from an electronic triplet ground state. The lifetime of the excited state is about 20–25 ns. Owing to the comparatively high saturation intensity in combination with the very low background fluorescence of the diamond lattice, N-V-centers may be employed as sources of single photons.

5.4.1.5 Further Spectroscopic Properties

NMR-Spectroscopy of Nanodiamond   NMR-spectroscopic examinations of nanodiamond turned out quite complicated. Most of all the in parts extremely long spin relaxation times τ lead to long measuring times. For high-grade diamonds of gem quality, τ may amount up to 3 days due to their low density of defects! For synthetic diamonds bearing several defects like nitrogen vacancy centers, the relaxation times are still observed to be about 1 s. For an explanation, one has to consider the low frequency of isotope ^{13}C and the associated weak dipole–dipole interactions between these nuclei as well as the long distance between the ^{13}C-spins and the nearest unpaired electron. For nanoscale diamond particles, on the other hand, rather short relaxation times of ca. 140 ms are observed as there is an efficient spin-lattice relaxation channel: the nuclear spins may couple to unpaired localized electrons resulting from the existence of unsaturated bonding sites.

Additional problems arise from cross-polarization experiments being rather complicated by the low number of protonated carbon atoms. ^1H-NMR-examinations are not very instructive due to the large width and the in parts extremely low intensity of signals.

Furthermore, it is usually a solid to be examined, for example by applying the MAS-technique, which means a significant broadening of signals. For the smallest diamond particles, the portion of surface atoms plays another important role. In comparison to the bulk material, they are placed in a distorted structure, so each magnetic nucleus is situated within a different electronic environment resulting

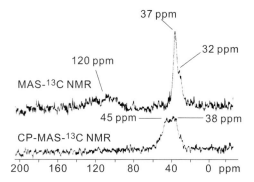

Figure 5.24 Solid-state ^{13}C-NMR-spectrum of nanodiamond obtained by means of the MAS-technique (© Elsevier 2000).

in a slight shift of the signal position. Summarized over the whole crystallite, this contributes to the broadening of ^{13}C-NMR-signals as well.

Two groups of signals are discernible in the ^{13}C-NMR-spectrum of nanodiamond samples. The highest intensity is observed for a peak at 37 ppm that is due to unprotonated, tetragonal crystalline carbon or, in short, to the diamond phase (Figure 5.24 top). The signal exhibits a shoulder at about 32 ppm, which also arises from tetragonal carbon that is, however, situated in distorted domains of the lattice structure like, for instance, close to defects. Signals of protonated carbon atoms are not observed at all as obviously their portion is too low to be detected at the sensitivity available. At about 120 ppm, the ^{13}C-NMR-spectrum reveals another signal (Figure 5.24 top). It corresponds to sp^2-hybridized carbon that is found, for example, in the outer layers of a nanodiamond particle. The signal largely disappears after reducing the portion of graphitic carbon by an acid treatment. Samples that have been subject to an oxidative treatment show an additional broad signal at about 170 ppm which is assumed to arise from carbonyl groups on the particle surface.

Information about protonated carbon atoms may be obtained from cross-polarization experiments. In a CP-MAS-^{13}C-NMR-spectrum of nanodiamond, two types of protonated carbons are observed (Figure 5.24 bottom). The signal at ca. 38 ppm presumably corresponds to CH_2-groups, whereas the peak at 45 ppm is assigned to CH-groups. Oxidized samples exhibit an additional signal at 71 ppm which is traced back to the formation of C–OH-groups.

ESR-Spectroscopy The ESR-spectroscopy provides information on the existence of unpaired electrons. As mentioned in Section 5.2 they play an important role both for the surface properties and in the crystal lattice of nanodiamond.

For bulk diamond the g-value is determined to be 2.0029, which is quite close to the value of a free electron. Samples thermally freed from their surface covering show spin densities of about 7×10^{18} spins per gram of material. This value decreases to as little as 2.2×10^{18} spins per gram for diamond hydrogenated at

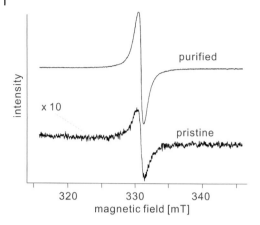

Figure 5.25 ESR-spectra of pristine detonation diamond (bottom) and of the same after the removal of graphitic carbon (top) (© Elsevier 2000).

800 °C. Hence it is obvious that in bulk diamond the major part of unsaturated bonding sites is found on the particle surface. But still a significant portion of free spins is localized inside the crystal lattice. Altogether, when assuming the conventional crystal faces to represent the outer structure, the number of free spins observed is markedly below that of free bonding sites possible in theory. This is because a part of the dangling bonds is saturated by surface reconstruction to the effect of a reduced spin density.

For nanodiamond, a g-value of 2.0027 is determined, and a much higher spin density of 10^{19}–10^{20} spins per gram is observed (Figure 5.25). These figures, although seemingly high on first sight, still represent just one to maximal 10 spins per primary particle at an average size of 5 nm. Again the spin density observed is by far lower than that conceivable for completely uncovered particles.

Contrary to bulk diamond, the spin density is increased by sample purification because the treatment, for example, with concentrated oxidizing mineral acids, removes the graphitic layer from the surface of the nanoparticles (Section 5.3.4). In this way new, unsaturated bonding sites are generated. A part of the spin density, however, is localized in the crystal lattice for nanodiamond as well, and again nitrogen centers and other defects give rise to the unpaired electrons.

5.4.2
Electronic Properties of Nanodiamond

Examining the electronic properties of nanodiamond turns out to be rather complicated by the variable sample quality and the associated differences in the electronic structure, so in this chapter, just some important aspects of electronic properties and structure will be discussed. For a more detailed presentation refer to the respective original literature given in the appendix.

Contrary to the bulk material there are also sp^2-hybridized regions in nanodiamond. In particular, they are situated on the particle surface. This graphitic portion

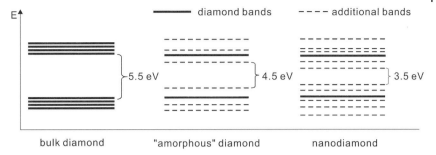

Figure 5.26 Comparison of surface band structures for bulk diamond and nanodiamond. The extra levels result from additional surface states (refer to the text). For amorphous diamond and for the nanomaterial, only the diamond levels on the edges of valence and conduction band are shown for clarity reasons (© Elsevier 1999).

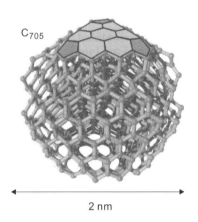

Figure 5.27 Model representing the structure of a diamond particle about 2 nm in size with a partly graphitized surface. A part of this shell is highlighted for better visibility. At the corners of this section the five-membered rings inducing curvature can be seen (© Springer 2005).

has large influence on properties like the electric conductivity. The latter increases by 12 orders of magnitude when the sp^2-portion of a sample is raised from 0% to 50%. Furthermore, the field emission properties are affected by the graphitic partial structures (Section 6.4.2.2).

Another factor is the small size of the particles, which leads to a considerable part of carbon atoms being situated on their surface. Owing to the special bonding situation, another band structure is observed here than for the bulk material. Most of all, the so-called surface states play a role in this context (Figure 5.26). Considering, for example, two model diamonds measuring 3 or 4 nm reveals that 73% or 57% of the carbon atoms, respectively, are situated in a distance of not more than 0.25 nm from the particle surface (Section 5.4.1.5). This is in good agreement with the estimate on the core region of nanodiamond particles as derived from NMR-studies. These indicate that just about 30% of the atoms are found in a normal sp^3-bonding situation, while the remaining 70% experience a marked influence of the particle surface. Thus a typical nanodiamond particle may be considered as a three-dimensional, sp^3-hybridized diamond core with a two-dimensional surface

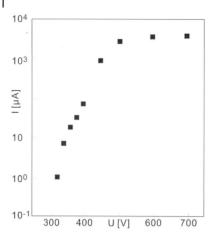

Figure 5.28 Current–voltage characteristic of field emission from a nanodiamond particle on a silicon substrate doped with nitrogen. The turn-on field intensity (attainment of a 0.1 µA current) is 3.2 V µm^{-1}, and at 5 V µm^{-1} a current density of about 95 mA cm^{-2} is attained (© Elsevier 2000).

layer made up from modified sp^3- and sp^2-carbon and from unpaired electrons (Figure 5.27).

Particularities like unpaired electrons at unsaturated bonding sites play a role for the electronic properties too, of course. However, a spin density of only 10^{19}–10^{20} spins per gram is determined from respective measurements (Section 5.4.1.5), which corresponds to a one-digit number of spins per nanodiamond particle. It results from a strong tendency toward saturation by the formation of π-bonds. In doing so, surface states rather graphitic in character are formed that cause, among other effects, also an electric conductivity (see below).

The electronic properties of the diamond core strongly resemble those of the bulk material. The bandgap in particular has a similar width of ca. 5.5 eV. According to quantum chemical calculations, there is no reason to expect a quantum confinement for real nanodiamond. Otherwise there would be a dependence of the bandgap on the particle dimensions, which is a well-known effect for silicon and germanium. For nanodiamond particles, on the other hand, it only sets in at diameters less than 2 nm, and significant changes would supposedly become observable only below 1 nm. Such tiny diamonds, however, are not stable and consequently have not been examined by now. The nanodiamond materials commonly employed exhibit particles measuring more than 3.5 nm, so any eventual change to the bandgap is virtually irrelevant. The values actually observed now are in parts less than 3.5 eV, which can be attributed to the existence of states within the diamond bandgap (Figure 5.26).

The electronic properties of nanodiamond are drastically altered by the presence of sp^2-carbon and possible impurities. Defects of these kinds give rise to additional energy levels within the bandgap that are or might be occupied by electrons (Figure 5.26). For this reason diamond samples with a partially graphitized surface exhibit a pronounced p-type surface conductivity with a concentration of holes ranging up to 10^{13} cm^{-2}. The conductivity of the sample is further increased by nitrogen lattice defects and by surface doping (also refer to Section 6.4.2).

The electronic structure of a nanodiamond sample may be examined by various spectroscopic techniques. Depending on the choice of method, the information obtained from doing so corresponds to different depths of the lattice. The Auger spectroscopy, for instance, has a low penetration depth, so it may serve to determine the situation of π-electrons on the surface. In the C1s-loss spectrum, on the other hand (Section 5.4.1.3), no π-transitions are observed as this method chiefly yields information on the second to seventh layer of atoms below the surface.

For various samples of nanodiamond, conductivity measurements have also been made. The degree of graphitization plays a major role here for the magnitude of resistance. On suitably purified nanodiamond particles, there are initially just small, incoherent π-systems, and the conjugation is anything but pronounced. The predominant sp^2-structures are rather dimeric because the individual crystal facets are very small and any reconstruction can only take place within these confined areas. The specific resistance is about $10^9 \Omega\,cm$. It drops to less than $0.3\,\Omega\,cm$ once the sample has been subject to a controlled thermal treatment. The sp^2-portion of the surface is known to be increased this way as completely graphitized carbon onions are the final products of heating nanodiamond particles (Section 4.3.5.3). Stopping this conversion in time by choosing a sufficiently low temperature and short duration of the treatment provides carbon nanocomposites consisting of an sp^3-core and a graphitic shell. The conductivities measured for a specific material then increase with the shell's coherence and its degree of graphitization.

The field emission properties as well are markedly influenced by the portion of graphitic carbon present. Tips coated with nanodiamond particles yield stable emission currents of $95\,mA\,cm^{-2}$ already at field intensities of ca. $5\,V\,\mu m^{-1}$, which is only explicable by assuming graphitic domains to exist on the diamond surface (Figure 5.28). Nanodiamond is an attractive material for field emission applications ("cold" cathode) due to its beneficial properties. These include, among others, a good thermal conductivity, a low electron affinity, and the very tightly bound crystal structure in the particles' core. As compared to a continuous film, a coating of nanodiamond provides the additional benefit of interrupted crystal structures, which increases the density of the emission current.

5.4.3
Mechanical Properties of Nanodiamond

The distinct mechanical characteristics of diamond are based on its lattice structure and electronic properties. It stands out for the highest hardness ever measured for a natural material, for large moduli of bulk and shearing and for a high scratch-resistance. Dislocations are little mobile in its lattice, and the material features a very high surface energy contributing to the hardness as well.

The existence of covalent bonds is an essential prerequisite for a great hardness of an element. A three-dimensional, orthotropic lattice with cubic symmetry may be formed from a minimum number of four bonds arranged at an angle of $109.47°$ (tetrahedral angle) relative to each other. In the periodic table, these conditions are first met for carbon with its four valencies in the diamond modification. The distribution of charges along the bonds is inhomogeneous. The electrons are preferably situated about 1/4 and 3/4 of the bond length, which contributes to a high modulus of shearing because the electron correlation is most pronounced in this state.

Metals are highly plastic as dislocations are rather mobile in their lattice, due to the metallic bonding situation (lattice of cation bodies with an electron gas). Hence individual layers are easily displaced one against the other. In the case of diamond (just like silicon and germanium, too) with its localized covalent bonds, such a displacement would comprise a number of fundamental changes to the bonding situation at the dislocated site. The process includes both cleaving old covalent bonds and forming new ones, which altogether requires much more energy than a displacement in the metal lattice with its delocalized bonding situation. It turned out that the activation energy corresponds to twice the width of the material's bandgap, which in the case of diamond calls for a very low mobility of dislocations indeed. This is why diamond as well as silicon and germanium are hard, brittle materials, whereas metals like silver or alkali metals are rather ductile.

The large surface energy of diamond distinctly complicates the cleavage of the crystal lattice with respect to most directions in space. The (111)-plane constitutes the crystallographic face with the lowest surface energy after all, so a cleavage preferably takes place in perpendicular to this facet.

The mechanical properties of bulky diamond crystals discussed so far largely hold true for nanodiamond as well because its crystal structure features the same, covalently bound lattice. Down to a certain level, especially the bulk modulus and the shearing strength are virtually unaffected by particle size. However, nanodiamond materials usually do not represent entirely pure sp^3-phases, but contain additional amorphous sp^3-material and graphitic structures on their surface. Furthermore, they exist as comparatively large agglomerates in the solid state. These peculiarities also cause certain changes in the mechanical properties observed. The hardness of nanodiamond is a little reduced for the over-representation of softer crystallographic faces and due to the almost spherical shape and the associated surface effects. For primary nanodiamond particles, studying these characteristics is complicated by the existence of the tightly bound aggregates. Further information on the mechanical properties of diamond materials may be found in Section 6.4.3.

Depending on the dimension of particles or agglomerates, respectively, nanodiamond can be applied either as abrasive (grinding or polishing agent) or as lubricant. The critical diameter is about 100 nm here. Below this value, the ability to reduce friction is dominant, while larger particles or agglomerates show abrasive effects. The latter characteristic definitely comes to the fore with larger diamond particles in the three-figure range of nanometers. On the surface of the substrates treated this way, scratches with a depth corresponding to the particle size are observed. Ever smaller particles allow for a reduction of the frictional coefficient *in vacuo* to as little as 0.01 (nanodiamonds 50 nm in diameter, measured vs. SiC). The good lubricating properties are due to the formation of a coherent film of single diamond agglomerates that altogether feature a low surface roughness. The size of primary particles is found to be largely irrelevant for the effect. The things rather to be considered are the structure of layers and the dimensions of the agglomerates actually present. Superficial functional groups containing oxygen contribute to the reduction of friction too by preventing adhesion.

5.5
Chemical Properties

5.5.1
Reactivity of Nanodiamond

In comparison to bulk diamond, nanodiamond particles are distinctly more reactive. This may be explained by the larger number of defects and by a markedly enlarged surface. Both effects increase the number of potential sites for the attack of a reagent, thus facilitating chemical modifications of nanodiamond particles. These include not only a functionalization of the surface, but also a conversion into other forms of carbon as discussed in Section 5.5.3. Due to the defective structure and to the presence of small graphitic domains on the particle surface, these transformations as well proceed much easier here than with macroscopic diamond particles.

As mentioned in Section 5.2 on the structure of nanodiamonds, they possess certain, but not very large number of unsaturated bonds owing to the saturation of free valencies by π-bond formation. The residual radical centers are normally surrounded by sp^2-structures, so eventual reagents cannot access them freely. Hence this approach to surface functionalization does not bear the desired results.

The sp^2-hybridized domains on the surface exist in parts as bent, condensed aromatic structures, and in parts as isolated double bonds. Therefore, a possible strategy might comprise the application of typical reactions of olefins or aromatic compounds like, for instance, the Diels–Alder reaction or other cycloadditions, the alkylation, or the halogenation of aromatic compounds. However, there is a certain drawback to this approach: usually it is only van der Waals forces connecting the graphitic structures situated on the surface to the particle's actual core. Hence, the functionality is only attached to a kind of shell, and stability problems may arise, especially if the respective material is to be employed in mechanically demanding applications.

Another aspect of the surface structure of nanodiamond materials should be considered too: in many cases the surface of the particles or agglomerates is completely covered by adsorbates. These have to be removed before the functionalization itself can take place. The simplest way to do so is heating the sample, although the process demands great care as nanodiamond particles are quite easily converted into graphitic structures. Another approach consists in the use of reagents leading to a covalent functionalization that can modify the surface in a way to

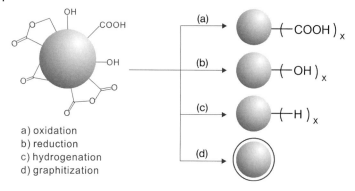

Figure 5.29 General strategies for an initial functionalization of diamond particles, including the homogenization of the surface.

disfavor adsorption, so the adsorbates are removed from the surface. Depending on the kind of substance adsorbed, suitable reagents may be applied, for example, fluorine, hydrogen, or concentrated mineral acids.

As discussed in Section 5.2.2, nanodiamond particles produced by detonation or shock wave synthesis exhibit a primary surface functionalization right from the preparation. It comprises a multitude of different groups, so it should be possible to make use of these functional groups. To ensure a reproducible quality of the secondary products obtained, however, a homogenization of the primary functionalities is required. There are several strategies to achieve this (Figure 5.29).

A suitably conducted thermal treatment, for instance, removes not only adsorbates, but also functional groups. At sufficient temperatures (usually >800 °C) *in vacuo*, the surface looses its functionalization, and a graphitization of the nanodiamond's outermost shell occurs. However, a thermal treatment still increases agglomeration, so a functionalization of single primary particles cannot be achieved in this manner so far.

Apart from the thermal homogenization, various chemical methods may serve as well to modifying the groups on the surface. Basically they can be classified to be reductive or oxidative reactions. In both cases the aim is to cover the particle surface as uniformly as possible with just one kind of functional group, so homogeneous products may be obtained upon further reaction.

An oxidation of the surface can be performed with a whole variety of reagents (Figure 5.30), the most commonly applied being concentrated mineral acids or mixtures thereof, hydrogen peroxide (often as a mixture with sulfuric acid), ozone or halogens (usually fluorine). Upon surface fluorination, C–F-bonds are generated that may serve as a starting point to a multitude of conversions like a halogen-metal exchange with subsequent reactions (Section 5.5.2). Still, as the halogenation is normally conducted on solid diamond samples placed in a tubular reactor, the nanodiamond particles obtained are inhomogeneously functionalized. The reaction with oxidants containing oxygen leads to an oxidation of both alkyl and

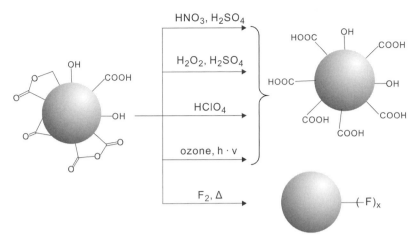

Figure 5.30 Oxidations on the diamond surface. Normally (except for the photochemical ozonization), the processes are conducted at high temperatures.

hydroxyalkyl groups. Tertiary hydroxyl groups are not affected under the usual conditions. From the oxidation various carbonyl groups like ketones, carboxylic acids, and lactones emerge, so the surface cannot exactly be termed homogeneous. There are, however, several methods to react specific types of carbonyl groups selectively, so a certain homogenization may still be achieved in subsequent conversions. In principle, carboxyl functions suit outstandingly well to further functionalizations of carbon materials (Section 5.5.2).

A reductive homogenization of the nanodiamond surface may be performed in different ways and with quite variable results. The surface hydrogenation must be conducted at elevated temperatures as a heterogeneous reaction in a tube reactor (Section 5.5.2.1). It must do without a catalyst, and so it requires very high temperatures (problems are graphitization and formation of amorphous carbon) and high pressure. Both functional groups and graphitic domains on the surface can be hydrogenated this way. Yet the reaction rarely proceeds completely, and again just the groups situated on the agglomerates' surface are affected. Following hydrogenation, the C–H-bonds formed may be employed, for example, for photochemical C–C-bonding reactions (Section 5.5.2.6).

The wet-chemical reduction aims for the transformation of existing carbonyl groups into hydroxyl functions. Therefore, complex hydrides (LiAlH$_4$) or borane are employed as reductant (Figure 5.31). In this way, it is possible to generate a diamond surface covered with hydroxyl groups, although they must be assumed to be a mixture of tertiary (existing beforehand), secondary, and primary variants. It is true now that they exhibit differing reactivities, but for further grafting the presence of just one kind of functional group is of considerable interest already, even if it is bound in different lattice structures. Like the carboxyl group obtained by oxidation, the hydroxyl group suits well to the attachment of further functional units.

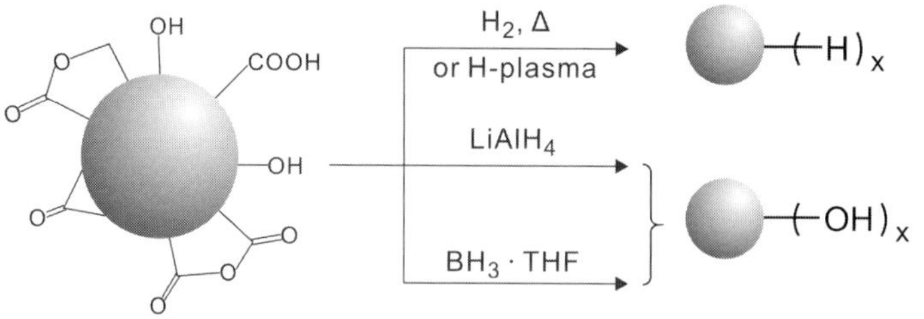

Figure 5.31 Reductions on the diamond surface.

A general problem arises regardless of the kind of functionalization chosen. With the particles existing as very tightly bound agglomerates (Figure 5.9), the products obtained must as well be thought of as agglomerates that are functionalized externally, but still consist of many primary particles. Subsequent deagglomeration then will consequently yield particles functionalized inhomogeneously. Some of them, for having been situated inside the agglomerate, will not be altered at all in comparison to the initial state. Therefore, a valid strategy has to start from destroying the agglomerates, and only then to react the primary particles generated in that first step. However, immediate reagglomeration occurs upon removal of the dispersing agent. Consequently, solid phase reactions like the hydrogenation or ozonization cannot be applied here, but any further step must be performed in the colloidal solution once prepared. Hence it would be sensible to have the surface reacted immediately after exposure. Thus a useful strategy consists in a simultaneous deagglomeration and functionalization of the nanodiamond particles.

5.5.2
Surface Functionalization of Nanodiamond

5.5.2.1 Hydrogenation

The hydrogenation is the most simple conversion of a nanodiamond surface. While this type of covering is the standard state for diamond films (Chapter 6), several problems arise when it comes to nanoparticles. The technique commonly employed for diamond film hydrogenation, that is, the reaction at very high temperatures in a stream of hydrogen, may be applied to the nanoparticles, too, yet it is always just powdery samples being reduced this way, which naturally leads to a certain inhomogeneity of the product. As a consequence there will always be unreacted functional groups inside the agglomerates, and destroying them afterward will provide unevenly hydrogenated particles. Sometimes hydrogen plasma is applied here to achieve higher concentrations of reactive hydrogen.

A reactor as commonly employed for the vapor deposition of diamond can also be used to perform experiments with hydrogen atoms generated on a hot filament that features enhanced reactivity. The IR-spectrum of diamond samples is altered significantly by the hydrogenation (Figure 5.32). The signals in the carbonyl region

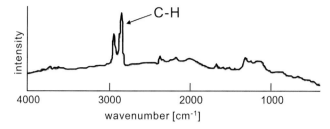

Figure 5.32 Reflection IR-spectrum of nanodiamond hydrogenated in a hydrogen stream (© RSC 2002).

vanish, and in the range of C–H-stretch vibrations characteristic of alkyl groups, new bands appear. Monitoring this process revealed that from about 900 °C on the surface may be considered entirely hydrogenated. A problem arises, however, because there is already appreciable graphitization taking place in this range of temperature.

A wet-chemical hydrogenation is hard to achieve. Most of all the separation of the catalyst (e.g., palladium on activated carbon) is very complicated. A reaction with zinc and hydrochloric acid in the sense of a Clemmensen reduction has not been described for nanodiamond either.

5.5.2.2 Halogenation

The halogenation constitutes another way of functionalizing the surface of nanodiamond. The usual reagents employed are fluorine or chlorine or, to a minor extent, also bromine, whereas, due to the insufficient reactivity, a direct iodination has not yet been reported. The reaction is performed in a tubular reactor heated to several hundreds of degrees (between 150 °C and 500 °C) where a mixture of elemental fluorine with hydrogen is passed over the sample. With gaseous chlorine, the reaction also takes place at elevated temperatures with the best results being obtained between 250 °C and 400 °C. Above this limit, the degree of chlorination decreases due to dissociation processes. As an alternative, the reaction with gaseous chlorine can be conducted photochemically making use of a mercury vapor lamp.

In a reaction with elemental fluorine (or chlorine), sp^2-hybridized carbon atoms are attacked as well as preexisting functional groups like, for example, the hydroxyl function are substituted (Figure 5.33). In detail, the reactions taking place include additions to double bonds, secondary aromatic substitutions and the destruction of aromatic systems by fluorine. Furthermore, there may be an exchange of hydrogen atoms and of OH-groups bound in different ways. Especially in the fluorination, hydrogen halides formed as intermediates can act as a catalyst, and contributions from radical processes may be assumed as well. The latter is particularly true for the photochemical chlorination. The halogenation restores an sp^3-state in formerly reconstructed regions of the surface, so altogether a homogenization of the bonding situation occurs.

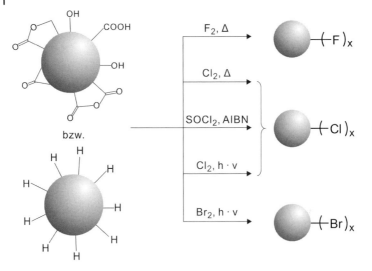

Figure 5.33 Methods for the halogenation diamond materials.

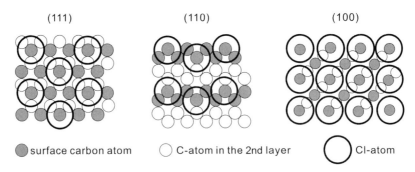

Figure 5.34 Covering of the distinct crystallographic faces of diamond with chlorine atoms. Owing to the irregular surface structure, however, these variants co-exist in reality (© Springer / Plenum Publ. 1978).

The chlorination of the surface can also be achieved by a controlled radical process at moderate temperatures. To this end, thionyl chloride is reacted in suspension with the previously hydrogenated nanodiamond in the presence of AIBN (azo-bis-*iso*-butyronitrile, Figure 5.33). A surface covering of ca. $3.5\,\mathrm{mmol\,g^{-1}}$ is attained, which almost comes up to a complete monolayer (Figure 5.34).

The surface bromination occurs as a radical process with elemental bromine that is reacted with the nanodiamond as solution in chloroform. By the action of light, bromine radicals are generated to attack on the surface. For the reduced reactivity of bromine radicals, however, just a partial bromination is achieved, and the degree of surface covering remains comparatively low at $0.87\,\mathrm{mmol\,g^{-1}}$.

In the reaction with fluorine, on the other hand, a high degree of conversion is achieved because it is small and very reactive. The respective surface covering is found to be about 3.8 mmol g^{-1}. The products obtained are distinctly hydrophobic. Hence they may be pushed from the aqueous into an organic phase where they are much easier to disperse than unprocessed nanodiamond. Still a complete destruction of agglomerates is not effected, but the product has an average particle diameter of 150 nm. Altogether the fluorinated particles and agglomerates of nanodiamond are attractive starting materials for further functionalization as the halogen atoms are easily substituted (Section 5.5.2.8).

5.5.2.3 Oxidation of Nanodiamond

One method of oxidizing nanodiamond surfaces has already been presented in Section 5.3.4 as a means of sample purification. The differing reactivity of graphitic and diamond carbon is utilized here. The unordered graphitic structures are oxidized and removed as gaseous products, whereas diamond particles are only modified on their surface.

Reagents and conditions usually employed include concentrated sulfuric, perchloric or nitric acid or mixtures thereof reacted at high temperatures and, occasionally, at elevated pressure. The oxidizing effect also leads to the conversion of methyl, hydroxymethyl, keto groups, etc., into carboxyl groups (Figure 5.30). After all, the surface covering amounts up to 3×10^{15} functional groups per square centimeter, corresponding to about 10 mmol g^{-1} (at a specific surface of ~200 m^2 g^{-1}). This value can be determined, for example, by titrating the acid groups with a base. Moreover, the IR spectrum reveals the expected changes. The surface of nanodiamond agglomerates modified like that becomes rather hydrophilic, and so they are readily dispersible in water. An oxidation of primary particles, on the other hand, is hardly possible this way. They sediment at low pH-values to form agglomerates again, so consequently, just the outer shell of these may be attacked.

An oxidation in air at elevated temperature is feasible, too. In doing so, the temperature must be kept below 500 °C to prevent decomposition. In the IR-spectrum a significant increase of the C=O-band at 1778 cm^{-1} is observed, while at the same time the signals of C–H-stretch vibrations at about 2900 cm^{-1} decrease.

Ozone is another oxidant available to modify nanodiamond. It can be applied by introducing it into a suspension or colloidal solution, yet the reaction is rather slow. The mechanism is presumably similar to that of the ozonization of adamantane or cyclohexane. Initially, a hydro peroxide is formed that decomposes to give a keto compound and an alcohol. The formation of the carbonyl group can be monitored by the increase of the IR-band at ca. 1740 cm^{-1}. In addition, the ozone reacts with graphitic domains on the nanodiamond surface in a [3+2]-cycloaddition type reaction. The ozonides initially formed decompose and diamond particles carrying oxygen-containing groups are generated. One of the advantages of ozonization is the possibility to work in solution. Therefore, primary particles as well can be functionalized this way.

5.5.2.4 Reduction of Nanodiamond

A hydrogenation as described in Section 5.5.2.1 is the simplest reduction of a diamond surface. However, it can also be reduced in other ways. The direct decarbonylation or decarboxylation of nanodiamond by heating it to 800 °C *in vacuo*, for example, yields a sample no longer carrying carbonyl groups on its surface. This is proven by IR-spectroscopy where a marked decrease of the carbonyl signal and a concomitant increase of C–H-band intensity are observed. Hydroxyl groups and adsorbed water are removed as well by this procedure. Heating to more than 900 °C must be avoided, however, as the graphitization of the diamond surface leading to the cleavage of all functional groups sets in at about 950 °C.

Besides reducing all functional groups to the level of hydrogenated nanodiamond, there are also ways of performing partial reductions (Figure 5.31). Upon reaction with borane (as THF-complex), lithium aluminum hydride, or $NaBH_4$, carbonyl functions are converted into hydroxyl groups. These can subsequently be employed to further functionalize the diamond surface (Section 5.5.2.8). Still the separation from byproducts, and in particular from inorganic compounds like aluminum hydroxide, poses considerable problems.

5.5.2.5 Silanization of Nanodiamond

Upon treating nanodiamond with alkoxy silanes, a condensation occurs between one or more alkoxy functions of the silane and the hydroxyl groups situated on the diamond surface. The resulting compounds feature a C–O–Si-bonding of the organic residues to the diamond surface (Figure 5.35).

A multitude of suitable silanes with terminally modified alkyl residues are easily available, so a wide variety of surface functionalizations can be achieved this way (Figure 5.35). In particular, it is the terminal modifications that may be employed for further functionalizations and for the attachment of, for instance, biologically active moieties. Furthermore, appropriately silanized nanodiamond suits to a covalent incorporation into composite materials. The exact way of the alkoxy silanes being connected to the diamond surface is still the subject of discussion. The individual units may be bound to the diamond by single or multiple condensation, to begin with, but in principle also a condensation with alkoxy groups of neighboring silanes or even with the groups of other diamond particles may be conceived. The latter process would account for the agglomerate dimensions increasing during silanization.

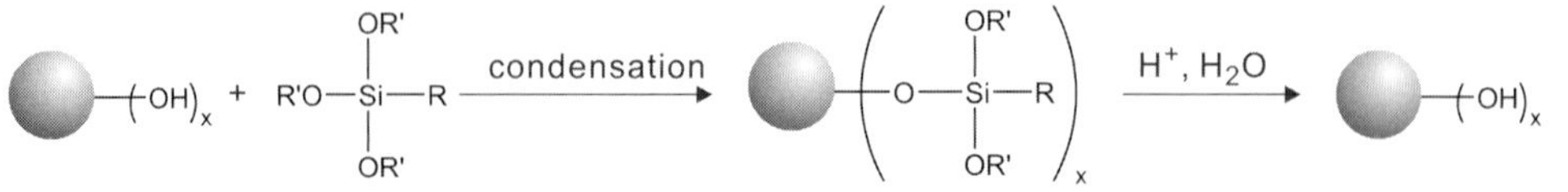

$R = -(CH_2)_n-CH_3, -(CH_2)_n-aryl, -(CH_2)_n-CH=CH_2, -(CH_2)_n-halogen, -(CH_2)_n-NH_2$ etc., $R' = Me, Et$

Figure 5.35 Silanization of the surface hydroxyl groups of diamond.

5.5 Chemical Properties

In many applications of functionalized nanodiamond, a very tight attachment of the functional groups to the surface is required. For silanized nanodiamond this is given only up to a point. Especially in an acidic environment, the C–O–Si-bonding is easily hydrolyzed and the starting material is recovered. Hence silanized nanodiamond can only be employed at operating conditions that ensure the stability of the C–O–Si-bond.

5.5.2.6 Alkylation and Arylation of Nanodiamond

There are several procedures leading to an alkylation or arylation of the diamond surface. These reactions are of special interest because, contrary to the pristine diamond, the products obtained exhibit a much less polar surface and hence should be dispersible in organic solvents.

Apart from the cycloadditions presented seperately in Section 5.5.2.7, reactions suitable to attach hydrocarbons directly to the diamond surface include the treatment with C-nucleophiles, the photochemical connection with ω-vinyl reagents or radical processes.

Starting from hydrogenated diamond surface (Section 5.5.2.1), a C–C-bond with the terminal vinyl group of a suitable reagent may be formed in a radical reaction under UV-irradiation (Figure 5.36a). This conversion is well-known already from the chemistry of hydrogenated semiconductor surfaces and has extensively been studied for diamond films as well (Figure 6.45, Section 6.5.2.6). Nanodiamond particles react in this manner as well, yet a homogeneous hydrogenation of the surface is by far more complicated here, and so side reactions on residual functional groups can occur as well.

10-Undecenyl compounds are commonly applied reagents here. The opposite end of the alkyl chain may carry a variety of (possibly protected) functional groups that can be employed in further reaction steps after C–C-bond formation. In principle, other (ω-1)-alkenes may be used as well. For successful radical generation, UV-radiation with a wavelength of less than 254 nm must be applied, so the efficiency of the mercury vapor lamps usually employed is limited at best. The radicals' high reactivity leads to a homogeneous covering of the particles as hydrogen atoms can be abstracted from virtually any position on the surface. A major advantage of this kind of functionalization is the residues being attached to the

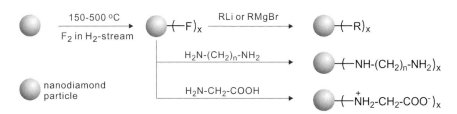

Figure 5.36 Alkylation of the diamond surface by reaction of fluorinated diamond with organometallic compounds. Besides alkylation, an amination of the diamond may be achieved as well by a reaction with amines.

diamond by C–C-bonds, so the connection established is stable at a variety of conditions.

Problems in the reaction of hydrogenated nanodiamond particles arise from the material's inhomogeneous surface structure. Due to the performance in solid state, functional groups situated inside the agglomerates can just insufficiently be attacked. For this reason homogeneously functionalized products can only be obtained once a uniform hydrogenation of the primary particles succeeds. Moreover, the application of hard UV-radiation may damage certain organic residues, and so just photochemically inert moieties can be employed as distal residue of the vinylic reagent.

Halogenated nanodiamond particles can be alkylated or arylated, respectively, by reaction with carbon nucleophiles (Figure 5.36b). Both fluorinated and chlorinated samples are suitable here. In principle, a reaction with brominated diamond would be possible too, yet by now no satisfactory covering of the diamond surface with bromine has been achieved. Suitable nucleophiles include organolithium as well as Grignard compounds, both of which react with the diamond surface by cleavage of the halogenide ion. This way, up to 10% of the superficial carbon atoms are alkylated or arylated. The degree of functionalization achieved increases with growing nucleophilicity of the organometallic reagent. To date, methyl, *n*-butyl, *n*-hexyl and phenyl groups have been described as organic residues of the organometallic compounds. Actually, however, it should be feasible with all available and sufficiently reactive organolithium and Grignard compounds. In comparison to pristine nanodiamond, the products obtained are distinctly more hydrophobic and, not unexpectedly, much easier to disperse in organic solvents. The "solubility" (actually better termed dispersibility) amounts to more than 50 mg/l in alcohols and to about 30 mg/l in THF, acetone, and chloroform. Altogether, dispersibility increases with a growing chain length of the alkyl residues and with a higher surface covering.

Another reaction successfully applicable to hydrogenated diamond surfaces is the radical alkylation with perfluorinated azo compounds. These are cleft photochemically to give nitrogen and perfluoroalkyl radicals. The latter then react with the diamond surface by hydrogen abstraction (Figure 6.37). The reaction progress may be monitored by IR-, XPS-, or ^{19}F-NMR-spectroscopy. Again, however, the situation is complicated by the inhomogeneous hydrogenation of the nanodiamond particles in agglomerates. Residual functional groups like OH or carboxyl units, if present, may react with the perfluoroalkyl radicals forming C–O–C_xF_y- or COO–C_xF_y-bonds, respectively (Figure 5.37). On hydrogenated diamond films, on the other hand, a complete surface hydrogenation can be achieved and this problem consequently does not arise.

5.5.2.7 Reactions on sp^2-Hybridized Domains on the Nanodiamond Surface

It has been mentioned before that, apart from functional groups, there are also domains on the particle surface that are characterized by an sp^2-hybridization of carbon atoms. The surface dimers forming on the (100)-plane are a well-known

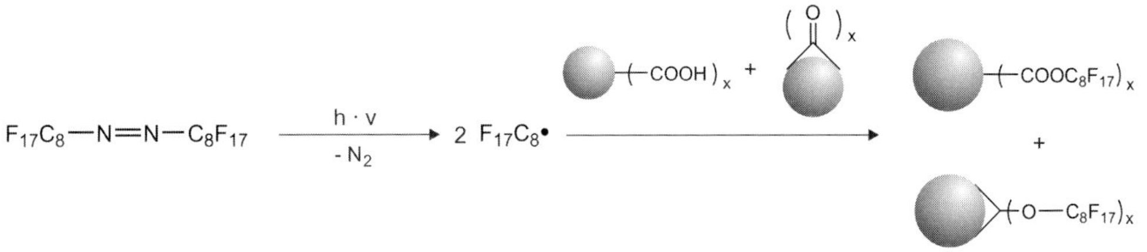

Figure 5.37 Reaction of the diamond surface with perfluorinated diazoalkanes.

example. They are distinguished by a double bond that is situated on top of the bulk phase. It suffers considerable strain as all bonding partners are situated out of its plane. Therefore the carbon atoms of the double bond are prehybridized toward an sp^3-situation, which means a weakening of the π-bond and a gain in reactivity. However, the individual crystal facets are very small in the case of nanodiamond particles as their shape is rather spherical. Hence unlike diamond films featuring an unaffected crystallographic orientation, a large number of neighboring and highly ordered surface dimers is not observed here. Still there are such π-bonded structures. They are situated especially at the steps between the individual small facets, so the respective reactions preferably take place at these positions. There are, however, also condensed systems extending over a markedly wider range on the particle surface. In parts they may feature aromatic character. Especially on the surface of those particles completely enclosed in an onion shell ("bucky diamond"), fullerene-like structures are observed (Figure 5.27). The presence of unsaturated structures on the diamond surface allows many reactions like the addition to double bonds or cycloadditions.

Many additions that can be performed on typical olefins are known to succeed on nanodiamond as well. The hydrogenation in a stream of hydrogen, for example, can be assumed to attack on existing double bonds, too. Halogens, and most of all the lighter species, react with double bonds in an analogous way even at very mild conditions. Still these reactions are always associated with a conversion of the functional groups existing on the surface as well, and so the effect of the addition to graphitic domains can hardly be examined alone.

5.5.2.8 Further Functionalization of Nanodiamond

A modification of functional groups already attached to the nanodiamond surface is of considerable interest for the development of new diamond materials for biomedical or mechanical applications.

Suitable initial functionalizations of the surface may be halogens as well as carboxyl or hydroxyl groups. Some examples of further modifications based on these groups will be given in the following. On diamond surfaces carrying halogen atoms, a nucleophilic substitution of the latter can take place. Suitable reagents include, among others, the carbon nucleophiles as presented in Section 5.5.2.6 on alkylation, but alcohols, amines, or ammonia may be reacted with chlorinated

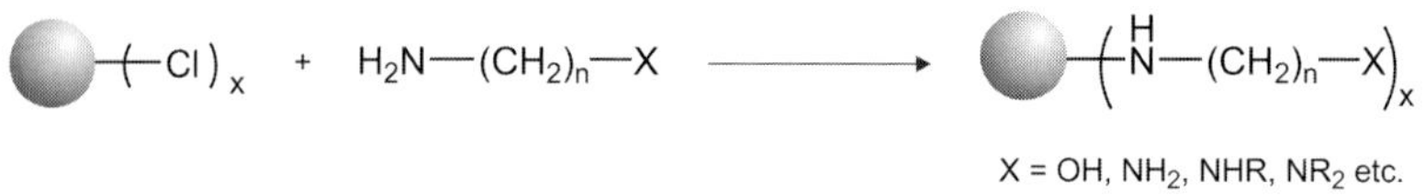

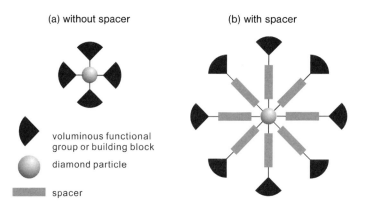

Figure 5.38 Reaction of halogenated diamond materials with amines.

Figure 5.39 Spacer concept for a more efficient grafting of functional moieties onto the surface of a diamond particle. Higher surface coatings can be achieved when using a space (b) than without this rigid moiety (a).

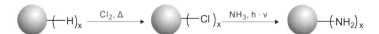

Figure 5.40 Amination of the diamond surface after intermediate chlorination.

diamond as well. The respective products bear ether structures derived from alcohols, or amines from a reaction with ammonia or amines (Figures 5.36 and 5.38). The amines or alcohols employed may carry further functional groups, so a reaction with other reagents becomes possible on these sites. This is of special importance for the attachment of larger units to achieve satisfactory surface covering (Figure 5.39). Suitable functionalities for this kind of transformation include amino, vinyl or amide groups as well as phenols and aldehydes.

The thermal conversion of chlorinated nanodiamond with gaseous ammonia does not yield definite results regarding the formation of an aminated material. The photochemical approach using a mercury vapor lamp, on the other hand, succeeds without doubt (Figure 5.40). The reaction can be monitored by the disappearance of IR- and XPS-signals associated to chlorine and by the emergence of bands typical of amino groups at about $1414\,cm^{-1}$ and above $3000\,cm^{-1}$. The direct amination of hydrogenated nanodiamond does not succeed, so this loop way has to be taken. The amino groups obtained then constitute valuable sites for the attachment of various functional units like, for example, amino acids. In this latter

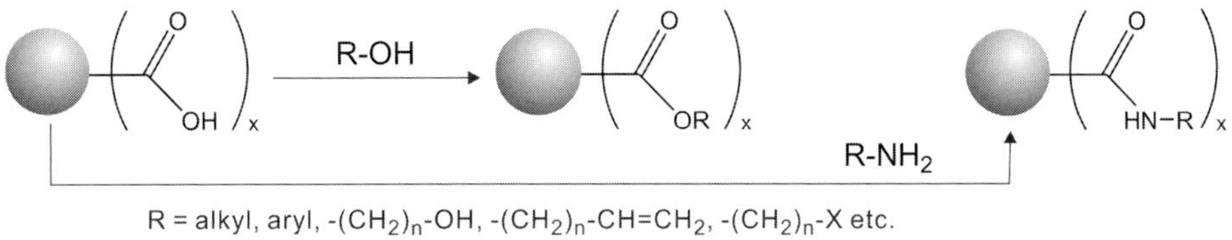

Figure 5.41 Further functionalization of surface carboxyl groups by reaction with alcohols or amines.

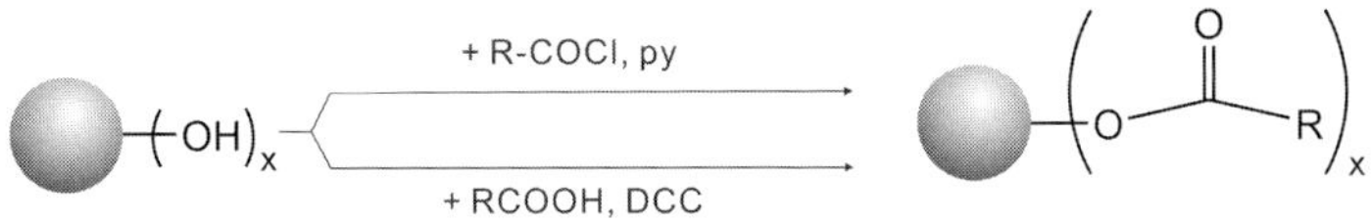

Figure 5.42 Reactions of hydroxyl groups on the diamond surface.

case, however, the aforementioned spacer concept is frequently applied as well to achieve sufficient degrees of covering.

Other reactions on chlorinated nanodiamonds include radical reactions with cresols and related alkylaromatic compounds. p-Cresol, for instance, allows for the introduction of terminal phenolic groups that may serve as valuable starting-point for the incorporation into polymers (Section 5.5.2.9) or for the attachment of biologically active moieties.

The carboxyl groups available from oxidation of the diamond surface also represent good anchoring sites for further functionalizations. Derivatives can be obtained by acid-catalyzed esterification as well as by a base-catalyzed formation of amides (Figure 5.41). Employing bifunctional alcohols or amines allows for subsequent grafting steps. In principle, the same compounds can be used here like in the modification of chlorinated nanodiamond with alcohols or amines.

There is, however, a certain drawback to nanodiamonds modified by esterification or amidation: at a variety of conditions, these groups are cleft again (e.g., by ester saponification), so derivatized nanodiamonds of this kind can only be employed in comparatively gentle environments.

Hydroxyl groups situated on the diamond surface can be reacted too, for example, in ester formations with acid chlorides. Again these esters may carry a variety of residues or feature multiple functionalities. A direct coupling of carboxylic acids is possible, too (Figure 5.42). Last, but not least, as mentioned in Section 5.5.2.5, the hydroxyl groups may serve to attach trialkoxy silanes.

The formation of O-acyl groups on the surface can also be realized starting from hydrogenated diamond. To this purpose, it is initially treated with a diacyl peroxide like dibenzoyl or dilauroyl peroxide. The reagent is thermally cleft to give two radicals that can generate radical centers on the diamond surface by hydrogen abstraction. These positions may then be reacted with a variety of

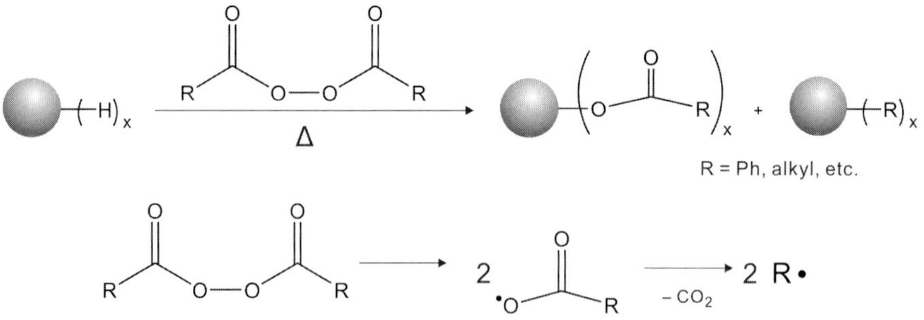

Figure 5.43 Reaction of hydrogenated diamond particles with radical initiators.

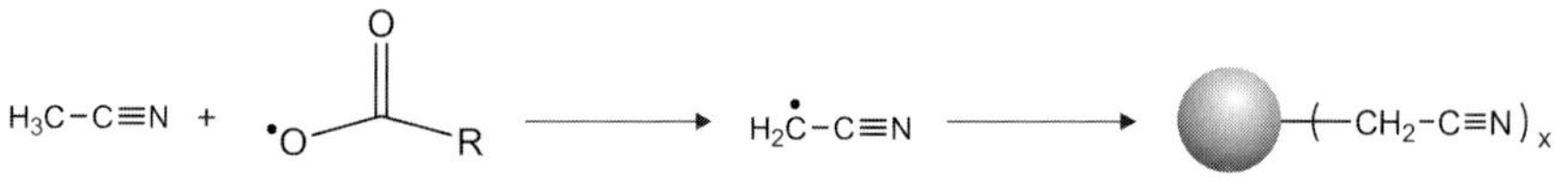

Figure 5.44 Reaction of acetonitrile with benzoyl peroxide and subsequent radical attack to the diamond surface.

compounds. Ester structures, for example, can be formed with carboxyl radicals generated *in situ* by reaction with the radical initiator present in the mixture (Figure 5.43).

In this way, it is possible to attach both aromatic and aliphatic carboxylic acids to nanodiamond. The reaction is conducted wet-chemically in a suitable dispersing agent like hexane, cyclohexane, THF, or DMF. In acetonitrile, a reaction occurs with the solvent itself. The radical initiator abstracts a hydrogen atom of the methyl group and the resulting alkyl is attached to a radical center on the diamond surface (Figure 5.44). This allows for establishing nitriles on the diamond that constitute valuable for further syntheses.

5.5.2.9 Composites and Noncovalent Interactions with Nanodiamond

Owing to its remarkable properties nanodiamond suits very well to being part of composite materials. In particular it is the small particle size, the hardness, the large chemical inertness, its nontoxicity and the high refractive index that may beneficially complement the properties of the polymer matrix. The latter may be connected to the diamond particles either by covalent bonding or by noncovalent interaction. Numerous examples of noncovalently bound composites have been reported in the literature (Section 5.6.1). Still the interaction with the matrix is by far more complex than discussed for the nanotubes and fullerenes. This is due to the more variable surface structure that features not only graphitized domains, but also a variety of polar and nonpolar functional groups.

Oxidized samples of nanodiamond carry many polar functional groups leading to a preferred interaction with polar compounds. The bonding is achieved via hydrogen bonds and electrostatic forces. Besides polymer materials, the noncovalent interactions suit also to immobilizing biomolecules like cytochrome c, poly-L-lysine, antibodies etc. on the particle surface. Numerous applications in the biomedical sector can be envisaged here. An interesting observation is made for the preparation of a nanocomposite with amphiphilic polymers like, for example, perfluorinated termini with a polar central section: with the polar groups on the diamond surface preferably interacting with the polymer's centerpiece, the resulting structure resembles a micelle that exhibits only nonpolar groups to the outside. As a consequence, such composites can easily be dispersed in non- or moderately polar organic media (e.g., 1,2-dichloroethane, THF).

From the hydrogenation or fluorination of a diamond material, a very hydrophobic surface results that may then enter into an exchange with rather nonpolar compounds. A connection via π-stacking, however, plays just a minor role because graphitic fragments are only found in small domains on the particle surface. In the case of thermally graphitized nanodiamond particles, on the other hand, the conditions largely resemble those observed for multiwalled nanotubes. The interaction of the π-electrons with the polymer molecule causes a stable noncovalent incorporation into the composite.

A covalent attachment of nanodiamond to various polymers is possible too and contributes to an improvement of mechanical properties. Again the specific kind and distribution of groups present on the particle surface play a decisive role. A most possibly homogeneous surface functionalization is a prerequisite for reproducible quality of the materials obtained. Provided a covalently bound composite can be prepared then by copolymerization, crosslinking or polymerization starting from the nanoparticles, the resulting products are usually characterized by a very homogeneous distribution of the diamond particles (Figure 5.45). To date, nanodiamond has already been incorporated into polyurethanes, epoxy resins, polymethacrylates, etc.

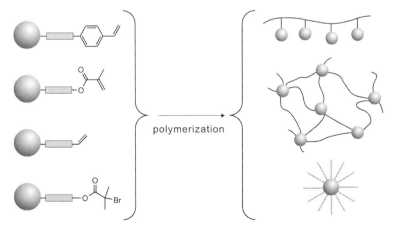

Figure 5.45 Strategies for the production of covalently bound diamond-polymer composites.

5.5.3
Transformations of Nanodiamond into Other Forms of Carbon

The conversion of nanodiamond into other forms of carbon represents a chemical transformation as well. In a range from a few up to several tens of nanometers the interconversion of different modifications is achieved quite easily (Section 5.2.3). Reactions leading to products as homogeneous as possible are of special interest in this connection. Hence an amorphization of the material as observed, for example, upon great mechanical stress does not constitute a transformation in the sense of this section. Another situation is given for the conversion of nanodiamond into carbon onions as described in Section 4.3.5. It can be induced by irradiation with high-energy electrons as well as by controlled heating. In both cases the conversion of sp^3-hybridized carbon proceeds from the outside into the core, and a hybrid material consisting of a diamond core with a graphitic shell is obtained as an intermediate (Figure 4.25). Further reaction yields a complete carbon onion inside of which no diamond can be detected anymore by any of the analytical methods described so far (Raman, XRD, EELS, HRTEM, etc.).

5.6
Applications and Perspectives

Owing to its properties, nanodiamond like the classical diamond is an attractive material for many applications. For the time being, however, just a limited number of industrial scale processes has really been established due to its inhomogeneity and the variable quality available from different suppliers. Pioneers in this area are the countries of the former Soviet Union where by now access has been made to various fields of application. The examples given herein comprise processes developed to an industrial scale already as well as such still operative on a laboratory scale. They include the preparation of composites and coatings, mechanical applications to reduce friction or to modify surfaces, uses in electro-deposition or biomedical applications.

5.6.1
Mechanical Applications

Nanodiamond as well features the hardness typical of a diamond in general. Therefore, it suits well to a use in polishing pastes to produce a precise finish of surfaces for electronic, optical or medical applications. Due to the small particle size, relief heights of just 2–8-nm are obtained, which means a considerable reduction of surface coarseness as compared to conventional polishing. Agglomerates of nanodiamond are also suitable to prepare polishing agents as individual primary particles may shear off from the agglomerates. Thus new cutting edges become available all the time. What is more, the presence of different crystallographic faces on the particle surface provides all degrees of hardness up to the maximum value of bulk diamond. Hence nanodiamond pastes can even be used to polish gem-quality stones. For treating metals or other materials, suspensions have turned out to be more stable and efficient. On the other hand, when nanodiamond is employed in polymer materials, it is usually applied as a powdery additive. The composite formation of PTFE (teflon) with 2% of nanodiamond is just one example here. In the resulting material, the coefficient of friction is decreased from 0.12 to 0.08.

The low frictional coefficient in general is another aspect rendering nanodiamond attractive for mechanical applications. Both surface coatings and composites can be made very wear resistant this way. For example, an admixture of 0.01% of nanodiamond in motor oil reduces the abrasion of moving parts by about 30% and increases the service life accordingly. The coating of magnetic recording devices (e.g., in hard-disk drives) as well experiences beneficial influence from the addition of nanodiamond. Higher numbers of revolution and, consequently, faster devices can be realized due to the wear resistance being doubled and the friction decreasing by about 20%. Moreover, the grain size obtained in the deposition of the ferromagnetic component is markedly reduced upon the addition of diamond, which means that the storage density of the respective medium is increased.

The preparation of composite materials in general is a very important application of the mechanical properties of nanodiamond. With many polymers like caoutchouc, polysiloxanes, fluoroelastomers polymethacrylates, epoxy resins, etc., composites with markedly improved mechanical characteristics have already been obtained from the noncovalent incorporation of nanodiamond by simple admixing during polymerization. The modulus of elasticity, the tensile strength, and the maximal elongation of the material all increase upon this modification. Depending on the basic polymer, just 0.1–0.5% (w/w) of nanodiamond are required to achieve this effect (Table 5.3). Polymer films can also be reinforced by the addition of nanodiamond. For a teflon film with ca. 2% of nanodiamond added, for example, friction is reduced at least 20%, and scratches inflicted by mechanical means are only half as deep as in neat teflon.

A possible reason for this enhancement of material properties may be the higher degree of crosslinking observed in the polymers mixed with nanodiamond. Hence an incorporation of suitably functionalized nanodiamond particles should lead to even more favorable properties and to a more homogeneous distribution of

Table 5.3 Effect of diamond added to polymers.

Polymer	Modulus of elasticity (MPa)	Tensile strength (MPa)	Maximum elongation
Fluoro elastomer	8.5 (at 100% elongation)	15.7	280%
Nanodiamond composite[a]	92	173	480%
Polysiloxane	19 (at 100% elongation)	52	730%
Nanodiamond composite[a]	53	154	1970%
Isoprene caoutchouc	7.7 (at 100% elongation)	20.5	No data given
Nanodiamond composite[a]	12.3	28.2	No data given

a) These are composites with the diamond noncovalently bound into the matrix.
Acc. to O. A. Shenderova, V. V. Zhirnov, D. W. Brenner, *Crit. Rev. Solid State Mater. Sci.* **2002**, *27*, 227–356.

diamond in the matrix. Especially the uptake of mechanical load by the diamond particles should occur yet more efficiently in a covalently bound, homogeneous composite. First publications on this topic justify the anticipation of interesting materials.

In electroplating, nanodiamond can be employed in the plating baths to provide markedly improved mechanical properties of the metal coating. The metal layer generated this way usually contains about 0.5% of nanodiamond (corresponding to ca. $0.2\,\mathrm{g\,m^{-2}}$ at a film thickness of 1 mm). This additive increases the corrosion resistance, the microhardness, and the scuff resistance as well as cohesion and adhesion. At the same time, the porosity of the coating and the frictional coefficient are reduced. Such improvement of film characteristics has been observed for a variety of metals, including chromium (already applied on an industrial scale), gold, silver, platinum, copper, aluminum and nickel. The fact that the noble metals form significantly harder films this way is of particular interest for electronic applications. The process of coating can be performed in the conventional apparatus by simply adding the nanodiamond. The service life of, for instance, casting moulds and gear components chrome-plated this way is markedly prolonged due to the effects described above.

5.6.2
Thermal Applications

The high thermal conductivity can be employed for nanodiamond applications as well. It is possible to prepare, for example, heat-conducting pastes. The material demand is only $1{-}10\,\mathrm{g\,m^{-2}}$ here. Another positive effect of using the nontoxic nanodiamond powder is to avoid the customary, very poisonous paste of beryllium oxide in some of these applications.

Nanodiamond is a suitable additive also for other coolants. It has been known for long that the addition of nanoparticles may overproportionally improve the thermal conductivity of cooling media. In general, an increase of up to 40% may be achieved. In the case of nanodiamond, an addition of only 0.3 vol% to cooling oils for large transistors causes a 20% growth of thermal conductivity. This effect serves to prevent the formation of hot zones inside the coolant and the consequent destruction of the transistor.

5.6.3
Applications as Sorbent

Materials like activated carbon have been known for long to possess good adsorptive properties, especially when they exhibit a large active surface. The specific surface of nanodiamond ranges up to $300\,\mathrm{m^2 g^{-1}}$, which should render it attractive for this type of application, too. It may adsorb up to four times its own weight of water, so it is a suitable siccative in certain areas of application. Yet it is not only water, but also other substances (like with biological origin) that can be adsorbed to the nanodiamond surface. Certain proteins, for instance, may be extracted from a serum or pollutants removed from solutions. The operative range can be extended still by functionalization of the diamond particle as then a specific bonding to the functional units presented on the large surface occurs.

Furthermore, nanodiamond is suitable to applications in liquid chromatography. A directed modification of surface polarity and adsorptive properties is feasible here by functionalization of the particles. Apart from this versatility, the nanodiamond material also stands out for another advantage: the large mechanical resistance and the small particle size allow a use in high-pressure applications, which is where the best separating power is achieved.

5.6.4
Biological Applications

Due to its low toxicity and large chemical inertness, nanodiamond is an attractive material for biological applications. Biological moieties may be attached by either covalent or noncovalent bonding. Most publications available to date describe the noncovalent adsorption of proteins, antibodies, enzymes, viruses, etc. In some cases, the surface of the diamond particles is pretreated, for instance, by coating it with L-polylysine or cellobiose, before the adsorption of the biologically active unit itself is effected. The diamond hybrid particles thus obtained suit to the preparation of transporting vehicles for vaccines or pharmaceutical actives or to the controlled release of genes, etc., inside living cells. In this regard then, the small particle size is rather beneficial.

Moreover, an application on biochips for the determination of certain proteins in a serum has been presented. The attachment takes place, for instance, on an immobilized enzyme with the read-out of information normally being enabled by fluorescence labeling. Besides an external fluorescence label, it is also possible to employ the inherent luminescence of lattice defects in the diamond itself (Section 5.4.1.4). It is especially nitrogen defects that may be detected this way, that is, by fluorescence microscopy (Figure 5.46). Defective nanodiamond particles with surface functionalization may also be employed as fluorescence label in *in vivo* experiments, so a system complementary to the metal chalcogenide quantum dots usually applied is available here. The nanodiamond adducts in these processes are characterized by their small particle size, stable fluorescence, and (at least according to current knowledge) by their nontoxicity.

5.6.5
Further Applications and Perspectives

A multitude of potential further applications opens up to nanodiamond materials in a variety of technological areas. Examples hitherto described include, among others, the preparation of field emitters for display uses. Current research focuses, for instance, on utilizing the lattice defects and the resulting fluorescence as well as the unpaired spins. More applications are expected to emerge in the field of scratch-resistant transparent coatings. Moreover, it should be possible to realize

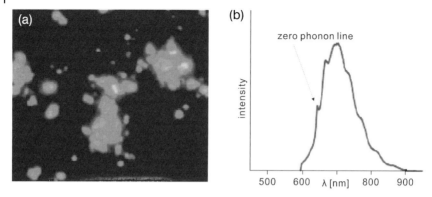

Figure 5.46 (a) N-V-defect induced fluorescence of nanodiamond inside living cells (bright spots); (b) corresponding fluorescence spectrum of the diamond nanoparticles under test (© ACS 2005).

coatings with determined electrical properties by using suitably doped nanodiamond particles.

Another field of possible applications can be envisaged in biology and medicine. First studies indicate that nanodiamond has some physiological effects that might be employed for the treatment of certain diseases. The research on this topic, however, is in its very beginnings, so for the time being, no definite statements can be made. Owing to the small particle size and the lack of any proven toxicity, however, nanodiamond can still be expected to be an attractive target for the development of transporting vehicles in controlled drug delivery to living cells.

A considerable advantage of nanodiamond consists in the material being available at reasonable prices today. High-grade samples are traded for as little as 2–5 Euro per gram. With increasing demands, the price would go down even further because mass production can easily be established. Hence nanodiamond has a clear advantage over nanotubes and fullerenes regarding an industrial scale application, and the realization of first actual examples seems possible in the near future.

5.7
Summary

Nanodiamond consists of particles showing a size in the range of nanometers. A distinction has to be made between materials containing very small particles (d ~ 4 nm) on the one hand and larger nanodiamond particles on the other. Very small particles tend toward agglomerate formation, which is favored by the presence of graphitic carbon as well as by the functional groups situated on the particle surface. The preparation can be achieved in different ways (Box 5.1)

Box 5.1 Preparation of nanodiamond.

- *By detonation:* A mixture of explosives with a negative oxygen balance is detonated inside a closed container. Due to the prevailing conditions, the resulting soot contains a large portion of nanoscopic diamond particles. These are purified by treatment with concentrated mineral acids that remove both metallic and graphitic impurities.

- *By shock synthesis:* A carbon material is converted into diamond by the action of a shock wave generated, for example, by a detonation or a projectile. This procedure is employed, for instance, to prepare polycrystalline microdiamond with primary particles measuring in the range of nanometers.

- Nanoscale diamond can further be generated by milling of larger diamond particles or by a transformation of other forms of carbon that is induced by irradiation or heating.

Regarding the structure of nanodiamond, a distinction has to be made between the diamond core that usually features a cubic lattice and, on the other hand, the surface. Depending on particle size, the portion of surface atoms may amount to as much as 50%. Generally the surface structure plays a major role for the material properties observed. At particle dimensions of more than 2 nm, the bandgap of nanodiamond corresponds to that of macroscopic diamond. Quantum effects are not observed, but there are interband states that can be attributed to partial surface graphitization and lattice defects, respectively.

Box 5.2 Physical and chemical properties of nanodiamond.

- *Surface structure:* In theory, there should be dangling bonds. Actually, however, these are saturated by partial graphitization or by the attachment of functional groups. The most abundant species are hydroxyl, carboxyl, and lactone groups and, to a lesser extent, alkyl functions.

- *Physical properties:* Nanodiamond stands out for its great hardness and surface conductivity, for field emission characteristics and for the possible fluorescence of defect centers. Spectroscopic examinations revealed both the band structure and the structural properties.

- *Chemical properties:* The reactivity is governed by the groups present on the surface. Therefore, a covalent attachment of various molecules and structures (e.g., bioactive moieties) can be achieved in different ways.

Owing to its interesting properties, there is a multitude of conceivable applications with some of them already being realized on industrial scale. These include the use as polishing agent, as additive in polymers, and as adsorbent as well as the employment in electroplating. Furthermore, intensive studies have been made regarding applications in electronics (e.g., as field emission source), for biological purposes, and in composites for scratch-resistant coatings.

6
Diamond Films

Its remarkable properties render diamond a coveted material, yet many applications require a certain shape of the respective device and hence an extensive machining of the initial workpiece. When starting from natural or artificial diamond, this turns out to be an arduous and expensive task due to the material's enormous hardness and friable character. The size of available diamonds further limits possible applications to very few fields like the use as grinding and polishing agent. Particular complications arose, however, in the preparation of thin layers with a predetermined shape. Consequently, it has been rather soon after the first generation of artificial diamond that procedures allowing for the production of diamond films were sought for.

6.1
Discovery and History of Diamond Films

It has already been mentioned in Chapter 1 that a method for the artificial production of diamond had been presented in the early 1950s. It was characterized by the application of high pressure and temperatures and by the employment of certain transition metals as catalyst. At about the same time, W. G. Eversole of the Union Carbide company had the idea that a diamond synthesis should also be feasible without using high-pressure apparatus. Initially he did extensive research on the decomposition of carbon monoxide. In doing so, he examined the system of CO, CO_2, and elemental carbon (both as graphite and diamond) for its kinetic and thermodynamic properties and drew the conclusion that it should be possible to generate diamond from carbon monoxide. In 1952, and thus 1 year *earlier* than the high-pressure synthesis at ASEA in Sweden or even three years before the publication of the General Electric results, he was the first ever to verifiably succeed in the synthesis of an artificial diamond material.

These results have never been published in any scientific journal though, and just a few patents bear testimony to the successful efforts. Most of his work was only laid down in internal reports. In 1958 then, Eversole published a patent describing the synthesis of diamond films by decomposition of hydrocarbons at ambient pressure. Still this procedure required the introduction of diamond crys-

Carbon Materials and Nanotechnology. Anke Krueger
Copyright © 2010 WILEY-VCH Verlag GmbH & Co. KGaA, Weinheim
ISBN: 978-3-527-31803-2

tallites to serve as nucleation centers, whereas without this addition, no formation of diamantoid carbon was observed. At first, these results were largely disregarded, and only much later, when the potential of these layers for future applications had been recognized, the research activity on the preparation of diamond films experienced nothing short of a boom.

The investigations on nanodiamond were influenced by numerous factors. At first, there had been unbelief among many scientists, who generally considered devious the idea of synthesizing diamond as a metastable phase at normal pressure. Even in 1954, and thus a year after the first preparation of diamond at low pressure, Neuhaus argued that from a theoretical point of view any effort on such a synthesis had to remain fruitless because diamond formation at normal pressure was downright excluded. This is all the more astounding as it was already known that a large variety of elements like sulfur could crystallize in a form thermodynamically unstable at normal conditions, only provided that a suitable method like sudden cooling of a melt was applied. Due to the arguments brought forth, the reports on the successful synthesis of diamond at normal pressure were met with large skepticism nevertheless.

In addition, the efforts were hampered by the political situation. In Cold War times, the communication between researchers in the USA and Western Europe on the one side and Russian scientists on the other was limited at best. Therefore in the mid-1960s, the invention and optimization of diamond synthesis at normal pressure and without the addition of diamond in large parts took place independently in the USA as well as in the Soviet Union. (Angus and Deryagin may be named as important workers in this field.) This is an example of how detrimental the effect of political circumstance on scientific progress might be. In the early 1980s then some Japanese publications attracted attention. They reported on the first successful syntheses of films with procedures still in common use today. In 1981, the first hot filament CVD was described, only to be followed by the radio frequency and the microwave plasma CVD methods in 1982. At first only low growth rates were attained, and a deposition on substrates other than diamond was rather complicated. Still satisfactory results could be achieved by varying the carbon source and by adding large amounts of hydrogen (an idea that had already been employed with good results in Russia) as well as by modifying the apparatus used.

For long time, it was unclear how to rationally plan the synthesis of diamond at metastable conditions. In 1973, Wilson published an article presenting some essential concepts that substantially contributed to the further development, for example, in Japan. He summarized the requirements to be met by a procedure that might be capable of diamond deposition. These include, firstly, the generation of single carbon atoms in an excited state that hold enough energy to form bonds like in diamond, and secondly, the excited atoms must feature a sufficient lifetime or an increased reaction rate for growth actually to occur. In the same paper, he also presented the solution for the first problem by suggesting various methods of generating energy-rich carbon particles like the dissociation of carbon compounds, electric discharge, electron bombardment, shock action, and X-ray- or

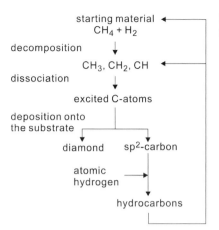

Figure 6.1 Scheme of the CVD process for the production of diamond films.

UV-irradiation. The second problem (a sufficient life-time) was solved only in 1981 by Spitsyn *et al.* who introduced the *in situ* generation of atomic hydrogen, which lead to a considerable increase in growth rates. Figure 6.1 shows a reaction scheme.

Since that time, synthetic diamond films have developed into an important high-tech product employed for many purposes. In comparison to other forms of diamond, the most attractive difference is the facile generation of diamond coated workpieces in almost any desired shape. The preparation of thin layers, for example, for electronic applications, became possible as well only after the development of CVD methods.

6.2
Structure of Diamond Films

6.2.1
General Considerations on the Structure of Diamond Films

Diamond films are a mono- or polycrystalline layer of diamond with a thickness (usually) in the range of micrometers and a comparatively wide dilation regarding the other two directions in space. Thus a – generally coherent – film is obtained. It can either be deposited on a substrate or can exist as a homoepitaxially grown layer or a free-standing structure.

Single crystalline diamond films may, upon suitable preparation, be distinguished by a well-defined orientation of the crystal lattice throughout the film. Hence they exhibit a very homogeneous surface structure corresponding to one of the crystallographic faces of cubic diamond (Figure 6.2a). In many cases the (100)-plane is observed, whereas the (111)- and especially the (110)-face are less frequent. The outermost layer of atoms constitutes the phase boundary against

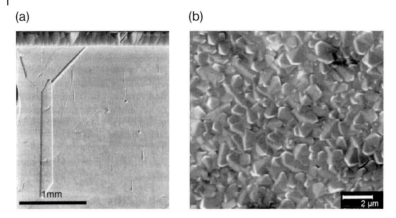

Figure 6.2 (a) Monocrystalline diamond film, (b) polycrystalline diamond film (© Elsevier 2005).

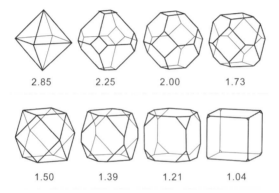

Figure 6.3 The shape of crystals depends on the growth parameter α.

the environment. To avoid unsaturated bonding sites, it has to undergo certain structural modifications that lead to interesting surface geometries (see below). In general, either a surface reconstruction or a functionalization at the so-called dangling bonds may occur.

Polycrystalline diamond films, on the other hand, do not consist of a coherent layer with homogeneous orientation, but of individual small crystals each exhibiting faceted surfaces themselves (Figure 6.2b). Again the (100)- and the (111)-plane constitute the most frequent crystallographic faces. The morphology of diamond films is determined by the growth rates of these planes. Their ratio is expressed by the so-called growth parameter α, which is a measure of the relative growth rate: $\alpha = \sqrt{3} \cdot v_{100}/v_{111}$ (with v_{hkl} being the growth rate of the respective crystallographic face). The shape of crystallites becomes more cubic with smaller α. In general, cubic, octahedral, and cuboctahedral species are observed. Figure 6.3 shows some idealized crystal shapes and the corresponding growth parameters.

In the process, a high concentration of carbon in the gas phase of the reactor favors high α-values, whereas increasing the substrate temperature or reducing the pressure leads to decreased α-values and, consequently, to a film structure characterized by cubic crystallites.

The orientation of the crystallites relative to the substrate plane is also governed by the growth parameter: In the end, only those crystallization nuclei with their largest growth rate aligned in perpendicular to the substrate surface will survive because at progressing growth, these crystallites suppress nuclei preferably growing in other directions. Hence all the resulting crystallites will exhibit the same respective orientation. Thus both the shape of crystallites and the texture of the diamond films are determined by the relative growth rates of crystal faces.

Contrary to single crystalline diamond films, the polycrystalline layers contain a much larger portion of sp^2-hybridized carbon. This can be attributed to the large number of grain boundaries and has considerable influence on the electronic properties of these films (Section 6.4.2).

The so-called "ultrananocrystalline diamond" (UNCD) is a special form of polycrystalline diamond films consisting of nanoscopically small individual crystallites with diameters of 3–10 nm (Figure 6.4). Owing to this fine structure, the portion of carbon atoms being part of grain boundaries is in its turn markedly increased as compared to microcrystalline diamond films, which again influences properties like the surface conductivity.

Apart from the types of diamond films mentioned so far, further films of sp^3-hybridized carbon have been described as well. Some of them feature amorphous structures and a large content of hydrogen. Within the scope of this book, these phases will be discussed briefly (Section 6.2.4), whereas further references may be found in the numerous review articles and in the original literature (Chapter 8).

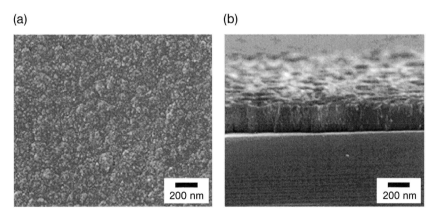

Figure 6.4 Nanocrystalline diamond film, SEM-image: (a) top view, (b) side view (© AIP 2006).

6.2.2
The Surface Structure of Diamond Films

On the phase boundaries in touch with the environment, the surface atoms of the diamond film are situated. Their valencies are saturated to a limited extent only as all of them lack at least one bonding partner on one side. For the (100)-face there are even two unsaturated bonding sites per atom extending into the surrounding space. Such a large number of dangling bonds is rather unfavorable from an energetic point of view, and so the surface has a strong tendency toward saturating these bonding sites. This may be effected either by passivation with external bonding partners or by a so-called surface reconstruction. In the latter case, additional bonds are formed between the atoms on the surface. Depending on the crystal face presented to the outside, different structural patterns are energetically favored in this process. It generates surface states with their wavefunctions entirely localized inside the surface, which gives rise to a two-dimensional surface band structure and characteristic surface properties.

The lattice structure of diamond films largely corresponds to that of cubic bulk diamond with a lattice constant of $a = 3.567$ Å and with the unit cell containing eight carbon atoms. The distance of proximate neighbors is 1.545 Å. The most important crystallographic faces of diamond are the (111)- and the (100)-plane. Both of them occur on polycrystalline diamond films or may be generated intentionally by chemical vapor deposition. The structure of the surface then depends not only on the crystallographic orientation, but also on the specific covering of the film with foreign atoms. In the following, some of the possible surface structures will be discussed in detail. In doing so, an uncovered diamond surface will be considered as well as one covered with model atoms (hydrogen and oxygen in this case).

6.2.2.1 Structure of the (111)-Plane

The (111)-face represents the natural cleavage plane of bulk diamond, whereas in diamond films its importance stands back a bit behind that of the (100)-face. In theory, two (111)-surfaces can be conceived with one of them featuring a single unsaturated bonding site per superficial carbon atom and with the other exhibiting methyl-like structures with three dangling bonds. The latter, however, is much less favorable from an energetic point of view, so only the reconstruction actually observed and resulting in a (111)-face with one free valency per C-atom will be discussed here (Figure 6.5).

The (111)-diamond plane is not stable without further modification, so in its uncovered state, it undergoes a complex reconstruction process. Looking at the (111)-face in top view reveals a striking similarity to a graphene layer. The surface atoms are connected to the atoms of the layer underneath in chair-form, six-membered rings, and in projecting these in perpendicular to the crystal plane results in six-membered rings with dimensions corresponding to those in the graphite lattice. This is suggestive of a strong tendency toward graphitization, which is confirmed experimentally indeed. Still saturating the surface with hydro-

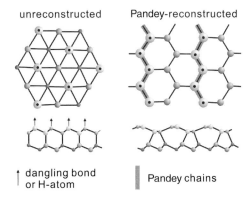

unreconstructed   Pandey-reconstructed

↑ dangling bond or H-atom

Pandey chains

Figure 6.5 Reconstruction (right) of the (111)-face of a diamond film in top view and side view. The traces marked in gray correspond to the Pandey chains.

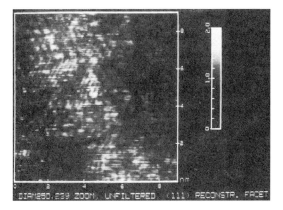

Figure 6.6 Atomic force microscopic (AFM) image of a reconstructed (111)-diamond face. The orientation of individual domains at an angle of 120° is evident (© APS 1993).

gen atoms is as well a suitable means of passivating the (111)-face, which will be detailed farther below.

In the case of an uncovered (111)-face a reconstructed surface with a 2×1-geometry is formed. This process is known as Pandey reconstruction and generates rows of connected carbon atoms on the surface and in the layer underneath (the low-lying atoms of the six-membered rings mentioned above, Figure 6.5). These Pandey-chains may be oriented in one of three possible directions on the surface ([$\bar{1}$01], [1$\bar{1}$0], [01$\bar{1}$]). What is actually observed in experiments are patterns of lines arranged at an angle of 120° relative to each other, corresponding perfectly well to the theoretical prediction. (Figure 6.6). The distance between surface atoms in a reconstructed structure amounts to no more than 1.43 Å, which is quite close to the value known for graphite (1.425 Å). In the course of the Pandey reconstruction, the dangling bonds turn into π-bonds and the symmetry of the surface is markedly reduced. Just one mirror plane is conserved throughout reconstruction. The Pandey chains are situated on the surface with a respective distance of 4.37 Å, so there are only weak interactions between individual chains.

What, on the other hand, does the surface look like after a saturation with bonding partners? The covering with hydrogen, which is also the process most commonly observed in experiment, suits very well to passivating the (111)-plane of diamond. Each carbon atom on the surface exhibits exactly one unsaturated bonding site. These are arranged at a respective distance allowing for an attachment of hydrogen atoms to every free valency (Figure 6.5). Therefore, the surface can be stabilized in a nonreconstructed state, thus also preserving its symmetry of course. The distance between the carbon atoms saturated with hydrogen measures 2.52 Å.

Covering the surface with oxygen instead of hydrogen, on the other hand, requires conserving of the 2×1-geometry of a reconstructed surface. The oxygen atoms presumably act as the bridges of the Pandey chains, yet detailed experimental studies on the matter are pending. Furthermore, saturation with other foreign atoms or with larger structures might be conceived. Due to the dense packing of carbon atoms in the diamond lattice, however, the available space is rather limited, so generally a complete saturation cannot be achieved with these bonding partners. This is true not only for the (111)-face of diamond, but also for any other face that may be saturated by reconstruction or passivation. More details on this topic may be found Section 6.5.

6.2.2.2 Structure of the (100)-Plane

In diamond films made by vapor deposition, the (100)-plane is by far more important than in the bulk material. It may be grown on substrates as well as homoepitaxially, and its properties are known in quite some detail. The top view on the unsaturated (100)-face of a diamond reveals the cubic symmetry of the lattice (Figure 6.7). The plane itself belongs in the point group C_{2v}. The side view shows

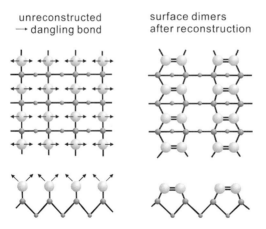

Figure 6.7 Reconstruction (right) of the (100)-face of a diamond film in top view (top) and side view (bottom).

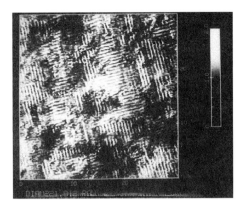

Figure 6.8 Atomic force microscopic image of a reconstructed (100)-diamond face. The ordered structure of the surface dimers is evident (© APS 1993).

a number of carbon atoms protruding from the lattice. These carry the unsaturated bonding sites with each surface atom holding two dangling bonds. The interval between carbon atoms on the (100)-face is 2.523 Å, which corresponds to the second closest distance in the diamond lattice.

The two dangling bonds can be saturated by bond formation with adjacent atoms. This process gives rise to surface dimers that are characterized by the presence of π-bonds. Owing to the larger distance, these are less stable than those on a (111)-face. The dimers are oriented along the [011]-direction, resulting in a reconstructed surface with a 2×1-geometry (Figure 6.8) and belonging to the point group C_{2v}. The distance between individual dimers is 2.52 Å, which allows for an exchange between them as well. Consequently, the electronic structure, initially only split into π- and π^*-orbitals, evolves into a surface band structure with a bandgap of ~1.3 eV. The occupied levels, however, are situated in the bulk diamond's valence band alone, thus constituting electronically inert states.

What happens now upon the saturation of the surface with foreign atoms? In principle, a hydrogenation of the surface may, as described in Section 6.2.2.1, saturate all dangling bonds of superficial carbon atoms. Due to the close proximity of individual atoms on the (100)-face, however, there is not enough room for a complete covering of the nonreconstructed surface, and the attachment of hydrogen occurs on the reconstructed face instead. The process can formally be considered an addition of H_2 to a π-bond (Figure 6.9) that does not alter the symmetry of the surface. Contrary to the two H-atoms hypothetically required in the case of a nonreconstructed surface, each superficial carbon atom actually carries exactly one hydrogen atom.

Oxygen, on the other hand, can saturate both free valencies at once. Hence a covering of the nonreconstructed surface is possible here, which means that the 1×1-geometry is conserved. In principle, there are two conceivable modes of attaching oxygen to the (100)-face of diamond: Firstly, the formation of ether structures

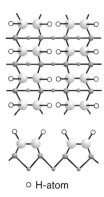

Figure 6.9 A (100)-face saturated with hydrogen in top view (top) and side view (bottom).

○ H-atom

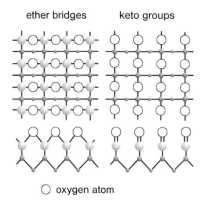

○ oxygen atom

Figure 6.10 A (100)-face saturated with oxygen in top view (top) and side view (bottom).

with a bridging arrangement of oxygen between two carbon atoms, and secondly, the generation of superficial keto groups upon both of a carbon atom's dangling bonds reacting with the same oxygen atom (Figure 6.10). It has not yet been fully elucidated which one of these arrangements is actually favored, but the majority of experimental and theoretical results available to date points to a preference of the ether structure.

6.2.2.3 Structure of the (110)-Plane

The (110)-plane is observed less frequently than those described before. It exhibits zig-zag lines of surface atoms resembling the structure of Pandey chains on a (111)-face (Figure 6.11). Only the distance of 3.57 Å between the chains turns out to be markedly smaller than on the latter. This crystallographic face can be stabilized without reconstructing the surface. With the distance between dangling bonds being just 1.545 Å, π-interactions can evolve directly. Consequently, and contrary to the (111)- and the (100)-face, the (110)-plane can remain in a 1×1-geometry even in its uncovered state.

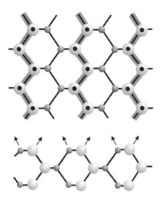

Figure 6.11 The structure of the (110)-face of a diamond film in top view and side view. The traces marked in gray correspond to the π-chains on the surface.

6.2.3
Defects and Doping of Diamond Films

It is very rare that the structure of a diamond film is really perfect and free of defects. Usually, there are many structural defects and impurities arising from the conditions of preparation. In some cases, however, a deliberate contamination with selected materials or certain structural features are even aimed at to make use of associated properties (e.g., electronic or optical characteristics).

The first defective structures (in the shape of dangling bonds) are already found on the surface of a diamond film. They exert considerable influence on the material's properties. These unsaturated bonding sites are especially concentrated in regions with many edges like crystallographic steps or other structures deviating from a plane morphology. Normally, several dangling bonds are situated here in direct proximity. On the undisturbed surface, on the other hand, there are also isolated free bonding sites observed being embedded in the layer of adsorbate. With their orbitals being just singly occupied, they constitute an amphoteric element that can act as both electron donor and acceptor. At a high density of dangling bonds, this leads to a graphitization of the diamond surface.

Within the lattice as well numerous defects exist. They include vacancies, dislocations, stacking disorders, and twinning. Normally, these defects are already generated during nucleation of the crystallites, which means that the density of defects in a diamond film can be influenced by scrupulously controlling the conditions of preparation.

Twin structures are very often observed in diamond films. Most frequently they are (111)-twins with the two neighboring crystals sharing a common [111]-axis (Figure 6.12). They are then tilted at an angle of 180° or 60° against each other, respectively. At the origin of the twinning, a boat-form six-membered ring of carbon occurs in the lattice (contrary to the chair-form rings usually present in cubic diamond, Figure 6.12). Twin structures exert substantial influence on the morphology of a diamond film. They are responsible for reentrant corners as well as for interpenetrating particle structures (Figure 6.13).

6 Diamond Films

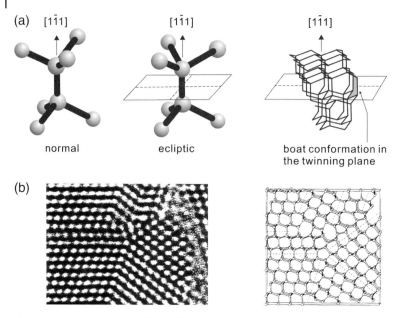

Figure 6.12 (a) Generation of twins by an ecliptic arrangement of C-atoms. Within the twinning plane the six-membered rings are boat-shaped. (b) Electron microscopic image and scheme of a multiple twin (© Taylor & Francis 2001).

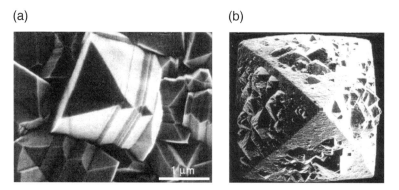

Figure 6.13 Effects of twinning: reentrant corner (a) and particle interpenetrations (b) (© Elsevier 2003).

Stacking disorders preferably appear on the (111)-plane, too. They are enclosed between two twinning planes. At the extreme case of a twin plane being inserted after every double layer, *Lonsdaleite*, the hexagonal modification of a diamond, is obtained (Section 1.2.2). Dislocations are also frequently observed in CVD-diamond films. They are formed during film growth (Section 6.3). In comparison to dislocations in silicon or germanium, these defects are significantly less mobile

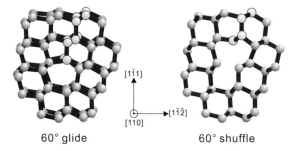

Figure 6.14 Examples of dislocations in a diamond lattice.

due to the carbon's characteristic ability of double-bond formation. The free bonding site in the center of the dislocation is presumably saturated as sp^2-carbon (Figure 6.14).

The grain boundaries of individual crystallites constitute another highly defective structural element. Owing to the saturation of the surface and other lattice imperfections the portion of sp^2-carbon is particularly high here. Actually the grain boundaries can rather be considered as an sp^2/sp^3-hybrid structure.

In addition to these structural defects, diamond films also contain numerous impurities. Contrary to the bulk diamond with nitrogen constituting the largest portion of foreign elements, hydrogen is the most frequent impurity in CVD diamond. This is due to its preparation from mixtures of methane and hydrogen. At the conditions prevailing (e.g., plasma or high temperatures), the latter also exists in an atomic state, thus facilitating its incorporation. Experiments with deuterated species revealed that the incorporated hydrogen does not originate from the methane, but from the hydrogen gas. For highly defective diamond films, the concentration of hydrogen atoms amounts to 10^{20}–10^{21} cm^{-3}, whereas it is less than 10^{19} cm^{-3} for high-quality films. Furthermore, hydrogen may be incorporated by directed proton implantation, which allows for achieving local concentrations of up to 5%.

The hydrogen is mainly situated in the defective or amorphous parts of the film, that is, on the grain boundaries and at dislocations, and it forms complexes with other lattice imperfections like steps, edges or impurities. The formation of boron–hydrogen interactions, for example, is responsible for the disappearance of the electronic states initially inserted into the bandgap by the presence of boron, which means a passivation of the acceptor. Similar effects also arise from hydrogen forming complexes with other impurities, and so the presence and concentration of hydrogen in a diamond film has significant influence on its electronic and optical properties.

The afore-mentioned boron plays a central role in the doping of diamond. It is the only acceptor available to date that is easy to incorporate into the crystal and at the same time provides satisfactory properties. The boron gives rise to a low acceptor level 0.37 eV above the valence band maximum, resulting in a p-doped material with semiconducting or, at higher boron concentrations, quasimetallic characteristics (Section 6.5.4). The dopant may originate from the gas phase (e.g., from diborane B_2H_6) as well as from solid sources like a support of boron nitride and is introduced by common techniques (coevaporation, implantation, etc.). The probability of boron being incorporated depends on the crystallographic orientation. The most easily it is achieved in the (111)-planes, followed by the (110)- and the (100)-planes. Boron atoms can easily substitute for carbon on its lattice positions as both species exhibit similar dimensions.

The same is true for nitrogen that can be incorporated into the diamond lattice as well (Figure 5.3). Interstitial nitrogen, on the other hand, or nitrogen aggregated within the lattice as known for type Ia bulk diamond is rarely ever observed. The lack of aggregates is due to the short growth period in CVD preparation. Hence the nitrogen atoms do not have enough time for migrating through the lattice. This dopant is a so-called deep donor that introduces into the diamond's bandgap a donor level at 1.7 eV below the conductance band. This is sufficiently low to prevent a facile excitation of electrons into the conductance band. Therefore, nitrogen is not a suitable dopant to obtain an n-semiconductor.

The somewhat bigger phosphorous, on the other hand, is a donor that introduces a level at 0.5–0.6 eV below the conductance band. Thus an n-semiconductor arises from this kind of doping. At first, the incorporation of phosphorous posed serious problems as it is poorly soluble in carbon. By now, however, an *in situ* process has been established that directly introduces the dopant into the lattice during CVD production of the diamond film. Still, owing to its larger diameter, phosphorous does not fit in well as compared to boron or nitrogen, which causes a slight lattice distortion and, consequently, an additional alteration of the diamond's properties.

Apart from those mentioned above, there are other contaminations known for diamond films, too. Oxygen and silicon are frequently observed with the latter being incorporated into the lattice in a substitutional way as well. Oxygen and further impurities like sulfur or metallic defects have not yet been studied in sufficient detail to give a final statement on their arrangement in the lattice and on their influence on the diamond properties. Calculations indicate that boron, nitrogen, phosphorous, and silicon are the only dopants suitable for a stable incorporation into the lattice.

6.2.4
Structure of Further Diamond-Like Film Materials

Besides the proper diamond films obtained by vapor deposition, the same method also allows for the preparation of further, similar materials. These include the so-called a-C:H- and a-C-phases that are alternatively termed "diamond-like carbon (DLC) films."

When ionized hydrocarbons hit the surface of a substrate, they can deposit there. It is noteworthy that the structure of the resulting film material does not depend on the kind of hydrocarbon employed, but on the energy of the ions' impact. Usually, energies around 100 eV are applied. These energy-rich $C_mH_n^+$-ions can be generated, for instance, in an RF-plasma. After decomposing into atomic carbon (which goes without problems at the prevailing conditions), they are able to insert into C–H-bonds and thus to attack on hydrocarbon structures already deposited on the surface. By doing so, they give rise to terminal triple and double bonds. Finally, in the course of film growth, a balance evolves between events stabilizing the film (bond formation) and destabilizing factors (strain due to distorted structures). This equilibrium is maintained by suitable incorporation of hydrogen atoms and multiple bonds.

The films obtained consist of an unordered network of sp^3-hybridized carbon atoms (hence **a-C:H** like amorphous) containing additional hydrogen atoms (hence a-C:**H**) and sp^2-carbon atoms. The latter are sometimes observed to coalesce into small clusters. The presence of π-bonds is confirmed by the small bandgap of only 0.5–2.5 eV that originates from $\pi \rightarrow \pi^*$-transitions.

The properties of a-C:H-phases vary strongly with their hydrogen content. The higher it is, the more transparent the films become. At the same time, however, they assume a rather soft, hydrocarbon-like consistency. Upon tempering a-C:H-films at 400 °C, the portion of sp^2-carbon increases, giving rise to a significant gain of electric conductivity. Altogether the a-C:H-phases assume an intermediate position between diamond and hydrocarbons.

Furthermore, it is possible to prepare diamond-like a-C-films. These do not contain hydrogen at all and consequently consist of sp^3- and sp^2-carbon alone. They have a low density of just 2–3 g cm^{-3}. Owing to the lack of hydrogen, the saturation of dangling bonds can only be effected by π-bonding. Depending on the method of production, the portion of sp^2-carbon and hence the material's properties can quite accurately be controlled. Normally, these films are largely homogeneous, and diamond crystallites are not observed.

6.3
Preparation of Diamond Films

Today, the preparation of diamond films can be performed on a large scale in commercially available apparatus. In this section, some important methods of depositing diamond layers on various substrates or as free-standing films are presented.

6.3.1
CVD Methods for the Preparation of Diamond Films

The chemical vapor deposition has developed into the leading method for the preparation of thin diamond films. It is mainly characterized by a precipitation of carbon from the gas phase onto a substrate. Applicable sources of carbon include methane, acetylene, or ethylene, which are normally admixed with a current of hydrogen. The latter, in an atomic state, turned out to be essential for an efficient production of high-quality diamond films. Actually, atomic hydrogen is generated *in situ* from

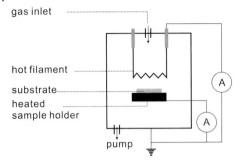

Figure 6.15 Scheme of an HF-CVD apparatus (hot filament chemical vapor deposition).

the hydrogen added to the gas phase. Suitable substrates include carbon, and especially diamond itself, as well as foreign materials like metals or silicon. Different kinds of diamond films with in parts strongly varying quality are obtained depending on the specific substrate. The temperature of the substrate should be 500–1200 °C as outside of this range, other forms of carbon-like graphite or diamond-like carbon (DLC) are deposited. The type of reactor chosen for the diamond synthesis also plays a role for the kind and quality of the resulting film. In the following, a few methods used in the chemical vapor deposition of diamond films will be presented. The growth mechanism is discussed in some detail in Section 6.3.2.

6.3.1.1 Hot Filament CVD

The core of such an apparatus is the filament (usually a tungsten wire) that is positioned in short distance above the substrate and heated to more than 2000 °C. On this wire occur both the generation of atomic hydrogen and a partial decomposition of the hydrocarbon (commonly methane) into excited fragments. Figure 6.15 shows a schematic representation of the setup. Normally, the substrate is heated too, thus enabling an exact temperature control. Furthermore, a positive potential may be applied (bias-enhanced CVD), which increases the rates of nucleation and film growth. This is because electrons driven by the bias between substrate and filament are extracted from the latter and bombard the substrate. This effect favors dissociation of the hydrocarbon on the surface and thus improves the growth parameters of the film.

The essential advantage of hot filament CVD lies in its simplicity. Diamond films can be deposited on various substrates without large effort pertaining to apparatus – the only thing required is a vacuum chamber to conduct the experiment in. However, the poor homogeneity and purity of the diamond films obtained is problematic. Due to its geometry, the filament acts as a linear source of reactive species above the substrate. As a remedial measure, a filament net expanding over the entire deposition zone may be employed. This leads to a more homogeneous distribution of reactive fragments. The impurities in diamond films usually originate from the filament. It decomposes due to the enormous thermal stress and the reaction with gaseous carbon atoms, and so fragments of the wire may be

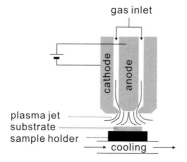

Figure 6.16 Scheme of the plasma jet method for chemical vapor deposition.

incorporated into the diamond film. Tungsten, for example, grows brittle upon the formation of carbides, which may, in some cases, cause a filament failure within shortest time. Moreover, the filament characteristics change over time at continuous deposition, which means that also the quality of the obtained diamond film varies to the same extent.

6.3.1.2 CVD at Simultaneous Electric Discharge

Another simple setup for the CVD preparation of diamond films employs a concomitant electric discharge between two electrodes to activate the species involved in the reaction. Either an arc or a glow discharge is generated between the electric poles with the substrate being attached to one of the electrodes. Owing to the arcs and the bombardment with electrons, the substrate is heated to about 800 °C and the hydrocarbon employed as carbon source decomposes. Sometimes the substrate heating is so strong that external cooling is required to keep within the range of temperature suitable for diamond deposition. At a current density of about 4 A cm^{-2} and with a voltage of 1000 V applied, a growth rate of ca. 20 µm h^{-1} and a nucleation rate of 10^8 cm^{-2} can be achieved.

A modification of the arc discharge method is realized in the so-called DC-arc jet (plasma jet). In this case, the electrodes are arranged in a way to form a sort of nozzle for the reactant gases. The cathode encloses the anode in a certain distance, and the gas mixture is led through the resultant gap. It partly decomposes between the electrodes before it hits the cooled substrate where the diamond film is deposited then (Figure 6.16). In this manner, an accurate control of the deposition zone is achieved, yet the results are highly dependent on the nozzle geometry and on a very constant reactant flow.

6.3.1.3 Microwave CVD

The hydrogen required for the deposition of a diamond film can furthermore be generated in the plasma arising from microwave irradiation at a frequency of 2.5 GHz. The radiation can act from the side of the apparatus (Figure 6.17) or from its top, either way with a plasma zone evolving inside the reactor. The latter is a vacuum chamber held at a pressure of 5–100 Torr/0.1–19.3 psi. The substrate is positioned inside the plasma zone and heated directly there. At the same time, it

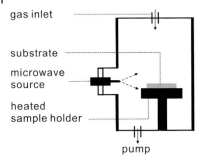

Figure 6.17 Scheme of the microwave-assisted plasma method.

Table 6.1 Typical parameters of CVD methods.

Method	Temperature (°C)	Pressure (Torr) (psi)	Growth rate ($\mu m\,h^{-1}$)
MW-CVD	800–1000	5–100 (0.1–1.9)	0.1–5
HF-CVD	700–1000	0.1–50 (0.002–1)	0.08–0.1
DC-arc-jet (plasma jet)	~800	500–700 (9.7–13.5)	80–1000
Combustion method (Section 6.3.4)	800–1000	760 (14.7)	140

is in intensive contact with the reactive species formed in the plasma zone from the gas mixtures fed from the top of the reactor. Inside the plasma the electron density is particularly high, which leads to a large rate of decomposition and hence to the nucleation of diamond on the substrate. The growth rates range about 0.1–5 $\mu m\,h^{-1}$. In some cases, an additional magnetic field is coupled to stabilize the plasma zone and thus to obtain better control of the deposition area.

Like all methods presented here, the microwave CVD has its pros and cons. Counting among the advantages is the good reproducibility of the deposition results obtained. Moreover, it turned out beneficial that the substrate may be cooled or heated independently of the plasma-generating source of energy. Last, but not least, the method has spread widely because there is commercially available a variety of microwave CVD apparatus yielding good results.

Still the inhomogeneity of the plasma has adverse effects on the outcome of MW-CVD. It causes an uneven decomposition of the hydrocarbon and an inhomogeneous distribution of the reactants in the gas phase. At the worst this will lead to an uneven deposition of the diamond film on the substrate with both the thickness and the quality of the film potentially being affected.

Table 6.1 summarizes some essential parameters of common CVD apparatus. For the time being, however, a final judgment on which of the setups provides the best results cannot be given. Therefore the most suitable method of diamond film deposition must be selected for each individual case depending on the specific requirements (high rate, high quality, good reproducibility, etc.).

6.3.2
Growth Mechanism of Diamond Films

Considering now that diamond is the thermodynamically most stable modification of carbon only at extreme pressures and temperatures, the latent question arises of how it may be formed in the first place at pressures so low and at comparatively moderate temperatures as well. However, diamond is metastable at standard conditions, and only a high activation barrier prevents it from spontaneously converting into graphite. As a matter of fact, it is this very aspect being utilized in the preparation of diamond films: once diamond crystallites have grown to sufficient size, they are protected from conversion by this large energetic barrier. The formation of such crystallites consequently has to take place not under thermodynamic, but under kinetic control, as it is known for the generation of metastable modifications of other elements as well.

At room temperature, graphite is by $2.9\,\mathrm{kJ\,mol^{-1}}$ more stable than diamond, corresponding to a difference of $0.03\,\mathrm{eV}$ per carbon atom. The atoms' thermal energy is of a similar order in the respective range of temperatures, so the energetic difference between graphitic and diamantoid carbon is no insurmountable barrier here. One only has to see for the first precipitating atoms to deposit as sp^3-carbon. Therefore, the growth process of a diamond film may be divided into the phases of (i) nucleation and (ii) actual growth.

During the nucleation phase, the reactive species generated from the gaseous source of carbon deposits on the substrate. Whether or not sufficient nucleation occurs here depends on the carbon being soluble in the support material or easily forming carbides with the substrate. The first case in particular means an impediment to the formation of nuclei, but if neither of these characteristics is very pronounced, a system supersaturated with regard to carbon is fast to evolve on the substrate's surface. This leads to the deposition of first tiny carbon nuclei (usually CH_x). They are frequently sp^2-hybridized, which, without further influences, would give rise to a precipitation of graphitic material (Figure 6.18). At the same time, however, atomic hydrogen, generated *in situ* from the molecular hydrogen present in the mixture of reactant gases, settles on the surface. The hydrogen radicals are very reactive and quickly enter into conversions with the sp^2-clusters on the substrate. The latter are hydrogenated with new C–H-bonds being formed on their surface. As an effect, eventual sp^2-carbon atoms are transformed back into sp^3-carbon, and the resulting species C_yH_x are more stable than the respective sp^2-clusters. The bound hydrogen passivates the cluster surface that consequently cannot graphitize just like that. The clusters then grow laterally and vertically by the addition of further CH_x-fragments that each of them replace a hydrogen atom on the surface (Figure 6.18). From a critical size of the crystal nucleus on, the energy barrier for the sp^3-cluster transforming into a graphitic object becomes too high, and so a spontaneous conversion can no longer take place due to this kinetic inhibition.

In the subsequent growth phase, the film accumulates further CH_x-units to keep expanding from the nucleus. It can grow laterally as well as vertically until a closed

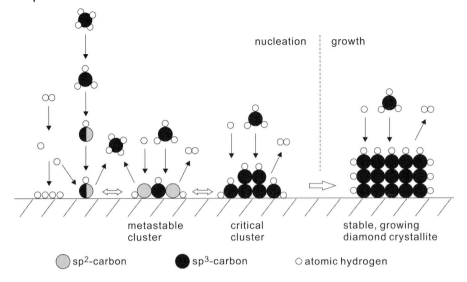

Figure 6.18 The mechanism of diamond deposition from the gas phase on a substrate.

film on the substrate surface is obtained. After having been adsorbed to the surface of the substrate, the activated CH_x-fragments can move by surface diffusion until finally being attached to the already existing clusters. Throughout this process the surface is protected from graphitization by the adsorbed hydrogen atoms. These maintain the sp^3-hybridization of surface atoms and prevent a surface reconstruction that would lead to π-bond formation if dangling bonds were present (Figure 6.18). If some graphitic material should emerge nevertheless, it is selectively attacked by the hydrogen radicals and converted into hydrocarbons that return into the gas phase. Here they may be decomposed once more to give reactive species for further reaction.

Suitable carbon sources include all gaseous and volatile hydrocarbons and their derivatives like, for example, alcohols and ethers. The substances used most frequently are methane, acetylene, and ethylene, yet methanol or ethanol can be employed as well. These are added at low concentrations (e.g., 5 : 95, in parts even less) to a current of hydrogen. Sometimes the reactant gases are additionally mixed with oxygen that exerts a similar function like the atomic hydrogen. Most of all it reacts preferably with graphitic material, establishing thus a low portion of sp^2-carbon in the film and high growth rates.

For the substrate, various metals and nonmetals are applicable. The most suitable of course is diamond itself that allows for homoepitaxial growth. The resulting diamond film crystallographically aligns to the support. In this way, highly ordered diamond films with defined orientation may be prepared (Figure 6.2a). Still diamond is rather expensive for a substrate (all the more if a specific surface is required), and it is strictly limited in size as well. On other materials, heteroepitaxial growth can be achieved, although severe complications are frequently met in doing so. This is due to the different lattice constants that cause a distortion at the phase boundary. Some elements, for example, copper or iridium, exhibit rather similar interatomic distances, so heteroepitaxial films are comparatively easy to prepare. For silicon, on the other and, the aforementioned distortional effect is observed. Usually, without meticulous control of experimental conditions, polycrystalline films consisting of many small, randomly oriented crystallites are obtained (Figure 6.2b).

Still, altogether, silicon wafers and silicon carbide suit very well to a use as substrate unless they are covered by amorphous silicon or SiO_2. These have inhibiting effects, so a deposition of diamond on such a support does not occur. Metals like tungsten or steel are suitable substrates as well. Depending on their interaction with carbon they can be divided in three groups: Metals belonging in the first group readily form carbides (Mo, W, Ti, etc.), while those of the second group do not give this reaction, but carbon shows significant solubility in them (Rh, Pt, Pb, etc.). The third group consists of elements neither forming carbides nor solving perceptible amounts of carbon (Au, Ag, Cu). By now one has succeeded in depositing diamond films of high to acceptable quality on all three types of substrates.

Generally the nucleation rate and the quality of the resulting film are observed to vary with the substrate as these are governed by the tendency toward carbide formation and by the solubility of carbon. Besides the nucleation rate, the adhesion of the diamond film to the support is a crucial parameter for the application of these structures. Owing to its extraordinary tribological properties, diamond adheres just poorly to many metals, for example, to copper. One way to overcome this problem is to deposit an intermediate layer like one of graphite, for instance, to achieve better adhesion. Substrates that are expensive or complicated to handle may themselves be condensed as thin films onto an inert support, only then to be covered with the diamond film. This setup is realized, for example, in the growth of diamond on iridium with the latter being bound as thin layer on a sapphire surface.

Nonmetals like α-Al_2O_3, $SrTiO_3$, MgO, or Si_3N_4 may be employed as substrates, too. Important parameters to be considered when selecting the material include its resistance to the temperatures (usually several hundreds of degrees) prevailing during deposition and to other effects like plasma, flames, or arc discharge. Hence, for the time being, it has not been possible to deposit diamond on common polymers, although this might open up a wide range of applications to diamond-coated materials.

However, not only the choice but also the pretreatment of the substrate play an essential role for the nucleation rate and the quality of the diamond films obtained. Significantly better results are observed upon mechanically pretreating the surface. A polishing with diamond powder or with another abrasive, for instance, causes many scratches that then constitute preferred nucleation centers. Where diamond grinding powder is employed, the remnant tiny diamond fragments are assumed to serve to the same purpose. This effect can also intentionally be produced by depositing suitable centers of crystallization on the substrate. These include,

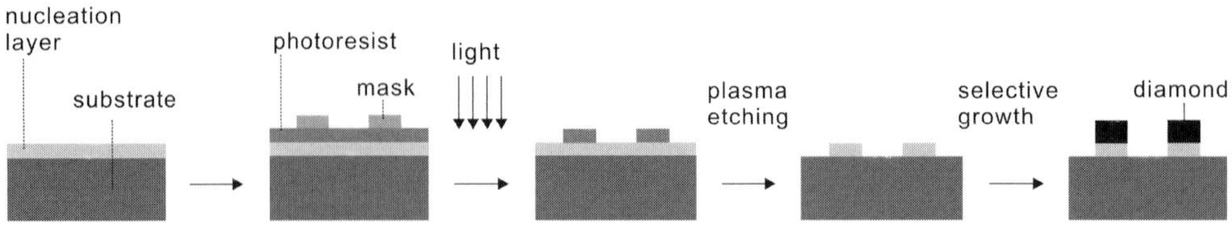

Figure 6.19 Preparation of structured diamond films by way of a masking procedure (© MRS 2001).

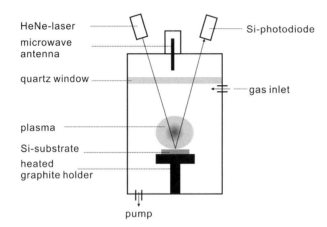

Figure 6.20 Preparation of UNCD-films by microwave-CVD.

among others, nanodiamond (Chapter 5), but Buckminster fullerenes (Chapter 2) have proven their worth for seeding as well.

By a patterned coating with an inhibiting layer, achieved for example by way of a mask, it is possible to directly deposit diamond only in selected areas on the substrate. Subsequent removal of the inhibitor provides structured diamond films (Figure 6.19). These are of considerable interest especially for electronic applications.

6.3.3
Preparation of UNCD

Yet another diamond film material drew much attention lately. Namely it is the so-called UNCD. It consists of extremely small crystallites measuring just 2–5 nm across. Their properties are largely determined by the grain boundaries.

Gruen and co-workers first succeeded in the respective preparation by modifying a typical microwave CVD method (Figure 6.20). In addition to the carbon source

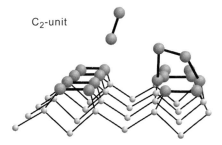

Figure 6.21 Mechanism for the deposition of nanocrystalline diamond films.

and a certain amount of hydrogen, the gas current is admixed with argon as an inert diluent gas. With an increasing content of argon in the mixture, the resultant crystallites turn out to be ever smaller – this effect is infinitely variable. Applicable carbon sources include methane, acetylene, anthracene, or C_{60}. Actually the latter gave the initial reason to start investigating into this method. It had turned out that Buckminster fullerenes do not vaporize as atoms, but in the shape of C_2-fragments, which led to the assumption that a completely new mechanism of vapor deposition might arise here. Indeed the carbon dimers upon reaction were observed to insert into the π-bonds of a reconstructed diamond surface (Figure 6.21). Hence it is obvious that growth can also take place without a saturation of the dangling bonds with hydrogen. In the case of methane, the reactive dimers are formed from intermediate acetylene. The dimer formation takes place in a plasma. Collisions with excited argon species contribute to the process, too.

Contrary to normal CVD, the low saturation of the diamond surface with hydrogen prevents the evaporation of small clusters in the shape of hydrocarbon-like structures, which in conventional diamond films leads to an accumulation of larger crystallites. In the case of UNCD, it would be more appropriate to consider a "survival of the smallest," as the particular mechanism favors the formation of small clusters. Some of the C_2-units insert just one of their carbon atoms into the superficial π-bonds. Hence, the other C-atom becomes available here to the nucleation of another crystallite. This, in spite of being infinitely small at first, does not evaporate owing to the lack of hydrogen, and a new grain boundary arises at this position. In comparison to normal CVD methods, the nucleation rate of $10^{10}\,\mathrm{cm}^{-2}$ is by several orders of magnitude larger. Given the constant supply of carbon with the gas current, this leads to the formation of a vast number of very small crystallites.

On the surface of the individual particles constituting the film, a rehybridization to sp^2 occurs as the remaining free bonding sites have to be saturated. The grain boundaries are two to four atomic layers thick. With about 10% of the carbon atoms in the film being situated in these boundaries, significant differences arise for the physical properties of UNCD as compared to conventional microcrystalline films. These are evident especially from the surface conductivity that solely originates from the existence of rehybridized grain boundaries.

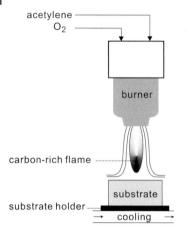

Figure 6.22 Production of diamond films by combustion of acetylene.

6.3.4
Further Methods of Diamond Film Production

The major part of diamond film production is actually covered by the CVD procedures presented in Section 6.3.1. Still there are a few other methods worth mentioning that suit to the generation of diamond films as well. These include, for instance, the flame combustion method. The respective apparatus essentially consists of a modified welding torch burning hydrocarbons at normal pressure (Figure 6.22). The carbon source most commonly applied here is acetylene, but the combustion of ethane, methane, ethylene, or methanol yields diamond films as well. The gas current is mixed with oxygen to support the combustion, but for being rich in hydrocarbon, the mixture does not burn up, and molecules of hydrocarbon still exist in the deposition zone.

The substrate is positioned in the flame cone (the so-called feather). The deposition of diamond precedes best here due to the high concentration of atomic hydrogen originating from the decomposition of hydrocarbon in other zones of the flame. With the latter being extremely hot (up to 3000 °C), the substrate inevitably has to be cooled to the desired temperature, usually ranging from 800 to 1200 °C. By way of the flame combustion method, the deposition of diamond is achieved at large growth rates. Moreover, the method is attractive for routine use due to its simple and flexible setup that does without expensive vacuum equipment. Still a certain portion of nitrogen is incorporated into these films. It originates from the surrounding atmosphere and enters into the flame by diffusion.

Another method was presented in the mid-1970s. Diamond films can be precipitated here by bombarding a substrate with carbon ions. These are extracted from an arc discharge taking place between carbon electrodes. *In vacuo* ($\sim 10^{-6}$ Torr/$\sim 1.9 \times 10^{-8}$ psi), the resulting positive ion beam is accelerated to 50–100 eV in an electric field between outlet and substrate. This energy does not suffice for an implantation of the ions into the substrate material, but it causes a high surface mobility of the carbon atoms. Their kinetic energy can be adjusted by varying the bias applied to the substrate. The resultant films are islands measuring up to 5 μm across consisting of particles 50–100 Å in diameter. The growth rate of 20 μm h^{-1} is quite satisfactory. Suitable substrates include silicon crystals, steel, or even glass.

6.4
Physical Properties of Diamond Films

The physical properties of diamond films largely correspond to those of the macroscopic material. The only significant differences to bulk diamond arise from surface defects and from a possible doping. The spectroscopic properties are employed to characterize the diamond films obtained, to evaluate their quality and, where applicable, to identify defects and impurities. In the following, the main attention will be directed just to those features differing from the bulk properties of diamond. Further aspects are also discussed in Section 5.4 on the physical properties of nanodiamond that shares some characteristics with the so-called ultrananocrystalline diamond in particular.

6.4.1
Spectroscopic Properties of Diamond Films

From examining the spectroscopic characteristics of diamond films, many insights into their structure may be obtained. Most of all the Raman spectroscopy, XRD and electron energy loss spectroscopy (EELS) provide valuable information. Other methods like IR-spectroscopy and XPS shed light on the surface structure. These techniques are supplemented by microscopy methods, for example, by AFM and STM, so altogether the morphology of the films' surface can be studied in quite some detail.

6.4.1.1 Infrared and Raman Spectroscopy

Infrared spectra of diamond films are usually recorded in reflection mode, while freestanding films can also be examined in transmission of course. The spectra obtained give indication of eventual surface groups as well as of foreign atoms in the diamond lattice that show characteristic vibrations. The functional groups situated on the surface that can be detected by IR-spectroscopy have already been discussed in Section 5.4.1.2. For diamond films, it is of particular interest to determine the situation on the surface (presence of C–H-bands or of carbonyl and ether absorptions, Figure 6.23a). Furthermore, the infrared spectrum of diamond films exhibits bands assignable to inherent diamond vibrations. The single phonon mode is not observed here as the corresponding vibration does not induce a dipole, thus rendering the signal symmetry forbidden. The two and three phonon bands, on the other hand, are seen at 1670–2500 cm^{-1} and at 3700 cm^{-1}, respectively. With impurities present in the diamond lattice, the crystal symmetry changes and the now permitted single phonon band appears at 1000–1400 cm^{-1}.

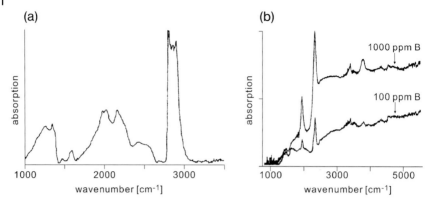

Figure 6.23 IR-spectra of diamond films without doping (a) (© Amer. Vac. Soc. 1992) and of such doped with boron (b) (© AIP 1995).

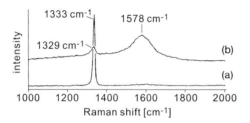

Figure 6.24 Raman spectra of diamond films containing little (a) or more sp²-carbon (b) (© MRS 1996).

Additional peaks in the spectrum arise from defect atoms. A doping with isolated nitrogen atoms, for instance, results in a band at about 1130 cm^{-1}, while for isolated boron atoms signals are observed at 900–1330 cm^{-1} and in the range about 2800 cm^{-1}. Especially the latter dopant, which is present in many films interesting for example, for electrochemical applications, gives rise to many characteristic bands in the IR-spectrum. This is indicative of different types of boron defects (Figure 6.23b).

It has already been illustrated in the previous chapters that Raman spectroscopy suits outstandingly well to characterizing different types of carbon materials. In particular, the fact of sp²- and sp³-carbon showing clearly distinguishable signals renders the Raman technique a powerful tool for determining the phase purity as well as other structural features both in bulk phase or in grain boundaries.

Depending on the type of carbon present, the bands are found at different wavenumbers. Diamond is characterized by its first-order band at 1332 cm^{-1} that is attributed to the vibrations of the interpenetrating cubic lattices. According to the quality and structure of the film under examination, further signals of sp²-hybridized carbon can be observed as well as peaks arising from unordered areas (Figure 6.24). These signals are localized at the same positions as given before in Sections

Table 6.2 Typical Raman signals of diamond films.

Wavenumber (cm^{-1})	Assignment
1140	Amorphous sp^3-carbon (T-band)
1315–1325	Hexagonal carbon (*Lonsdaleite*)
1332	Diamond signal
1355	D-band, microcrystalline graphite, unordered sp^2-carbon
1500	Amorphous sp^2-carbon
1580	Graphite (G-band)

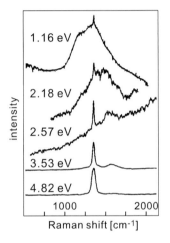

Figure 6.25 Dependence of Raman signal intensities from the excitation wavelength (1.16–4.82 eV, © Royal Soc. 2004).

3.4.5.1, 4.4.1.1 and 5.4.1.1. Table 6.2 provides an overview on the most important signals and their origin.

For the sp^2- and sp^3-portions, the relative signal intensities are observed to depend strongly on the wavelength of the excitation laser. Generally, the relative sensitivity of the diamond signal is much higher at short wavelengths than at long-wave excitation. This is due to the Raman scattering cross section of sp^2-carbon decreasing at short excitation wavelengths. Upon UV-excitation then, also the D-band of unordered material vanishes and the photoluminescent background of the spectrum markedly decreases. Therefore an excitation at, for example, 254 nm (4.88 eV) or 228 nm (5.44 eV) suits outstandingly well to prove the existence of diamond in a given film. Testing a diamond film for the absence of sp^2-hybridized carbon, on the other hand, requires an excitation in the infrared region (Figure 6.25).

The characterization of ultrananocrystalline diamond with its large portion of grain boundaries benefits as well from the application of Raman spectroscopy as the content of sp^2-material in a sample can be determined rather exactly this way. In addition to the aforementioned dependency on the excitation wavelength, the

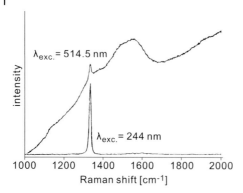

Figure 6.26 Raman spectrum of an ultrananocrystalline diamond film at different excitation wavelengths (© Elsevier 2000).

position and shape of Raman signals is also affected by the dimensions of individual crystallites. With decreasing particle size, for instance, the diamond signal shifts toward lower wavenumbers and its shape becomes broader. Upon excitation – outside the UV-region, however, the diamond band normally cannot be detected at all besides the signals associated to the grain boundaries, and so one has to rely on UV-Raman spectroscopy for the characterization of UNCD-films (Figure 6.26).

The signal position for diamond films is further dependent on possible strain within the film. These distortions result from interactions with the substrate. Defects and foreign atoms then give rise to additional Raman bands. The signal at 1344 cm^{-1} originating from an isolated nitrogen defect may be mentioned here as an example. For this defect, the band was found to result from a vibration of the surrounding carbon atoms, while the nitrogen atom itself remains in place. Boron doping also causes a change of the Raman spectrum. There are, among others, two additional signals at 500 and 1230 cm^{-1}. However, these become visible only at excitation wavelengths above the UV-region. Altogether the influence of doping on the Raman spectrum of a diamond film is hard to quantify as the observable changes rather depend on the actual number of charge carriers than on the absolute amount of foreign atoms (with the two figures being different in a majority of cases).

6.4.1.2 Optical Properties of Diamond Films

The optical properties of diamond films have been studied quite extensively, too. In the spectral range from about 220 to 1000 nm pure diamond does not show any absorption at all. This transparency renders it an attractive material for spectroscopic applications like as windows or lens systems in spectrometers.

The material's bandgap absorption only sets in at less than 220 nm. Yet the existence of defects such as foreign atoms or sp^2-carbon (e.g., on a reconstructed surface or at grain boundaries) generates further states within the bandgap so radiation of larger wavelength may be absorbed as well (Figure 6.27). With the related absorption occurring in the visible range of the spectrum, a coloration

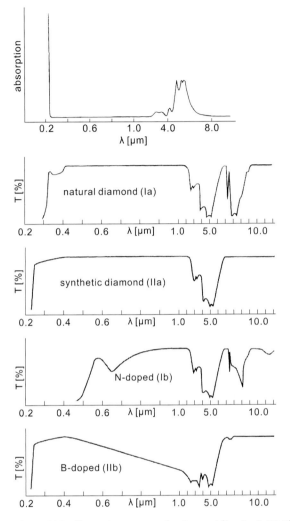

Figure 6.27 Absorption spectra of a diamond film (top) (© Elsevier 1993) and optical properties of various diamond materials (bottom) (© Wiley Interscience 1994).

characteristic of the specific defect arises, like yellowish-green or blueish for diamonds doped with nitrogen or boron, respectively. Besides the absorption of defects like π-bonds, the grain boundaries also act as scattering centers for the light. The high refractive index of diamond plays an additional role, for example to the effect of films with sp^2-components and small crystals featuring a dark gray color.

The luminescence characteristics of diamond films are also strongly affected by the presence of defects. Although it requires considerable energies, an excitation across the entire bandgap is possible. It may be effected by (laser) light, by an

6 Diamond Films

Table 6.3 Luminescence of lattice defects in diamond.

Energy (eV)	Wavelength (nm)	Name, assignment
4.582	270	Interstitial carbon
2.985	415	N_3-signal, N_3-V (V = vacancy)
2.85	435	Blue band A (broad band), dislocations
2.2	563	Green band A (broad band), boron(?)
2.156	575	Single N-V
1.682	737	Silicon
1.673	741	GR1, neutral vacancies V

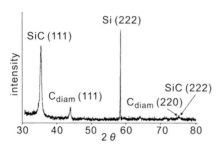

Figure 6.28 XRD-spectrum of a diamond film on silicon substrate (© AIP 1992).

electric field, or by the action of an electron beam. The phenomena observed are termed photoluminescence, electroluminescence, and cathodoluminescence, respectively. The spectra obtained from these methods strongly resemble each other. The sharp peaks observed to appear next to the broad bands in the luminescence spectra originate from foreign atoms incorporated into the lattice (also refer to Section 5.4.1.4). Table 6.3 collects some of the more frequent defects in the diamond lattice and the corresponding wavelengths of luminescence.

Furthermore, there are broad bands that can each be assigned to a range of transitions. Particularly worth mentioning among them are the so-called blue band A at ca. 435 nm (~2.85 eV) that arises from dislocations, and the "green band A" at about 563 nm (~2.2 eV) presumably originating from boron doping.

6.4.1.3 XRD, XPS, and EELS of Diamond Films

Examining diamond films by X-ray diffraction (XRD) provides important information on the structure of the material at hand. The diamond lattice can unequivocally be detected by way of its characteristic reflexes (Figure 6.28). Furthermore, it is possible to give a statement on the film's purity based on the absence or presence of signal related to sp^2-carbon. The average particle size of individual crystallites can be inferred from the width of the diamond signals, which is relevant especially for nanocrystalline diamond films.

However, the quality control of diamond films by means of X-ray diffraction has largely been replaced by Raman spectroscopy. It allows for a much more sensitive

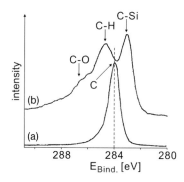

Figure 6.29 XPS-spectra of a diamond film: (a) on the film's surface, (b) at the phase boundary with the substrate (© AIP 1994).

structure determination of diamond materials and their possible impurities, providing at the same time a better spatial resolution.

The X-ray photoelectron spectroscopy (XPS) constitutes another method that suits well to testing the purity of diamond films. Here it is in particular the material's surface being examined as only photoelectrons from the uppermost layers of atoms can emit from the sample. With each element showing characteristic signals here, even the doping with foreign atoms may be determined. Hence XPS is a suitable means of surface analysis for diamond films. Especially the superficial structures like reconstructions or functional groups may be identified by means of their characteristic XPS-signals. The C(1s)-signal seen in Figure 6.29, for example, can be deconvoluted into various components assignable to carbon atoms in different bonding situations. The main constituent is a signal at 284.5 eV that belongs to carbon atoms bound to hydrogen or other carbon atoms. Furthermore, there are C–O- and C=O-bonds present on diamond films resulting in signals at 285.8 and 286.9 eV. The existence of double bonds as present, for example, in surface dimers is documented by a π–π*-signal at 289.2 eV. Apart from the signals corresponding to the C(1s), those arising from eventual other elements present in the film are of interest too, of course. From the O(1s)-spectrum, for instance, the existence of the C–O-bond as already detected in the C(1s)-spectrum can be confirmed by a signal at 532.8 eV. For diamond films functionalized with substituents that carry characteristic elements, successful bonding can be demonstrated by XPS as well. An F(1s)-spectrum obtained from a perfluoroalkylated diamond film, for example, revealed the expected signals of fluoroalkyl groups (Figure 6.38).

Furthermore, XPS suits in determining the bonding situation between the substrate and the first layer of the diamond film, provided the diamond growth is interrupted at an early stage of nucleation. It could be shown this way that a layer of carbide is formed between the diamond film and many substrates like, for example, silicon (Figure 6.29). For metallic substrates, a thin layer of graphite was found upon which the diamond starts growing then. In this early stage of growth also hydrocarbons have been observed on the surface, which is indicative of the mechanism of nucleation (Section 6.3.2). The influence of oxygen on

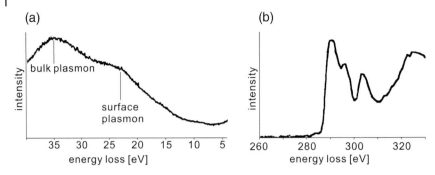

Figure 6.30 (a) Low-loss spectrum (© Elsevier 2005) and (b) core-loss spectrum of a diamond film (© Nature Publ. Group 1993).

diamond nucleation could also be examined by XPS. It turned out, for instance, that the presence of oxygen suppresses the formation of a graphite film on metal supports and that at the same time the nucleation rate of diamond decreases significantly.

The EELS can serve in particular in determining the bonding situation within a carbon material. As shown before in the chapters on nanotubes, carbon onions, and nanodiamond, the signals of sp^2- and sp^3-hybridized carbon each assume typical positions. When examining diamond materials, both regions of the EEL-spectrum provide valuable information. For the low-loss region from 0 to 50 eV, the signals especially worth mentioning are the one corresponding to the σ-plasmon of the bulk at 30–34 eV and that of the surface plasmon at 23 eV (Figure 6.30). A π-plasmon signal appearing at 6–7 eV indicates the presence of double bonds. It is not observed for diamond films with saturated surface.

In the spectrum's core-loss region the presence of graphitic material in a sample can be inferred from the position of the K-edge. The signal corresponding to the transition from a 1s-orbital into an unoccupied π*-2p-orbital at 285 eV is only observed for graphitic carbon. The signal arising from the C 1s→σ*-2p-transition of sp^3-hybridized carbon sets in only at 290 eV. The existence of sp^2-hybridized domains, situated, for example, on the phase boundary between substrate and diamond film, can even be examined at spatial resolution by means of the STEM-technique. It provides an insight into the ways of bonding in different regions of the diamond film. Defects like dislocations influence the signal positions and consequently can be detected in the loss spectrum, too. The effect is strongest here if the electron beam is oriented in perpendicular to the line direction of the defect.

6.4.2
Electronic Properties of Diamond Films

A strong incentive for the fast development of technologies for the production of high quality diamond films arose from their prognosticated outstanding electronic properties. Diamond is considered by many to be *the* material of the future for

6.4.2.1 Electric Conductivity of Diamond Films

Diamond as a pure material is a very good insulator. Owing to its large bandgap of 5.46 eV, it is virtually impossible to generate charge carriers by thermal excitation. Even at 700 K, the bandgap is still 5.34 eV wide. The intrinsic density of charge carriers of $10^{-27}\,\text{cm}^{-3}$ is so small that just a single electron–hole pair would exist in a diamond the size of the earth! The resistance of diamond as determined experimentally is about $10^{16}\,\Omega$. Hence, in the case of a doped diamond material, intrinsic conductivity will not affect the electronic characteristics of respective devices.

The value of $10^6\,\Omega\,\text{cm}$ found when first measuring the specific resistance of diamond films was much less than what had been expected. It could be increased to 10^{13}–$10^{14}\,\Omega\,\text{cm}$ by tempering, but exposing such a treated diamond film to a hydrogen plasma brought the value back to $10^6\,\Omega\,\text{cm}$ again. It quickly became obvious, thus, that certain structures on the surface of the film were responsible for its conductivity being markedly higher than that of bulk diamond.

The so-called surface conductivity was studied extensively. For films saturated with hydrogen it is about 10^{-4}–$10^{-5}\,\Omega^{-1}$, arising from a concentration of charge carriers of ca. $10^{13}\,\text{cm}^{-2}$ on the surface of the film. The charge carriers were identified to be not electrons, but holes accumulated on the surface. What is the reason now for this depletion of electrons in the regions close beneath the surface? The saturation with hydrogen is essential to the conductivity found, but actually the high values observed only turn up once the film has been in contact with air. Upon this contact, a layer of adsorbate largely consisting of water is formed. This now has been postulated to play a role in the accumulation of holes in the surface of the film by way of reaction (Eq. 6.1). It occurs on the film and generates elemental hydrogen by consuming electrons from the diamond material, thus accounting for the aforementioned electron depletion.

$$2H_3O^+ + 2e^- \rightarrow H_2 + 2H_2O \tag{6.1}$$

The driving force behind this reaction is the difference between the *Fermi* level E_F in diamond and the chemical potential μ_e of electrons in the aqueous adsorbate layer. As long as $\mu_e < E_F$, electrons exit from the diamond surface in the course of reaction (Eq. 6.1) taking place. Once $\mu_e = E_F$, equilibrium is achieved. Thus a space charge is generated in the uppermost layer of the diamond film, which leads to the so-called surface band bending (Figure 6.31).

However, this is only true for hydrogenated surfaces. With the diamond featuring a different surface structure, other effects come to the fore. Reconstructed surfaces, for instance, bear π-bonds that may be arranged in various ways on the main crystallographic faces (Section 6.2.2). The surface dimers present on the (100)-plane, for example, cause the respective orbitals to split into a π- and a π*-

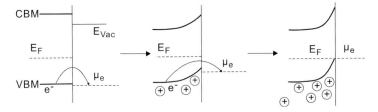

Figure 6.31 The surface conductivity of diamond films arising from surface band bending (© APS 2000).

orbital. However, the bandgap is still 1.3 eV wide here, and with the π-orbitals being situated inside the valence band, a charge transfer with the diamond in the core of the film does not occur. Consequently, no surface band bending is observed here. For the (111)- and the (110)-faces, the situation is different. In this case there is a chain-like arrangement of π-bonds, and metallic behavior is expected to show along these chains. For the time being, however, the experimental results are contradictory in parts, and a conclusive statement cannot be made.

Nanocrystalline diamond films also frequently show remarkable conductivity that is primarily attributable to the large portion of atoms situated in grain boundaries. The sp^2-hybridized carbon atoms generate electronic states within the diamond's bandgap. These lead, for instance, to the so-called hopping conductivity, that is, a conductance arising from a successive "hopping" from one state to the next.

Doping a diamond film with foreign atoms causes drastic changes of its electronic properties. By substituting a part of the carbon atoms by other elements, the insulator or semiconductor with wide bandgap may turn into a good semiconductor or may even attain conductivity comparable to metals. This feature has been discovered quite early in the history of diamond films, and numerous experiments on doping have been performed ever since. The production of doped films is effected either by codeposition or by ion implantation (in the latter case with limitations arising from possible damage done to the diamond film). Diffusion doping as frequently applied to other materials like silicon cannot be employed here due to the low diffusivity of most elements in diamond. Upon successful doping, the additional electronic states introduced into the bandgap of diamond enable the material's conversion into a semiconductor with reduced bandgap. Consequently, the charge carriers can now be excited at room temperature already.

First experiments on p-doping with aluminum did not yield very encouraging results, but then boron was found to suit outstandingly well to the p-doping of diamond. Boron is an acceptor introducing a level situated 0.37 eV above the valence band maximum. Depending on the concentration of the dopant, the resulting films are found to behave like a classical semiconductor or, from a certain concentration on, to show metal-like characteristics. Hence diamond films doped with boron may be employed as material for electrodes used in electrochemical examinations (Section 6.6).

However, the assumption of nitrogen in an analogous way being the best element for n-doping could not be confirmed. It is true that nitrogen is a donor inserting a deep level at 1.7 eV below the conduction band minimum, but for the time being, its incorporation into the lattice and the resulting electronic characteristics do not allow for an effective application. Just recently then, one has succeeded in the incorporation of phosphorous into the diamond lattice and in studying its influence on the electronic properties. Diamond films doped this way actually represent n-semiconductors. The phosphorous inserts a level at 0.5 eV below the conduction band minimum. Upon a doping of up to $10^{19}\,\text{cm}^{-3}$, the charge carriers are thermally activated like in a classic semiconductor. Above this value, and up to the maximal doping of about $5 \times 10^{19}\,\text{cm}^{-3}$ then, hopping conductivity is observed. By now, one has succeeded in combining diamond doped with boron and with phosphorous, respectively, into p–n-elements and in characterizing their properties.

6.4.2.2 Field Emission from Diamond Films

The field emission characteristics of diamond films are of considerable interest, too (Figure 6.32), for example, for the production of displays. It is a whole variety of its properties rendering diamond an ideal material for this purpose. Apart from the low electron affinity, it is especially its great hardness, its favorable chemical behavior, the high breakdown field strength, the good heat conductivity, and last, but not least, the large bandgap, which altogether enable working at high output and high temperatures.

Due to the wide bandgap, the conduction band of diamond approximates the vacuum level. Consequently, electrons excited into the conduction band may leave the diamond's surface as there is no significant potential difference between conduction band and vacuum level. In hydrogenated diamond films, the conduction band is even observed to exceed the vacuum level, resulting in a negative electron affinity of the respective film. This causes the emission of excited electrons from the film to occur all the easier.

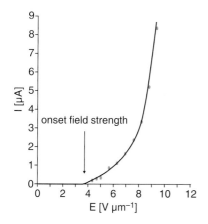

Figure 6.32 Field emission from a diamond film. Plotted here is the strength of the emission current vs. the intensity of the field applied.

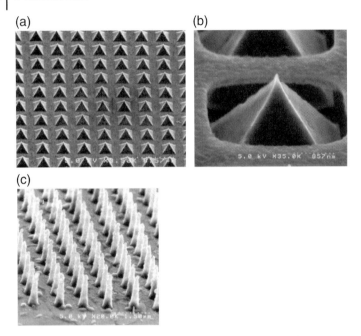

Figure 6.33 Diamond emitter tips; (a) array, (b) with gate, (c) diamond nanotips (© Elsevier 2005).

For materials not subject to a special pretreatment, however, the emission properties are found rather unfavorable. The drawbacks include an inhomogeneous emitter structure, a random distribution of emitting sites on the surface of the film, and a more or less inconsistent emission behavior. This reflects in the measured field strength required for an onset of emission partly being up to $10\,V\,\mu m^{-1}$. By enhancing the structure of the film's surface, this value can be brought down to $5\,V\,\mu m^{-1}$ (Figure 6.33), and doping the diamond film may lower the onset field strength to as little as $2.3\,V\,\mu m^{-1}$. Naturally, n-doping is more efficient in doing so than p-doping. Furthermore, there is an influence by the specific substrate chosen as through this the subsequent supply of electrons must come to pass.

6.4.3
Mechanical Properties of Diamond Films

Diamond has been well-known for long for its outstanding mechanical properties like great hardness and low frictional coefficient. Therefore, it was self-suggesting to examine the respective characteristics of diamond films as well. Actually the results are found to vary depending on the morphology of the film and, for polycrystalline materials, on the particle size.

Like in any material, the mechanical properties of diamond films have chemical reasons. The basic effect underlying friction is the making and breaking of

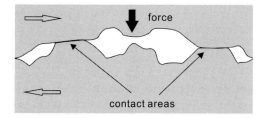

Figure 6.34 Friction between two rough surfaces: The forces act at the contact sites.

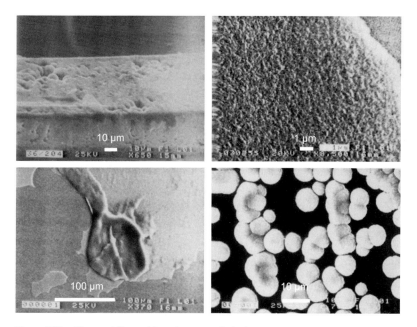

Figure 6.35 Diamond films with various morphologies.

bonds between touching surfaces, whereas the cleavage and re-formation of bonds on the individual surfaces is the reason for wear. To either process, the so-called dangling bonds on the film's surface are of particular relevance. These unsaturated bonding sites may be generated by bond breaking, or they are uncovered by the desorption of adsorbates. Their strong tendency toward saturation has to be considered the reason for interphase bond formation.

The actual abrasive properties of a given material are largely governed by its surface roughness as force transmission only occurs at the actual contacting sites (Figure 6.34). There is now a wide variation of coarseness and texture among diamond films obtained from different methods of preparation (Figure 6.35). The tips of individual crystallites protruding from the surface act as preferred contact sites. These may be very small depending on the size and shape of particles constituting a polycrystalline film, and so the load resulting at these positions may be extreme in parts. The entire force resting on the tips alone may result in plastic deformation or fracture. Consequently, the crystallites' edges and tips are worn by impact and shearing. There are several factors influencing friction that depend on the types of film and counterpart chosen. Important contributions include a so-called plowing of the crystallites' hard tips in the opposite friction surface as well as a rising of pointed structures from the two faces moving against each other (an effect particularly pronounced in diamond). Furthermore, there are shearing forces acting on microcontacts and plastic or elastic deformations of the crystallites' tips. The plowing plays a relevant role only if the friction takes place on a much softer counter face. After a certain run-in period under mechanical load, a tribological equilibrium evolves with the mechanical parameters and wear remaining constant then.

Measuring the tribological properties of diamond films turns out to be rather complicated for several reasons. For running such an experiment between two diamond objects, the tip or ball of the tribometer has to be coated with diamond too, which constitutes a technological challenge. Using softer tips instead may in some cases lead to a material transfer from the tip to the diamond surface under test, resulting in a modified surface with altered mechanical properties. The surface coarseness is also complicated to measure and hard to adjust. Finally, problems in the diamond film's adhesion to the substrate constitute another parameter to be controlled.

Depending on their quality and environmental conditions there is a wide variation of the frictional coefficient determined on diamond films. A value of 0.1 is obtained in air, while in a vacuum it increases to 0.9. This drastic change is due to the formation of dangling bonds which in its turn results from the adsorbate layer being removed *in vacuo*. This desorption causes both friction and wear to increase. A marked decrease in friction is observed, on the other hand, upon coating the surface with a suitable agent like, for instance, saturated hydrocarbons. A similar effect is achieved by the partial or complete graphitization or fluorination, respectively, of the film's surface. Ultimately, all these measures serve to passivate the unsaturated bonding sites, so no interphase bonds can be formed anymore.

The temperature shows an influence on the tribological characteristics too. Heating *in vacuo* promotes the desorption of the passivating surface layer, which, according to the above, increases friction. If exposed to air, reactions may occur with elements or compounds present in this environment. Oxygen and water in particular as well as hydrogen readily react with the emerging unsaturated bonding sites. In the presence of oxygen, there is even peril of the material burning up, which initially leads to a marked increase in abrasion.

The reconstruction of the surface plays an important role especially for monocrystalline diamond films, but polycrystalline materials are affected too. The π-bonds already existing on such diamond faces largely facilitate a graphitization, which manifests in a strong reduction of the friction coefficient at very high temperatures (when graphitization assumes a significant extent).

The mechanical properties of UNCD (Section 6.3.3) are dominated by a multitude of small, pointed crystallites protruding from the surface. Altogether, the latter is much more homogeneous than in microcrystalline diamond films, leading to a decrease in friction at otherwise comparable surface structure. Furthermore, the portion of sp^2-carbon situated in the grain boundaries is significantly higher for UNCD than for normal bulk diamond. This too alters the frictional properties as the graphitic material takes up part of the load.

The attrition of a diamond film is promoted by a high portion of sp^2-material. The grain boundaries in UNCD represent predetermined breaking points and facilitate fracture of the film. The comparatively low elasticity of diamond contributes to its friability as well. Furthermore, sp^2-carbon atoms are more easily oxidized at contact with air, leading to material losses of the film which thus might become porous upon strong wear.

Interesting results were found when measuring the Young modulus. Nanocrystalline films prepared with many nucleation centers feature a Young modulus of ca. 1000 GPa, which is rather close to the value observed for free-standing, polycrystalline diamond or microcrystalline diamond films. Films prepared with less nuclei, on the other hand, clearly show the resulting surface morphology to exert negative influence on the mechanical properties. The Young modulus is only 500 GPa here.

The characteristics of monocrystalline diamond films are much more clearly defined. Still polycrystalline films are employed in most cases as the high price interferes with large scale application of the monocrystalline material. Even for thin layers there is no significant change to the essential characteristics of diamond. For this reason as well as to save further material, it is a common practice to employ coated substrates with a film the thickness of micrometers spread on their surface (Section 6.6.1). The endurance of such films against mechanical stress is essentially influenced by two factors: Firstly, by delamination (peeling off) of the film from the substrate, and secondly by normal, gradual wear.

The problem of poor adhesion to the substrate may be mitigated by the choice of an appropriate material. Especially metals with the ability to form carbides are suitable for this purpose. However, carbide formation alone is not sufficient to obtain films reliably adhering to the respective substrate. Films deposited on molybdenum, for instance, tend toward spontaneous delamination upon stress. Hence the film should be sought to bear the least possible stress to achieve good adhesion. The so-called stress is a quantity hard to assess for being influenced by a multitude of parameters like the thickness of the film, problems related to the grain boundaries, lattice mismatch, and thermal strain.

Silicon is one of the most suitable substrates. It is able to form a carbide phase, thus allowing for a stable attachment. Silicon substrates in general exhibit just little stress, so a delamination of the deposited film is not promoted. Usually, an even and well-adhering film with favorable morphology is desired. There are now numerous ways to achieve this. They include a scratching of the surface with a fine-grained, hard material (like diamond) to generate a large number of nucleation sites, or the seeding with crystallization centers, for example, nanodiamond. Still these have to be sufficiently small to achieve a homogeneous film.

6.4.4
Thermal Properties of Diamond Films

It has been known for long that diamond is the material with the largest thermal conductivity (although CNTs in axial direction show even higher values, Section 3.4.6.2). Even good heat conductors such as silver or copper come off badly when compared to diamond. The thermal conductivity of the latter passes through a maximum of 175 W cm^{-1} K^{-1} at 65 K (about 1/30 of the Debye temperature). In the range of temperatures about 300 K as relevant for practical applications, the value is still 15–30 W cm^{-1} K^{-1}.

The conduction of heat in metals and nonmetals takes place by different mechanisms. In metallic conductors, the freely moving electrons effect the transport of heat, which leads to metals with high electric conductivity (e.g., silver or copper) being good heat conductors as well. In semiconductors, on the other hand, no free electrons are available to transport the heat. In this case, phonons take over the transmission of energy. Depending on the intensity of interaction of the phonons with the environment (e.g., phonon scatter on lattice defects, phase boundaries, or impurities) the measured thermal conductivities are usually below those found for metals. Diamond with its rigid, tightly bound lattice, on the other hand, is a very apt material for the transport of heat by way of phonons, which leads to an extremely high thermal conductivity.

Actually, however, first experiments on (low quality) diamond films produced disappointing results. The thermal conductivities found were in the range of 10 W cm^{-1} K^{-1}, that is, about half of the expected value. This is due to the presence of numerous defects in the films. Most of all a contamination with foreign atoms plays a role here (for bulk diamond, the conductivity of type Ia containing nitrogen as well is observed to be less than that of type IIa). Furthermore, there are defects (e.g., dislocations) and the grain boundaries. All of these cause a scatter of phonons and hence interfere with the transport of heat. This is confirmed by experiments that show thermal conductivity to lower upon decreasing crystallite dimensions, lower thickness of films, and increasing portions of sp^2-carbon. High-quality diamond films with a lower density of defects, on the other hand, may feature values of distinctly more than 20 W cm^{-1} K^{-1}.

6.5
Chemical Properties of Diamond Films

6.5.1
Considerations on the Reactivity of Diamond Films

The chemical properties of diamond films are determined by the kind of surface they present. The bulk phase, as far as it might be considered such for a thin film, has no significant influence as it does not participate in the reaction. The latter rather takes place on the surface of the film or of the individual crystallites. Only in some fluorinations the fluorine atoms are observed to infiltrate deeper into the material. Contrary to nanodiamond (Chapter 5), the portion of surface atoms is markedly smaller in a diamond film. Therefore, relative to the mass of diamond employed, even a complete covering of this surface with functional groups inevitably yields a lower concentration of these units. On the other hand, the functionalization of diamond films coated onto substrates is normally intended to modify their surface characteristics, for example, for a use in electronic or tribological applications or as sensing devices. Hence no disadvantages arise here – provided the covering as such is sufficiently dense. The crystallographic orientation of the surface is relevant, too: Depending on which face is functionalized, there is a variation of the distance between surface atoms and, consequently, of the highest achievable covering of this face.

As mentioned before in Section 6.2.2, a so-called reconstruction occurs on a noncovered diamond surface and results in the formation of π-bonds. Depending on the crystallographic orientation of the surface, these are arranged in different ways and consequently show different reactivity in chemical conversions. Generally the π-bonds existing on a diamond surface are distinct from typical alkene-type double bonds for a lower bonding strength (40–80 kJ mol^{-1} as compared to 250 kJ mol^{-1}) and for a bending of the other bonds from the plane of the double bond. Hence, in comparison to a typical alkene, the π-bond is much more reactive.

For the time being, most studies have been conducted on the (100)-face or on nanocrystalline diamond films presenting a mixture of different faces. Owing to the saturation of dangling bonds by reconstruction processes, a nonfunctionalized diamond surface enters into reactions typical of double bonds. These include, among others, the addition of hydrogen and halogens, cycloadditions, the reaction with ammonia, or oxidations. A multitude of other possible reactions is known and will be discussed further in more detail. Both covalent and noncovalent modifications of the diamond surface have been reported. A polymer coating has been achieved as well as the attachment of biologically active moieties, and the influence of the modifications on the band structure of the diamond film has been studied.

Doping the diamond film also affects its surface reactivity as the foreign atoms alter the electron affinity, thus either hindering or facilitating an attack by nucleophiles or electrophiles, respectively. Especially redox reactions like in electrochemistry (Section 6.5.4) are influenced by a doping with donors or acceptors. Generally, however, the reactions mentioned in the following sections should be considered taking place on untreated as well as on n- or p-doped diamond films, although certain differences may exist for the reactivity of doped films. A more detailed description of the doping effects is given in Section 6.4.2.

Reactions on a diamond film inevitably take a heterogeneous course as a diamond film obviously cannot be brought into solution. Therefore, the reactions are usually conducted in a gas current or by immersion into a reagent solution. This often results in considerable reaction times, but then, however, the fact of all reaction centers being situated on the nonporous surface of the film allows for an easy access to all of them, and so complete conversions may be achieved.

6.5.2
Covalent Functionalization of Diamond Films

Depending on the method of preparation, the surface of a diamond film is either functionalized already, for example, covered by hydrogen atoms, or it exhibits an array of so-called surface dimers (π-bonds arising from reconstruction). The latter case is mostly found for samples that have been subject to a secondary thermal treatment to remove their initial, usually inhomogeneous surface functionalization.

Before a diamond surface is actually reacted with larger organic moieties, there is often a basic functionalization being done. Suitable reactions include the hydrogenation, halogenation, amination, or oxidation. Normally these are conducted in a current of the respective gas with the additional application of either high temperatures, plasma, or high energy irradiation of light to achieve satisfactory degrees of conversion. The attachment of heteroatoms or -groups is effected by addition to the existing π-bonds, thus turning sp^2-carbon atoms into sp^3-centers. In the following sections, some reactions leading to a covalent functionalization of diamond surfaces will be discussed in more detail. In doing so, no distinction will be made between microcrystalline and nanocrystalline diamond films as in principle the reactions take a rather similar course.

6.5.2.1 Hydrogenation of Diamond Films

The hydrogenation of diamond films takes place by the addition of hydrogen to the double bonds on a reconstructed surface. To this end, gaseous hydrogen is employed in a flow reactor or directly in a CVD apparatus (Section 6.3.1). The process is conducted either at high temperatures (>850 °C) or in hydrogen plasma at elevated temperatures. The resulting hydrogenated diamond films feature a decidedly hydrophobic surface that suits well to further functionalization, for example, by reaction with vinylic species or by halogenations and oxidations. A sufficiently long reaction time provided, the surface is completely covered with hydrogen. Hence, in this manner a homogeneous initial functionalization of the diamond film is achieved. This is of particular importance for the production of electrodes and sensors (also refer to Sections 6.5.4 and 6.6.3). Besides the functionalization there is also another interesting aspect to the hydrogenation of diamond films: In chemical vapor deposition of diamond, it conserves the sp^3-hybridization of surface atoms. These would otherwise be passivated by the formation of sp^2-centers, thus interfering with or even preventing the growth process (Section 6.3).

6.5.2.2 Halogenation of Diamond Films

Halogenating the surface of a diamond film is of interest for several reasons. Firstly, diamond films covered with, for example, fluorine are expected to show attractive properties like a decreased coefficient of friction, and secondly the carbon–halogen bond is a suitable point of attack for further functionalization.

The fluorination of diamond surfaces has been reported most frequently. It can be effected by a variety of fluorinating agents and conditions like elemental fluorine at elevated temperatures, atomic fluorine in a plasma or xenon difluoride, and gaseous CF_4 being reacted with the diamond film in a microwave plasma reactor. These methods suit best to the modification of unfunctionalized surfaces. On diamond films already hydrogenated, on the other hand, the reaction often remains incomplete so no homogeneous functionalization is achieved here. In these cases a suitable alternative was found in the use of perfluorinated alkyl iodides. After condensing a film of these substances onto the diamond at low temperatures (~120 K), they may be decomposed by irradiation of energy-rich light or X-rays. The resulting perfluoroalkyl radicals then react with the diamond surface. In a first step, a diamond surface covered with covalently bound perfluoroalkyl groups is formed. By subsequent heating, these surface groups are removed under direct covering of the diamond with fluorine atoms. The conversion can easily be monitored by, for instance, XPS. The F(1s)-signal of CF_3- and CF_2-groups appearing at first vanish upon heating, while at the same time there is a significant increase in the F(1s)-signal at 686 eV corresponding to fluorine atoms bound directly to the diamond surface. The most suitable reagent to be employed for this method is perfluorobutyl iodide, but trifluoromethyl iodide can be used as well. The ratio of fluorine to superficial carbon atoms in the fluorinated final product amounts to 0.6 or ca. 0.2 upon the use of C_4F_9I or CF_3I, respectively. With other methods, values of up to 0.75 can be achieved for previously unfunctionalized films. Thus, in no case a complete monolayer is achieved.

The properties of fluorinated diamond films differ from those of unfunctionalized or hydrogenated films. The hydrophobic character, for instance, increases, while a significant decrease in the frictional coefficient is observed. It turns out further that the usable potential window of fluorinated diamond electrodes is extended by up to 2 V into the negative range, so an overall range of more than 5 V is available here. In comparison to conventional materials used for electrodes, this is an exceptionally large value (Section 6.5.4). On first sight, the morphology of fluorinated diamond films does not differ much from that of other diamond films. However, a direct comparison of surfaces after fluorination reveals that up to a certain degree an etching of crystallites must have occurred as the shapes observed are markedly rounded. Hence, some material losses must be considered to take place during fluorination.

The chlorination of diamond films succeeds by the photochemical reaction of gaseous chlorine under irradiation with a 400-W-mercury vapor lamp for 36 h. Still the XPS-spectrum of chlorinated diamond films always shows the signal of oxygen besides the one corresponding to chlorine. This is assumed to result from a secondary reaction of the chlorinated diamond surface with adsorbed water, leading to the formation of hydroxyl groups. A bromination or iodination of diamond surfaces appears to be complicated for the low reactivity of these elements. Actually, there are no reports known to date reporting on diamond films with this kind of functionalization.

The amination of a diamond surface succeeds by reacting it with gaseous ammonia under irradiation with high-energy light. In a radical reaction, NH_2-groups are generated on the diamond. These may very well be employed for further functionalizations. Chlorinated diamond films may as well be aminated by reaction with gaseous ammonia.

6.5.2.3 Oxidation of Diamond Surfaces

Among other reasons, the oxidation of diamond surfaces is of considerable interest because there are significant differences when compared to the functionalization of silicon. In most cases, a method employed for the modification of silicon surfaces may as well be applied to diamond films. Concerning the oxidation, however, the results differ to quite some extent. On the one hand this is due to diamond not forming a bulk oxide, or in other words, not covering itself with a layer of oxide like silicon does. Furthermore, carbon has a much higher tendency toward forming carbonyl structures, that is, carbon–oxygen double bonds. This ability is just weakly pronounced in silicon.

A variety of reagents may be employed to oxidize the diamond surface. Again a distinction has to be made between noncovered, reconstructed surfaces on the one hand and those already prefunctionalized (e.g., hydrogenated) on the other. Noncovered diamond films are even oxidized by water upon heating in high vacuum ($\sim 10^{-7}$ Torr/$\sim 1.9 \times 10^{-9}$ psi). The initial products are hydroxyl groups that transform into ether and carbonyl structures in the further course of reaction. This process takes place even with the traces of water usually present in a CVD apparatus, so diamond films after having been prepared by vapor deposition always carry on their surface not only hydrogen, but also a certain portion of functional groups containing oxygen.

The reactions with ozone or molecular oxygen at elevated temperatures ($\sim 500\,°C$) both yield oxidized diamond films as well, yet the products differ in their surface conductivity. It is generally reduced by the sp^2-centers reacting with the oxidant. Still it is conserved to some extent in the reaction of polycrystalline diamond films with ozone, whereas with molecular oxygen it is completely lost. The partial preservation of conductivity upon ozonization can be explained by a charge transfer from the O_3/OH^--redox couple into the diamond film. The IR-spectra of diamond samples treated with oxygen or ozone show a characteristic signal of carbonyl functions at $1731\,cm^{-1}$ that coincides very well with the respective signal of a C=O-stretch vibration in adamantanone ($1732\,cm^{-1}$, Figure 6.36). Hence an analogous bonding of the carbonyl function to the diamond surface may be assumed.

Concentrated oxidizing acids (sulfuric, perchloric or nitric acid, or mixtures thereof) are also suitable to oxidize noncovered as well as prefunctionalized diamond materials. Both hydroxylated and carboxylated surfaces can be generated this way. However, characterizing exactly the kind of carbonyl function with either XPS or IR-spectroscopy is rather complicated. An actual example is the reaction with so-called piranha water, that is, a mixture of concentrated sulfuric acid and

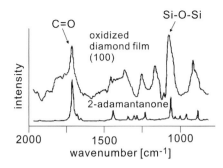

Figure 6.36 Reflection IR-spectra of adamantanone (bottom) and of an oxidized diamond film (top) (© ACS 2003).

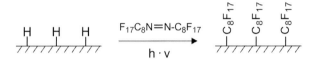

Figure 6.37 Photochemical reaction with perfluoroazoalkanes.

hydrogen peroxide. Upon ~10 min action, the formation of about 5×10^{-12} mol cm^{-2} of hydroxyl groups on the diamond surface is observed, which corresponds to a covering of 0.5–1% of a monolayer. This value may be comparatively small (e.g., ~7×10^{-10} mol cm^{-2} are achieved on gold surfaces functionalized with thiol groups), but it is absolutely sufficient to further modify the surface with larger units.

6.5.2.4 Radical and Photochemical Reactions on Diamond Surfaces

Apart from the photochemical halogenation, various other radical reactions are known to take place on diamond films. Their initiation has been reported to be achieved either photochemically or by other methods. An example has already been mentioned with the fluorination of diamond films by perfluoroalkyl iodides. Omitting the subsequent heating of the sample provides a diamond surface covered with perfluoroalkyl groups. A similar reaction occurs with the corresponding perfluoroalkyl azo compounds. These decompose upon irradiation with high-energy light (60 W Hg low-pressure lamp, λ = 185 nm). Under nitrogen cleavage, they are converted into perfluoroalkyl radicals that then attack the diamond surface. To this end, the diamond film with its substrate is first immersed into a solution of the respective azo compound (e.g., $F_{17}C_8$–N=N–C_8F_{17}) in a perfluoroalkane employed as solvent, and then irradiated. This reaction can also take place on hydrogenated diamond films (Figure 6.37). In this case, a perfluoroalkyl radical cleaves a hydrogen atom from the surface. In a second step the now unsaturated bonding site reacts with another radical to give the alkylated product. Surface groups containing oxygen yield the respective ester or ether structures (Figure 5.37).

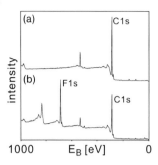

Figure 6.38 XPS-examination of pristine (a) and perfluoroalkylated diamond films (b) (© ACS 2004).

The reflection IR-spectrum of the resultant film does not show the typical signals of C–H-stretch vibrations known for hydrogenated diamond films, and a new band corresponding to a C–F-stretch vibration is observed at 1142 cm^{-1}. In XPS the presence of fluorine can easily be detected (Figure 6.38). The contact angle of water, increasing from 81° to 118°, also indicates a marked hydrophobization. The friction coefficient of a perfluoroalkylated diamond film is 0.1, while for the starting material it is 0.2.

Diazirine compounds as well may photochemically be reacted with diamond surfaces. In this case even blue or near UV-light suffices to generate the reactive carbene species (Section 6.5.2.5). If a suitable functionalization of the diazirine is provided, a complex surface modification can easily be effected this way. Diazirines functionalized with terminal maleimides, for instance, have been used to covalently immobilize sugars on the diamond surface (Figure 6.39). The reaction of maleimide with, for example, thiogalactose is employed here to establish a connection with the diazirine. The sugar derivative thus immobilized on the diamond surface is still biologically active as evident from the persisting ability to interact specifically with suitable reagents. Diamond films modified this way can consequently be employed in bioassays intended to detect certain substances in sera or other liquids. However, the production of such bioassays requires a reduction of the diamond's hydrophobicity as only a suitable functionalization enables an effective interaction with the usually quite polar biologically active compounds. Further examples of photochemical reactions on diamond films are given in Section 6.5.2.6.

The reaction with aryl, acyl, or aroyl radicals, generated from the respective peroxides by thermal cleavage, leads to an arylation or acylation of the diamond surface (Figure 6.40). Substitutions performed on the aromatic rings then allow for further modifications of the surface and for consecutive reactions.

Aryl radicals may also be prepared from the respective diazonium salts by electrochemical reduction. The radicals obtained from this single electron transfer then react with the diamond surface, arylating it in the course of the process. Varying a bit with the reagent chosen, the resultant surface covering is about 13% of a monolayer. Depending on the substitution pattern of the aromatic compound, the most different functionalizations of diamond films can be achieved this way. For example, doubly *meta*-chlorinated or *para*-nitrated diazonium salts can be

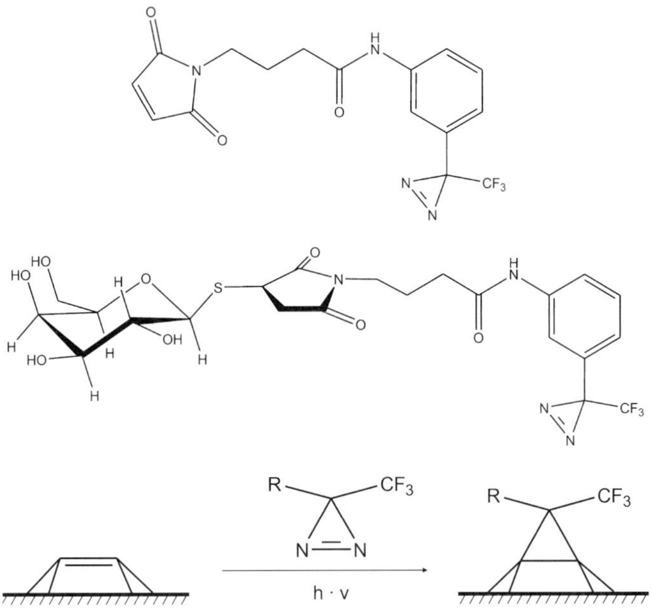

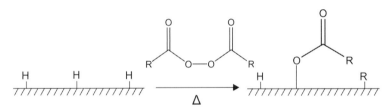

Figure 6.39 The photochemical reaction of diamond films with functionalized diazirines leading to cyclopropane formation (e.g., refer to the text).

Figure 6.40 Reaction of hydrogenated diamond surfaces with radical reagents.

employed. The latter then may directly be converted into the respective anilines by electrochemical reduction, which serves to a significant increase in the diamond's hydrophilic character.

The reaction with aryl radicals of this kind has found an interesting application in the direct polymerization of vinylic monomers like styrene, methyl methycrylate (MMA), or hydroxyethyl methacrylate (HEMA) on the diamond. Furthermore, the covalent attachment of an initiator molecule allows for an atom transfer radical polymerization (ATRP) to take place immediately on the diamond surface and results in a covalently bound composite material (Figure 6.41).

The major advantage of an arylation or alkylation lies in the very strong covalent bonding of the functional group to the diamond surface by way of a C–C-bond, which also reflects in the high stability of diamond films functionalized in this manner.

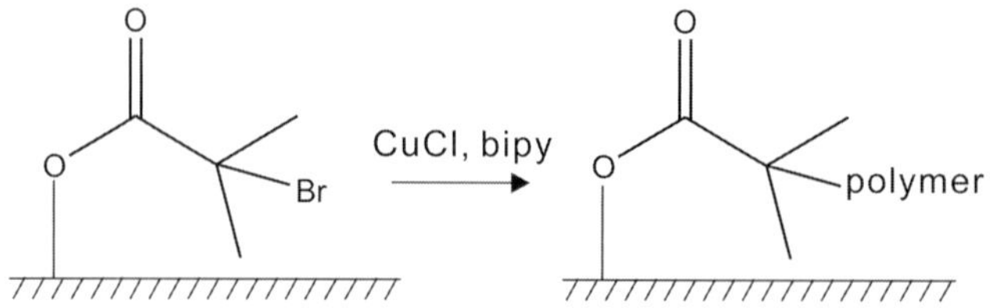

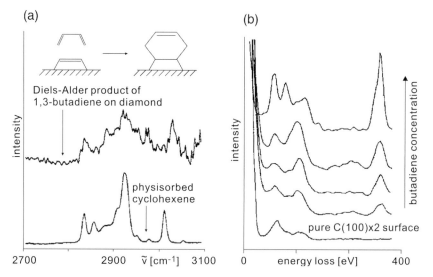

Figure 6.41 Atom transfer radical polymerization (ATRP) on diamond surfaces.

Figure 6.42 The reaction of diamond surfaces with 1,3-butadiene: (a) reflection IR-spectra of a reacted diamond and, for comparison, of physisorbed cyclohexene (© Elsevier 2001), (b) EEL-spectra of diamond films reacted with different concentrations of 1,3-butadiene (© Oyo Butsuri Gakkai 1999).

6.5.2.5 Cycloadditions on Diamond Surfaces

The presence of π-bonds on a diamond surface passivated by reconstruction suggests the possibility of cycloadditions to unfunctionalized diamond films. Actually, these diamond surfaces are observed to react with several typical reagents for cycloaddition. In comparison to a typical alkene double bond, the superficial π-bond exhibits a lower bonding strength because it is something in between a double bond and a diradical in character (Section 6.5.1). The bending of the adjacent covalent bonds projecting toward the bulk phase also contributes to the consequent increase in reactivity. Usually the double bonds on a reconstructed surface are not conjugated, and so the superficial π-bonds always act as ene-component. In a reaction with 1,3-butadiene, for example, the only product observed results from the Diels–Alder reaction that yields a cyclohexene structure bound to the diamond (Figure 6.42). In electron micrographs these structures are found to be arranged in the rows typical of reconstructed (100)-faces. Other dienes

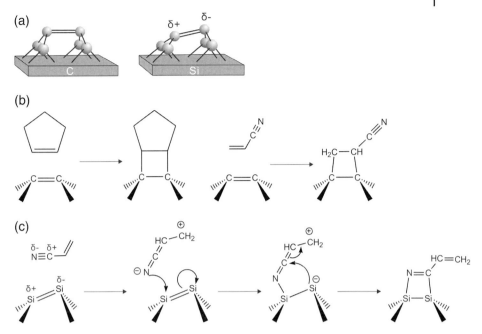

Figure 6.43 (a) The structure of surface dimers for diamond and silicon, (b) reaction of surface dimers on diamond with cyclopentene or acrylonitrile, (c) mechanism for the reaction of silicon surfaces with acrylonitrile.

may be employed as well in this [4+2]-cycloaddition. Contrary to silicon, the thermally forbidden formation of the [2+2]-adduct does not occur (see below).

With alkenes, on the other hand, the [2+2]-cycloaddition to surface dimers is observed even though strictly speaking it is thermally forbidden. It has been reported to take place, for instance, with cyclopentene, ethylene and acrylonitrile. Still, altogether, the [2+2]-cycloaddition to diamonds is much harder to be effected than that to surfaces of silicon or germanium. This is due to the comparatively wide bandgap of diamond (carbon: 350 kJ mol^{-1}, silicon: 110 kJ mol^{-1}, germanium: 140 kJ mol^{-1}): the excitation of electrons from the π- to the π*-orbital required for the reaction is hardly possible here. Consequently, the reactivity as compared to silicon is found to be reduced by a factor of 1000. A functionalization of diamond surfaces can be achieved this way nevertheless, only provided the reaction times are extended for long enough.

What is more, the asymmetric structure of the surface dimers in silicon and germanium causes a polarization of the double bond. Hence may occur a nucleophilic attack on the π*-orbital and the formation of a π-complex from the attacking, electron rich alkene and the electron-deficient end of the surface dimer. The subsequent addition leading to the final product is easy to take place then (Figure 6.43). Obviously this is not the concerted and symmetric mechanism typical of pericyclic reactions, which is also why the prohibition of a thermal reaction is by-passed.

For the reaction of diamond, a nonconcerted, radical mechanism is assumed as well, yet the formation of an intermediate π-complex is not observed as the symmetric surface double bond is not polarized (Figure 6.43). The lack of polarity becomes obvious for acrylonitrile: in comparison to the addition to silicon, another regioselectivity is found for diamond. In the first case, the reaction takes place on the C≡N-triple bond, while on a diamond surface it is the C=C-double bond forming a cyclobutane structure.

Further cycloadditions include the 1,3-dipolar cycloadditions as well as the respective [2+1]-reactions. The latter have already been mentioned in the reaction of carbenes photochemically generated from diazirines (Section 6.5.2.4). Experimental examples for [3+2]-cycloadditions have not yet been reported. Still theoretical considerations gave rise to the assumption that such reactions with, for example, azides, diazomethane or other classical 1,3-dipoles should readily take place on the surface dimers. The calculations further revealed that the cleavage of nitrogen under formation of the respective azacyclopropane as known for "normal" alkenes should also take place for the [3+2]-adducts bound to diamond. For the reaction with ozone (Section 6.5.2.3) it has not yet been clarified whether or not an ozonide is initially formed by [3+2]-cycloaddition, only then to be transformed into the carbonyl compound.

6.5.2.6 Further Reactions on Functionalized Diamond Films

After having achieved the basic functionalization of a diamond film it is possible to attach much more complex structures. Hydrogenated, halogenated, aminated and oxidized diamond films are all suitable for these reactions that will be illustrated by some typical examples in the following.

The reaction of surface carboxyl groups can lead to either esters or amides. The first step in both cases is to generate acid chlorides on the diamond surface by treating it with thionyl chloride. This intermediate then can easily be reacted with the respective alcohols or amines. In this manner it has been possible, for instance, to establish on a diamond film a thymidine ester that could be connected to a strand of DNA in a subsequent ligase-mediated reaction (Figure 6.44). Hydroxyl

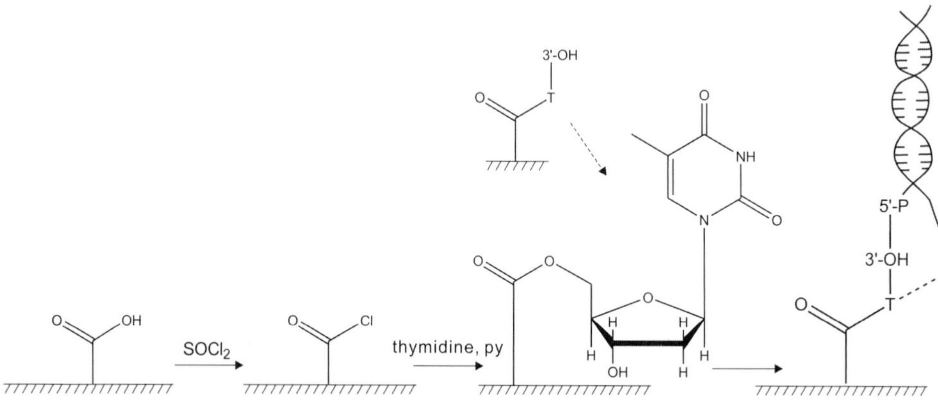

Figure 6.44 Attachment of DNA-oligonucleotides to diamond surfaces via a thymidine ester.

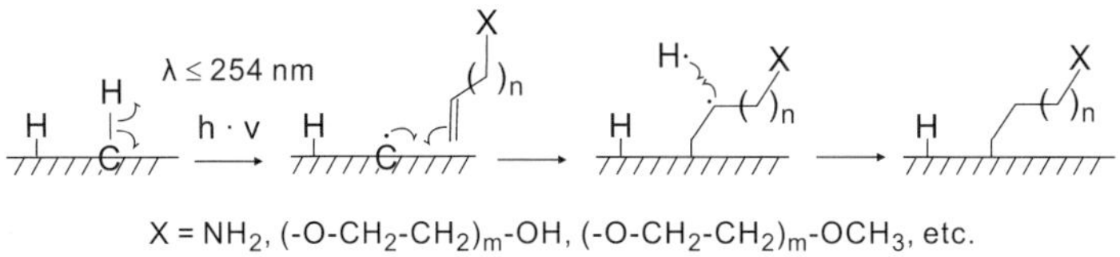

$$X = NH_2, (-O-CH_2-CH_2)_m-OH, (-O-CH_2-CH_2)_m-OCH_3, \text{etc.}$$

Figure 6.45 Mechanism for the photochemical reaction of hydrogenated diamond films with ω-vinyl species rendered suitable for further reaction by carrying terminal functional groups.

groups (obtained, e.g., from the action of "piranha water") can also be converted into esters. In this case, a suitable carboxylic acid or its carboxylate is immobilized on the diamond surface with the aid of dicyclohexylcarbodiimide (DCC). Biotin bound this way can be made visible by its specific interaction with fluorescence-labeled streptavidine.

In addition to the aforementioned reactions on hydrogenated diamond films (e.g., Sections 6.5.2.2 and 6.5.2.4) there is yet another valuable reaction having been described for diamond films covered with hydrogen. It is, in resemblance to an analogous process on silicon surfaces, the photochemical connection with terminal vinyl groups. To this end, the diamond film is covered with a suitable vinyl reagent and irradiated with energy rich UV-light ($\lambda \leq 254$ nm). It is a known fact that in hydrogenated diamond films, electrons may be excited across the wide bandgap into unoccupied levels by short wave UV-irradiation.

The reaction mechanism of superficial C–H-bonds is based on this excitation of electrons from the valence band that causes a cleavage of hydrogen radicals. The subsequent reaction with the vinyl reagent leads to a saturation of the diamond surface. The radical center now situated on the second carbon atom is in its own turn saturated by scavenging a hydrogen radical (Figure 6.45). The resultant alkyl derivative is covalently bound to the diamond by way of a C–C-single bond which, in comparison to other modes of attachment, is particularly stable against external influences. Depending on the reagent chosen, the surface covering achieved is about 30% of a monolayer. A variety of protected carboxylic acids and amines has been employed as difunctional vinylic reagent. After deprotection these are suitable to attach biologically active moieties. This may be considered as an important step toward the production of operative diamond based sensors and bioassays.

Usually derivatives of ω-undecene are employed for the photochemical C–C-bond formation. Among them there have also been derivatives modified with oligoethylene glycol that confer by far better hydrophilic properties (Figure 6.45). The terminal group may be a carboxylic acid protected as trifluoroethyl ester or an amino group protected with Boc or trifluoroacetamide (TFA). Subsequent deprotection then provides carboxyl or amino groups attached to the diamond surface via alkyl chains. Further reacting these films with linker units and a subsequent covalent attachment of DNA fragments allow for the production of diamond films functionalized with DNA. They are significantly more stable than the analogous silicon or gold films. Moreover, 1-chloro-5-hexene has been photochemically connected with diamond films. Here as well the linking of biological structures can

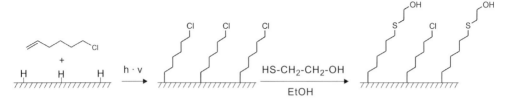

Figure 6.46 Photochemical reaction of hydrogenated diamond films with 1-chloro-5-hexene, followed by a substitution of the terminal chlorine atoms.

be achieved by a reaction with 2-mercaptoethanol. In this case the hydrophilicity and hence the surface covering may be adjusted by controlling the degree of conversion with the alcohol (Figure 6.46).

6.5.3
Noncovalent Functionalization of Diamond Films

Owing to their extremely hydrophobic character, diamond surfaces altogether hardly tend toward an unspecific adsorption of polar organic molecules. Therefore functionalized diamond films suit very well to sensor applications, for example, for biomolecules. The specific adsorption then only takes place at the respective positions predetermined by the functionalization.

Treating a diamond film in a plasma with aerial oxygen or reacting it with oxidizing acids largely increases its surface hydrophilicity by introducing functional groups that contain oxygen. On diamond films modified this way, now polar substances can be also adsorbed. The better bonding results from stronger interactions being possible by way of hydrogen bonds and van der Waals forces. The noncovalent immobilization of active antibodies against salmonella and staphylococci may serve as example here. The functionality of the structures fixed this way was proven in a test that shows the respective bacteria to be bound specifically to those regions that have noncovalently been functionalized with the matching antibody. Cytochrome c as well may be immobilized on hydrophilic diamond surfaces by noncovalent interaction. It is an excellent probe for an electron transfer between adsorbate and diamond (Section 6.6.3).

6.5.4
Electrochemistry of Diamond Films

First reports on the electrochemical properties of diamond date from the year 1983, and from the mid-1980s on, numerous extensive studies have been performed on the electrochemistry on diamond electrodes. The first electrodes were made from diamond prepared by CVD processes. It featured a certain electric conductivity due to lattice defects. Therefore it was possible to determine the capacity and the photoresponse as well as the voltage/current characteristic.

Why now is diamond so attractive to be used as material for electrodes? Firstly, it is a very stable material under both mechanic and chemical aspects, so it can even be employed in highly aggressive media. Secondly, it features favorable electrochemical properties like a very wide potential window (see below) and a low background current. Furthermore, it is resistant to the so-called fouling and does not form oxides passivating its surface. Hence it may be employed as a sensor or in electrosynthesis (Section 6.6.3). The low sp^2-content causes an inert behavior in many media here. For example, high-quality diamond electrodes are stable even in a melt of KCl/LiCl at 450 °C.

Electric conductivity of the bulk phase generally provided (caused either by doping or by defects), the electrochemical properties of the deposited diamond film are largely governed by its surface. A hydrogenated diamond electrode, for instance, is very hydrophobic and impairs the electrode reaction of species to be adsorbed, for example, in the cathodic or anodic generation of hydrogen or oxygen, respectively. This is why in aqueous solution no decomposition of water is observed over a wide potential range (Figure 6.47). The only increase in the electrode's hydrophilicity arises from anodic polarization and a concomitant adsorption of oxygen.

Still the electrochemical behavior of diamond electrodes is influenced by more parameters than the surface termination alone. Most of all it is a doping of the diamond phase that provides the required electric conductivity. Usually boron-doped electrodes are employed. In this case, the electric conductivity increases with the content of boron – the resulting properties range from an isolator at low

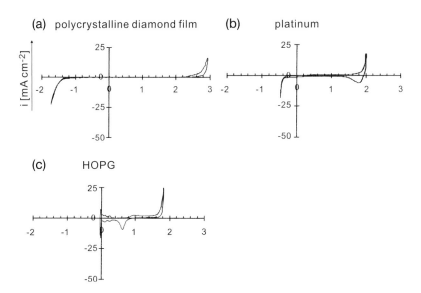

Figure 6.47 Current–voltage characteristic of different electrode materials (with the abscissa scaling being in volts): (a) diamond, (b) platinum, and (c) highly ordered graphite (HOPG) (© Electrochem. Soc. 1996).

contents to a quasi-metal at high values. The boron atoms insert an acceptor level into the bandgap at 0.37 eV above the valence band maximum (Section 6.2.3). At high contents of boron (>10^{20} cm^{-3}) an overlapping dopant level is formed and the specific resistance drops from about 10^4 to 10–1000 Ω cm: A p-doped electric conductor has been generated. n-Type conductivity can be achieved by a doping with nitrogen or phosphorous. Both elements represent donors inserting a donor level at 1.7 eV or 0.6 eV below the conduction band minimum, respectively (Section 6.2.3). Nitrogen doping, however, is only feasible for ultrananocrystalline diamond films.

Apart from a doping of the bulk phase, the electronic band structure may also be influenced by processes on the phase boundary between the diamond film and its environment. For example, solutes like, for example, O_2, CO_2, or H_2 contained in the water film adsorbed on the diamond surface may cause an electron transfer from or into the diamond electrode according to their chemical potential. With the latter being situated below the Fermi level of the electrode, electrons will be transferred to the adsorbed film until the same level is attained for both the diamond's Fermi level and the potential of the surface film (Figure 6.31). In doing so a p-doped space-charge zone arises directly in the diamond surface. At opposite conditions the electron transfer takes its course into the electrode and an n-doped zone is formed.

Irradiating the diamond surface with energy-rich light or with X-rays may as well affect the electronic properties. This effect is termed the "photoresponse" of the diamond electrode. Depending on the energy of the radiation employed it is either electrons from states within the bandgap or, sufficient energy provided, such from the valence band being excited. Owing to the wide bandgap in diamond, the latter can only be achieved by high-energy radiation. If the excitation of valence band electrons is possible, it will give rise to an electron–hole pair that is separated in the space-charge zone. The minority charge carrier (the electrons in the case of B-doped diamond) migrates to the surface of the film, while the major species (the holes in boron-doped material) move into the bulk phase. If now the additional charge carriers supplied to the electrode's surface are consumed (e.g., by a redox reaction occurring there), a photo current results. With no consumption taking place, on the other hand, the surface capacity is charged and a photo potential arises. This reduces the drop of potential across the space-charge zone, and the corresponding bending of the band structure in the film's surface is decreased (Section 6.4.2.1). The sign of the photo potential provides information on the kind of majority charge carrier present.

Figure 6.48 illustrates the correlation of electron energy and electrode potential (measured vs. standard hydrogen electrode) and some examples of important redox couples with their respective potentials to enable a direct comparison between the band structure of diamond and the electrode potential. The latter is related to the electron energy according to $e \cdot E = 4.44$ eV $+ \varepsilon$ with e being the charge of the electron (i.e., −1), E(V) being the electrode potential and ε(eV) being the energy of an electron unaffected by surface forces. Thus, both the maximum of the valence band and the minimum of the conduction band may be deduced from electrochemical measurements, and the position of the Fermi level is also com-

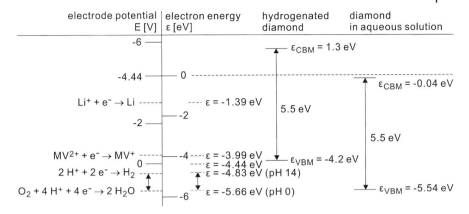

Figure 6.48 Comparison of electrode potentials (measured vs. standard hydrogen electrode) to electron energies for various redox couples (MV: methylviologen, CBM: conduction band minimum, VBM: valence band maximum; © Elsevier 2004).

paratively easy to determine this way. The structure of the diamond surface plays an essential role for the values obtained here. If it is covered with oxygen (e.g., in aqueous solutions), the maximum of the valence band shifts toward a more positive potential. At the same time, the minimum of the conduction band (about 1.3 eV for hydrogenated diamond) drops to -0.04 eV. Usually boron-doped electrodes with a degree of doping higher than 10^{19} cm^{-3} are employed in such examinations as well as in sensor applications to achieve sufficient conductivity. More details on specific uses are collected in Section 6.6.3.

6.6
Applications and Perspectives

Contrary to the "new" carbon materials presented so far, diamond films are a product that is already being employed with many variants in large scale applications. Especially the coating of tools and fast turning parts represents a considerable market, but there are also electronic uses and synthetic optical windows on offer. Some of the important fields of application are presented in more detail below.

6.6.1
Mechanical Applications

Owing to remarkable mechanical properties like great hardness, scratch resistance and a low frictional coefficient in combination with a high tolerance for aggressive environmental conditions, diamond-coated objects have been developed soon after the first preparation of diamond films. Today the respective products are employed in many technical applications.

Figure 6.49 Examples of cutting and drilling tools coated with CVD diamond (© L. Schäfer).

A distinctive feature of diamond films is their wear resistance, which in a self-suggesting manner led to diamond-coated tools being developed. Usually a primer layer of, for example, silicon carbide is applied to achieve a good adhesion between the diamond film and the tool support and thus to ensure a long service life of the coated object. Otherwise delamination might occur, which would mean immediate failure of the tool. Exemplary tools are such for drilling or cutting like precision saw-blades or rock-drills (Figure 6.49), moreover diamond coatings are suitable for long-lived grinding tools.

Diamond films can also be employed as a scratch-resistant coating on substrates. The same resistance in combination with the aforementioned low friction may further be employed to protect mechanical parts stressed by fast rotation like gearings, bearings, and axles. For this kind of application high demands are made on the fit and on the precision of the surface. In the case diamond films, however, there are certain problems in producing such objects. Due to the great hardness it is first and foremost a polishing of the surface being hardly achievable by mechanical means. By now the so-called chemical polishing has been developed here, which in principle is based on an etching process. A new field of application emerged in the range of high-speed hard disk drives where a layer of diamond enabled faster rotation and thus shorter access times. In this case not only the mechanical stability, but also the high thermal conductivity play a role as even at high speed it prevents a heating of the storage device.

The combination of the diamond films' beneficial mechanical properties with their biocompatibility (Section 6.6.3) renders them an ideal material for coating implants and prostheses. The wear of such parts is markedly decreased, and diamond hardly evokes rejection reactions of the surrounding tissue.

6.6.2
Electronic Applications

It has already been mentioned in Section 6.4.2 that diamond films feature some outstanding properties predestinating them for a use in electronic components.

Some of these applications have already been realized, and a part of them is even in commercial use. The respective diamond films may or may not be doped. One of the major problems in the application of diamond components is contacting them. The contact points are normally generated by applying metals with a large electron work function like gold or platinum, so in parts very high barriers arise at these positions. They mean a serious limitation to the components' operability.

An attractive aim for the application of diamond films is the construction of field emission displays using diamond emitters as electron source. To this end, several techniques have been developed in recent years. They focus on generating a structured diamond surface that exhibits a controlled distribution of emitters (Figure 6.33). Apart from display applications, diamond films can also be employed to produce high-power electron sources. Altogether the use of nanocrystalline diamond films, for example, as coating on a substrate, turned out to be more beneficial for the emission characteristics, as in microcrystalline films emission only occurs close to crystal edges. In nanocrystalline films, on the other hand, the more even distribution and the higher density of emitting sites lead to better emission properties. They also obey a somewhat different mechanism of emission with a continuous supply of electrons being provided by the graphitic domains on the grain boundaries. These act as conduction channels, and subsequent tunneling into a diamond crystallite enables emission into the vacuum.

Classical semiconductor applications of diamond films can be realized as well. These include, for example, field effect transistors (FET) and diodes. The latter have already been constructed as an all-diamond device operating stably at temperatures of up to 1000 °C. The substrate employed is a type-Ib diamond on which a layer of UNCD p-doped with boron is deposited. The diode is completed by subsequent coating with a monocrystalline, undoped layer and, finally, with an n-layer containing nitrogen. In some cases then, the device is etched in a way partly exposing the p-layer. At 20 V, the rectification achieved with this diode comprises 10 orders of magnitude with only very little leakage current flowing in the opposite direction.

Field effect transistors as well have successfully been realized with diamond. In this case, hydrogen-terminated diamond is used. It meets essential requirements for the production of a surface channel FET by providing in its surface a high density of charge carriers ($>10^{13}\,cm^{-2}$) with just a thin layer ($\sim 10\,nm$) for them to dwell in. The small dimensions of the layer containing the charge carriers allows for a thickness of the gate of less than 20 nm and for a better control over the current flowing. Moreover H-terminated diamond films feature a very low Schottky barrier, which is favorable for contacting the device. Recently a specially designed FET has been presented. The metal gate electrode is replaced by an electrolyte solution here (Figure 6.50). The diamond films employed may be boron-doped or rendered surface-conductive by saturation with hydrogen. The whole setup stands out for being operative even at harsh environmental conditions, tolerating, for instance, pH-values from 1 to 13.

Another application of diamond films, which actually is in commercial use already, has been presented with the surface acoustic wave (SAW) devices. Their

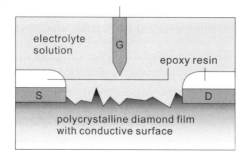

Figure 6.50 Scheme of a field effect transistor with an electrolyte as the gate (S: source, D: drain, G: gate).

function is based upon the high velocity of surface waves on a suitable material, which is combined with a piezoelectric material. Among other applications, they serve as frequency filters and resonators in mobile phones or TV tuners. Owing to its high elasticity diamond allows for extreme surface velocities of up to 10 000 m s^{-1}, so a 5-GHz filter may be constructed at a 2-µm distance between electrodes. Usually the diamond material is processed together with the piezoelectric material (e.g., zinc oxide) required for wave generation to become a sandwich-like structure. However, these devices must exhibit very precise diamond surfaces, which makes high demands on the polishing of the diamond wafers employed. For the time being, frequency filters covering the range from 1.8 to 4 GHz are available on the market, and further high-frequency applications in communication techniques are sure to follow in the future.

6.6.3
Chemical, Electrochemical, and Biological Applications

As discussed in Section 6.5, there are many ways of chemically modifying the surface of diamond films. This allows for the attachment of various structures like biomolecules. Diamond films stand out, among others, for the chemical inertness of their unfunctionalized surface, for good biocompatibility and low unspecific adsorption and for their tunable surface hydrophilicity. Owing to these beneficial properties, diamond has substantial advantages over other materials commonly employed in bioapplications like silicon, gold, or metal oxides. It suits perfectly well to the construction of biosensors, medical implants and to the development of electronic devices based on organic matter. Diamond-coated implants (Section 6.6.1) cause much less fibroses than analogous parts treated with silicon. Furthermore, the long-term stability in body fluids like blood or serum is significantly increased for diamond as compared to other materials.

By now an immobilization on diamond films succeeded for several biologically active structures like DNA-fragments, oligonucleotides, peptides, enzymes, etc. All of them were proven to retain their activity also in the bound state by showing their respective interaction in suitable test systems. Apart from biological moieties, also catalysts or other reagents may be connected to diamond films. Many studies on functionalized diamond films aim for the development of a so-called lab on a chip – a kind of diagnostic center placed on a small, diamond-coated plate that is covered with the required specific structures. Detection can be achieved, for instance, by measuring an electric current at the normally boron-doped diamond chips.

Diamond also features particular advantages for a use as electrode material. It is very stable in a mechanical and largely inert in a chemical way and, due to the high hydrogen overvoltage, provides a wide potential window for the examination of many analytes. Furthermore, it shows favorable electrochemical characteristics like fast response times, a wide linear range, low background current, and a fast stabilization of the current. Hence it is a suitable material for electrodes in electroanalyses, that is, in measuring an electric signal (voltage, current or charge) related to the concentration of the analyte that is oxidized or reduced in a reaction inducing the respective current. Normally boron-doped diamond films with a hydrogenated surface and coated onto a support of, for example, Si, W, or Mo are employed. In this way the required electric conductivity of the electrode is achieved. The content of boron amounts to about $10^{19}\,cm^{-3}$, resulting in a specific resistance of less than $0.1\,\Omega\,cm$. Ultrananocrystalline diamond films also suit well to coating electrodes. Even at a low thickness of layers a complete and dense covering of the support is achieved, and the coating of unevenly shaped electrodes is easier.

Numerous examples of diamond electrodes being employed in electroanalytical systems have been reported to date. The toxic azide ion, for instance, can very sensitively be detected by oxidizing it to nitrogen ($2\,N_3^- \rightarrow 3\,N_2 + 2\,e^-$). In this case, the low background current is beneficial for the method's sensitivity. Also nitrite, metal ions, aliphatic polyamines, etc., all have successfully been quantified with diamond electrode systems. Determining the content of NADH of a solution constitutes an interesting example here: other electrode materials irreversibly adsorb the products arising from NADH oxidation, which results in much worse electrode performance regarding stability and sensitivity. With diamond, this adsorption does not occur that easily, and the electrode characteristics are found to improve. Other biologically active substances like histamine, serotonin or uric acid may be electroanalyzed, too. Diamond films modified by covalent bonding or by physisorption are also suitable for electrochemical analysis. In these cases, sensitivity is increased by an additional, selective bonding of the analyte to the electrode favored by a specific interaction. By now, several electrode systems for laboratory use are commercially available.

6.6.4
Further Applications of Diamond Films

Numerous other fields of diamond film application are known, and more are sure to emerge in the future. One potential application makes use of the diamond films' high thermal conductivity, which by far exceeds that of any other bulk material. Hence, diamond is suitable for heat dissipation, for example, in electronic circuitry. For instance, a surface may be coated with a layer of CVD diamond which then effects the removal of heat. Another conceivable setup would be using a diamond-coated material for a plate bar supporting the electronic devices. In other ranges of application also free-standing diamond films might be employed for heat dissipation as they feature sufficient mechanical stability even at very low film thickness.

The optical properties as well render diamond an attractive material. Owing to its transparency over a wide range of radiation wavelengths, windows and lenses for various optical apparatus can be produced. In particular, systems equipped with diamond windows may be employed in difficult environments like in space because diamond is much more stable than the materials commonly employed for such optical systems. One possible application is the use as infrared window, for example, in rockets or planes, but also in spectrometers, with diamond replacing the conventional materials like ZnS or ZnSe. Only at a wavelength of 5 μm, diamond windows exhibit absorption of their own. Diamond is also suitable for a use as X-ray window where it can serve as a substitute for the highly toxic beryllium. With its atomic mass being rather low as well and with the feasibility of preparing extremely thin layers (~1 μm), it is possible to make highly transmissive windows that exceed the parameters of beryllium windows with their usual thickness of about 8 μm.

6.7
Summary

Diamond films may be polycrystalline or monocrystalline layers. In polycrystalline films, the diameter of particles is either on the range of micrometers, or they measure just a few nanometers across (UNCD). The preparation is mostly achieved by deposition from the gas phase (CVD methods). A variety of gaseous hydrocarbons like methane serve as carbon source. Film formation only occurs in the presence of atomic hydrogen that must be generated *in situ*, for example, in a plasma, on a hot filament, or in a flame. The deposition takes place on a substrate heated at 800–1200 °C.

Box 6.1 Preparation of diamond films.

- In microwave-CVD (MWCVD) the plasma is generated by microwave irradiation.
- In HF-CVD (hot filament CVD) the decomposition of starting material is effected by a hot wire.
- The combustion method employs the deposition from an oversaturated flame.
- UNCD is obtained from adding argon to the gas mixture in an MWCVD process.

Diamond films exhibit different characteristics depending on their morphology and mode of preparation. They feature great hardness and a low frictional coefficient. Doping can turn the wide gap semiconductor (bandgap ~5.5 eV) into a semiconductor or even into a material with a conductivity comparable to that of metals (e.g., upon boron doping).

The surface can be chemically modified. Basically it is either hydrogenated or reconstructed, depending on its degree of hydrogen covering. In the latter case, π-bonds are present on the surface that may be attacked by various reagents. They can enter, for instance, into Diels–Alder reactions with suitable dienes. By reacting superficial C–H-bonds with terminal alkenes, a C–C-bonding with the respective substituents is achieved in a photochemical reaction. A multitude of other possible conversions has been reported too, so altogether the surface may be functionalized covalently or noncovalently in many ways.

The electrochemical properties of diamond films are very promising. Films doped for better conductivity feature a wide potential window, and owing to their stability and fast response times, etc., they suit very well to a use as electrode material in electroanalysis. Diamond films with their surface being suitably modified further suggest themselves for the analysis of biological material. Due to a low unspecific adsorption, interactions will occur only at those positions carrying the respective structure. This is of considerable interest for the development of the so-called lab on a chip.

Further applications of diamond films lie in a mechanically resistant coating of components and implants. Moreover, the use in electronic devices seems near at hand now, and one may well assume that diamond films will at least partially replace silicon in this field. The optical qualities of diamond render free-standing films an ideal material for windows in spectroscopic apparatus, etc. Reliable methods of preparation and a controllable doping have been established, which altogether allow for a large scale, commercial application of diamond films.

7
Epilog

The scope of this introductory text does not allow for a full appreciation of any recent development in the field of carbon materials as otherwise the book would have attained unreasonable extent. Therefore, after studying the present text, it is advised to consult the respective original literature and reviews on special findings and on further kinds of carbon materials, e.g. graphene. A multitude of substances and principles is waiting to be discovered there. Moreover, these publications provide many hints on and conceptions of which direction the development will assume. So what now will the future be like in the field of carbon materials?

Diamond materials and related substances already hold an important position concerning surface coatings and electronic devices. In this case, the development producing ever smaller and more efficient components will continue. Diamond might once completely replace silicon for being more stable and biocompatible than the latter, and it may be possible to establish electronic devices even at the phase boundary to living systems. A nanostructuring of diamond surfaces as well as the directed doping or chemical functionalization will multiply the potential utilization of diamond in both everyday applications and the high-tech segment.

Yet high hopes are also laid on the carbon nanotubes. Preparing them in sufficient amounts of satisfying purity constitutes a formidable challenge that is mastered, however, to an ever increasing degree. Once it becomes possible to selectively obtain a certain structural variant, there will be virtually no limit anymore to applications in electronics and sensing. Large electronic devices consisting of carbon alone – the proverbial computers made of diamond and nanotubes – come near at hand.

Still it is not only the promising field of potential applications that inspires nanotube research – another strong motivation is the hope of discovering ever more tube-related species like, for instance, Y-junctions, peapods, or the tori shown in Figure 7.1. They might give an idea of a structural variety that has nothing of the sort among the other elements. The question of whether this range has been fully investigated already cannot conclusively be answered as new variants to the known modifications are continually being found. Moreover, it is never for certain if not completely different structures might be realizable besides those discovered so far. One trend heads toward carbon objects with a more three-dimensional structure. Different types of super"-nanotubes with the constituent

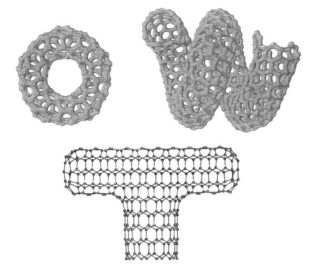

Figure 7.1 Tori (© RSC 1995), helices from controlled synthesis (© APS 1997), and T-junctions are advanced developments of carbon nanotubes.

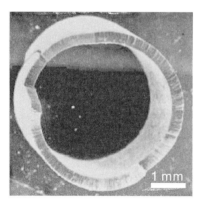

Figure 7.2 A tube made of carbon nanotubes. The individual CNT are arranged radially. The "supertube" may, for example, serve as filter material (© Nature Publishing Group 2004).

tubes being radially arranged around a superaxis or forming the "bonds" of the superstructure have been prognosticated and in parts even confirmed by experiment already (Figure 7.2).

Growing interest is also put in three-dimensional structures made up from fullerenes. Interesting properties like, for example, magnetism may be expected in particular for such objects with the carbon atoms' degree of hybridization ranging between sp^2 and sp^3 and featuring both concave and convex domains. However, it is still a long way to go for the aim of preparing a sort of Mackay crystal consisting of polymerized fullerenes (Figure 7.3). Another trend leads toward developing photosensitive fullerene systems to exploit solar energy. Basic

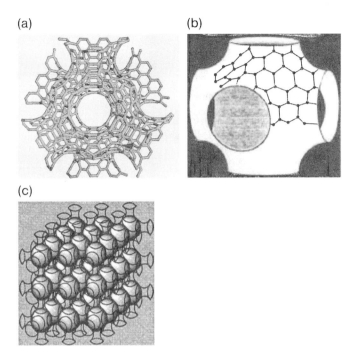

Figure 7.3 The structure of the three-dimensional Mackay crystals consists of convex and concave elements. (a) A scheme illustrating the structure of such crystals, while structural fragments of these idealized C_{60}-polymers are shown in (b) and (c) (© RSC 1995).

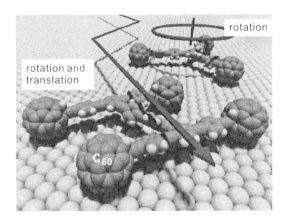

Figure 7.4 With the tip of an atomic force microscope, the so-called fullerene car may be steered across a surface (© ACS 2005).

research will bring forth new syntheses for small fullerenes ($n_C < 60$), and maybe it will be possible one day actually to prepare the smallest possible fullerene C_{20} in visible amounts.

Interesting results in the field of surface manipulation (Figure 7.4) can also be envisaged to arise from the arrangement of nanotubes, fullerenes, or tiny diamond particles on surfaces or from the direct addressing of certain positions on carbon structures. The analytical laboratory on a chip, nanotube-based catalysts, highly efficient fuel cells or a luminous display with low power consumption illustrate the enormous potential of carbon research.

Altogether, there are countless possibilities and there is hardly any limit to creativity, and so we should be curious what kind of surprises element carbon still holds in store for us.

8
Further Readings and Figure References

8.1
Further Readings

For a deeper understanding of carbon materials, it is advisable to gather information on further properties and applications from more specialized literature. The choice of respective readings given below is a small selection of the available publications and reviews. There is a variety of other excellent reviews and monographs providing a more detailed insight into the subject as well; still for reasons of space they could not be mentioned here. Although there are (for reasons of readability of such an introductory text) no direct references to the original literature in the text, there are, however, ample examples for detailed experimental and theoretical studies given in the figures of Section 8.2.

Carbon in General

Dresselhaus, M.S., Dresselhaus, G., Sugihara, K., Spain, I.L., and Goldberg, H.A. (1988) *Carbon Fibers and Filaments, Springer Series in Materials Science*, vol. 5, Springer, Berlin.

Greenwood, N.N. and Earnshaw, A. (2005) *Chemistry of the Elements*, 2nd edn, Elsevier Butterworth Heinemann, Amsterdam.

Hazen, R.M. (1999) *The Diamond Makers*, Cambridge University Press, Cambridge.

Huheey, J., Keiter, E., and Keiter, R. (1995) *Anorganische Chemie*, 2nd edn, W. de Gruyter, Berlin.

Jenkins, G.M. and Kawamura, K. (1976) *Polymeric Carbons – Carbon Fiber, Glass and Char*, Cambridge University Press, Cambridge.

Robertson, J. (1986) *Adv. Phys.*, 35, 317–374.

Wiberg, N., Wiberg, E., and Hollemann, A.F. (2007) *Lehrbuch der Anorganischen Chemie*, 102th edn, W. de Gruyter, Berlin.

Gmelin Handbook of Inorganic and Organometallic Chemistry, 8th edn, (1967–1978) Springer Verlag, Heidelberg.

Harlow, G.E. and Danies, R.H. (eds) (2005) *Elements*, 1, 67–108.

Fullerenes

Cozzi, F., Powell, W.H., and Thilgen, C. (2005) *Pure Appl. Chem.*, 77, 843–923.

Dai, L. and Mau, A.W.H. (2001) *Adv. Mater.*, 13, 899–913.

Diederich, F. and Gómes-López, M. (1999) *Chem. Soc. Rev.*, 28, 263–277.

Dresselhaus, M.S., Dresselhaus, G., and Eklund, P.C. (1996) *Science of Fullerenes and Carbon Nanotubes*, Academic Press, London.

Echegoyen, L. and Echegoyen, L.E. (2001) The electrochemistry of C_{60} and related compounds, in *Organic Electrochemistry*, 4th edn (eds H. Lund and O. Hammerich), Marcel Dekker, New York, pp. 323–340.

Fowler, P.W. and Manolopoulos, D.E. (1995) *An Atlas of Fullerenes*, Oxford University Press, New York.

Guldi, D.M. (2002) *Chem. Soc. Rev.*, 31, 22–36.

Guldi, D.M., Rahman, G.M.A., Sgobba, V., and Ehli, C. (2006) *Chem. Soc. Rev.*, 35, 471–487.

Hirsch, A. (1994) *The Chemistry of the Fullerenes*. Thieme, Stuttgart.

Hirsch, A. (ed.) (1999) *Top Curr. Chem.*, 199, 1–246.

Hirsch, A. and Brettreich, A. (2005) *Fullerenes: Chemistry and Reactions*, Wiley-VCH Verlag GmbH, Weinheim.

Kadish, K.M. and Ruoff, R.S. (eds) (2000) *Fullerenes: Chemistry, Physics, and Technology*, Wiley-Interscience, New York.

Kleineweischede, A. and Mattay, J. (2003) Photochemical reactions of fullerene and fullerene derivatives, in *CRC Handbook of Organic Photochemistry and Photobiology* (eds W. Horspool and F. Lenci), CRC Press, Boca Raton, pp. 28-1–28-42.

Kroto, H.W., Mackay, A.L., Turner, G., and Walton, D.R.M. (eds) (1993) *Phil. Trans. R. Soc. A*, 343, 1–154.

Langa, F. and Nierengarten, J.-F. (eds) (2007) *Fullerenes – Principles and Applications*, RSC Publishing, Cambridge.

Lu, X. and Chen, Z. (2005) *Chem. Rev.*, 105, 3643–3696.

Martin, N. (2006) *Chem. Commun.*, 2093–2104.

Nakamura, E. and Isobe, H. (2003) *Acc. Chem. Res.*, 36, 807–815.

Nierengarten, J.-F. (2004) *New J. Chem.*, 28, 1177–1191.

Osawa. E. (ed.) (2002) *Perspectives of Fullerene Nanotechnology*, Kluwer Academic Publishers, Dordrecht.

Powell, W.H., Cozzi, F., Thilgen, C., Hwu, R.J.-R., and Yerin, A. (2002) *Pure Appl. Chem.*, 74, 629–695.

Carbon Nanotubes

Andrews, R., Jaques, D., Qian, D., and Rantell, T. (2002) *Acc. Chem. Res.*, 35, 1008–1017.

Avouris, P. (2002) *Acc. Chem. Res.*, 35, 1026–1034.

Balasubramanian, K. and Burghard, M. (2005) *Small*, 1, 180–192.

Banerjee, S., Hemraj-Benny, T., and Wong, S.S. (2005) *Adv. Mater.*, 17, 17–29.

Bianco, A., Kostarelos, K., Partidos, C.D., and Prato, M. (2005) *Chem. Commun.*, 571–577.

Coleman, J.N., Khan, U., and Gun'ko, Y.K. (2006) *Adv. Mater.*, 18. 689–706.

Dai, H. (2002) *Acc. Chem. Res.*, 35, 1035–1044.

Dekker, C. (1999) *Phys. Today*, 52(5), 22–28.

Dresselhaus, M.S., Dresselhaus, G., and Eklund, P.C. (1996) *Science of Fullerenes and Carbon Nanotubes*, Academic Press, London.

Dresselhaus, M.S., Dresselhaus, G., and Avouris, Ph. (eds) (2001) *Carbon Nanotubes: Synthesis, Structure, Properties, and Applications*, Top. Appl. Phys., vol. 80, Springer, Berlin.

Dresselhaus, M.S., Dresselhaus, G., Charlier, J.C., and Hernández, E. (2004) *Phil. Trans. R. Soc. A*, 362, 2065–2098.

Endo, M., Hayashi, T., Kim, Y.A., Terrones, M., and Dresselhaus, M.S. (2004) *Phil. Trans. R. Soc. A*, 362, 2223–2238.

Harris, P.J.F. (1999) *Carbon Nanotubes and Related Structures*, Cambridge University Press, Cambridge.

Harris, P.J.F. (2009) *Carbon Nanotube Science – Synthesis, Properties and Applications*, Cambridge University Press, Cambridge.

Joselevich, E. (2004) *Chem. Phys. Chem.*, 5, 619–624.

Meyyappan, M. (ed.) (2005) *Carbon Nanotubes: Science and Applications*, CRC Press, Boca Raton, FL.

Nikolaev, P. (2004) *J. Nanosci. Nanotechnol.*, 4, 307–316.

Oberlin, A. and Endo, M. (1976) *J. Cryst. Growth*, 32, 335–349.

Ouyang, M., Huang, J.L., and Lieber, C.M. (2002) *Acc. Chem. Res.*, 35, 1018–1025.

Reich, S., Thomsen, C., and Maultzsch, J. (2004) *Carbon Nanotubes*. Wiley-VCH Verlag GmbH, Weinheim.

Tasis, D., Tagmatarchis, N., Bianco, A., and Prato, M. (2006) *Chem. Rev.*, 106, 1105–1136.

Willner, E. and Katz, I. (2004) *Chem. Phys. Chem.*, 5, 1084–1104.

Carbon Onions

Banhart, F. (1999) *Rep. Prog. Phys.*, 62, 1181–1221.

Kuznetsov, V.L. and Butenko, Y.V. (2005) Nanodiamond Graphitization and Properties of Onion-Like Carbon, in *Synthesis, Properties and Applications of Ultrananocrystalline Diamond*, NATO Science Series, vol. 192 (eds D.M. Gruen, O.A. Shenderova, and A.Y. Vul'), Springer, Dordrecht, pp. S199–S216.

Terrones, M., Hsu, W.K., Hare, J.P., Kroto, H.W., Terrones, H., and Walton, D.R.M. (1996) *Phil. Trans. R. Soc. A*, 354, 2025–2054.

Tomita, S., Sakurai, T., Ohta, H., Fujii, M., and Hayashi, S. (2001) *J. Chem. Phys.*, 114, 7477–7482.

Diamond Films

Angus, J.C. and Hayman, C.C. (1988) *Science*, 241, 913–921.

Clare, T.L., Clare, B.H., Nichols, B.M., Abbot, N.L., and Hamers, R.J. (2005) *Langmuir*, 21, 6344–6355.

Dahotre, N.B., and Kichambare, P.D. (2004) Nanocrystalline Diamond, in *Encycl. Nanosci. Nanotechnol*, vol. 6 (ed. H.S. Nalwa), American Scientific Publishers, Stevenson Ranch, pp. S435–S463.

Fox, B.A. (1997) Diamond films, in *Thin Film Technology Handbook* (eds A. Elshabini-Riad and F.D. Barlow, III), McGraw-Hill Professional, New York, pp. 7-1-7-74.

Gruen, D.M. (2001) *MRS Bull.*, 26, 771–776.

Hamers, R.J., Coulter, S.K., Ellison, M.D., Hovis, J.S., Padowitz, D.F., Schwartz, M.P., Greenlief, C.M., and Russel, J.N., Jr. (2000) *Acc. Chem. Res.*, 33, 617–624.

Kobashi, K. (2005) *Diamond Films: Chemical Vapor Deposition for Oriented and Heteroepitaxial Growth*, Elsevier, Amsterdam.

Koizumi, S., Nebel, C.E., and Nesladek, M. (2008) *Physics and Applications of CVD Diamond*, Wiley-VCH Verlag GmbH, Weinheim.

Nebel, C.E. and Ristein, J. (eds) (2003) *Thin Film Diamond I, Semiconductors and Semimetals*, vol. 76, Elsevier Academic Press, Amsterdam.

Nebel, C.E. and Ristein, J. (eds) (2004) *Thin Film Diamond II, Semiconductors and Semimetals*, vol. 77, Elsevier Academic Press, Amsterdam.

Ristein, J., Maier, F., Riedel, M., Cui, J.B., and Ley, L. (2000) *Phys. Stat. Sol. A*, 181, 65–76.

Spear, K.E. and Dismukes, J.P. (eds) (1994) *Synthetic Diamond: Emerging CVD Science and Technology*, Wiley-Interscience, New York.

Sussmann, R.S. (ed.) (2009) *CVD Diamond for Electronic Devices and Sensors (Wiley Series in Materials for Electronic & Optoelectronic Applications)*, John Wiley & Sons, Ltd, Chichester.

Nanodiamond

Dahotre, N.B. and Kichambare, P.D. (2004) Nanocrystalline Diamond, in *Encycl. Nanosci. Nanotechnol*, vol. 6 (ed. H.S. Nalwa), American Scientific Publishers, Stevenson Ranch, pp. S435–S463.

Dolmatov, V.Y. (2001) *Russ. Chem. Rev.*, 70, 607–626.

Donnet, J.-B., Lemoigne, C., Wang, T.K., Peng, C.-M., Samirant, M., and Eckhardt, A. (1997) *Bull. Soc. Chim. Fr.*, 134, 875–890.

Gruen, D.M., Shenderova, O.A., and Vul', A.Y. (eds) (2005) *Synthesis, Properties and Applications of Ultrananocrystalline Diamond*, NATO Science Series, vol. 192, Springer, Dordrecht.

Obraztsova, E.D., Kuznetsov, V.L., Loubnin, E.N., Pimenov, S.M., and Pereverzev, V.G. (1996) Raman and Photoluminescence Spectra of Diamond Particles with 1–5 nm Diameter, in *Nanoparticles in Solids and Solutions* (eds J.H. Fendler and I. Dékány), Kluwer Academic Publishers, Amsterdam, pp. 485–496.

Shenderova, O.A. and Gruen, D.M. (2006) *Ultrananocrystalline Diamond: Synthesis, Properties, and Applications*, William Andrew Inc., Norwich.

Shenderova, O.A., Zhirnov, V.V., and Brenner, D.W. (2002) *Crit. Rev. Solid State Mater. Sci.*, 27, 227–356.

8.2
Figure References

Many figures in this book originate from other sources: The following figures were generously provided by scientists and companies. I gratefully acknowledge them:

1.4: http://ekati.bhpbilliton.com/docs/Koala.pdf, Ekati-Mine, Billiton Corp., Kanada; **1.10:** Dr. M. Ozawa, University of Kiel; **1.14:** Dr. A. Schwarz, University of Hamburg; **2.3:** Prof. M. Sumper, University of Regensburg; **2.6c:** Felix Köhler, University of Kiel; **3.1b:** Prof. F. Banhart, University of Sbrasbourg; **3.23b:** Prof. K. P. C. Vollhardt, Berkeley University, USA; **5.12:** Dr. Vlad Padalko, Alit Corp., Zhitomir, Ukraine; **5.23:** Dr. Fedor Jelezko, University of Stuttgart; **6.49:** Dr. Lothar Schäfer, Fraunhofer Gesellschaft, IST Braunschweig.

The following figures were reproduced from various original publications. The respective publishers are mentioned in alphabetical order, within these sections the entries are sorted by figure numbers.

The following figure was reproduced with kind permission of the American Astronomical Society:

4.8a: T. J. Bernatowicz, R. Cowsik, P. C. Gibbons, K. Lodders, B. Fegley, jr., S. Amari, R. S. Lewis, *Astrophys. J.* **1996**, *472*, 760–782.

The following figures were reproduced with kind permission of the American Association for the Advancement of Science (AAAS):

2.31: J. M. Hawkins, A. Meyer, T. A. Lewis, S. Loren, F. J. Hollander, *Science* **1991**, *252*, 312–313; **2.50c:** M. Saunders, *Science* **1991**, *253*, 330–331; **3.14:** A. Thess, R.

Lee, P. Nikolaev, H. Dai, P. Petit, J. Robert, C. Xu, Y. Hee Lee, S. G. Kim, A. G. Rinzler, D. T. Colbert, G. E. Scuseria, D. Tománek, J. E. Fischer, R. E. Smalley, *Science* **1996**, *273*, 483–487; **3.18, 3.33b**: K. Hata, D. N. Futaba, K. Mizuno, T. Namai, M. Yumura, S. Iijima, *Science* **2004**, *306*, 1362–1364; **3.40**: R. Krupke, F. Hennrich, H. v. Löhneysen, M. M. Kappes, *Science* **2003**, *301*, 344–347; **3.46**: P. Poncharal, Z. L. Wang, D. Ugarte, W. A. de Heer, *Science* **1999**, *283*, 1513–1516; **3.54**: S. Frank, P. Poncharal, Z. L. Wang, W. A. de Heer, *Science* **1998**, *280*, 1744–1746; **3.58**: M. J. O'Connell, S. M. Bachilo, C. B. Huffman, V. C. Moore, M. S. Strano, E. H. Haroz, K. L. Rialon, P. J. Boul, W. H. Noon, C. Kittrell, J. Ma, R. H. Hauge, R. B. Weisman, R. E. Smalley, *Science* **2002**, *297*, 593–597; **3.79b**: M. S. Strano, C. A. Dyke, M. L. Usrey, P. W. Barone, M. J. Allen, H. Shan, C. Kittrell, R. H. Hauge, J. M. Tour, R. E. Smalley, *Science* **2003**, *301*, 1519–1522; **3.110b**: J. Kong, N. R. Franklin, C. Zhou, M. G. Chapline, S. Peng, K. Cho, H. Dai, *Science* **2000**, *287*, 622–625; **3.113**: Y.-L. Li, I. A. Kinloch, A. H. Windle, *Science* **2004**, *304*, 276–278.

The following figures were reproduced with kind permission of the American Chemical Society (ACS):

2.27, 2.30: H. Ajie, M. M. Alvarez, S. J. Anz, R. D. Beck, F. Diederich, K. Fostiropoulos, D. R. Huffman, W. Krätschmer, Y. Rubin, K. E. Schriver, D. Sensharma, R. L. Whetten, *J. Phys. Chem.* **1990**, *94*, 8630–8633; **2.39**: Q. Xie, E. Perez-Cordero, L. Echegoyen, *J. Am. Chem. Soc.* **1992**, *114*, 3978–3980; **2.45b**: M. M. Olmstead, A. S. Ginwalla, B. C. Noll, D. S. Tinti, A. L. Balch, *J. Am. Chem. Soc.* **1996**, *118*, 7737–7745; **2.47**: T. Suzuki, Y. Maruyama, T. Kato, K. Kikuchi, Y. Achiba, *J. Am. Chem. Soc.* **1993**, *115*, 11006–11007; **2.74**: X. Lu, Z. Chen, W. Thiel, P. v. R. Schleyer, R. Huang, L. Zheng, *J. Am. Chem. Soc.* **2004**, *126*, 14871–14878; **3.20**: T. Guo, P. Nikolaev, A. G. Rinzler, D. Tomanek, D. T. Colbert, R. E. Smalley, *J. Phys. Chem.* **1995**, *99*, 10694–10697; **3.21**: K. Hernadi, L. Thiên-Nga, L. Forró, *J. Phys. Chem. B* **2001**, *105*, 12464–12468; **3.31b**: H. Hou, Z. Jun, F. Weller, A. Greiner, *Chem. Mater.* **2003**, *15*, 3170–3175; **3.34**: J. Gao, A. Yu, M. E. Itkis, E. Bekyarova, B. Zhao, S. Niyogi, R. C. Haddon, *J. Am. Chem. Soc.* **2004**, *126*, 16698–16699; **3.37**: C. A. Furtado, U. J. Kim, H. R. Gutierrez, L. Pan, E. C. Dickey, P. C. Eklund, *J. Am. Chem. Soc.* **2004**, *126*, 6095–6105; **3.38, 3.39**: M. E. Itkis, D. E. Perea, R. Jung, S. Niyogi, R. C. Haddon, *J. Am. Chem. Soc.* **2005**, *127*, 3439–3448; **3.45b**: M.-F. Yu, B. I. Yakobson, R. S. Ruoff, *J. Phys. Chem. B* **2000**, *104*, 8764–8767; **3.50b**: P. Avouris, *Acc. Chem. Res.* **2002**, *35*, 1026–1034; **3.56a**: C. A. Furtado, U. J. Kim, H. R. Gutierrez, L. Pan, E. C. Dickey, P. C. Eklund, *J. Am. Chem. Soc.* **2004**, *126*, 6095–6105; **3.60a**: C. Engtrakul, M. F. Davis, T. Gennett, A. C. Dillon, K. M. Jones, M. J. Heben, *J. Am. Chem. Soc.* **2005**, *127*, 17548–17555; **3.60b**: A. Kitaygorodskiy, W. Wang, S.-Y. Xie, Y. Lin, K. A. S. Fernando, X. Wang, L. Qu, B. Chen, Y.-P. Sun, *J. Am. Chem. Soc.* **2005**, *127*, 7517–7520; **3.83a**: K. A. S. Fernando, Y. Lin, W. Wang, S. Kumar, B. Zhou, S.-Y. Xie, L. T. Cureton, Y.-P. Sun, *J. Am. Chem. Soc.* **2004**, *126*, 10234–10235; **3.96b, c**: G. Korneva, H. Ye, Y. Gogotsi, D. Halverson, G. Friedman, J.-C. Bradley, K. G. Kornev, *Nano Lett.* **2005**, *5*, 879–884; **3.97**: J. Sloan, A. I. Kirkland, J. L. Hutchison, M. L. H. Green, *Acc. Chem. Res.* **2002**, *35*, 1054–

1062; **3.98b**: T. Okazaki, K. Suenaga, K. Hirahara, S. Bandow, S. Iijima, H. Shinohara, *J. Am. Chem. Soc.* **2001**, *123*, 9673–9674; **3.102**: L. Li, C. Y. Li, C. Ni, *J. Am. Chem. Soc.* **2006**, *128*, 1692–1699; **3.109**: J. R. Wood, M. D. Frogley, E. R. Meurs, A. D. Prins, T. Peijs, D. J. Dunstan, H. D. Wagner, *J. Phys. Chem. B* **1999**, *103*, 10388–10392; **4.17, 4.30**: M. Ozawa, H. Goto, M. Kusunoki, E. Osawa, *J. Phys. Chem. B* **2002**, *106*, 7135–7138; **4.31**: Q. L. Zhang, S. C. O'Brien, J. R. Heath, Y. Liu, R. F. Curl, H. W. Kroto, R. E. Smalley, *J. Phys. Chem.* **1986**, *90*, 525–528; **5.46**: S.-J. Yu, M.-W. Kang, H.-C. Chang, K.-M. Chen, Y.-C. Yu, *J. Am. Chem. Soc.* **2005**, *127*, 17604–17605; **6.36**: P. John, N. Polwart, C. E. Troupe, J. I. B. Wilson, *J. Am. Chem. Soc.* **2003**, *125*, 6600–6601; **6.38**: T. Nakamura, M. Suzuki, M. Ishihara, T. Ohana, A. Tanaka, Y. Koga, *Langmuir* **2004**, *20*, 5846–5849; **7.4**: Y. Shirai, A. J. Osgood, Y. Zhao, K. F. Kelly, J. M. Tour, *Nano Lett.* **2005**, *5*, 2330–2334.

The following figures were reproduced with kind permission of the American Institute of Physics (AIP):

1.20b: F. Tuinstra, J. L. Koenig, *J. Chem. Phys.* **1970**, *53*, 1126–1130; **3.29b**: M. Endo, Y. A. Kim, T. Hayashi, Y. Fukai, K. Oshida, M. Terrones T. Yanagisawa, S. Higaki, M. S. Dresselhaus, *Appl. Phys. Lett.* **2002**, *80*, 1267–1269; **3.31a**: S. Yang, X. Chen, S. Motojima, *Appl. Phys. Lett.* **2002**, *81*, 3567–3569; **3.52**: R. Saito, M. Fujita, G. Dresselhaus, M. S Dresselhaus, *Appl. Phys. Lett.* **1992**, *60*, 2204–2206; **3.114b**: C. Kim, Y. J. Kim, Y. A. Kim, T. Yanagisawa, K. C. Park, M. Endo, M. S. Dresselhaus, *J. Appl. Phys.* **2004**, *96*, 5903–5905; **4.26**: V. L. Kuznetsov, I. L. Zilberberg, Yu. V. Butenko, A. L. Chuvilin, B. Segall, *J. Appl. Phys.* **1999**, *86*, 863–870; **4.37b**: S. Tomita, T. Sakurai, H. Ohta, M. Fujii, S. Hayashi, *J. Chem. Phys.* **2001**, *114*, 7477–7482; **5.16a, 5.17b**: O. O. Mykhaylyk, Y. M. Solonin, D. N. Batchelder, R. Brydson, *J. Appl. Phys.* **2005**, *97*, 074302; **6.4**: Z. H. Shen, P. Hess, J. P. Huang, Y. C. Lin, K. H. Chen, L. C. Chen, S. T. Lin, *J. Appl. Phys.* **2006**, *99*, 124302; **6.23b**: J. W. Ager III, W. Walukiewicz, M. McCluskey, M. A. Plano, M. I. Landstrass, *Appl. Phys. Lett.* **1995**, *66*, 616–618; **6.28**: X. Jiang, C.-P. Klages, *Appl. Phys. Lett.* **1992**, *61*, 1629–1631; **6.29**: F. Arezzo, N. Zacchetti, W. Zhu, *J. Appl. Phys.* **1994**, *75*, 5375–5381.

The following figures were reproduced with kind permission of the American Physical Society (APS):

1.20a: M. P. Conrad, H. L. Strauss, *Phys. Rev. B* **1985**, *31*, 6669–6675; **3.6**: M. Fujita, R. Saito, G. Dresselhaus, M. S. Dresselhaus, *Phys. Rev. B* **1992**, *45*, 13834–13836; **3.31c**: S. Ihara, S. Itoh, J. Kitakami, *Phys. Rev. B* **1993**, *48*, 5643–5648; **3.59c**: G. Wagoner, *Phys. Rev.* **1960**, *118*, 647–653; **4.6**: M. I. Heggie, M. Terrones, B. R. Eggen, G. Jungnickel, R. Jones, C. D. Latham, P. R. Briddon, H. Terrones, *Phys. Rev. B* **1998**, *57*, 13339–13342; **5.16b**: J. W. Ager, III, D. K. Veirs, G. M. Rosenblatt, *Phys. Rev. B* **1991**, *43*, 6491–6499; **6.6, 6.8**: Th. Frauenheim, U. Stephan, P. Blaudeck, D. Porezag, H.-G. Busmann, W. Zimmermann-Edling, S. Lauer, *Phys. Rev. B* **1993**, *48*, 18189–18202; **6.31**: F. Maier, M. Riedel, B. Mantel, J. Ristein, L. Ley,

Phys. Rev. Lett. **2000**, *85*, 3472–3475; **7.1b**: M. Menon, D. Srivastava, *Phys. Rev. Lett.* **1997**, *79*, 4453–4456.

The following figures were reproduced with kind permission of the American Vacuum Society:

3.15: M. J. Bronikowski, P. A. Willis, D. T. Colbert, K. A. Smith, R. E. Smalley, *J. Vac. Sci. Technol. A* **2001**, *19*, 1800–1805; **6.23a**: K. M. McNamara, K. K. Gleason, C. J. Robinson, *J. Vac. Sci. Technol. A* **1992**, *10*, 3143–3148.

The following figures were reproduced with kind permission of Cambridge University Press:

1.13a: G. M. Jenkins, K. Kawamura, *Polymeric carbons – carbon fibre, glass and char*, Cambridge University Press, Cambridge **1976**; **3.10**: P. J. F. Harris, *Carbon Nanotubes & Related Structures*, Cambridge University Press, Cambridge **1999**.

The following figure was reproduced with kind permission of CRC Press:

3.13: M. Meyyappan (editors), *Carbon Nanotubes Science & Applications*, CRC Press, Boca Raton, FL, **2005**, p 83.

The following figure was reproduced with kind permission of the Electrochemical Society:

6.47: H. B. Martin, A. Argoitia, U. Landau, A. B. Anderson, J. C. Angus, *J. Electrochem. Soc.* **1996**, *143*, L133–L136.

The following figures were reproduced with kind permission of Elsevier:

2.44b: M. M. Olmstead, L. Hao, A. L. Balch, *J. Organometal. Chem.* **1999**, *578*, 85–90; **3.16b**: F. Lupo, J. A. Rodriguez-Manzo, A. Zamudio, A. L. Elias, Y. A. Kim, T. Hayashi, M. Muramatsu, R. Kamalakaran, H. Terrones, M. Endo, M. Rühle, M. Terrones, *Chem. Phys. Lett.* **2005**, *410*, 384–390; **3.19b**: H. W. Zhu, X. S. Li, B. Jiang, C. L. Xu, Y. F. Zhu, D. H. Wu, X. H. Chen, *Chem. Phys. Lett.* **2002**, *366*, 664–669; **3.28a**: C. J. Lee, J. H. Park, J. Park, *Chem. Phys. Lett.* **2000**, *323*, 560–565; **3.28b**: Y. F. Li, J. S. Qiu, Z. B. Zhao, T. H. Wang, Y. P. Wang, W. Li, *Chem. Phys. Lett.* **2002**, *366*, 544–550; **3.30b**: T. Yamaguchi, S. Bandow, S. Iijima, *Chem. Phys. Lett.* **2004**, *389*, 181–185; **3.33a**: K.-H. Lee, K. Baik, J.-S. Bang, S.-W. Lee, W. Sigmund, *Solid State Commun.* **2004**, *129*, 583–587; **3.35**: R. E. Morjan, V. Maltsev, O. Nerushev, Y. Yao, L. K. L. Falk, E. E. B. Campbell, *Chem. Phys. Lett.* **2004**, *383*, 385–390; **3.36**: D.-C. Li, L. Dai, S. Huang, A. W. H. Mau, Z. L. Wang, *Chem. Phys. Lett.* **2000**, *316*, 349–355; **3.59a**: M. Kosaka, T. W. Ebbesen, H. Hiura, K. Tanigaki, *Chem. Phys. Lett.* **1994**, *225*, 161–164; **3.59b**: M. Kosaka, T. W. Ebbesen, H. Hiura, K. Tanigaki, *Chem. Phys. Lett.* **1995**, *233*, 47–51; **3.61b**: T. Pichler, M. Sing, M.

Knupfer, M. S. Golden, J. Fink, *Solid State Commun.* **1999**, *109*, 721–726; **3.95b**: L. Jiang, L. Gao, *Carbon* **2003**, *41*, 2923–2929; **4.3c, 4.4**: Q. Ru, M. Okamoto, Y. Kondo, K. Takayanagi, *Chem. Phys. Lett.* **1996**, *259*, 425–431; **4.8b**: P. J. F. Harris, R. D. Vis, D. Heymann, *Earth Planet. Sci.* **2000**, *183*, 355–359; **4.11**: T. Cabioc'h, J. P. Rivière, M. Jaouen, J. Delafont, M. F. Denanot, *Synth. Metals* **1996**, *77*, 253–256; **4.15**: W. A. de Heer, D. Ugarte, *Chem. Phys. Lett.* **1993**, *207*, 480–486; **4.19**: D. Ugarte, *Chem. Phys. Lett.* **1993**, *209*, 99–103; **4.20**: S. Tomita, M. Fujii, S. Hayashi, K. Yamamoto, *Chem. Phys. Lett.* **1999**, *305*, 225–229; **4.21**: L.-C. Qin, S. Iijima, *Chem. Phys. Lett.* **1996**, *262*, 252–258; **4.32**: T. Cabioc'h, A. Kharbach, A. Le Roy, J. P. Rivière, *Chem. Phys. Lett.* **1998**, *285*, 216–220; **4.33**: E. D. Obraztsova, M. Fujii, S. Hayashi, V. L. Kuznetsov, Yu. V. Butenko, A. L. Chuvilin, *Carbon* **1998**, *36*, 821–826; **4.34**: S. Tomita, A. Burian, J. C. Dore, D. LeBolloch, M. Fujii, S. Hayashi, *Carbon* **2002**, *40*, 1469–1474; **4.35**: W. A. de Heer, D. Ugarte, *Chem. Phys. Lett.* **1993**, *207*, 480–486; **4.36**: S. Tomita, M. Fujii, S. Hayashi, K. Yamamoto, *Chem. Phys. Lett.* **1999**, *305*, 225–229; **4.37a**: R. Selvan, R. Unnikrishnan, S. Ganapathy, T. Pradeep, *Chem. Phys. Lett.* **2000**, *316*, 205–210; **4.41**: P. Redlich, F. Banhart, Y. Lyutovich, P. M. Ajayan, *Carbon* **1998**, *36*, 561–563; **5.2**: T. L. Daulton, D. D. Eisenhour, T. J. Bernatowicz, R. S. Lewis, P. R. Buseck, *Geochim. Cosm. Acta* **1996**, *60*, 4853–4872; **5.14**: Y. Q. Zhu, T. Sekine, T. Kobayashi, E. Takazawa, M. Terrones, H. Terrones, *Chem. Phys. Lett.* **1998**, *287*, 689–693; **5.17a**: E. D. Obraztsova, M. Fujii, S. Hayashi, V. L. Kuznetsov, Yu. V. Butenko, A. L. Chuvilin, *Carbon* **1998**, *36*, 821–826; **5.24**: J.-B. Donnet, E. Fousson, L. Delmotte, M. Samirant, C. Baras, T. K. Wang, A. Eckhardt, *C. R. Acad. Sci. Fr.* **2000**, *3*, 831–838; **5.25**: A. I. Shames, A. M. Panich, W. Kempinski, A. E. Alexenskii, M. V. Baidakova, A. T. Dideikin, V. Yu. Osipov, V. I. Siklitski, E. Osawa, M. Ozawa, A. Ya. Vul', *J. Phys. Chem. Solids* **2000**, *63*, 1993–2001; **5.26**: H. Hirai, M. Terauchi, M. Tanaka, K. Kondo, *Diamond Relat. Mater.* **1999**, *8*, 1703–1706; **5.28**: D. He, L. Shao, W. Gong, E. Xie, K. Xu, G. Chen, *Diamond Relat. Mater.* **2000**, *9*, 1600–1603; **6.2a**: T. Teraji, M. Hamada, H. Wada, M. Yamamoto, K. Arima, T. Ito, *Diamond Relat. Mater.* **2005**, *14*, 255–260; **6.2b**: J. Schwarz, K. Meteva, A. Grigat, A. Schubnov, S. Metev, F. Vollertsen, *Diamond Relat. Mater.* **2005**, *14*, 302–307; **6.13a**: E. Blank in C. Nebel, J. Ristein (editors), *Thin Film Diamond I, Semiconductors and Semimetals*, Vol. 76, Elsevier, Amsterdam **2003**, p 59; **6.13b**: E. Blank *in* C. Nebel, J. Ristein (editors), *Thin Film Diamond I, Semiconductors and Semimetals*, Vol. 76, Elsevier, Amsterdam **2003**, p 61; **6.26**: Z. Sun, J. R. Shi, B. K. Tay, S. P. Lau, *Diamond Relat. Mater.* **2000**, *9*, 1979–1983; **6.27a**: A. T. Collins, *Physica B* **1993**, *185*, 284–296; **6.30a**: S. Michaelson, A. Hoffman, *Diamond Relat. Mater.* **2005**, *14*, 470–475; **6.33**: W. P. Kang, J. L. Davidson, A. Wisitsora-at, Y. M. Wong, R. Takalkar, K. Subramanian, D. V. Kerns, W. H. Hofmeister, *Diamond Relat. Mater.* **2005**, *14*, 685–690; **6.42a**: J. N. Russell, J. E. Butler, G. T. Wang, S. F. Bent, J. S. Hovis, R. J. Hamers, M. P. D'Evelyn, *Mater. Chem. Phys.* **2001**, *72*, 147–151; **6.48**: J. C. Angus, Y. V. Pleskow, S. C. Eaton *in* C. Nebel, J. Ristein (rditors), *Thin Film Diamond II, Semiconductors and Semimetals*, Vol. 77, Elsevier Academic Press, Amsterdam **2004**, p 99.

The following figure was reproduced with kind permission of the Institute of Physics (IOP):

4.10: Y. Shimizu, T. Sasaki, T. Ito, K. Terashima, N. Koshizaki, *J. Phys. D.: Appl. Phys.* **2003**, *36*, 2940–2944.

The following figure was reproduced with kind permission of the Japanese Chemical Society (JCS):

2.32: K. Kikuchi, N. Nakahara, M. Honda, S. Suzuki, K. Saito, H. Shiromaru, K. Yamauchi, I. Ikemoto, T. Kuramochi, S. Hino, Y. Achiba, *Chem. Lett.* **1991**, 1607–1610.

The following figures were reproduced with kind permission of the Materials Research Society (MRS):

3.32: X. Chen, W. In-Hwang, S. Shimada, M. Fujii, H. Iwanaga, S. Motojima, *J. Mater. Res.* **2000**, *15*, 808–814; **4.38**: A. Romanenko, O. A. Anikeeva, A. V. Okotrub, V. L. Kuznetsov, Y. V. Butenko, A. L. Chuvilin, C. Dong, Y. Ni, *Mat. Res. Soc. Symp. Proc.* **2002**, *703*, 259–264; **6.19**: D. M. Gruen, *MRS Bull.* **2001**, *26*, 771–776; **6.24**: C. D. Zuiker, A. R. Krauss, D. M. Gruen, J. A. Carlisle, L. J. Terminello, S. A. Asher, R. W. Bormett, *Mat. Res. Soc. Symp. Proc.* **1996**, *437*, 211–218.

The following figures were reproduced with kind permission of the Nature Publishing Group:

2.48: M. Saunders, H. A. Jiménez-Vázquez, R. J. Cross, S. Mroczkowski, D. I. Freedberg, F. A. L. Anet, *Nature* **1994**, *367*, 256–258; **3.1a**: S. Iijima, *Nature* **1991**, *354*, 56–58; **3.22**: M. Endo, M. Muramatsu, T. Hayashi, Y. A. Kim, M. Terrones, M. S. Dresselhaus, *Nature* **2005**, *433*, 476; **4.1**: *Nature* **1992**, *359*, cover picture 22.10.1992; **4.27**: H. W. Kroto, K. McKay, *Nature* **1988**, *331*, 328–331; **6.30b**: D. A. Muller, Y. Tzou, R. Raj, J. Silcox, *Nature* **1993**, *366*, 725–727; **7.2**: A. Srivastava, O. N. Srivastava, S. Talapatra, R. Vajtai, P. M. Ajayan, *Nature Mater.* **2004**, *3*, 610–614.

The following figure was reproduced with kind permission of the Neue Schweizerische Chemische Gesellschaft (Switzerland):

2.45a: H. B. Bürgi, P. Venugopalan, D. Schwarzenbach, F. Diederich, C. Thilgen, *Helv. Chim. Acta* **1993**, *76*, 2155–2159.

The following figures were reproduced with kind permission of the Oyo Butsuri Gakkai (Japan):

3.61a: R. Kuzuo, M. Terauchi, M. Tanaka, Y. Saito, *Jpn. J. Appl. Phys.* **1994**, *33*, L1316–L1319; **6.42b**: Md. Z. Hossain, T. Aruga, N. Takagi, T. Tsuno, N. Fujimori, T. Ando, M. Nishijima, *Jpn. J. Appl. Phys. (2)* **1999**, *38*, L1496–L1498.

The following figures were reproduced with kind permission of the Royal Society London:

2.11: H. W. Kroto, D. R. M. Walton, D. E. H. Jones, R. C. Haddon, *Phi.l Trans. R. Soc. A* 1993, *343*, 103–112; **3.56c**: M. S. Dresselhaus, G. Dresselhaus, J. C. Charlier, E. Hernandez, *Phil. Trans. R. Soc. A* 2004, *362*, 2065–2098; **4.7**: M. Terrones, W. K. Hsu, J. P. Hare, H. W. Kroto, H. Terrones, D. R. M. Walton, *Phil. Trans. R. Soc. A* 1996, *354*, 2025–2054; **6.25**: S. Prawer, R. J. Nemanich, *Phil. Trans. R. Soc. A* 2004, *362*, 2537–2565.

The following figures were reproduced with kind permission of the Royal Society of Chemistry (RSC):

2.28: J. P. Hare, T. J. Dennis, H. W. Kroto, R. Taylor, A. W. Allaf, S. Balm, D. R. M. Walton, *J. Chem. Soc., Chem. Commun.* 1991, 412–413; **2.70**: S. Yoshimoto, E. Tsutsumi, O. Fujii, R. Narita, K. Itaya, *Chem. Commun.* 2005, 1188–1190; **3.56b**: Q. Li, I. A. Kinloch, A. H. Windle, *Chem. Commun.* 2005, 3283–3285; **3.84b**: J. Sun, L. Gao, M. Iwasa, *Chem. Commun.* 2004, 832–833; **3.99**: D. A. Britz, A. N. Khlobystov, K. Porfyrakis, A. Ardavan, G. A. D. Briggs, *Chem. Commun.* 2005, 37–39; **3.112**: A. Bianco, K. Kostarelos, C. D. Partidos, M. Prato, *Chem. Commun.* 2005, 571–577; **5.32**: T. Tsubota, O. Hirabayashi, S. Ida, S. Nagaoka, M. Nagata, Y. Matsumoto, *Phys. Chem. Chem. Phys.* 2002, *4*, 806–811; **7.1a, 7.3**: H. Terrones, M. Terrones, W. K. Hsu, *Chem. Soc. Rev.* 1995, *24*, 341–350.

The following figures were reproduced with kind permission of Springer:

1.12b: M. S. Dresselhaus, G. Dresselhaus, K. Sugihara, I. L. Spain, H. A. Goldberg, *Graphite Fibers and Filaments*, Springer Ser. Mater. Sci. 1988, *5*, p 4, p 11; **2.73**: V. Blank, S. Buga, G. Dubitsky, N. Serebryanaya, M. Popov, V. Prokhorov *in* E. Osawa, *Perspectives of Fullerene Nanotechnology*, Kluwer Academic Publishers, Dordrecht 2002, p 227; **3.11a**: J.-C. Charlier, S. Iijima *in* M. S. Dresselhaus, G. Dresselhaus, Ph. Avouris, *Top. Appl. Phys.* 2000, *80*, 65; **3.11b**: J.-C. Charlier, X. Blase, A. De Vita, R. Car, *Appl. Phys. A* 1999, *68*, 267–273; **3.96a**: D. Ugarte, T. Stöckli, J. M. Bonard, A. Châtelain, W. A. de Heer, *Appl. Phys. A* 1998, *67*, 101–105; **3.107**: S. Uemura *in* E. Osawa, *Perspectives in Fullerene Nanotechnology*, Kluwer Academic Publishers, Dordrecht 2002, p 60; **5.22**: A. E. Aleksenskii, V. Yu. Osipov, N. A. Kryukov, V. K. Adamchuk, M. I. Abaev, S. P. Vul', A. Ya. Vul', *Techn. Phys. Lett.* 1997, *23*, 874–876; **5.27**: J.-Y. Raty, G. Galli, *in* D. M. Gruen, O. A. Shenderova, A. Ya. Vul' (editors), *Synthesis, Properties and Applications of Ultrananocrystalline Diamond*, NATO Science Series, Vol. 192, Springer, Dordrecht 2005, p 15–24; **5.34**: E. P. Smirnov, S. K. Gordeev, S. I. Kol'tsov, V. B. Aleskovskii, *J. Appl. Chem. USSR* 1978, *51*, 2451–2456.

The following figures were reproduced with kind permission of Taylor & Francis:

4.18: M. S. Zwanger, F. Banhart, *Phil. Mag.* **1995**, *72*, 149–157; **5.20**: L. A. Bursill, J. L. Peng, S. Prawer, *Phil. Mag. A* **1997**, *76*, 769–781; **6.12**: D. Dorignac, S. Delclos, F. Phillipp, *Phil. Mag. B* **2001**, *81*, 1879–1891.

The following figure was reproduced with kind permission of John Wiley & Sons:

2.29: M. S. Dresselhaus, G. Dresselhaus, P. C. Eklund, *J. Raman Spectrosc.* **1996**, *27*, 351–371.

The following figures were reproduced with kind permission of Wiley Interscience:

2.15b: J. C. Grossman, C. Piskoti, S. G. Louie, M. L. Cohen, A. Zettl *in* K. M. Kadish, R. S. Ruoff, *Fullerenes*, Wiley Interscience, New York 2000, p 898; **2.46**: H. Shinohara *in* K. M. Kadish, R. S. Ruoff, *Fullerenes*, Wiley Interscience, New York 2000, pp. 272–273; **6.27b**: K. E. Spear, J. P. Dismukes, *Synthetic Diamond: Emerging CVD Science and Technology*, Wiley Interscience, New York, **1994**, S. 405.

The following figures were reproduced with kind permission of Wiley-VCH:

1.19: U. Schwarz, *Chem. Unserer Zeit* **2000**, *34*, 212–222; **3.26b**: S.-H. Jung, S.-H. Jeong, S.-U. Kim, S.-K. Hwang, P.-S. Lee, K.-H. Lee, J.-H. Ko, E. Bae, D. Kang, W. Park, H. Oh, J.-J. Kim, H. Kim, C.-G. Park, *Small* **2005**, *1*, 553–559; **3.48, 3.49**: E. Joselevich, *Chem. Phys. Chem.* **2004**, *5*, 619–624; **3.50a**: S. Reich, C. Thomsen, J. Maultzsch, *Carbon Nanotubes*, Wiley-VCH, Weinheim **2004** (adapted from M. Machón, S. Reich, C. Thomsen, D. Sánchez-Portal, P. Ordejón, *Phys. Rev. B* **2002**, *66*, 155410); **3.51**: E. Joselevich, *Chem. Phys. Chem.* **2004**, *5*, 619–624; **3.57**: A. Hartschuh, H. N. Pedrosa, J. Peterson, L. Huang, P. Anger, H. Qian, A. J. Meixner, M. Steiner, L. Novotny, T. D. Krauss, *Chem. Phys. Chem.* **2005**, *6*, 577–582; **3.77b**: S. Banerjee, M. G. C. Kahn, S. S. Wong, *Chem. Eur. J.* **2003**, *9*, 1898–1908; **3.88b**: Z. Wei, M. Wan, T. Lin, L. Dai, *Adv. Mater.* **2003**, *15*, 136–139.

The following figure was reproduced with kind permission of World Scientific (Singapore):

3.53: P. Lambin, J.-C. Charlier, J.-P. Michenaud *in* H. Kuzmany, J. Fink, M. Mehring, S. Roth (editors), *Progress in Fullerene Research*, World Scientific, Singapore **1994**, p 130–134.

The following figures were taken from the encyclopedia Wikipedia:

1.2: Hope diamond in National Museum of Natural History. Picture taken in April 2004. Permission Source: English Wikipedia, original upload 29 May 2004 by

Kowloonese; Permission is granted to copy, distribute and/or modify this document under the terms of the GNU Free Documentation License, Version 1.2 or any later version published by the Free Software Foundation; with no Invariant Sections, no Front-Cover Texts, and no Back-Cover Texts. **1.17a**: *rough_diamond.jpg* produced by the U.S. Geological Survey, in Wikipedia "*Diamant*"; **2.2.b**: Expo '67 American Pavilion (now the Biosphère), by R. Buckminster Fuller, taken by Montréalais. http://en.wikipedia.org/wiki/Image:Mtl-biosphere.jpg, Permission is granted to copy, distribute and/or modify this document under the terms of the GNU Free Documentation License, Version 1.2 or any later version published by the Free Software Foundation; with no Invariant Sections, no Front-Cover Texts, and no Back-Cover Texts.

Index

a

abrasive 424ff., *see also* polishing agents
absorption spectroscopy
– carbon onions 290, 316f.
– diamond 24
– diamond films 413f.
– fullerenes 59ff., 65f.
– nanodiamonds 358
– nanotubes 209ff., 221
acids, *see also* piranha water
– nanodiamond purification 335, 349
– nanotube purification 171f.
– reactions with graphite 27
activated carbon 15f., 31
addition reactions, *see also* cycloaddition
– fullerenes, energetically favored sites 68
– multiple, regioselectivity 68ff.
– nomenclature 69f.
– nucleophilic 93ff., 231, 376
– template-directed 71, 97
aggregates, nanodiamond 338ff., 350f., 370
allotropes 6ff., 33, 125
aluminum, as type IIb diamond impurity 9
aminofullerenes 94f.
anisotropy
– graphite 8, 22
– nanotubes 213, 216, 276
annulenes
– electronic characteristics 195f., 201
– source of carbon nanostructures 33, 48, 160, 297
anthracite 16
arc discharge
– carbon onion formation 291ff.
– diamond film deposition 405
– endohedral metallofullerene preparation 82, 302

– fullerene synthesis 50
– nanotube growth mechanisms 180ff.
– nanotube synthesis 140ff., 150ff., 158
armchair nanotubes 126ff., 182f.
aromaticity
– fullerenes 39, 102
– nanotube 133ff., 218
arrays, nanotube 149f., 168ff.
atomic force microscopy (AFM)
– nanotube tips 267f.
– uses, for nanostructure studies 191, 395, 397, 413
azafullerenes, *see* heterofullerenes

b

ballistic electron transport 203f.
bamboo-like nanotubes 156f., 163f.
band structures 194ff., 352ff.
beltenes 162f.
Bingel-Hirsch reaction 95ff., 231f.
biological applications, *see also* medical applications
– bioassay 434, 439
– carrier systems (drug/vaccine delivery) 274f., 385
– DNA molecular recognition 273
– nanostructure toxicity 273, 385
– sensors 273ff., 446f.
boiling point 13
bonding, atomic 2, 6f.
– bond lengths 7, 38f., 43f.
– dangling bonds, reactivity 131, 218, 399
– nanostructure growth mechanisms 42, 309, 311
– surface reconstruction 333f., 394ff.
Boudouard equilibrium 144, 341
Brillouin zone 194, 196f., 199, 353
bromofullerenes 92f.

Carbon Materials and Nanotechnology. Anke Krueger
Copyright © 2010 WILEY-VCH Verlag GmbH & Co. KGaA, Weinheim
ISBN: 978-3-527-31803-2

buckminsterfullerene (C_{60}) 33, 411
– crystalline lattice derivatives 64f., 116f.
– solubility 57f.
– structure 37ff.
bucky diamonds 334, 337f., 377

c

C_{20} 40
C_{36} (fullerene structures) 46f., 117f.
C_{60}, see buckminsterfullerene
C_{70}
– combustion synthesis 49f.
– solubility 59
– structure 43
C_8K 28
cage structures
– analagous to fullerenes 33ff.
– curvature 37f.
– mathematically possible 40
calixarenes 113f.
carbenes 98, 230f.
carbides 25f., 409, 427, see also silicon carbide
carbon, as element
– abundance 1
– allotropes 6ff., 33, 125
– chemical properties 24ff.
– electronic structure 6f., 20
– history (uses and discoveries) 1ff., 33ff.
– isotopes 2f., 21
– phase diagram 20f., 63f.
carbon black, see soot
carbon clusters
– dangling bonds 42, 309, 311
– structures 45f., 310ff., 337f.
– surface nucleation 407f.
carbon fibers 15
– manufacture, from polyacrylonitrile (PAN) 11
– structure 11f., 124
– uses 31, 123, 246
carbon nanotubes (CNTs), see also multiwalled nanotubes; nanotube chemistry; single-walled nanotubes
– applications 266, 271ff.
– array construction 149f., 168ff.
– chemical reactivity 217ff.
– composite materials 246ff., 264ff., 275ff.
– discovery 123f.
– electronic properties and applications 194, 202ff., 267ff., 279f.
– geometric variants 163ff., 451f.
– growth mechanisms 180ff.
– mechanical properties 190ff.
– production 140, 159ff.
– purification 148, 171ff.
– solubility and dispersion 187ff.
– spectroscopic properties 206ff.
– structural classification 126f.
– thermal properties 216f.
carbon nanohorns (SWNH) 165f.
carbon onions
– chemical properties 321ff.
– faceted nanoparticles 289f.
– growth mechanisms 307ff.
– occurrence 290f., 316
– physical properties 313ff.
– preparation methods 291ff., 305f.
– production by carbon transformation 298ff.
– structural arrangement 283ff.
– uses and applications 324ff.
carbonados 4, 19, 345
carborundum, see silicon carbide
carbyne structures 10
catalysts
– activity of carbon onions 326
– co-catalyst method, in CVD 159
– deposition techniques 155f., 168
– floating 144ff., 154, 156, 165
– nanotubes as support 277f.
chaoite 10
chemical vapor deposition (CVD) 147ff., 154ff., 158ff., 391, 403f.
– growth mechanisms, nanotube 185f.
– hot filament 155, 404f.
– low pressure (LPCVD) 292f.
– microwave assisted 155, 347f., 405f., 410f.
– plasma jet 405
– plasma-enhanced (PECVD) 147, 154f., 171, 293f.
chiral nanotubes 126f., 136f.
chlathrates 114
chlorofullerenes 91f.
chondrites, see meteorites
circumtrindene II 52
Clar-structure 134f.
coal, deposits and formation 16
cocrystallization 80f.
coke 14
complexation
– nanotube scaffold templates 264ff.
– organofullerenes 112ff.
conductivity, electrical, see also doping
– carbon onion agglomerates 320f.
– diamond 23, 421
– diamond films 421ff.

– graphite   22, 27
– nanodiamonds   363ff.
– nanotubes   193, 202ff.
– percolation threshold   254
coordination complexes (metal–nanostructure)   77ff., 236
copolymerization   249
corannulene   33f., 48
crown ethers   113f.
cup-stacked nanotubes   164f.
cycloaddition
– [2+1]   98f., 438
– [2+2]   99f., 437f.
– [3+2]   99ff., 232f.
– [4+2]   102f., 233f., 436f.
cyclodextrins
– fullerene complexation   113f.
– nanotube complexation   265f.
cyclophanes   53f.

d

debundling (nanotube)
– assessment, by fluorescence   190
– effect on reactivity   222
– formation and dispersion   187f.
dendrimers   109ff.
density of state distributions   194ff.
density, carbon allotropes   10, 22f.
detonation synthesis   330, 340ff.
diamond, *see also* diamond films; nanodiamonds
– artificial synthesis   5f., 19f., 329, 389
– chemical reactions   29
– industrial uses   30
– jewel manufacture   18f.
– lattice structures   9f., 394
– mining production   4f., 17f.
– natural deposits   3f., 17
– physical properties   22ff., 365f., 421
diamond films
– chemical reactivity   428ff.
– covalent chemical reactions   430ff.
– defects and doping   399ff.
– derivatives and composites   438ff.
– development, history of   389ff.
– electronic properties   420ff., 440ff.
– growth mechanisms   407ff.
– physical properties   413, 424ff.
– preparation methods   403ff., 412f.
– spectroscopic properties   413ff.
– structural types   391ff.
– surface atomic geometry   394ff.
– uses and applications   420f., 443ff.
diamond-like carbon (DLC) films   402f.

diazonium salts   238f.
Diels–Alder reaction   102, 233f., 436f.
dodecahedrane ($C_{20}H_{20}$)   45, 90
doping
– diamond   20
– diamond films   401f., 414, 422f., 441ff.
– fullerene   40
– nanodiamond   332f.
– nanotube   229, 252, 256
double-walled nanotubes (DWNTs)
– fluorination   227
– production methods   158f., 170, 183, 261

e

electrochemical reactions   434f., 440ff.
electrodes, diamond   447
electron affinity, fullerenes   67, 72ff., 93
electron donor-acceptor systems   40, 58, 84f., 265
electron energy loss spectroscopy (EELS)
– carbon onions   317f.
– diamond films   420
– nanodiamond   357f.
– nanotubes   215f.
electron spin resonance (ESR) spectroscopy
– carbon onions   318f.
– nanodiamonds   361f.
– nanotubes   212f.
electronic components
– composite materials   111
– use of diamond   420f., 444ff., 451
– use of nanotubes   451
electronic properties
– conductivity   22, 27, 202ff., 364f., 421ff.
– field effect   269ff.
– field emission   181, 204ff., 268f., 365, 423f.
endofullerenes
– metallofullerenes   82ff., 260, 302
– nomenclature   36f.
– nonmetallic   86f.
endohedral chemistry   67, 256ff.
epoxides   76, 237f., 261f.
epoxy resins   251
exohedral fullerene modifications   67

f

faceted carbon nanoparticles
– atomic structure   289f.
– formation   300, 303f., 320
– transformation to carbon onions   308
field emission   181, 204ff., 365, 423f.
– display applications   268f., 385, 445
floating catalysts   144ff., 154, 156, 165

fluorescence
- diamond   24
- fullerenes   59, 61
- isolated nanotubes   190, 210ff.
- nanodiamond   332, 359f., 385
fluorination
- diamond films   431
- fullerenes   90f.
- graphite   26f.
- nanodiamond   368, 373
- nanotubes   227ff.
frictional coefficient   383, 426
fuel cells   278
fullerene chemistry
- cycloaddition reactions   98ff.
- electrochemistry   72ff.
- halogenation   90f., 117f.
- hydrogenation   87ff.
- inorganic reactions   77ff.
- nucleophilic addition   93ff.
- photochemistry   99, 103ff.
- polymers and composites   107ff.
- reactions with radicals   105ff.
- redox reactions   74ff.
- regiochemistry   68ff., 97
- size and reactivity   68, 117f.
- supramolecular complexes   112ff.
fullerenes, *see also* endofullerenes; fullerene chemistry; heterofullerenes
- applications   118ff., 452ff.
- carbon onion shell sizes   286
- chemical characteristics   66ff.
- crystal forms   64ff., 116f.
- discovery   33ff.
- growth mechanisms   41ff.
- natural occurrence   47f.
- nomenclature   36f.
- physical properties   57ff., 65f.
- preparation methods   48ff.
- separation and purification   54ff.
- structure   37ff., 43ff.
functionalization
- asymmetric   225f.
- covalent derivatization   233, 238ff., 377ff.
- micelle incorporation   111, 116
- phase stability   120
- solubilization   119, 188ff., 223, 241
- surface film immobilization   111f., 120, 225f., 440

g

gemstones   3f., 18f.
giant fullerenes   44f., 288

glassy carbon   15, 31, 299
- structure   12
Goldberg polyhedra   286ff.
graphite
- artificial   14f.
- chemical reactions   26ff.
- electrolytic conversion to nanostructures   305f.
- in fullerene production methods   50ff.
- natural deposits   13f.
- physical properties   21ff.
- production and processing   14ff.
- structures and bonding   7f.
- uses   30
- vaporization, in nanotube synthesis   140ff.

h

halogenation, *see also* fluorination
- diamond films   430ff.
- fullerene derivatives   90ff.
- fullerene precursors   52f.
- nanodiamonds   371ff.
- nanotubes   227ff.
hardness, of diamond   23, 365f.
- scratch/wear resistance   383, 443f.
heat capacity
- graphite   22
- nanotubes   216
heterofullerenes
- nomenclature   36
- structure   46f.
- synthesis   56f.
hexagonal diamond, *see* Lonsdaleite
hexogen   342
highly ordered pyrolytic graphite (HOPG)   15
HiPCo process   144ff.
Hückel transition states   42, 194
hydrocarbons
- flame combustion   412
- partial combustion   49f.
- pyrolysis   48f., 154, 156
hydrofullerenes
- stability   89f.
- synthesis   87ff.
hydrogen storage   120, 166, 257, 278f.
hydrogenation reactions
- diamond films   396f., 430
- fullerenes, experimental methods   87ff.
- nanodiamonds   334, 369ff.
- nanotubes   226f.
hydrothermal purification (nanotube)   172f.

i

infrared (IR) spectroscopy
- carbon onions   314
- diamond films   413f.
- fullerenes   61f.
- nanodiamonds   354ff., 370f.
- nanotubes   175f., 207ff.

intercalation compounds   27ff., 74ff., 245f., 255f.

ion bombardment   294ff., 402f., 412f.

ionization
- energies, of carbon   20f.
- in endohedral metallofullerenes   84f.
- potential, of fullerenes   72

isolated pentagon rule (IPR)   41

isomers, fullerene
- chiral   44, 55f., 66
- regioisomers, of addition products   68ff., 97
- stability prediction   41
- theoretical   40

isotopes
- ^{13}C depletion, in artificial diamond   20
- naturally occurring   2f., 21

j

jewels, *see* gemstones

k

Kekulé structure   135
Kimberlite pipes   4, 17

l

lab on a chip technologies   385, 446f.

lasers
- ablation   142ff., 153f.
- irradiation methods for carbon onions   306

lattice defects   332f., 399ff., 416ff.

LCAO (linear combination of atomic orbitals)   198f.

Lonsdaleite (hexagonal diamond)   341, 346
- structure   9f., 400

lubrication, *see* tribological applications

m

Mackay crystal   452f.
magneto-resonance tomography (MRT)   120
mechanical properties
- carbon onions   325
- diamond films   424ff.
- graphite   22
- nanodiamond   365ff., 382ff.
- nanotubes   190ff.

medical applications, *see also* biological applications
- antioxidants   119f.
- charcoal   2
- gene therapy   275
- imaging methods   120, 385f.
- targeted drug delivery   274f.
- tumor therapy   118

melting points   13, 63

metals
- as nanotube impurities   171ff.
- catalysts, in nanodiamond synthesis   348
- catalysts, in nanotube synthesis   141ff., 183ff.
- diamond film substrates   409
- electroplating, with nanodiamond inclusion   384
- fullerides   74ff.
- metallofullerenes, endohedral   82ff., 260, 302
- nanocluster deposition   245, 257ff., 278
- nanotube-reinforced complexes   254f.
- organometal fullerene complexes   77ff.
- substrates for carbon ion bombardment   294ff.

meteorites
- evidence of cosmic carbon onions   290f.
- nanodiamond particles   290, 330f.

micelles   111, 116, 188, 264

modulus of elasticity (Young's modulus)
- anisotropy, in graphite   22
- diamond films   427
- nanotube measurements   191f.

monocrystalline films   391f., 427

multiwalled nanotubes (MWNTs)
- comparison with carbon onions   283, 289, 322
- electronic properties   201f.
- helical (hMWNTs)   166ff.
- production methods   150ff., 181f.
- structure   135ff.
- tube interactions   137ff.

n

nanodiamonds
- agglomeration   338ff., 351, 370
- chemical reactions   370ff.
- derivatives and composites   377ff.
- discovery   329f.
- formation inside large carbon onions   323f., 346f.
- occurrence   290, 331
- physical properties   351, 362ff.
- purification and dispersion   335, 349ff.

– size and stability 329, 336ff.
– structure 332ff., 362ff.
– surface reactivity 334f., 367ff., 376f.
– synthesis 340ff., 347f.
– thermal transformation 303f., 310, 382
– uses and applications 345, 382ff.
nanohorns, carbon (SWNHs) 165f., 279
nanoonions 293ff.
nanotube chemistry
– endohedral inclusions 256ff.
– intercalation compounds 245f., 255f.
– metal coordination compounds 236
– noncovalent attachments 240ff., 249f.
– open ends and caps 224f.
– polymer composites 246ff.
– radicals and photochemistry 234f.
– redox reactions 220ff.
– sidewall covalent attachments 226ff., 237ff., 248f.
– supramolecular complexes 263ff.
nitrenes 98, 230f.
nitrogen
– impurity of diamond 9
– vacancy (N-V) centers 332f., 359f.
nuclear magnetic resonance (NMR)
– carbon onion spectra 319
– fullerene spectra 62f.
– nanodiamond spectra 360f.
– nanotube spectra 214f.

o

optical properties, see also spectroscopic properties
– diamond 24, 416ff., 448
– graphite 22f.
orbitals
– atomic 6f.
– delocalized π-electrons 7f., 39f., 133ff.
– frontier orbital symmetry 198f.
– molecular 67f., 85
– nanotube bundling 187
– π-stacking 240, 242ff.
osmylation 80, 234, 237
oxidation reactions
– diamond film 397f., 432f.
– fullerenes 76f.
– nanodiamond surface 368f., 373
– nanotubes 220ff.
oxidative cutting/purification 173, 176f., 220f.
oxides
– fullerene 76
– nanotube ozonides 232
– of carbon 25

p

Pandey reconstruction 395f.
peapod formation 114f., 259ff.
Peierls distortion 195
peptides, attachment to nanotubes 233
photochemistry
– diamond films 433ff., 439f.
– fullerenes 99, 103ff.
– nanotubes 234, 237
photoresponse, electrode 442
picotubes 161
piranha water
– diamond film oxidation 432f.
– nanotube cutting 176f.
polishing agents 18, 345, 366, 382, 409
polyacetylene, electronic characteristics 195
polyacrylonitrile (PAN)
– carbon fiber manufacture 11
– nanotube incorporation 276
polyanilines (PANI) 252
polycrystalline films 392f.
polycyclic aromatic hydrocarbons (PAHs)
– fullerene preparation 48f.
– soot nucleation centers 10
polymers
– composites, with nanodiamonds 380f., 383f.
– composites, with nanotubes 246ff., 264ff., 275ff.
– fullerene incorporation 107ff.
– nanotube wrapping 189f., 240ff.
polymethacrylates (PMMA) 251f., 265
polyphenylenevinylidenes (PmPV) 241f., 253f.
polystyrene 253
porphyrins
– fullerene complexation 113
– nanotube attachment 244f.
purification processes
– centrifugation 173, 350
– chemical impurity removal 171ff., 335, 349
– chromatography and filtration 54ff., 173, 178
– cutting (nanotubes) 176f.
– electrochemical separations 178ff.
– evaluation of purity 173ff.
– size sorting and separation 177f., 350f.
pyrene derivatives 242ff.
pyrolysis
– aromatic, cobalt catalysed 159ff.
– ferrocene catalysed 146f., 157
– flash vacuum (FVP) 52f., 297

– hydrocarbons 48f., 154
– organic compound decomposition 146f., 156ff.
pyrolytic carbon 15

r

radial breathing mode (RBM) 180, 206ff.
radicals
– "sponge" character of fullerenes 105ff.
– in alkyl/aryl attachments 235, 238, 375f., 434f.
radiocarbon dating 3, 21
Raman spectroscopy
– carbon onions 314f.
– diamond films 414ff.
– fullerenes 62
– nanodiamonds 351ff.
– nanotubes 176, 180, 206ff.
redox potentials, fullerenes 72ff.
reduction reactions
– fullerenes 74ff.
– nanodiamond surface 369f., 374
– nanotubes 222f.
regiochemistry 68ff., 97
resistive heating, fullerene synthesis 51

s

Schlegel diagrams 37, 70
semiconductors, nanotube, see also doping
– explanation of properties 194ff.
– field effect applications 269ff.
– purification 178ff.
sensors
– chemical 120, 271ff.
sensors
– physical 271
shock synthesis, nanodiamond 344ff.
shungite 48
silanization, nanodiamond 374f.
silicon carbide
– carbon nanostructure production 306, 348
– in artificial graphite production 2, 14f.
– production and uses 26, 444
single-walled nanotubes (SWNTs)
– aromaticity 133ff.
– closed, cap structure 132f.
– dissolution–precipitation model 182f.
– graphene sheet geometry 127ff.
– metallic and semiconductor characteristics 194ff.
– nanohorns (SWNHs) 165f.
– open ended 131f.
– production methods 140ff.
snowball mechanism 310f.

solar cells 119f., 452
solubility
– fullerenes 54f., 57ff.
– nanotubes 171, 187ff.
soot
– explosive detonation product 341ff.
– industrial production 16
– structure 10f.
– thermal transformation 298ff.
– transformation by irradiation 300ff.
– uses 30f.
Soxhlet chromatography 55
spectroscopic properties, see also optical properties
– absorption spectra 59ff., 209f., 316f., 358, 416f.
– electron energy loss spectra (EELS) 215f., 317f., 357f., 420
– electron spin resonance (ESR) 212f., 318f., 361f.
– fluorescence/luminescence spectra 61, 210ff., 359f., 417f.
– infrared spectra 61f., 208f., 314, 354ff., 413f.
– nuclear magnetic resonance (NMR) spectra 62f., 214f., 319, 360f.
– Raman spectra 62, 206ff., 314f., 351ff., 414ff.
– X-ray diffraction 315f., 356f., 418
– X-ray photoelectron spectra (XPS) 419f., 434
spirocyclization 43
spiroid carbon particle growth 310ff.
Stone-Wales rearrangements
– nanostructure defects 45, 192f., 204, 219
– strain energy redistribution 42, 46, 288, 308
surface acoustic wave (SAW) devices 445f.
surface film immobilization
– fullerene attachment 111f., 120
– nanotube attachment 225f.
surfactants 188, 263f.
symmetry
– description, for nanotubes 127ff.
– exohedral adducts 71
– icosahedral, in fullerenes 38, 62
– rotational, in carbon onions 284ff.

t

template-directed synthesis
– carbon onions 296
– fullerene derivatives 71, 97
– nanotubes 161f., 169f., 180

tensile strength
- anisotropy, in graphite   22
- nanotubes   190ff.
- polymer-NT composites   246ff., 275f.
thermodynamic properties
- carbon onions   319f.
- diamond   23, 384, 428, 447f.
- fullerenes   63f.
- graphite   21f.
- nanotubes   216f.
thermogravimetric analysis (TGA)   175
topological resonance energy (TRE)   134
transformations (phase transition)   337
- high pressure–high temperature (HPHT)   19, 340
- irradiation, electron beam   300ff., 304f., 308f.
- thermal conversion   298ff., 303f.
transistors   179, 269ff., 384, 445
tribological applications
- carbon onions   325
- nanodiamonds   366f.
twin structures   331, 399f.

u
ultrananocrystalline diamond (UNCD)
- preparation methods   410f.
- structure and properties   393, 415f., 427
ultrasound, in nanotube debundling   188

v
van der Waals interactions   7f., 113, 137, 240
van Hove singularities   200f., 206

w
water, *see also* solubility
- detonation coolant   343
- hydrothermal nanotube purification   172f.
- medium for carbon onion formation   291ff.
- nanodiamond surface adsorption   335, 355f., 384
- supercritical, carbon transformation   348
water gas equilibrium   341
windows, diamond   448
wrapping (nanotube)   189f., 240ff.

x
X-ray diffraction
- carbon onions   315f.
- diamond films   418
- nanodiamonds   356f.
X-ray photoelectron spectroscopy (XPS)   419f., 434

y
Young's modulus, *see* modulus of elasticity

z
zeolites, as nanotube templates   162, 169f.
zig-zag nanotubes   126f., 136
zone-folding model   206f.

Paper Cutout DIY Kit

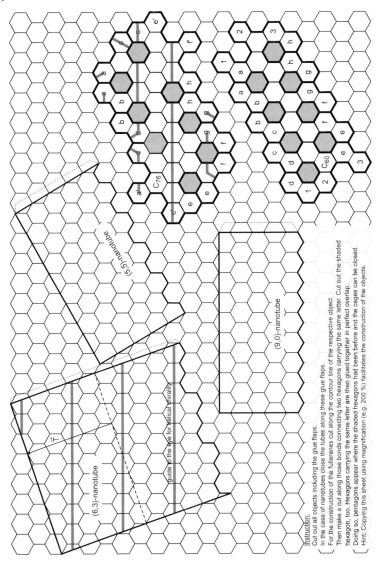

Carbon Materials and Nanotechnology. Anke Krueger
Copyright © 2010 WILEY-VCH Verlag GmbH & Co. KGaA, Weinheim
ISBN: 978-3-527-31803-2